Drawing the Line

ALSO BY MARK MONMONIER

How to Lie with Maps

Mapping It Out:
Expository Cartography for the
Humanities and Social Sciences

Maps with the News:
The Development of
American Journalistic Cartography

Technological Transition
in Cartography

Drawing the Line

Tales of
Maps and
Cartocontroversy

Mark
Monmonier

Henry Holt and Company
New York

Henry Holt and Company, Inc.
Publishers since 1866
115 West 18th Street
New York, New York 10011

Henry Holt® is a registered
trademark of Henry Holt and Company, Inc.

Library of Congress Cataloging-in-Publication Data
Monmonier, Mark S.
Drawing the line: tales of maps and cartocontroversy/
Mark Monmonier. — 1st ed.
p.   cm.
Includes bibliographical references and index.
1. Cartography.   2. Deception.   I. Title.
G108.7.M65   1994   94-16945
912 — dc20   CIP

ISBN 0-8050-2581-2

Henry Holt books are available for special
promotions and premiums. For details contact:
Director, Special Markets.

First Edition — 1995

Designed by Paula R. Szafranski

Printed in the United States of America
All first editions are printed on acid-free paper. ∞

3   5   7   9   10   8   6   4   2

For Bernard V. Gutsell,
founder and editor of *Cartographica,* friend,
and fellow traveler on the cartographic prairie

Contents

Acknowledgments

Numerous people helped make this book possible. Among the many informants acknowledged in the Notes section, I am particularly grateful to Lee Cobb, Allegany County administrative office; Denise Cote-Hopkins, Cortland County Low-Level Radioactive Waste Office; John Dean, New York Senate Energy Committee; Ute Dymon, Kent State University; Debra Levine, New York Joint Legislative Task Force on Demographic Research and Reapportionment; Frances Robotti, Fountainhead Publishers; and John Snyder, formerly of the U.S. Geological Survey.

Mike Kirchoff in the Syracuse University Cartographic Laboratory photographed the facsimile illustrations and provided helpful advice on FreeHand. Dorcas MacDonald and her interlibrary loan staff saved me at least a week on the road, and Mary Anne Waltz and the staff of Bird Library's map and government documents department helped me locate examples and check facts. The Maxwell School's Appleby-Mosher Research Fund provided travel support.

I owe special thanks to editor-in-chief Bill Strachan, who thought my fascination with controversial maps matched nicely his eagerness to publish an "introduction" to geography for the general reader, and to assistant editor Darcy Tromanhauser, copy editor Carole Berglie, and the staff of Henry Holt and Company.

Always understanding, my wife, Marge, and daughter, Jo, kept our five cats at bay while I was working.

DeWitt, New York
May 1994

Drawing the Line

Introduction

As powerful tools of persuasion in science and public affairs, maps have had a remarkable effect on our view of the world, our health, and the impact of our votes. At the root of their power is our frequently unquestioning acceptance of cartographic messages. Even folks who are routinely suspicious of written text equate maps with fact and fail to realize that no map is capable of including all information or telling all possible stories. In fact, the process of mapmaking requires cartographers to limit content in order to create a readable map and so allows them to manipulate their audience with the information they choose to include. This combination of power and subjectivity has repeatedly put maps at the center of controversy.

My goal here is to lay out the territory of map controversy by exploring the ways maps are used to convince people and by examining how a map can play various roles as a contest, prize, or stratagem. The contest might be to control real estate, represented by the map's boundary lines and labels, or to influence beliefs, values, or self-esteem. Or the map itself could be the prize, as in contests over whose language should be used to name places and physical features. In such cases maps often become ideological symbols: Do you accept my view of the world, my explanation for a geographic pattern, or my brand of cartographic representation, rather than that of my opponent? Controversy can also arise when maps are used as debating strategies, their convincingly crisp lines and labels effectively concealing an argument's tenuous assumptions.

As an exploration of the versatility of maps in science and politics, this book is an extension of my earlier book, *How to Lie with Maps*, which I modeled after Darrell Huff's classic *How to Lie with Statistics*. My goal in *How to Lie with Maps* was to promote a healthy, informed skepticism about maps through a series of simple hypothetical examples. In *Drawing the Line*, I would like to go a step further and take a tour of the most striking examples of map controversy. Chosen both from famous historical incidents and from the modern day, these narratives involve real people, places, and events. I look first at a number of national and international controversies and then investigate several local contests involving political influence, access to information, and environmental health.

You might already be familiar with some of these stories. Perhaps you once heard the argument that the moon came from the Pacific Ocean, or you saw maps showing that the continents were once one giant land mass. Maybe you recall news stories that used maps to show potential dumping sites for nuclear waste or ones about legal battles over new legislative districts. You might recall an election ballot missing the familiar name of a legislator or House member forced to run in another district. Possibly you remember an effort by local clergy to persuade the school board to buy new maps—weird maps with emaciated-looking renderings of Africa and South America. If any of these examples is familiar, you have been influenced by maps and deserve to know how.

Focusing on maps, people, and institutions, my approach is more a "carto-anthropology" than a cartographic history. Few of my examples come from the old, rare, ornate maps favored by map collectors, and social science methodology plays a far smaller role than simply looking at maps, talking with map users, and collecting information about things cartographic. I deliberately avoid critical theory, semiotic jargon, and other academic baggage that might get in the way of the story. Instead, I write about maps and the people and organizations that use (and sometimes abuse) them. Where helpful and practicable, I include photofacsimiles that let the maps speak for themselves.

Many of the tales that follow involve various forms of what I now call *mapism*. I don't know who coined the term, but I first heard the word in May 1991, while giving a live talk-show interview on Wisconsin Public Radio. A caller suggested *mapism* to describe the ill-founded but unshakable belief that a specific world-map projection is vastly superior to all others. Analogous to racism, the term implies the same mix of ignorance, prejudice, and arrogant conviction that leads seemingly honest, well-meaning people to assert the superiority or inferiority of a racial or ethnic group. And like racism, mapism can be subtle as well as blatant. The pervasiveness of one kind of map—the Eurocentric Mercator projection—in America's classrooms is a good example of subtle mapism. A less subtle example is the aggressive advocacy of another—the Peters projection—as an antidote not only for the widespread misuse of the Mercator world map but also for cultural and economic oppression of less developed nations. By ignoring established principles of map design, the Peters projection earned the suspicion, if not the scorn, of most geographers and mapmakers. After all, we cartographers have a pervasive mapism of our own, based on our principles, techniques, rules, guidelines, and standards—what we like to think of as truth. At odds with these beliefs, the Peters map provoked controversy.

With several forms of mapism, the Peters controversy is a fitting focus for chapter 1. The news media, which like a good fight, reported the story off and on throughout the 1980s. Reporters found lots of eager sources: an enigmatic German historian, numerous Third World advocates and UN officials, a liberal Christian publisher, and an assortment of academic cartographers who suddenly found themselves on the wrong side of political correctness. Like a number of these scholars, I disputed Peters's claims and defended the mapmaking community against his charges. Hardly an unbiased observer, I occasionally enjoyed showing the deficiencies of the Peters map and gloated over the historical ignorance of its advocates. Even so, I must also applaud these zealots' highly effective challenge to the misuse of Mercator's map in countless classrooms, textbooks, atlases, and other serious publications. Clearly, geographers and cartographers have been far too tolerant of a mapism that magnifies the size — and presumably the importance — of Europe and North America at the expense of Africa, Latin America, and tropical Asia.

Chapter 2 moves from the size of continents to the naming of places. Language is a powerful tool for both asserting ownership and expressing contempt. For centuries mapmakers in the employ of kings and presidents have claimed territory by renaming towns and physical features. Subtle cartographic statements that deny more legitimate, long-standing claims of native peoples are common on maps of the Canadian north, Eastern Europe, and the West Bank. Place-names on maps can also enshrine an earlier, less enlightened generation's ethnic insults, as older U.S. Geological Survey maps attest. In exploring the use and misuse of language on maps, chapter 2 also examines recent efforts to purge both the landscape and its cartographic record of racial epithets.

Chapter 3 narrates a different kind of cartographic attack on ethnic sensitivity. Its focus is Yale University's once-famous Vinland map, which challenged both the significance of Columbus's "discovery" of the New World and the ethnic pride of Italian Americans. This controversy lasted eight years, from Yale's provocative, well-publicized announcement on October 11, 1965—the day before Columbus Day—to the university's grudging admission in early 1974 that its map was a forgery. Yet news reports from the 1960s reveal little skepticism about the Vinland map's authenticity: because the map looked believable, the media reported it as fact.

That maps tend to be believed until they are discredited is a subtle yet rampant form of mapism. This often undeserved trust explains why lawyers enthusiastically introduce maps as evidence in boundary disputes. But as chapter 4 observes, a map presented at a trial typically requires the testimony of an expert witness to certify its authenticity and explain its relevance. Attorneys involved in boundary litigation thus recruit historians and cartographers to find and interpret maps that favor their case as well as to rebut and discredit the maps of their opponents. In order to further clarify their arguments and score points in the news media, lawyers occasionally prepare new composite maps that combine evidence from several sources and incorporate persuasive titles and labels. Although inadmissible as evidence, this cartographic posturing can encourage a particular outcome. As rulings by the U.S. Supreme Court and the International Court of Justice demonstrate, a map showing competing claims can provide a rhetorically useful base for a workable compromise.

In chapter 4 I also look at the extension of terrestrial boundaries offshore and the corresponding maps that justify claims to fishing grounds, submarine mineral rights, or shares of Antarctica. The UN Convention on the Law of the Sea and the 1959 Antarctic Treaty are important, cartographically inspired international attempts to deal peacefully with new conflicts arising from technological advances in fishing, mining, and transportation. Like so many tools dependent on the goodwill of the people who use them, maps can be instruments of compromise and peace as well as means of conquest and subjugation.

Forums of scientific, political, and scholarly opinion are as receptive to cartographic rhetoric as courts of law, and perhaps more so. Chapter 5 explores this through the careers and cartographic propaganda of two important theorists: Alfred Wegener (1880–1930), the father of continental drift, and Halford Mackinder (1861–1947), whose concepts of "Heartland" and "World Island" had a profound effect on foreign policy in Germany during the 1920s and 1930s and in America during the 1940s and 1950s. Wegener's and Mackinder's astonishingly successful world maps illustrate the deep respect among both physical and social scientists for carefully authored, highly generalized cartographic models. The map's role in promoting the theory of continental drift is an intriguing saga about the discovery, refinement, and marketing of scientific ideas. Equally revealing is the story of how maps help geopoliticians organize data, devise theory, and influence foreign policy. Maps that trigger a sense of fear, invincibility, or righteousness can be powerful weapons in international contests for diplomatic, military, economic, and cultural influence.

Maps are also effective weapons in domestic partisan politics. Every ten years, after the Bureau of the Census counts heads, each state becomes a battleground for legislative control. Reapportionment can be especially bitter when below-average population growth costs a state seats in Congress. Even when a single party controls the reapportionment of a state's legislative and congressional districts, population shifts can put two incumbents in the same district or force a white incumbent to face a powerful challenger in a new African-American or Hispanic district. Minority communities like a single politician to represent the entire neighborhood, and county leaders resent any partition that dilutes their influence and prestige. Because of the diverse ways legislative and congressional boundaries can be reconfigured to meet the "one person, one vote" goal, no reapportionment plan can satisfy everyone. Nowadays, though, the process is comparatively open: public hearings encourage people to air their complaints and suggestions, and sophisticated geographic information systems allow unprecedented experimentation with the size, shape, and ethnic composition of districts. And, as I show in chapter 6, a gerrymandered district might be fairer and more efficient than the geometrically compact districts favored by courts and political scientists. Traditional approaches to redistricting create an illusion of equality that ignores the disproportionate power of incumbents. However open and efficient, the map-based process of creating political districts is a form of mapism that can distract attention from the need for election reform and a fuller use of public referenda.

Chapter 7 investigates another map-based political process, the siting of landfills, incinerators, hazardous-waste treatment facilities, nuclear power plants, and other environmentally obnoxious facilities. Siting decisions become intensely political when neighboring counties and towns try to divert noxious facilities into someone else's backyard. Much of the debate turns on cartographic evidence that one location is "best" or another "even better." Heated arguments can arise based on hastily collected environmental data and maps of questionable reliability. Besieged decision makers often resort to an electronic geographic information system (GIS) that promises to make the data respectable and the siting process scientifically objective. Because the GIS can frighten off opposition and strengthen those controlling the information, advocates of "open government" argue persuasively for citizen access. Not surprisingly, appeals for "electronic democracy" receive the enthusiastic support of GIS developers, who gain both sales and status.

Geographic information systems and electronically generated maps are equally important elements in chapter 8, which looks more broadly at their use in environmental monitoring and emergency management. Among the challenges of GIS is the development of efficient methods for using maps to communicate information about hazards and risks to nontechnical people. Poorly conceived maps can inhibit appropriate responses to a wide range of hazards as well as thwart orderly evacuation and disaster relief. As pilots and military commanders are well aware, misleading maps are themselves an intolerable hazard.

Central to each of these controversies is the fact that no map is a thoroughly objective, value-neutral device for describing distances and locations. While a map might appear accurate and relevant, there usually is more to the story than one map can conveniently communicate. As several of my examples demonstrate, maps can work at more than one level simultaneously and hold different meanings for different users and in different contexts. A map that symbolizes an idea or theory provides a banner to carry and rally around, for instance, and an especially controversial map makes a good smoke screen to focus debate on certain issues and away from others. Moreover, even when nonbelievers ridicule or attack a map, its notoriety helps legitimize the debate. Despite the power of maps to persuade or distract, many examples here prove the map need not be an invincible weapon, especially if the intended victim is sufficiently skeptical to challenge the aggressor's data, interpretation, or choice of presentation. In almost all cases, the map that is presented is but one of a large number of cartographic viewpoints, some of which differ radically from others. After all, like other artists, map authors select what suits them and ignore what doesn't. Militant informed skepticism is the citizen's best defense against cartographic harassment by bureaucrats or ideologues.

The Peters Projection Controversy

Look at the Peters map in Figure 1.1 and then look at a globe, and you'll understand why professional cartographers were appalled by the good press this odd-looking picture of the world was getting in the early 1970s. The Peters worldview is one of a class of map projections called *rectangular* because its parallels and meridians are straight lines intersecting at right angles to form a grid of rectangles. But on a globe, as Figure 1.2 illustrates, the grid cells next to the poles are spherical triangles, whereas those near the equator are more nearly square. In contrast, the rectangular cells on the Peters map have a pronounced north-south elongation near the equator and an equally obvious east-west elongation at the poles. In the words of Arthur

Robinson, a prominent cartographic educator, the resulting "land masses are somewhat reminiscent of wet, ragged, long winter underwear hung out to dry on the Arctic Circle." Robinson's sarcasm had a point, because most maps produced by professional cartographers for atlases, textbooks, and wall maps more accurately represent continental shapes. The Peters map, which periodically catches the fancy of news editors and reporters, is simply a poor map.

Left-leaning and religious publications have been particularly enthusiastic about the Peters projection. For example, shortly after the National Council of Churches (NCC) began distributing the Peters map in the United States in 1983, *Christianity Today* ran a story with the headline "A New View of the World" below a picture captioned "West German historian and map maker Arno Peters has redrawn the world." The picture showed Peters sitting at a drafting table and holding a bow compass. According to the reporter, "Peters first presented his map in Germany ten years ago, and its influence has been growing in schools and institutions there." In its "Readings" section, *Harpers* included a half-page example of the Peters map below the title "The Real World," and reported the sale of eight million copies of "the new map" in eighty-five nations. *Science 84,* a popu-

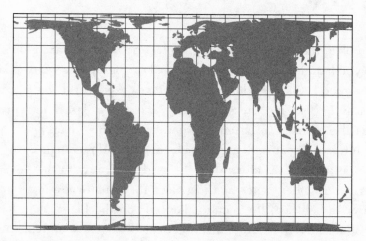

Fig. 1.1. The Peters projection, also called the Gall-Peters projection.

Fig. 1.2. Perspective view of a globe.

lar magazine published by the prestigious American Association for the Advancement of Science, was more cautious. Despite the title "A Fairer View of the World," its story quoted U.S. Geological Survey projection expert John Snyder, who said, "[Peters's] map isn't any better than similar maps that have been in use for 400 years." Nonetheless, Peters and his supporters were getting the attention they sought as the news media eagerly reported claims and counterclaims about a controversial map pitting science against religion.

Peters's claim of originality rankled the cartographic establishment. Despite generous use of the word *new* in press releases from the NCC, the World Council of Churches, and the UN, the Peters projection wasn't new at all. Over a hundred years earlier, in 1855, an Edinburgh clergyman named James Gall had devised an identical projection. Figure 1.3 shows the projection as Gall first published it in the initial volume of the *Scottish Geographical Magazine.* Gall called his map the *orthographic equal-area projection* because its grid lines intersected at right angles and it preserved the relative sizes of continents. He invited mapmakers to use this and his other mathematical inventions freely. "All that I would ask," he wrote, "is that, when they be used, my name be associated with them."

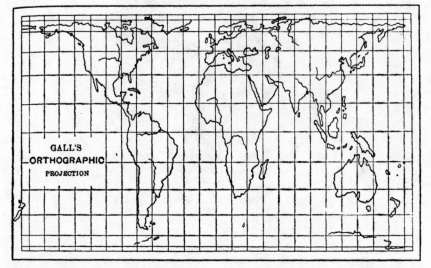

GALL'S ORTHOGRAPHIC PROJECTION.

EQUAL AREA.   PERFECT.

For Physical Maps, chiefly Statistical.

GALL'S
ORTHOGRAPHIC
PROJECTION

Fig. 1.3. The "Peters projection," as published by Gall in 1885. (From James Gall, "Use of Cylindrical Projections for Geographical, Astronomical, and Scientific Purposes," *Scottish Geographical Magazine* 1[1885], pp. 119–123.)

Gall's orthographic equal-area projection never impressed Gall as much as its reincarnation impressed Peters. Gall described two other cartographic creations in his 1885 paper, the Stereographic and Isographic projections that also carry his name. With more openness than Peters, Gall acknowledged that while "we may obtain comparative area with mathematical accuracy" by using the orthographic projection, in so doing "we must sacrifice everything else." In fact, he clearly preferred another of his innovations. "For general purposes," Gall wrote, "the Stereographic is best of all; for though it has none of the perfections of the others, it has fewer faults."

In rebutting Peters's propaganda, professional cartographers pointed with a mixture of scorn and thinly disguised glee to Peters's apparent ignorance of Gall's invention. To make the connection with the map's first inventor as well as undermine Peters's credibility, they took to writing about the "Gall-Peters projection." And privately it became the "unmitigated Gall-Peters projection."

The most noteworthy asset of Gall's orthographic projection—or the Peters projection, I'll call it for simplicity—is a property called *equivalence*, whereby a map portrays correctly the relative areas of land masses, countries, and other chunks of the world. The Peters map does this by enlarging tropical areas with a north-south expansion that makes Africa and South America look tall and thin, as well as by reducing higher-latitude areas with a north-south compression that makes Canada and northern Asia look short and fat. As a result, the rectangular cells in Figures 1.1 and 1.3 have the same relative areas on the map as they have on the globe.

But Gall was not the first to create an equivalent-area, rectangular projection. Credit for this discovery belongs to the mathematician Johann Heinrich Lambert, who in 1772 proposed what cartographers now call Lambert's *cylindrical equal-area projection.* Lambert's solution to the problem of preserving relative area was far simpler than Gall's. He began with a Plate Carrée projection, literally a "plane chart," on which an evenly spaced grid of squares makes a degree of latitude equal a degree of longitude (Figure 1.4). To adjust for area distortion, which increases toward the poles, Lambert devised a scaling formula to alter the spacing of parallels so that areas on the flat map have the same relationships as areas on a globe. As Figure 1.5 demonstrates, Lambert's formula provides a generally accurate representation of shape along the equator at the expense of enormous distortion near the poles.

Eighty-three years later Gall mollified this severe compression of shape by creating a more balanced solution, with two belts of low shape distortion centered on parallels at 45° north and 45° south. Poleward of these lines he maintained a north-south compression, while between them he induced a north-south expansion. But the key to preserving area lies in Lambert's relative spacing of the parallels: stretch Gall's projection horizontally to twice its width and you have

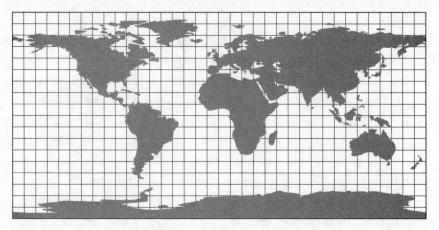

Fig. 1.4. The Plate Carrée (plane chart) projection.

Lambert's projection, and compress Lambert's to half its original width and you have Gall's.

Although Gall did a better job than Lambert in minimizing distortion of shape, cartographers willing to bend grid lines a little have done even better. An early solution is the *sinusoidal projection,* which creates the Christmas-tree-ornament outline in Figure 1.6 by curving the meridians inward toward a central meridian. Older and more obscure in origin than Lambert's projection, the sinusoidal projection provided a framework for world and continental maps as early as 1570. Convergence of the meridians allows this equal-area projection to have evenly spaced, straight-line parallels, useful for showing relative latitude. Like other *pseudocylindrical* projections, the sinusoidal

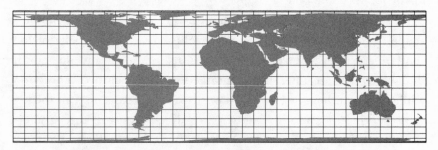

Fig. 1.5. Lambert's cylindrical equal-area projection.

has perpendicular axes of low distortion along both the equator and central meridian. Distortion is modest in the Tropics and around the central meridian but extreme in the four corner regions well removed from both axes. Seldom used as a world map because of extreme shearing in these corners, an appropriately centered sinusoidal offers a good representation of a single hemisphere.

Let me skip the rogues' gallery of better or merely different pseudocylindrical projections and jump to my favorite, Goode's *homolosine equal-area projection*, shown in Figure 1.7. A geography professor at the University of Chicago, J. Paul Goode wanted a world map that preserved relative area with only minimal distortion of the continents. His ingenious map reminds me of a globe painted on a spherical orange, from which the skin is peeled and pressed flat. Like an orange peel, Goode's projection is torn, or interrupted, but these interruptions occur only over water in order to preserve the integrity of land masses. As described in his classic paper, published in 1925 in the *Annals of the Association of American Geographers*, Goode's projection divides the world into six lobes—two in the Northern Hemisphere and four in the Southern Hemisphere. Each lobe has its own central meridian, around which a locally centered equal-area pseudocylindri-

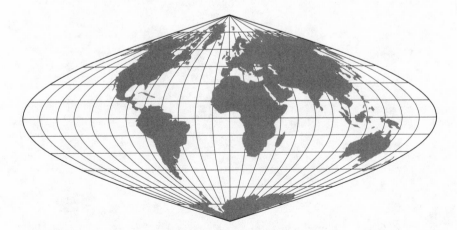

Fig. 1.6. The sinusoidal projection.

cal projection keeps distortion low. But to provide an even better representation of shape, Goode divided each lobe into two sections at latitudes 40°44' north and south, and used a sinusoidal projection on the equatorial side and a Mollweide projection on the polar side. The Mollweide projection offers lower distortion of shape in higher latitudes because its meridians converge to a point less sharply than on the sinusoidal. Goode's composite map projection splices its twelve pieces together seamlessly to not only preserve relative area—as the Peters map does—but provide a truer, more globelike portrayal of continents and countries.

How, then, did Peters convince journalists that his map was new, better, and beneficial? Like a clever high school debater, he found an easily refuted counterproposal, or "straw man"—in this case, the well-known Mercator projection. Most news stories instigated by his press releases included at least one illustration comparing the true relative areas of the Peters map to the gross enlargement of mid-latitude and polar areas on the Mercator. *Christianity Today*, for instance, used the Peters-over-Mercator comparison in Figure 1.8 to show that Mercator's worldview makes Europe's 3.8 million square miles look larger than South America's 6.9 million square miles. A second illustration—the pair of Mercator maps in Figure 1.9—highlighted two

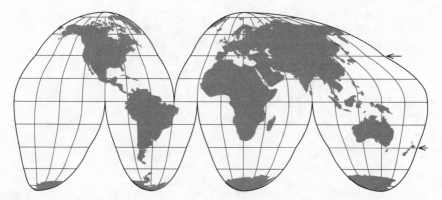

Fig. 1.7. Goode's homolosine equal-area projection. Arrows point out division of lobes into equatorial and polar sections.

additional flaws: Greenland looks larger than China (a country more than four times as large), and the relatively industrialized "North" appears larger than the less economically advanced "South." *Harpers*, *Science 84*, and *Mother Jones* followed the lead of the NCC press releases, the latter craftily positioning a much-reduced Mercator map in the lower-right corner of the Peters map, to illustrate a story titled "Cutting the Old World Down to Size."

However inappropriate on general reference maps, distortion of area is not a topic the press and the public generally get worked up about. So to make their straw man more diabolical and newsworthy,

rect sizes. To do so, it enlarges and elongates most Third World countries at the expense of the northern hemisphere, particularly Europe. That's exactly what Peters, a historian from Bremen, West Germany, had in mind when he drew the map.

Another German, Gerhard Kramer, first drew the more familiar Mercator map in 1569. (Kramer Latinized his last name to "Mercator.") His map produced severe size distortions in some countries because he located his own homeland, Germany, in the center of the map. It

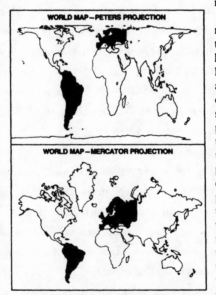

The Peters projection, top, reflects the actual sizes of land masses. The Mercator map, bottom, makes Europe appear larger than South America. It is actually smaller.

although it distorts shapes in order to preserve accurate geographical relationships.

"In our epoch, relatively young nations of the world have cast off the colonial dependencies and now fight for equal rights," he says. "It seems important to me that the developed nations are no longer placed at the center of the world, but are plotted according to their true size." His work is perhaps more of a contribution to world politics than it is to cartography, since its shape distortion renders it unsuitable for use in navigation. *(cont.)*

Fig. 1.8. Comparison of the Peters and Mercator projections. (From *Christianity Today*, 17 February 1984, p. 39. © CHRISTIANITY TODAY, 1984. Used with permission.)

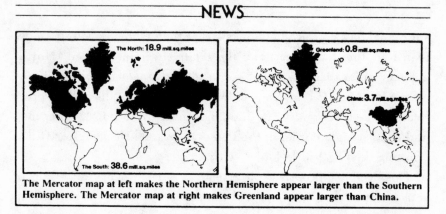

The Mercator map at left makes the Northern Hemisphere appear larger than the Southern Hemisphere. The Mercator map at right makes Greenland appear larger than China.

Fig. 1.9. Examples of areal misrepresentation by the Mercator projection. (From *Christianity Today*, 17 February 1984, p. 40. © CHRISTIANITY TODAY, 1984. Used with permission.)

Peters and his disciples drew attention to the Mercator projection's "Eurocentric" bias. As one press release quoted in *Harpers* said, the Mercator map "distorts the world to the advantage of European colonial powers." Peters's anticolonial, North vs. South angle also caught the attention of editors at *Christianity Today*, which ran its report under the subtitle "A German Map Maker Says the Third World Has Suffered Long Enough from a Distorted Map." *Science 84* quoted Ward Kaiser, director of the NCC's Friendship Press, who agreed with Peters that the Mercator map "portrays a colonialist and racist mentality."

Peters also attacked Mercator maps that showed a world centered vertically on a horizontal axis through Europe, the U.S.S.R., and the United States. (Eliminating Antarctica while retaining Greenland, as many mapmakers did, moves the vertical center of the Mercator world map well north of the equator, to about 50° north or more.) Ignoring the perhaps questionable rationale of omitting an embarrassingly inflated, uninformative, and awkward portrayal of Antarctica, Peters saw a more cynical, propagandistic motive for putting the industrialized North across the middle of the cartographic stage. "In our epoch, relatively young nations of the world have cast off the colonial dependencies and now fight for equal rights," *Christianity*

Today quoted him as saying. "It seems important to me that developed nations are no longer placed at the center of the world, but are plotted according to their true size."

Peters and the NCC were dead right in attacking the misuse of the Mercator projection on general-reference world maps for which it was never intended. Most projections serve a narrow range of tasks and regions, and the Mercator map is useful primarily to navigators, who have little interest in comparing the size of continents. A sailor or aviator who wants to find the direction to a distant destination need only draw a straight line from where he is to where he wants to go. This *rhumb line* represents an easily followed, constant-direction course between the two points. The rhumb line intersects all meridians at the same angle, and the navigator can quickly read his bearing with the aid of a protractor. Figure 1.10 illustrates this process for a voyage from Honolulu to San Francisco. A protractor placed on the rhumb line calls for an initial bearing 60 degrees east of North—an enormously valuable piece of information for the sailor who can't see the destination and has no landmarks or signposts for guidance.

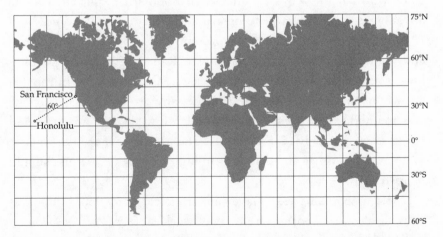

Fig. 1.10. On the Mercator projection, a straight line is a rhumb line, which shows the constant geographic direction and bearing between two points.

In addition to its usefulness in practical navigation, the Mercator projection provides acceptably accurate shapes and relative areas on highly detailed maps of small areas. For this reason, most of the U.S. Geological Survey's large-scale topographic maps are based on a variation called the Transverse Mercator projection. Because these maps cover areas smaller than twenty-five square miles, scale variation and area distortion are neither noticeable nor troublesome. But when a single small-scale map sheet covers a much larger region, such as a continent or hemisphere, areal distortion becomes both obvious and obnoxious, and other projections provide better representations of shape, area, distance, or direction. These alternatives each offer specific advantages. For example, regional maps showing airline routes often employ the Gnomonic projection, which can portray shortest-distance great-circle routes as straight lines.

Peters was not the first to object to using the Mercator projection for general purposes. Its misuse has long dismayed cartographic educators, although they have attributed people's misplaced reverence for it more to ignorance and expediency than to colonialism and racism. In 1938, for example, prominent cartographic designer Erwin Raisz noted in his classic textbook *General Cartography* that "the Mercator world map enjoys an unmerited popularity," but cited as its advantages straightforward construction, easy use of latitude and longitude, and more space for representing Europe's crowded cities and political boundaries. According to Raisz, "Every cartographer knows how difficult it is to represent the small countries of Europe on a world map; consequently, he welcomes any projection that exaggerates the higher latitudes." To explain is not to excuse, however, and Raisz asserted firmly that because of the Mercator map's "erroneous impressions of areas and distances in high latitudes . . . its use should be restricted."

Most mapmaking texts have similar caveats. In 1953, in the first edition of the important reference work *Elements of Cartography,* Arthur Robinson observed that because it has benefited mariners since 1569, Mercator's projection is "one of the most famous projections ever devised," but he also warned readers that "it is of little use for purposes other than navigation." State Department geographer S. Whittemore Boggs was even more vehement when he wrote in a 1947 issue of *Scientific Monthly* that "use of the Mercator projection for world maps should be abjured by authors and publishers for all purposes." Calling attention to the dangers of misusing maps intended for other purposes, Boggs argued that "no man ever saw or will ever see a world that has much resemblance to the Mercator map, and the misconceptions it has engendered have done infinite harm."

This warning was not a closely held secret. In a widely read laical introduction to mapmaking, David Greenhood discussed the merits and deficiencies of the projection, and speculated that Mercator's map "has been a favorite schoolroom map because it shows almost the whole world simply—in a single continuous panel [and] does not seem to distort shapes." After issuing his own injunction, Greenhood quoted a much earlier alert from British cartographer G. J. Morrison, who in 1902 warned that "people's ideas of geography are not founded on actual facts but on Mercator's map."

So how valid is Peters's complaint about widespread misuse of the Mercator map? In 1984, British cartographic scholars Iain Bain and Peter Lawrence addressed this question by looking for Mercator maps of the world among the collection of twelve reference atlases at *The Geographical Magazine* editorial offices and the twenty school atlases collected for an international cartographic exhibition. Surprisingly, they found not one, and concluded that "as a world map [the Mercator] seems to have disappeared from British atlases." Moreover, equal-area projections were so numerous as to "now dominate world maps." I conducted a similar nonsystematic survey of American reference works and likewise failed to detect a Mercator world map among my nonscientific sample of recent, expensively produced commercial atlases from major cartographic publishers such as Hammond, National Geographic, and Rand McNally. But a broader search revealed that the Mercator projection was alive and well, if not thriving, among wall maps, inexpensive outline maps distributed free for various promotional purposes, and cheap atlases and encyclopedias occasionally sold in supermarkets. Like most consumers in a free-market system, map buyers often get what they pay for.

Although not the rule, such poorly conceived maps and atlases betray cartography's status as more an occupation than a profession. While mapmakers with collegiate or other formal training in the subject obviously know the advantages and limitations of the Mercator projection, many people become mapmakers by less rigorous routes. Most maps in newspapers and magazines, for instance, are drawn by persons trained in commercial art, and the design and production staffs of commercial map and atlas publishers often reflect a variety of backgrounds, including commercial art, engineering drawing, and military mapping. Among these uncredentialed mapmakers, an arcane or unknown cartographic theory can be less important than tradition, history, and the Mercator projection's enormous name recognition.

Mercator's reputation as a cartographic pioneer might explain the popularity of his projection among the interior designers and graphic artists responsible for prominent cartographic displays. In his deconstruction of Peters's map, Arthur Robinson grimly conceded a number of highly visible and inappropriate uses of the Mercator map, including the backdrops for the ABC and NBC evening news programs and in briefing rooms at the State Department and the Pentagon. In 1991 the U.S. Postal Service used the Mercator projection for the centerpiece map of an unusually large sheet of ten stamps commemorating the fiftieth anniversary of World War II. The rationale for choosing a Mercator view surely lies in tradition, aesthetics, nostalgia, or some other milieu with values very different from those of cartographic theorists.

Although the Mercator projection is not the only alternative to his own creation, Peters devised a "catalogue of attainable map qualities" that conveniently dismisses Goode's interrupted projection and other nonrectangular attempts to portray the curved earth on a flat map. In his book *Die Neue Kartographie* (*The New Cartography*), Peters described ten criteria for an ideal world map, the second of which, "fidelity of axis," requires a rectangular network of meridians and parallels. Projections on which the meridians converge toward the poles lack "fidelity of axis" because "the direction to the north is different at different points on the map [so that] orientation becomes more difficult." A composite map such as Goode's world map in Figure 1.7 violates this condition by pasting together locally centered pseudo-cylindrical projections. According to Peters, Goode's and similar equivalent projections "have fidelity of area, but they bought this quality at the price of abandoning important qualities of Mercator's map and could therefore not supplant it."

Except for equivalence, which he termed "fidelity of area," Peters's other criteria are equally dogmatic or vague. In addition to fidelity of area, Goode's map possesses only three other "attainable qualities": fidelity of scale, totality, and clarity. Fidelity of scale refers to area, not distance, and merely reflects fidelity of area. (Because flat maps distort some, if not all, distances, "fidelity of distance" is possible only on a globe.) Totality refers to a map's ability to portray the entire world—a Mercator projection cannot portray the polar regions completely and thus lacks totality. Clarity refers to not deforming "by extreme distortion any of the countries, continents and seas portrayed" and can be recognized by a map's "harmonic proportions, practical dimensions, ease of comprehension and of construction of the necessary co-ordinates for its grid system." Goode's map qualifies because of minimal distortion of continental land masses, but Peters claimed his projection did as well. ("Extreme distortion" apparently is in the eye of the beholder.) In a checklist comparing eight projections, Peters rated his own map a perfect ten but gave none of the others a score higher than four.

Goode's projection fails on the Peters test because of six rather fuzzy requirements with impressive labels: fidelity of axis, supplementability, fidelity of position, proportionality, universality, and adaptability. As noted earlier, Goode's map lacks fidelity of axis because its meridians converge and thus don't all point northward in the same direction. It also lacks supplementability, which would allow a mapmaker "to detach a section from the left hand side and to reattach it to the right." But Peters rules unfairly against Goode's map on fidelity of position, whereby "all points which exist at an equal distance from the equator are portrayed as lying on a line parallel to the equator." The parallels in Figure 1.7 are all straight lines, parallel to the equator, and clearly fulfill this requirement. Equally questionable is the claim that Goode's map lacks proportionality, which requires "longitudinal distortion along [the map's] upper edge as great (or as small) as along its lower edge." In Figure 1.7, both poles are points, with zero longitudinal distortion. Only Peters's own map meets the last two criteria. Especially arcane is universality, a quality that "permits the construction of grid systems for maps of each section of the Earth's surface as well as for a global map, and . . . permits the portrayal of all contents of a map for all applications." Huh? I won't even bother challenging this babble or the equally puzzling quality of adaptability, whereby a projection "can cope with specialist requirements of general map contents."

Among the first to challenge Peters's "catalogue" was the German Cartographical Society, whose response was neither kind nor temperate. A sister organization in Britain published an English translation of the statement signed by the nine top scholars who constituted the German society's board of directors. They charged that Peters's criteria not only stack the deck in his favor but "contradict the findings of mathematical cartography and arouse doubts regarding the author's objectivity." Because "known scientific terms are being used and manipulated in a falsifying way with subjectively altered meanings, . . . subjective value judgments are being presented as map qualities in a misleading way, . . . [and] the catalogue displays violations of the elementary rules of logic" in the Peters test, the German cartographers concluded "it is totally unsuitable for assessing world maps." They not only attacked the Peters map's portrayal of shape as "totally unsuitable for representing the Countries of the Third World," but noted that "each of the old and new equal-area world maps represents a non euro-centric view of the world." Equally disturbing to them was Peters's one-size-fits-all solution to cartographic problems: "To make publicity for just one map of the world—which exhibits a rigid grid and numerous errors—as being the standardised, all-purpose map of the world—amounts to an anachronism in our day and age." They concluded in no uncertain terms that "the Peters map [not only] conveys a distorted view of the world [but] is by no means a modern map and completely fails to convey the manifold global, economic and political relationships of our times."

Because the controversy focused as much on Peters and his tactics as on his projection, a brief biographical sketch is in order.

A half-day scrounging in the library turned up little. *Who's Who in the World* lists Arno Peters only in its 1978–79 and 1980–81 editions. Identical entries describe him as a historian and geographer, born in Berlin on May 22, 1916—and thus in his late sixties when the National Council of Churches began touting his map in the early 1980s. Educated at Friedrich-Wilhelm University in Berlin, Peters received a Ph.D. in 1945. His first four children—Anja, Axel, Anita, and Aribert—were probably named in alliterative honor of himself and his first wife, Anneliese. His fifth child, Sabine, is probably his daughter with Birgit Francke, whom he married in 1959. Peters lived in Bremen, West Germany, and was affiliated with the Institute for World History. He joined the institute in 1974 and became its director a year later.

Peters's critics fill in a few of the gaps. His academic training focused on journalism, history, and art. Several sources note a bit ominously that his doctoral dissertation, titled "Film as a Means of Public Leadership," suggests an early appreciation and understanding of the tactics and strategy of media propaganda. Between 1958 and 1964, Peters worked as a journalist for a socialist magazine, and in 1967 he described his projection at a meeting of the Hungarian Academy of Science.

His critics also mention—perhaps a bit too pointedly—the 1952 "history textbook scandal" in which Peters got the U.S. High Commissioner in occupied Germany to back a highly leftist world history text. According to an Associated Press story published in the *New York Times* on November 15, 1952, Peters received $47,600 from a U.S. government program to encourage and support new textbooks for German schools. But what the Americans got, apparently, was a text "written by a Communist and heavily loaded with Red propaganda." After a West German official pointed this out—"What's going on here?" his telegram asked—embarrassed U.S. officials condemned the book's omissions and biases, and started pointing fingers. They had failed to run a security check on Peters because of "high endorsements from eminent German educators," the Associated Press reported, and before officials could confiscate all copies of the textbook, Peters "ran off a large printing of his own for sale in many German towns."

The AP dispatch reflects much of the cold-war, anti-Communist hysteria of the early 1950s. Three days later, though, a less strident Reuters story confirmed the basic facts that the U.S. High Commissioner had given Peters the subsidy for a book with "alleged Communist leanings" and distributed 9,200 copies to U.S. cultural centers throughout West Germany. But the Reuters account also mentions a press conference at which Peters and his wife and coauthor, Anneliese, denied they were Communists and defended their book as free of prejudice. Indeed, the Communists seemed even less pleased with their writing than the Americans. According to the Peterses, an East German publisher had rejected their book as "so remote from the Socialist understanding of history that we have no use for it here."

In *The New Cartography*, Peters traced his interest in world maps back to his controversial world history text, in which cartography was an important key to awareness of other nations and people. In his words, "the quest for the causes of arrogance and xenophobia has led me repeatedly back to the global map as being primarily responsible for forming people's impression of the world." Existing world maps were "worthless for an objective representation of historical situations and events," he wrote, so "after taking careful stock of the current situation I came to conclusions which were different from those reached by the increasingly scientific methods of modern cartography."

Where and how thoroughly Peters studied cartography is not known. In *The New Cartography*, he called himself a "historian with geographical leanings." However limited his own formal training in the field, Peters had an in-house source of cartographic expertise: his son Aribert, who published a number of respectable papers on map projections. By 1983, when Peters published *The New Cartography*, he was aware of Lambert's equal-area projection but not Gall's. He mentioned Lambert's map once in the caption of a figure showing seven equal-area world maps, including Goode's. "Ever since the Mercator map has disfigured our geographic view of the world," he wrote, "cartographers have taken pains to overcome this Europe-centered view of the world," yet he rejected the efforts of these others, who omitted qualities of Mercator's map that only he retained.

By the end of the decade Arno Peters still had not acknowledged Gall's development of the map now widely known as the Peters projection. I never had an opportunity to talk with Peters and ask him about this, but in March 1989 British geographer Peter Vujakovic did in an interview arranged by Peters's publisher (Longman) while he was visiting England to promote the *Peters Atlas of the World*. When Vujakovic asked the "G question," Peters apparently expressed a mixture of ignorance and skepticism. According to Vujakovic, "He claims not to have been aware of the Gall variation until recently and has not as yet seen documentary evidence."

Asking how Peters discovered or devised the Peters projection is perhaps no more relevant than asking how Stephen King gets his ideas. Far more important is how Peters, a noncartographer, was so enormously successful in promoting his projection. Part of the answer seems to lie in his career as a journalist, when he learned the art of writing press releases, holding press conferences, and exploiting channels markedly different from the scientific journals and technical conferences commonly used by his critics. Indeed, Peters's skill as a publicist explains much of the wide acceptance of his invention among the media and Third World advocates.

Like many persistent ideological crusaders, Peters attacked in widely spaced assaults, usually tied to his new publications. The first major offensive began in 1973 with a press conference in Bonn, where Peters announced his "Orthogonal Map of the World." According to a report in the Manchester *Guardian,* nearly 350 reporters attended and Peters distributed a slick brochure "Der Europa-zentrische Charakter unseres geographischen Weltbildes und seine Überwindung." (I like the German better than its milder, rather quirky translation, "The Europe-centered Nature of Our Geographical Picture of the World and Its Conquest.") Peters's earliest Anglophone critic, Derek Maling, called this handout "a remarkable example of sophism and cartographic deception." The booklet "normally . . . would not merit any mention" in the Royal Geographical Society's *Geographical Journal,* where Maling roundly panned it, but serious treatment of the projection by newspapers and the technical journal *Kartographische Nachrichten* demanded "some authoritative statement about its utility."

Peters's campaign probably would have fizzled without endorsement by the World Council of Churches, Christian Aid, UNESCO, UNICEF, and numerous other anticolonial, prodevelopment organizations eager to promote his North vs. South attack on the Mercator world map. Early adoption by these high-profile groups in turn encouraged some lower-profile agencies to use the Peters projection. To study the extent of adoption of the Peters projection in Britain, Peter Vujakovic sent questionnaires to Christian Aid, Oxfam, and fifty less prominent development and relief organizations. Forty-two editors or publications directors replied, and thirty-six of these reported using maps in their publications. Among the group using maps, twenty-five had adopted the Peters projection as their principal world map. Cartographers rarely had a voice in these decisions: only one organization among the thirty-six agencies using maps in their publications sought the advice of a cartographic professional. Two forms of the projection were common. More typical were the small, page-wide thematic maps on which authors either portrayed distributions (such as famine or per capita income) or identified specific locations (such as missions or refugee centers). But equally noteworthy were the even smaller, highly simplified renderings of the projection as a design element in book covers, newsletter nameplates, and logos.

Vujakovic asked the twenty-five map users to list the advantages of the Peters projection. Their most common response was the map's ability to show true relative areas (48 percent), followed by its distinctive, thought-provoking appearance (36 percent), the elimination of a Europe-centered worldview (32 percent), and a fairer representation of Third World nations (24 percent). One respondent found the projection useful as "a political statement in itself."

Vujakovic traced Christian Aid's decision to use the Peters projection to Pamela Gruber, an employee impressed by a German version she had seen in New Zealand. Gruber convinced Christian Aid to print and distribute a colored wall map. Although production costs precluded translation of the German text within the map, English-language text outside the map described both the projection's benefits and Christian Aid's activities. Published in 1977 and distributed widely throughout Britain, the Christian Aid map attracted the attention of many relief and development agencies, and encouraged further adoptions by the Centre for World Development Education, the United Nations Association (UK), and the World Development Movement. Like religions and revolutionary movements, the Peters projection depended on both conspicuous conversions and champions like Pamela Gruber.

Peters's most influential advocate was Willy Brandt, chancellor of the Federal Republic of Germany from 1969 to 1974, and winner of the Nobel Peace Prize in 1971. Brandt chaired the Independent Commission on International Development Issues, launched in 1977 with a generous grant from the Dutch government. The commission's goal was to study and recommend solutions for economic and social disparities among nations. Its first report, "North–South: A Program for Survival," displayed the Peters projection on the bottom half of its front cover with eye-catching contrast among its white background, red land masses, thin black grid of meridians and parallels, and thick, black meandering line separating the more developed North from the less developed South. (This visually intriguing line ran relatively directly from left to right across the U.S.-Mexican border, through the Mediterranean Sea, and along the Soviet-Chinese border. Upon reaching the Pacific, though, it turned south to exclude Japan, and then swung west and south of Australia and New Zealand.)

The Brandt Commission extended its endorsement of the Peters map beyond the cover of its report. A combined explanation and credit note filled the upper half of the book's copyright page directly opposite the table of contents. "The map on the front cover is based upon the Peters Projection rather than the more familiar Mercator Projection," it announced. After a concise litany of the projection's "several innovative characteristics," the blurb addressed distortion. "The surface distortions that do appear," it noted a bit cryptically, "are distributed at the Equator and the poles; the more densely settled earth zones, it is claimed, appear in proper proportion to each other. This projection represents an important step away from the prevailing Eurocentric geographical and cultural concept of the world." The map was used "courtesy of Dr. Arno Peters of the University of Bremen."

Another strong supporter was the West German government's Press and Information Office, which distributed press releases promoting the Peters projection. In 1977 one of these missives ironically appeared in the bulletin of the American Congress of Surveying and Mapping. Packing more punch and general information than most articles in the *ACSM Bulletin*, the strongly pro-Peters article used four illustrations with quaint, untranslated labels such as "Grönland" and "Sud-amerika" to demonstrate the Mercator map's more prominent flaws and indict its Eurocentric view. The account noted, almost apologetically, that Gerhard Kremer, a German cartographer with the Latinized name "Mercator," "had put his chosen homeland, Germany . . . in the center of the map together with all of Europe." To support its claim that Peters "has radically changed the world map," the story listed eight qualities "which hitherto had never been provided on one and the same map." After praising the Peters map lavishly, the article concluded, "It may surprise the layman and shame the expert that it was not a geographer or cartographer but a historian who created this new geographic scheme."

Despite this not-so-subtle provocation, neither the *Bulletin*'s noncartographer editor nor ACSM's noncartographer headquarters staff, which reviews and approves the house organ's content, saw anything amiss. Cartographers constituted only one-eighth of the organization's ten thousand members and were far less touchy and image-conscious than the land surveyors, who dominate ACSM. So the *ACSM Bulletin* ran the release, apparently verbatim, with the headline "Four Centuries After Mercator: Peters Projection—To Each Country Its Due on the World Map."

ACSM's cartographic members were not amused. In its very next issue, the *Bulletin*'s "What They Say" section published objections from Arthur Robinson and John Snyder under the heading "American Cartographers Vehemently Denounce German Historian's Projection." Robinson attacked both the projection and its originator. "Map projections are fascinating for many reasons," he observed, "not the least of which is the way people, such as Dr. Arno Peters, who know little of the subject, regularly devise something new and wonderful." But while some naïve discoveries "pass into oblivion because the originators have the good sense to check out the idea with the cartographer knowledgeable about projections," others, like Peters, "don't have such good sense, don't realize what they don't know, put it forward, and end up looking ridiculous." Snyder was less personal but equally affronted: "For Peters's promoters to declare that this is the first world map projection of consequence since Mercator is ridiculous and insulting to dozens of other inventors over the years who have done a better job with much more innovation and much less fanfare." A closing editorial note attempted to soothe ACSM's irate members by concluding— far too optimistically, it turned out—that "our eminent cartographers have debunked the projection and seem to have 'laid it to rest' forever in no uncertain terms!"

With the active support of the National Council of Churches, in 1983 Peters launched a publicity campaign similar to his first. This second wave featured two publications: an English-language version of the Peters map of the world and his book *The New Cartography*, published in a bilingual German-English edition. The book is attractive in design and provocative in content. A gold grid of vertically elongated rectangles decorates a dark green cloth cover nearly as large as a small atlas. Its 164 pages include 135 maps, many collected in blocks of four or six for ready comparison. On the copyright page, the credit line, "This work was commissioned and sponsored by The United Nations University," implies endorsement by the UN. (The university is an autonomous institute set up by the General Assembly in Tokyo in 1975.) The first section, "The Emergence and Development of Cartography," traces the evolution of cartographic thought from pre-Christian times to the Second World War. In the next section, "Taking Stock," Peters attacks the "outdated theory" of the professional cartographer and "strips away" ten myths about what a world map projection should or might do. His list of myths includes several widely accepted cartographic principles, which would make a nearly complete outline for a course on map projections. In the third and final section, titled "The New Cartographic Categories," Peters presents two other lists: his ten "attainable map qualities" discussed earlier and ten "attributes of the New Cartography." A two-page conclusion announces, "The revolutionary character of the new cartography lies in its defeat of the ideologies which hitherto have stamped all world maps." Cartographic scholars unable to find their names in the eleven-page index could fume at the bibliography's revoltingly short list of a mere twenty-eight entries, five by Peters himself.

Throughout his book, Peters attacked cartography's shortsighted subservience to religion, mathematics, pragmatism, and colonialism. "Since Mercator produced his global map over four hundred years ago for the age of European world domination," he wrote, "cartographers have clung to it [and tried] to render it topical by cosmetic corrections." His opinion of world maps created after the thirteenth century, when philosophers, historians, and theologians ceded map production to professional cartographers, was even worse. This new profession, he argued, devised an arcane mathematical theory and complex vocabulary to thwart popular understanding of its craft. Mathematical cartographers "who wished to retain Mercator's map with its pleasing shapes and its clarity of portrayal" coined the term *conformality* to describe this property and justify widespread use of this and other non-equal-area projections. While this is true, they did so principally to promote detailed maps of small areas, for which distortion of area is insignificant. In Peters's mind, though, "This error led to the stagnation of the development of cartography, a discipline which has affected our global concept right to the present day."

Peters's attack on the mapmaker's products and theories angered cartographers, who not only challenged his rhetoric but attacked his solutions. Derek Maling, an early critic, was a specialist in both map projections and cartometry, a subfield concerned with making precise measurements on maps. His skepticism about Peters's flagrant and inflammatory claims led to doubts about Peters's skill as a mapmaker. After carefully measuring the spacing of grid lines on a 1974 version of Peters's map and comparing these results with precise spacings computed using Peters's own formula, Maling found discrepancies "too large and too regular to be accounted for by random errors of measurement, paper deformation or small uncertainties in the location of the standard parallels and scale of the map." The map displaced some parallels as much as four millimeters, a huge distance in the world of cartometry. In an article in *Kartographische Nachrichten*, published in Peters's home court, Maling concluded, "Peters' Projection is not equal-area." Later versions of the map by more skilled drafters presumably addressed these drawing errors.

In a similar vein, geographer Phil Porter and projection expert Phil Voxland found Arno Peters "a thoroughly confused cartographer" and sarcastically asked that he not be confused with Peter Arno, the well-known cartoonist for *The New Yorker* magazine. In addition to attacking the projection's "disfigurement of Africa" and pointing out a number of inconsistencies in *The New Cartography,* Porter and Voxland demonstrated that Peters's remedy to Europe-centered maps was seriously flawed: because the earth's land area is centered around Europe, the Peters projection—like any map centered at the intersection of the equator and the Greenwich meridian (0° north or south, 0° east or west)—is itself Europe-centered. A better solution, they argued, would be an equal-area projection that puts India and China in the middle of the cartographic stage (centered at 45° north, 90° east) and highlights population, not land area.

But Peters clearly was playing to a different audience—a much larger audience of victims and sympathizers eager to overthrow a tyranny that was at once graphic, ideological, and economic. His revolution supplemented existing mathematical theory with "practical, aesthetic and didactic" concepts of the new cartography so that an "egalitarian map [could now] demonstrate the parity of all the peoples of the earth." It is not surprising that his claims found receptive ears among development agencies and religious groups.

Peters's prime advocate in North America was Ward Kaiser, an ordained minister, religious writer, and director of the National Council of Churches' Friendship Press. Kaiser was not only the North American publisher of Peters's work but his press agent, spokesperson, and committed defender. Any American newspaper or magazine reporter who wanted Peters's viewpoint usually talked to Kaiser. In early 1984, for instance, a *USA Today* story titled "The Global Perspective: Third World Gains" repeated Kaiser's pitch that "every teacher who really wants to help his or her students understand the real world ought to have this map." Kaiser, like Peters, deplored the Mercator map, which, he told *Mother Jones*, "makes the predominantly white-dominated areas of the world seem more important than they are." In his interview for *Science 84*, he argued that the Mercator projection "portrays a colonialist and racist mentality."

Kaiser seemed more eager than Peters to inject race and social consciousness into the debate and less inclined to pursue a losing argument with cartographers. "I will not claim the Peters Projection is the only map that ever ought to be used," he told *Mother Jones*, and in response to cartographer John Snyder's 1988 article in *Christian Century*, Kaiser recognized "a common interest: to restore the Mercator to its honorable and original use as a navigator's tool and replace it with a more suitable map or maps." Two months earlier, Snyder—a socially conscious Quaker—had been equally conciliatory in reflecting that "at least Peters's supporters are rightly communicating the fact that the Mercator should not be used for geographical purposes."

Snyder, who had focused his article on erroneous claims by Peters's advocates, was especially critical of Kaiser's booklet "A New View of the World," which the Friendship Press published in 1987 as a "handbook" to the Peters projection. He pointed out how Kaiser had turned a West German press release reprinted in the *ACSM Bulletin* into an endorsement. Kaiser's handbook asserted that "Support for Professor Peters' map has been forthcoming from a number of professional communities. Geographers and cartographers among these. Thus the American Congress on Surveying and Mapping could say . . ."—and at this point Kaiser quoted a particularly supportive passage in the controversial 1977 *Bulletin* article. But as Snyder noted, the *Bulletin's* editor had clearly and accurately attributed the entire piece to the Press and Information Office of the Federal Republic of Germany. Moreover, Kaiser claimed ACSM approval a second time, later in his book. When advising readers on how to pitch the "mathematical or scientific superiority" of the Peters projection to their neighborhood schools, churches, public libraries, and bookstores, Kaiser suggested, "You may wish to quote statements that recognize this breakthrough, such as those by the American Congress on Surveying and Mapping [and two prominent German academics], given in Chapter 1." Snyder wrote that "ACSM's 16-member board was so incensed by this false inference that it unanimously passed a resolution asking that this misuse of its name be retracted, since the ACSM has taken no official position on the projection."

The Friendship Press acknowledged ACSM's objection by hand-pasting into the front of the handbook a correction slip identifying the original source of the offending quote and noting that the ACSM has never "made an official statement of its own on the Peters projection." Left unreported, though, were the strongly negative opinions of Robinson, Snyder, and other ACSM members who publicly denounced the map.

Obviously aware of this controversy, Kaiser dismissed Peters's critics in the "Questions People Ask" section of his handbook. "These people tend to be professional geographers or cartographers [who] form the elite of their profession. . . . But one wonders if they have looked lately at how often the maps we common people turn to are Mercators." He conceded the problem of distorted shape on the Peters projection but countered: "We know that some distortion of shape is inevitable if three fundamental properties are to be attained: fidelity of area, fidelity of axis, and fidelity of position." By getting his readers to accept the first requirement and quickly slipping in the other two, he could safely conclude that the Peters map was the only acceptable alternative to the Mercator map. "Dr. Peters has chosen what we [call] 'Column A' goals," he wrote, ". . . rather than absolute fidelity of shape. I firmly believe that is a wise choice, and his map is eminently useful because of that."

Peters's cause, Kaiser asserted, is as much social consciousness as cartographic fidelity: "The Mercator map . . . lends support to the assumption—unfortunately all too common even now—that nationalism or ethnocentrism or even racism is all right: that it is grounded in geographical realities." Although subtle and often unintended by its users, he explained, the Mercator projection promotes racism by enhancing the portion of the world inhabited by whites. In contrast, the Peters map "takes the world's people seriously. Far from exalting a (white) minority while relegating other peoples to relative obscurity, it provides every nation its rightful place."

Kaiser encouraged readers to take up the cause. His chapter on "Teaching a New World Vision" suggested discussion topics for social studies teachers or converts willing to address groups concerned with peace and nonviolence, race relations, or "world-mindedness." In a section called "What Can We Do?" he encouraged individuals or groups to display the map in their homes, lobby teachers and school boards, ask TV stations to integrate the Peters map with their newscasts, and write articles for their local newspaper. "The transformation of the world begins in the transforming of our minds," he concluded, "and the renewal of our minds begins with the transforming of the images we entertain: images we hang on our walls and images we carry with us in our heads."

The third wave rolled across England in 1989, when the Longman Group published the *Peters Atlas of the World,* and reached the United States a year later, when Harper and Row distributed an identical edition. The atlas has three principal parts: an 87-page section with relatively detailed topographic maps each covering a different portion of the earth, a 92-page section with less detailed worldwide maps each addressing a specific theme, and a 40-page index of place names. Included in a pocket at the back is a large, folded copy of the Peters world map, which the publishers also distributed separately to promote the atlas. Text at the bottom of the sheet points out such innovations and assets as "The World in True Proportion for the First Time," "Instant Orientation," and of course, "Fairness to All Peoples." In his foreword, Peters condemns the Mercator view of the world and its "Eurocentric way of thinking," and hopes his atlas "can contribute to an understanding of the causes of the North-South divide and the tensions between East and West."

For his topographic maps, Peters maintained fidelity of area by dividing the world's land area into forty-three equal-size, somewhat overlapping sections and setting a constant-area scale (each square centimeter represents 6,000 square kilometers). A world map on the front endpaper also serves as an index to these section maps, each spread across two facing pages. The Arctic and the Antarctic each account for four sections—quarter slices of polar pies framed by the familiar grid of concentric parallels and converging meridians. For the other thirty-five sections Peters mollified north-south stretching by centering the projection for each section map on a local meridian and a local parallel. As a result, sectional maps covering portions of Africa and Latin America have roughly similar scales in the north-south and east-west directions, instead of the long underwear–like vertical stretching of his whole-world map. But because page size is fixed and area scale is constant, Peters had to divide some tradition-ally intact world regions such as Central America and the U.S.S.R. among two or more section maps.

The second section ("Nature, Man and Society in 246 Thematic World Maps") has an impressive and intriguing table of contents. Each of the forty-five two-page layouts containing from one to six-teen world maps examines a specific theme. After a few layouts addressing physical geography (e.g., "Continents 280 Million Years Ago," "Mountains," and "Vegetation"), the atlas lives up to its human-istic focus with sheets devoted to such themes as education, life expectancy, sport, nutrition, prostitution, urbanization, and child labor. Layouts treating inequality and the status of women under-score a concern with human-rights issues.

Reviews were mixed. As might be expected, cartographers writing for academic journals were critical of the atlas's portrayal of shape. But they also found fault with poor detail for congested areas, incon-sistency in the spelling of place-names, thematic maps without dates, and Peters's failure to identify the sources of his data. Reviewers writ-ing for less technically sophisticated readers were generally more enthusiastic. In *American Reference Books Annual*, eminent geographer

Chauncy Harris labeled the atlas "intriguing and challenging [and] entirely new" and praised its maps as "clean, uncluttered, legible, and graphic." *Library Journal*'s reviewer spotted a few inaccuracies (such as Largo, Florida, in the same population-size category as St. Petersburg) but felt that "such occasional lapses should not dissuade librarians from acquiring this pathbreaking atlas." In the critical camp, *Wilson Library Bulletin*'s reviewer noted that partitioning countries such as Kenya among two or three sectional maps thwarts visual estimates of relative size, and Britain's conservative weekly *The Economist* (which publishes its own world atlas) concluded that Peters "appears to view western liberalism as a moral absolute, a perspective at least as Eurocentric as Mercator's."

I too am intrigued by the atlas's provocative contents, suspicious of its scholarship, and irked by the dismal, droopy-looking shapes of its continents. But as I pore over these pages of fascinating maps, I am puzzled by Peters's obsession with land area on maps addressing human misery, human needs, and human hopes. Peters's claim of "fairness to all peoples" seems less accurate than "fairness to all acres" because small but populous nations such as Indonesia, with over 190 million people living in an area of about 735,000 square miles, are less conspicuous on Peters's maps than, say, Canada, with only 26 million people but over five times as much land. Readers are thus much less likely to notice Indonesia's relatively high infant mortality rate on a map that apportions cartographic clout by land area, not people.

If I were Peters, my atlas would include some cartograms—perhaps half my thematic maps would have a demographic base like Figure 1.11. Cartograms are neither new nor esoteric; for several decades, in fact, textbooks for geography undergraduates studying cartography have shown how thematic maps might be made fair to all people, all households, all children, all domesticated animals, all houseplants, or all tractors. Indeed, in the early 1980s, Michael Kidron and Ronald Segal impressed reviewers, scholars, educators, and informed citizens of all ideological persuasions with their provocative *State of the World Atlas*, published in color as a trade paperback and sold widely for about $10—a modest price for an atlas. Many of their maps were cartograms based on measures including nations' shares of military spending, scientific articles, electricity generating capacity, and voting strength in the International Monetary Fund. Although Europe- and North America–centered, these cartograms had an important humanistic message.

What can we make of Peters and the controversy over his quaint but wildly successful projection? To be sure, the debate provoked the mapping priesthood to argue a bit more loudly for more realistic views of the world in classrooms and public displays. In 1989, for instance, seven North American professional and educational organizations endorsed a resolution asking publishers, government agencies, and the news media not to use *any* rectangular projection on maps for "general purposes or artistic displays." Repeated exposure might "have a powerful and lasting effect on people's impressions of the shapes and sizes of lands and seas." The resolution mentioned

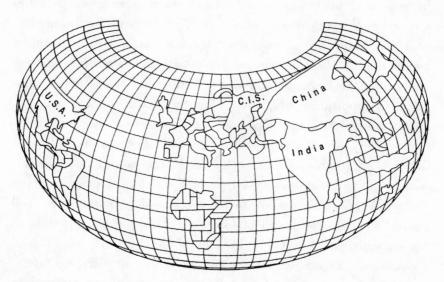

Fig. 1.11. A demographic base map based on 1980 populations. (© 1982 Mark Monmonier)

only one projection by name—the Mercator—but its authors quite pointedly observed that "other rectangular world maps proposed as replacements for the Mercator also display a greatly distorted image of the spherical earth." In a more proactive effort to inform editors, artists, students, and scholars, the American Cartographic Association's Committee on Map Projections has published three carefully written, well-illustrated booklets: "Which Map Is Best?" "Choosing a World Map," and "Matching the Map Projection to the Need."

Although cartographic scholars largely ignored Peters's arguments for fidelity of position and axis, he has won a few sympathizers for his argument that the type of map people commonly see conditions their view of the importance of places. Geographer Thomas Saarinen explored this idea by gathering 3,863 sketch maps by students from forty-nine countries. Although subjects typically centered local-area sketch maps near their homes, most of Saarinen's students centered their world maps on the Greenwich meridian. The prime exceptions were students in Australia, the West Coast of the United States, and other places near the edge of the traditional Europe-centered view. Saarinen observed that "a colonial mentality and Eurocentric image of the world still remains dominant in many places a quarter of a century after the end of the colonial era." Yet, while agreeing with Peters's diagnosis, Saarinen preferred improved geographic education focused on "modify[ing] old Eurocentric world images to fit a newer, more egalitarian concept" to the single-projection Peters prescription.

But can the Mercator map be blamed for the low geographic self-esteem Saarinen observed among Third World students? Not in the view of geographer John Pickles, to whom Europe-centered and north-up maps are effects, not causes. While agreeing that maps are not value free, Pickles chose to emphasize that "cartography shares and reproduces the values of the age." Among the various ways maps reflect values is their focus on political boundaries, property lines, physical features, and the built environment in contrast to their comparative neglect of public easements and the human environment.

Although cartography might be useful in promoting different, more egalitarian views of the world, maps themselves are more adept at reflecting existing images of the world than in forming or reshaping those images.

The late Brian Harley, a respected historian of cartography, interpreted the Peters controversy not as a Peters vs. Mercator debate about projections but as a Peters vs. cartographers struggle for authority and respect. Cartographers, Harley observed, see themselves as scientific defenders of an "ethic of accuracy." They don't regard maps as propaganda tools, however subtle, of government and science; they deny the effect of ideology on map design; and they don't appreciate challenges by outsiders. "The real issue in the Peters case," he wrote, "is power: there is no doubt that Peters' agenda was the empowerment of those nations of the world he felt had suffered an historic cartographic discrimination." The often irate, sometimes defensive, occasionally desperate tone of cartographers' objections to the Peters phenomenon underscores Harley's point.

The Peters controversy indicates not only broad ignorance about cartography but also the enormous persuasiveness of maps, even bad ones. Naïveté about maps and how they work explains the willingness of educated people—editors, reporters, religionists, and development officials—to take Peters seriously. Had these journalists and advocates received a firm foundation in both geography and geometry, the Peters phenomenon would have been impossible. But equally important is the power of even poor maps to organize and communicate information. Pardon the tautology, but if the Peters map were utterly misleading and totally useless, no one would have used it, not even Peters.

The Peters controversy also demonstrates that maps meet needs at different levels and play a variety of roles. For scientists and scholars, maps organize data and reveal patterns. For navigators and travelers, they aid in the selection of destinations and point out routes and landmarks. For utility companies and geologists, maps reveal underground facilities and rock formations. For state and local politicians and public administrators, they support land ownership, tax assessment, and land-use planning. For students, researchers, and newspaper readers, maps describe places and explain geographic relationships. But the role of many maps is purely symbolic or, in scholars' jargon, iconic — and in some cases ideological as well. Adoption of the Peters projection for a wall display, in a logo, for maps in a newsletter, or even for a T-shirt signaled that an individual or organization was socially conscious — cartographically inexact perhaps, but politically correct.

Whether an empathic emblem, political icon, or merely a cartographic illustration, the Peters projection enjoys the subtle benefit of a well-proportioned overall shape — specifically, the 1.57 ratio of its width to its height. In a 1974 lecture, Peters proudly observed that this ratio is very close to the "golden section" of 1.62 (for a thirteen-by-eight frame), which artists consider aesthetically pleasing. Hardly a prized asset in the technical world of the mathematical cartographer, the width-length ratio of his projection is indeed conveniently similar to those of most illustrations in newspapers and magazines. In contrast, Lambert's 1772 cylindrical equal-area projection (Figure 1.5) and Goode's interrupted projection (Figure 1.7) are much too elongated for the traditional page layout.

As a symbolic gesture, adopting the Peters map cost little in either money or inconvenience. The revolution of the Peters projection was largely visual, with little real effect on the daily routines of people, companies, governments, development agencies, or religious groups. Peters himself demonstrated this point when he proposed a far more disruptive change that attracted little if any attention or support. During a 1989 visit to Britain to promote his atlas, he recommended to the Royal Geographical Society that the zero meridian be moved from its present position through Greenwich to longitude 169° west — more or less through the Bering Strait. Adopting a new vertical axis not tied to any single nation would be a useful anti-imperialist measure, Peters suggested — but the enormous cost of revising time zones, relabeling maps, and disrupting lives worldwide would hardly be warranted by the ideological impact of a more equitable prime meridian. However revolutionary or well intended, this proposal went nowhere.

The world will survive the Peters projection, and might even become a slightly better place because of it. Maps are generally robust, and a unique projection at least encourages people to look at parts of the world they tend to forget. By challenging cartographers in the media, Peters inadvertently helped pull the map-projection genie all the way out of the bottle. Educators and publishers now are more fully aware of the Mercator map's distortions, and the news media are using an ever wider variety of world maps, thanks also to flexible, inexpensive mapping software. Among these users and makers of maps, at least, suggestions that a particular projection can be best, ideal, or even politically correct now seem dogmatic and silly. Indirectly the Peters projection has helped beat back some of the ignorance that allowed it to prosper.

• 2 •

Place-Names,
Ethnic Insults,
and Ideological
Renaming

Although some will still argue that the Eurocentric bias of the
Mercator projection is more imagined than real, geographic features named Nigger Lake or Chink Creek make any map blatantly
racist. Cartographic insults such as the toponym (place-name) at the
center of Figure 2.1 shock people, embarrass mapmakers, and make
maps controversial. Yet in less sensitive times mapmakers readily
adopted almost any place-name used and recognized by local people:
in the cartographer's mind, accurate representation was more important than the feelings of those defamed or offended. Without
present-day sanctions inhibiting local use of racially, ethnically, or
scatologically offensive feature names, the mapmaker perpetuated the

prevailing cultural landscape, however abhorrent. In most cases, an impolite, insensitive label was the only English-language name the feature ever had.

Offensive toponyms, never rampant, have become rarer still in recent decades because of the unfavorable attention they have received. Cartographers not only reject derogatory place-names as they make new maps but systematically remove them as they revise and update old ones. To many people, however, this geographic cleansing has been largely inadequate, as illustrated by later editions of the map in Figure 2.1 that merely renamed the feature Negro Lake.

Although they reflect the insensitivity of original namers and betray the cruel tolerance of racism by deceased cartographers, offensive toponyms also have inspired efforts to change local usage and eradicate cartographic insults. By calling attention to the cultural violence of past centuries, topographic maps can carry strong arguments for their own revision—especially when circulated beyond the locali-

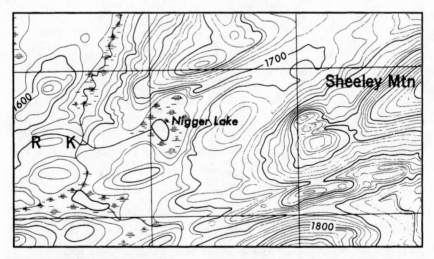

Fig. 2.1. The 1946 edition of the Canada Lake, N.Y., 7.5-minute topographic quadrangle map shows "Nigger Lake," a small, swampy body of water about one and a half miles south-southwest of the village of Canada Lake in Fulton County, New York. Map scale is 1:24,000, and the area shown is approximately 1.5 square miles. (U.S. Geological Survey)

ties they depict. The controversy can then multiply if residents insensitive to multicultural issues resist changes proposed by outsiders.

Names on maps have inspired a variety of cartographic controversies, nationalistic as well as racial and ethnic. Rather than focus on one or two specific incidents, this chapter examines three kinds of ideologically offensive or politically obsolete place-names. The first section looks at racially and ethnically insulting toponyms in North America. Although widely distributed geographically, derogatory place-names are nonetheless rare, and renaming them has been a slow and largely piecemeal process. The second section addresses widespread geographic renaming resulting from such political or military upheavals as the fall of communism in the Soviet Union and "ethnic cleansing" in the Balkans. A thorn in the side of commercial mapmakers such as Rand McNally and Hammond, these periodic cartographic revolutions are an accelerated version of America's efforts to rename its Jap Valleys and Nigger Hills. The final section explores the pragmatic and culturally sensitive effort in Canada and the United States to recognize place-names used and preferred by indigenous peoples.

Soiled Maps

Figure 2.2 represents a sampling of offensive feature names from maps drawn in the last few decades. The upper-left panel reproduces Chinks Peak in Idaho from a 1974 U.S. Geological Survey map. The name commemorates rather contemptuously the thousands of Chinese laborers who helped build the Union Pacific Railroad, which runs through Pocatello, five miles to the northwest. The panel to the right highlights Dago Gulch, shown on the 1989 edition of the Geological Survey's Greenhorn Mountain, Montana, quadrangle map. More than two hundred prospects and mine shafts scattered across the map reflect intensive gold and silver mining during the 1880s and 1890s. Someone probably named this small valley after a Hispanic prospector who lived nearby, and the name stuck. The lower-left panel shows Jap Valley, enlarged somewhat from the 1962 edition of the Price, Utah, regional topographic sheet. The Geological Survey began to remove offensive labels from its maps in the 1960s, and the 1966 edition of its Hayes Canyon, Utah, map sheet used the nonpejorative "Japanese Valley." In the panel at the lower right, the twin-peaked hill named Squaw Tits on the 1979 Big Horn, Arizona, topographic map simultaneously insults both women and North American Indians.

How common are such ethnically offensive place-names? I approached this question by examining the national index of the massive, eleven-volume *Omni Gazetteer of the United States of America,* published in 1991. As its preface acknowledges, the gazetteer's compilers relied heavily on the Geographic Names Information System, a government database cartographers identify by its euphonious acronym, GNIS (pronounced GEE-nis). Developed as a part of the Geological Survey's effort to computerize map production, GNIS contains not only the names of the nation's 55,000 topographic quadrangles but also official and variant labels for the approximately 1.5 million places and physical features these maps identify by name. As a late-1980s snapshot and summary of this huge electronic archives, the *Omni Gazetteer* distinguishes current official names, called "main entries" and printed in boldface, from "variants" printed in italics. Many derogatory names are variants—either no longer found on recent maps or destined for replacement when map sheets are revised. But like Chinks Peak, Dago Gulch, and Squaw Tits, most of the offensive geographic names I found are main entries, still lurking among the contour lines and hydrographic symbols.

Among the entries in the *Omni Gazetteer,* overtly derogatory names are rare yet conspicuous in their combined virulence and persistence. Some ethnic groups seem more or less disfavored than others. Although *China* is widely used as an adjective or prefix for more than a hundred instances of over forty different toponyms, such as China Creek, Chinamans Spring, and Chinatown, the derogatory form

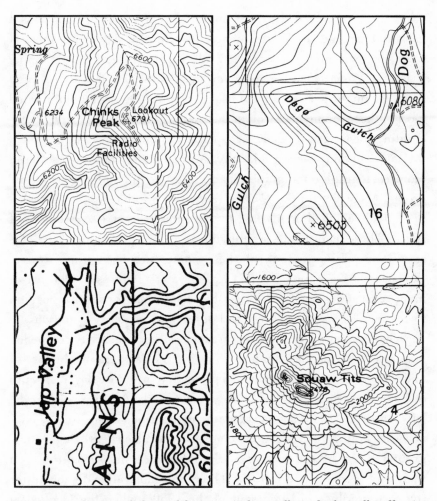

Fig. 2.2. A selection of physical features with racially and ethnically offensive names: Chinks Peak in Idaho (upper left), Dago Gulch in Montana (upper right), Jap Valley in Utah (lower right), and Squaw Tits in Arizona (lower right). (Scale of these excerpts from U.S. Geological Survey maps is approximately 1:100,000 for the lower-left panel and 1:24,000 for the others.)

Chink appears only three times as a main entry and only once as a variant. Hispanics are less fortunate: only three of sixteen instances of *Dago* are variants. In contrast, all eighteen features defaming Japanese Americans are now variants. Although the abbreviated form *Jap* might persist in local usage, mapmakers have rehabilitated most of these toponyms through the simple editorial expedient of substituting the longer, officially acceptable national adjective *Japanese.*

As the nation's largest minority, African Americans have suffered more cartographic indignities than any other nonindigenous ethnic group. According to the *Omni Gazetteer,* by the late 1980s cartographers had removed or marked for deletion at least 102 instances of *Nigger.* These names once identified features in thirty-four of the fifty states. Surprisingly, none of these names are listed for Georgia, Louisiana, Mississippi, South Carolina, and Tennessee. The list of names and states would be longer had I searched for second-position forms such as *Old Nigger Creek* and different derogatory synonyms.

What happened to all the features that ethnic sensitivity compelled mapmakers to rename? Figure 2.3 traces the cartographic history of one such place, a small northward projection into Port Bay, just south of Lake Ontario in Wayne County, New York. As illustrated in the left-hand panel, a 1943 government map identified the feature as Niggerhead Point. Town folklore connects the place with the Underground Railroad and tells of abolitionists who hid slaves there on their way to freedom in Canada. The center illustration demonstrates the cartographic enlightenment of the 1950s, when the Geological Survey renamed the feature Negrohead Point. Indeed, most of the *Omni Gazetteer's* 102 features listed with a *Nigger* variant now have a *Negro* main entry. The right-hand panel, reproduced from a 1977 quadrangle map published by the New York State Department of Transportation, suggests a second wave of renaming, after many African Americans rejected *Negro* in favor of *Black* in the mid-1970s. Perhaps to deal with this racial pejorative once and for all, the mapmakers borrowed the name of the road running north through the area and renamed the feature Graves Point.

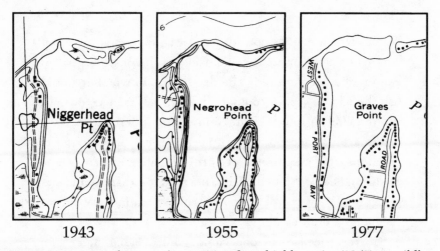

| 1943 | 1955 | 1977 |

Fig. 2.3. A geographic name's transition from highly racist (1943) to mildly racist (1955) to nonracist (1977). (Portions of 7.5-minute, 1:24,000-scale maps of the North Wolcott, New York, quadrangle. Left-hand map, enlarged from 1:31,680 to 1:24,000 for comparison, was produced by the Tennessee Valley Authority and the U.S. Geological Survey; the center map, a 1955 imprint of the 1953 edition, by the Geological Survey; and the right-hand map by the New York State Department of Transportation.)

Replacing offensive toponyms is the responsibility of the U.S. Board on Geographic Names. Established in 1890 by President Benjamin Harrison, the board is charged with standardizing names used for topographic maps and other federal publications and with resolving disputes when multiple names are in use. Although only federal agencies are bound by the board's decisions, state and local governments and commercial mapmakers recognize the advantages of standardization and usually conform. Three-quarters of the states have a names committee that works closely with the board. Proposals submitted to the board by individuals are referred to these committees for guidance. Although state committees may initiate their own proposals, ultimate authority resides in Washington.

The Board on Geographic Names is sensitive to names that offend citizens or embarrass the government. A U.S. Geological Survey handbook on federal cartography summarizes the board's policy:

Names which may be considered derogatory, obscene or vul-
gar, whimsical, or merely amusing or inconsequential are
continually encountered. The decision as to whether to publish the name as is, or to alter, abbreviate, disguise, or delete
is often subject to criticism. Names with a derogatory impli-
cation are not accepted for publication. Experience and the
reactions from map users have proved that good taste is much
easier to defend than vulgarity. If a solution that retains identification and yet avoids undue criticism cannot be devised,
the name is omitted.

Correct decisions concerning controversial names and
name changes must be reached before the maps involved are
published. All names of natural features are the responsibility
of the Board. . . .

Although the board will reject a new or replacement name it considers offensive, rehabilitating an established toponym requires a formal
request from a state or individual. Only twice has the board ordered
a general revision of racially offensive names. In 1963, at the initiation
of Secretary of the Interior Stewart Udall, the board approved
changing all instances of *Nigger* to *Negro,* and in 1967, again at the
request of the Interior Department, *Jap* became *Japanese.*

Because inspecting and editing 55,000 map sheets is a huge task,
these blanket changes took effect only when a map was revised or
reprinted. In some cases a map editor would drop a pejorative name
rather than change it, and after the mid-to-late 1970s, when *Negro*
became offensive, a *Nigger* label was deleted altogether. In 1982, the
Geological Survey applied the board's ruling electronically by substituting *Negro* and *Japanese* as main entries in the GNIS database.
According to Roger Payne, executive secretary of the Domestic
Names Committee, map users occasionally report instances of these
banned pejoratives on old stock sold by map dealers or on map sheets
not yet reprinted because of low demand.

The board changes other racially or ethnically offensive names when petitioned by an individual or state committee. Chinamans Spring in Yellowstone National Park thus became Chinese Spring because of a request by a Chinese American from Hawaii who was vacationing there and took offense. But any change requires a carefully researched proposal, and the substitute name must meet strict requirements. For example, features may not be named after living persons, and no proposal to name a feature after an individual will be considered within the first year of his or her death. In addition, the person for whom a feature is to be named should have had "either some direct association with the feature or have made a significant contribution to the area or State in which it is located," although an exception is made for "an individual with an outstanding national or international reputation." Because proposals from localities or individuals are referred to the state names committee (if the state has one), renaming can take several years.

Local opposition can block attempts to rename places, and a long-standing name often triumphs over civic leaders who find it offensive, embarrassing, or simply unattractive. Even risqué and kinky names have their backers, as in the case of Whorehouse Meadow, near Vale, Oregon. The area acquired its name after a local madam set up a tent city to accommodate horny cowboys. In 1968 a Bureau of Land Management employee with two young daughters proposed changing the name to Naughty Girl Meadow, and the board agreed. But the local newspaper and the Oregon Historical Society objected, and in 1983, a board less sensitive to colorful language reversed the decision.

Objections to the substitute name occasionally thwart efforts to eradicate racist toponyms. In 1990, for instance, Dorothy Igau opposed an NAACP proposal to rename Niggerhead Hill, a 1,300-foot peak on her cattle ranch forty miles north of Austin, Texas, after an African-American historian. Her family had owned the ranch since 1910, Mrs. Igau told a *New York Times* reporter. "I can see why it shouldn't be called 'Niggerhead,' but that's what everyone around here knows it as, and it's not bothering anyone." Although Igau was willing to accept "Warbler Hill" as a compromise, NAACP leaders objected and action was deferred.

When exposed in the news media, racist labels on maps often incite political reaction. In 1988, for instance, a simple press release from the New York Department of Environmental Conservation (DEC) triggered an executive order to find and eliminate all offensive toponyms. The story began when the DEC announced plans to release pheasants in a wetland area between Lake Champlain and Lake George — a wetland named Nigger Marsh. When a *New York Post* reporter asked for an explanation, DEC officials pointed to a U.S. Geological Survey topographic map of the area. Investigating further with local historians, the reporter traced the name back to the post-Revolutionary era, when a heroic black soldier named Prince Taylor helped lay out the area's earliest roads. Another story, from the 1830s, attributed the name to a black barge worker who was suffering from smallpox and put ashore to die. Despite these roots, this cartographic insult deeply offended Governor Mario Cuomo, who ordered state workers to fix the problem.

Executive Order 113 observed that derogatory place-names on official maps connotes governmental approval. To avoid this implied endorsement, the governor ordered that no map used or published by any state agency "shall contain any derogatory racial, ethnic or religious name or other epithet." State agencies were to flag derogatory names encountered while compiling new maps or revising old ones and to refer derogatory names to the State Committee on Geographic Names. The committee was to develop a review process "to identify and eliminate problem names from all State maps" and to promote "the prompt adoption of replacement names in consultation with the interested local community."

Curious about what had happened, I visited Albany in July 1992. In speaking with the State Museum official responsible for geographical names and examining the Historical Survey's file on derogatory place-names, I found little evidence of cartographic cleansing. The historian originally in charge of the effort had died suddenly in 1990, the Education Department had eliminated a number of positions in the State Museum, the remaining staff were overworked, and funding for research and travel had not been approved. Another historian had only recently been assigned to work on geographic names. Although several offensive names had been found, no recommendations for their deletion had yet been sent to the U.S. Board on Geographic Names. Without funding and follow-up, Mario Cuomo's words, however well intended, had little effect.

Despite minimal results, New York's derogatory names initiative raised the issue of whether some questionable toponyms are truly offensive or largely innocuous. In an example of bureaucratic overkill, DEC deputy commissioner and general counsel Marc Gerstman submitted a list of forty-five names that "the Department feels are offensive and should be considered for renaming." His list included thirty-eight features with obviously offensive names, but the remaining candidates for renaming seemed less compelling. Guinea Road and Spook Woods were ambiguous. Was Christian Falls really a religious epithet, or was it on the department's hit list largely to balance the inclusion of Jews Creek? Whom and in what way would the names Dingle Hole Road and Dingle Hole WMA (Wildlife Management Area) offend? Did Wappingers Falls, the name of a prosperous village south of Poughkeepsie, really offend Italian Americans? Even though the *Wap* in Wappingers might sound a bit like the pejorative *wop*, the village, the town, and a nearby creek are all named after the Wappinger Indians.

By November 1989, the State Committee on Geographic Names had devised a policy for accepting or rejecting questionable geographic names based upon a distinction between intent and impact. For example, although a place named Coon Hill to commemorate either an abundance of raccoons or an early settler named John Coon might reflect an acceptable intent, the perceived derogatory impact was clearly unacceptable. In the committee's eyes, "clear impact" was more important than intent—a name perceived as derogatory had to be changed even if the intent of the original namer was unknown or benign. Of course, an investigation of intent might conveniently suggest a straightforward nonderogatory alternative, such as Raccoon Hill or John Coon Hill.

The DEC's suggestion that the name Dingle Hole was in fact an obscure double entendre prompted research and discussion of its history. A footnote to a draft policy statement indicates a thoughtfully detailed analysis of whatever scatological meaning might be hidden here:

> Dingle Hole and Dingle Hole WMA do not appear at present to suggest derogatory intent or impact. To the Committee's knowledge the term "Dingle" or "Dingle Hole" does not refer to any ethnic group. If "Dingle" and its variants refer to sexual behavior and/or genitalia, the Committee is not aware of that fact.
>
> "Dingle" indeed can be easily traced to the Middle English of the 13th Century and describes a small wooded valley, a dell. A dingleberry is a shrub found primarily in the Southeastern United States. Thus Dingle Hole Road and Dingle Hole WMA may very well function as identifying names with non-derogatory intent. The Committee will investigate the possibility of derogatory and/or unacceptable sexual-reference impact before making a recommendation.

The committee apparently rejected the opinion of the game warden or environmental protection officer, who considered the name a crude reference to nonbotanical *dingleberries*, defined in *Newspeak: A Dictionary of Jargon* as "pieces of excrement clinging to a poorly cleansed anus." Just as well, I suppose.

In all of these examples, the decision to replace, suppress, or retain a geographic name is inherently and ultimately political. Renaming controversies commonly boil down to questions of who is offended and how severely, who has a vested interest in keeping the current name, and how much the local, state, or federal government is embarrassed by the status quo. Maps fuel these controversies by calling attention to embarrassing names, and groups or individuals assert their power by deciding for both the locality and the nation that a place-name is sufficiently offensive to be removed.

Place-Names and Geopolitics

Military conquest and political revolution can precipitate a toponymic slaughter far more extensive than bureaucratic skirmishes over offensive place-names. Because geographic names connote power and ownership, history shows that a victor eager to validate itself and replace an old regime will often erase toponyms commemorating the previous government's accomplishments, saints, and language. When local resistance makes political control questionable, relabeling the map is a convenient way to both assert and exaggerate the new regime's authority. If a region is to be annexed, new geographic names signify the new owner's intent to make the occupation permanent. Renaming also allows settlers to transplant part of their cultural and linguistic heritage—to make themselves at home by quickly and inexpensively redecorating the cultural landscape.

Meron Benvenisti, a former deputy mayor of Jerusalem and outspoken critic of the Israeli government's policy of expanding Jewish settlements in the West Bank, examined the deep significance of names on maps to both Israelis and Palestinians. In the epilogue to his insightful book *Conflicts and Contradictions*, Benvenisti focused on his father's life work, a new map of Israel with all place-names in Hebrew. Well before Israel became independent in 1948, the elder Benvenisti, a geography teacher, had been tracking down Hebrew place-names and plotting them on a set of maps of Palestine compiled by the British in the 1880s. Reconstructing the original Hebrew toponymy was a lengthy, laborious task requiring decades of field-work and archival research, but the effort was important to the Zionists, who had returned to their homeland and sought to become once again an indigenous people. Benvenisti senior considered his work a sacred trust, and when research could not produce a place or feature's original Hebrew name, he would create a descriptive name based on local history and geography. In most cultures, indigenous names are more often descriptive than commemorative.

Paradoxically, the new Hebrew map of Israel relied heavily on Arab maps of Palestine. When the Romans conquered the original Semite inhabitants in the first century B.C., they ordered the use of Roman place-names, which the indigenous Christianized Semites dropped after the Roman Empire withered in the fourth and fifth centuries. When the Arabs conquered Palestine in the seventh century, most Palestinians converted to Islam and translated their indigenous geographic names into Arabic. The Crusaders, who conquered the Islamized indigenous population several centuries later, also imported their own place-names. Citing toponyms like Château-Perelin and Montfort, Benvenisti described the Crusader map of the Holy Land as a transported map of southern France. But indigenous names are surprisingly persistent, and when the Crusaders withdrew, local usage restored Arabic versions of many of the original Semitic descriptors.

The 1970s and 1980s witnessed vigorous cartographic change throughout the West Bank, which Israel had captured from Jordan in the 1967 war. Figure 2.4, an excerpt of a map published in early 1984 by the Central Intelligence Agency, describes the expansion of Jewish settlements across the occupied West Bank, from Israel eastward to the Jordan River. Hebrew toponyms suggest a permanence only slightly diminished by the note "Israeli occupied—status to be determined" below the name *West Bank*. The CIA, which monitored the area from satellites, apparently used the label "site" to identify very new settlements not yet assigned an official name.

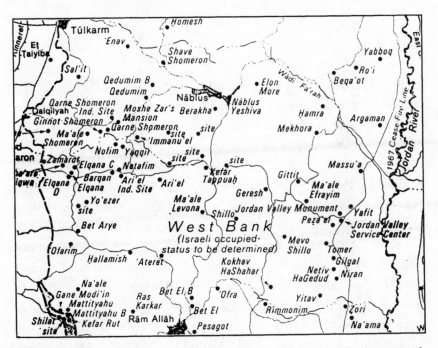

Fig. 2.4. Portion of CIA map titled "Israeli Settlements in the West Bank, September 1983."

As different views of a common terrain, the Israeli and Palestinian maps signify not only different cultural perceptions of the region but the bitter polarization of mortal enemies. By creating new settlements and constructed features, the Israelis have had a more profound impact than earlier regimes. As Benvenisti observed, "We have done more than create a paper empire. We have actually transformed the physical reality, built cities, drained marshes, made the desert bloom. We not only eradicated Arab place-names, we actually destroyed the places as well." The Palestinians deeply resent this, of course, and their research institutes in Beirut and refugee camps throughout the region feed the hunger for a restored Palestinian homeland by persistently promoting a pre-1948 map of Palestine that denies the existence of Jewish settlements, boundaries, and place-names. The conflict between Israelis and Palestinians has hardened to the point where, in Benvenisti's words, "maps cease being geographical and turn into an act of faith, a call for action, for revenge."

. . .

In the decades following World War II, Britain, France, the Netherlands, and other European countries with vestiges of an overseas empire withdrew from dozens of colonies in Africa, Asia, and Latin America. Pakistan illustrates the widespread geographic renaming typical of a former colony eager to assert its political and cultural independence. The British controlled the area from the early nineteenth century until 1947, when they separated Pakistan from India and granted both nations independence. As did the Indians, the Pakistanis promptly set about renaming their country's cities, streets, and administrative areas. Campbellpur thus became Attock City, Montgomery reverted to Sahiwal, and networks of streets with names like Brewery Road, Garden Road, and Mission Road now honor Liaquat, Tipu Sultan, Aurengzeb, and other Pakistani leaders. In contrast, physical features seldom required renaming because the British had readily accepted local names for mountains, passes, and rivers. Although British names persist for some peaks and coastal features, such as Black Mountain and Hawkes Bay, these comparatively inconspicuous exceptions might eventually fall to the demands of nationalists eager to obliterate all traces of colonial subjugation.

A few British town names have lasted as well, most notably Jacobabad, an administrative center founded in 1851 by General John Jacob, a colonial administrator who gained the admiration of local people. Jacob not only brought peace to an area plagued by marauding bands of thugs but initiated many public works and other local improvements. Ten thousand native mourners attended his funeral, and local residents persistently reject proposals to change the toponym that commemorates this respected historical figure. Clearly an exceptional case in Pakistan, Jacobabad illustrates the effectiveness of local opposition to changing a name that residents themselves consider ideologically appropriate.

· · ·

While civil war often prompts geographic renaming, as the case of Cyprus illustrates, secure factions can afford to be more tolerant than their opponents. Before civil war led to a partition of the island in 1974, about 120,000 Turkish Cypriots lived somewhat uneasily among a Greek Cypriot population four times larger. During the summer of 1974, a Greek junta overthrew the coalition government led by Archbishop Makarios and aroused Turkish fears of a forced merger of Cyprus with Greece. To protect the interests of the Turkish Cypriot minority, an army from Turkey invaded the northern part of the island. As many as 65,000 Turkish Cypriots moved north while up to 200,000 Greek Cypriots fled or were forced southward. The Turkish Cypriots proclaimed their independence in 1975, and although only Turkey recognizes their northern republic, the Turkish faction changed all Greek place-names to signify their control of the region. In contrast, Greek Cypriots condemned the succession as illegal but made no attempt to eliminate Turkish toponyms. The need to rename places is perhaps strongest among people insecure about their territory.

Strong shifts in political power or ideology and their corresponding changes in territorial boundaries become a recurrent problem for both map publishers and customers requiring up-to-date atlases and wall maps. Take Europe, for example. World War I and the Treaty of Versailles, German and Russian machinations in the mid-1930s, and World War II and its aftermath have all kept mapmakers repeatedly redrawing boundary symbols and updating geographic names. Even more boundary lines shifted when the recent fall of the Communist Party precipitated the reunification of Germany, the breakup of Yugoslavia and Czechoslovakia, and a perplexing restructuring of the Soviet Union, now known—but for how long?—as the Commonwealth of Independent States. Furthermore, the newly independent Soviet Socialist Republics often adopted new names or preferred spellings, as when Belorussia became Belarus, Moldavia became Moldova, and Tadzhikistan became Tajikistan.

Even before the Soviet Union bowed to free-market principles, ideological renaming allowed the government to posthumously purge its discredited leaders. In the late 1950s, for instance, Stalingrad became Volgograd, and in 1988 the city of Brezhnev became Naberezhnive Chelny (meaning "Dugout Canoes on the Riverbank"). Indeed, the list of Soviet cities renamed since the mid-1980s reads like a litany of Stalin's lieutenants and other party notables. Political renaming accelerated in 1990 as Mikhail Gorbachev promoted economic reforms, and in 1991, after years of delay attributed to Vladimir Lenin's prestige as the U.S.S.R.'s founder, Leningrad again became St. Petersburg. Because renaming almost always restored a place's pre-1917 name, a 1991 *Economist* article observed that "every time a city changed its name, the act symbolically turns the clock back 74 years."

"Ethnic cleansing" in the Balkan states is the most recent conflict that will profoundly alter modern maps of Europe. Differences run deeper than religion and alphabet for the religious and national groups in this region who are overcome by mutual hate and mistrust. Following World War II, Marshal Tito's dictatorial Communist government held together an ethnically diverse population of Serbs, Croats, Slovenians, Montenegrins, Macedonians, Albanians, Hungarians, Romanians, and Germans. Tito's death in 1980 and the worldwide weakening of socialism later in the decade encouraged various factions' hopes for independence. In 1991 and 1992 the union of Yugoslavian republics disintegrated amid declarations of independence by Bosnia-Herzegovina, Croatia, Slovenia, and Macedonia. In a genocidal offensive reminiscent of Nazi atrocities, Serbian Christians displaced tens of thousands of Bosnian Muslims. Serbs also claimed parts of nearby Croatia, expelled Croats, and expanded their territory. On a smaller but equally vicious scale, Croats drove Serbs out of Croatian districts in Bosnia and Croatia. As a result of murder, bombings, and forced evacuations, once heterogeneous settlements are now homogeneous, and the ethnic map of the Balkans is radically different. Although some argue the new ethnic distribution is more stable, the toponymy of the Balkans promises to remain fundamentally volatile as hostilities continue to intensify.

We can explore yet another kind of political renaming in Canada, where official policy accords English and French equal status and so requires both bilingual toponyms and bilingual maps. As a move to appease Québec successionists, "bilingualization" is most prominent at the federal level. In 1960, the Surveys and Mapping Branch of the Department of Energy, Mines and Resources began to produce partly bilingual topographic maps. Notations in the map sheet's margin, or collar, were printed in both English and French, as were large, important features such as "St. Lawrence River–Fleuve Saint-Laurent." The names of places and smaller features were written according to their most common local form—any Canadian sufficiently literate to use a map would surely know, for instance, that *lake* and *lac* are equivalent—and generic labels such as *golf course* and *hospital* were still written solely in English. But beginning in 1977, on map sheets covering Québec, labels for these generic features were converted from English to French. On all sheets, though, a glossary in the collar translates these generic labels into the other language, and unusual features such as mink farms are labeled in both languages. Because densely packed feature symbols thwart bilingual labeling, the *Atlas of Canada* and maps covering large areas at scales smaller than 1:1,000,000 are published in separate English and French editions.

Individual Canadian provinces are less committed to bilingualization. Only New Brunswick is officially bilingual, and the government of the Northwest Territories has officially recognized both English and Inuit, an aboriginal language spoken by the majority population. Elsewhere in Canada provincial maps are essentially unilingual—fiercely so in Québec, where language bills introduced by the Parti Québécois in the 1970s and early 1980s restrict the use of English in government, education, and commerce, as well as for public signs, maps, and place-names. As part of this crusade, the Commission de

Toponymie has been promoting the purity of the province's linguistic landscape by renaming populated places and physical features, usually by translating English names into French.

In contrast to the comparatively passive geographic-names authorities in Canada's other provinces, the Commission de Toponymie du Québec is actively prescriptive. In 1987, the commission published a massive gazetteer listing 93,379 officially recognized place-names, up by over 60,000 names from a provincial gazetteer published in 1969. A table classifying names by linguistic group observed that the French share had risen from 69 to 72 percent while the English share declined from 20 to 12 percent. Equally significant, the Amerindian share had grown from 4.7 to 7.5 percent and the Inuit share from 0.4 to 1.9 percent. Ironically perhaps, in asserting French linguistic hegemony, the Québec government has demonstrated an exemplary respect for the geographic names used by Cree, Inuits (Eskimos), and other native peoples. None of Canada's other nine provinces preserves and promotes indigenous place-names as effectively. By accepting native toponyms on its maps, Québec can remove symbols of English control as well as assert the supremacy of locally dominant languages in general.

Diverse motivations for political renaming in Israel, Pakistan, Cyprus, Eastern Europe, and Canada attest to the map's power as a symbol of ownership and control. Much of this power reflects the ease with which toponyms mark territory. Because names and other linguistic descriptors are to humans what smells are to animals, renaming is the human equivalent of a dog eagerly and systematically sprinkling bushes, trees, fence posts, and fireplugs. Maps complement public signs in encouraging the local population to accept official names; and by organizing a broad array of geographic names on a scaled-down representation of the landscape, a map can assert the existence of a nation as well as advertise its territorial claims to a wider audience of neighbors and world powers.

Going Native

Renaming a colony's geographic features can be as intimidating and haphazard as redecorating a home. Yet while a homeowner or tenant can conveniently ignore former occupants, the colonist usually finds exploitation simpler and more profitable than annihilation. Since the native people already will have named most places and physical features worth naming, the colonist who wants to hire or trade with them must assimilate their toponyms. Language differences can lead to mangled adaptations, however, and further modifications to improve pronunciation and spelling have distorted many native place-names beyond recognition. Because North American Indians, for example, relied largely on mental maps, cartographic folklore, and ad hoc drawings rather than a permanent, tangible, easily inherited cartographic record, many American Indian place-names not appropriated by the Europeans simply disappeared when the aboriginal inhabitants moved on, died, or were assimilated.

As novelist and language expert George Stewart pointed out in his classic 1945 study *Names on the Land,* the American colonists appropriated American Indian names in three different ways. Most commonly they merely listened, pronounced what they heard, and transcribed the result. But phonetic transfer is rarely precise because of poor pronunciation, bad hearing, and unfamiliar native sounds that might be transcribed in several different ways. Consequently, adapted place-names often have a variety of spellings — Winnipesaukee, for instance, has at least 132 variations. A second practice, which Stewart called *folk-etymology,* applies a familiar spelling or English-sounding word to part of the native name, as in Kingsessing and Westkeag. When folk-etymology could supply all parts of a native place-name, the result could look and sound very English, as with Lamington, Pompton, and Wantage. The third method translates the American Indian name into English and then substitutes a concise synonym. As an example, Clearfield, Pennsylvania, avoids the awkward literal translation of the native name meaning "left the face of the country as bare as though it had been cleared by a grub-axe."

European settlers sometimes shifted a name from one feature to another or applied it to several nearby features. In Michigan, for example, the Algonquian term *siskowit,* which refers to a kind of fish, now identifies not only the Siskiwit River but Siskiwit Falls, the Siskiwit Islands, and Siskiwit Mine. In the case of rivers, name shifting can have far-reaching effects. In Maine, for instance, the Penobscot River now carries all the way to the Atlantic the native name for a small stream north of Bangor. Farther west, the French created the convenient image of a vast, integrated river system by extending the Algonquin toponym *Mississippi* all the way from Minnesota to Louisiana. As further testimony to the power of naming, the name *Mississippi River* displaced dozens of local names used by natives living along its course.

Indigenous names are making a comeback. Although the U.S. Board on Geographic Names has long considered American Indian names a valuable part of the nation's cultural heritage, a policy adopted in 1990 deliberately promotes names derived from American Indian, Inuit, and Polynesian languages. For features on tribal lands, changing a nonnative name to an indigenous one is fairly easy, unless the current name is widely used and the state withholds support. For unnamed features off the reservation, the board eagerly accepts American Indian names that are "linguistically appropriate to the area." Especially suitable are distinctive names that describe the terrain, relate to the locale's natural history, or commemorate local folklore or a noteworthy incident.

The board does confront several impediments to a wider use of native place-names, however. In addition to its restriction that proposed names should be easy to pronounce and should not duplicate names used elsewhere within the state, the board sometimes finds that names one tribe prefers might be considered offensive by another. Furthermore, because the land holds deep religious signifi-

cance, American Indians are reluctant to reveal sacred names for profane display on printed maps.

Another impediment is the powerful politician who favors a feature's current name. Perhaps the most outrageous example is Representative Ralph Regula, an Ohio Republican who has used his
seniority and position on the House Interior and Insular Affairs
Committee to block renaming Mount McKinley. Regula's resistance
first appeared in 1975, when the Alaskan legislature petitioned the
Board on Geographic Names to officially endorse the Athabaskan
name *Denali,* meaning "large or great one," for North America's highest peak (20,320 feet). Although most geography and travel books
identify Denali as Mount McKinley—a name the board endorsed
hurriedly after President William McKinley was assassinated in
1901—Denali is a much older name and far more significant to local
residents. Alaskans have requested its restoration off and on since
1913, but before the board could act, Congress intervened on behalf
of McKinley. Does Regula's persistent introduction of bills asserting
congressional satisfaction with the name McKinley mean he is pointedly anti-Athabaskan? Not really, say associates; his district includes
much of the region that fellow Republican McKinley represented in
Congress before becoming president, and in Regula's mind Ohio history obviously supersedes Alaskan toponymy. Because the board
refuses to rule on a change also under consideration by Congress,
Mount McKinley's commemorative name is safe as long as Regula or
a successor keeps reintroducing the McKinley bill.

Official adoption of native place-names can be an important sign of
political recognition and cultural respect. Local toponyms have
become especially important in Canada's Northwest Territories,
where a small and widely scattered Inuit population is nonetheless far
larger than the English and French minorities, as well as in parts of
Québec where Cree outnumber other groups. In the 1980s Canada
began to adapt the official map to local usage by substituting indigenous place-names for the ones imposed by European explorers and

the Hudson's Bay Company. In 1987, for example, the municipality of Frobisher Bay, the principal center on Baffin Island, became Iqaluit, which means "place of fish" in Inuktitut. In northern Québec many geographic names evolved in two stages: first from English to French, and then from French to Inuktitut. George River, for instance, became Port-Rouveau-Québec in 1965, and Kangiqsualujjuaq in 1981. Alan Rayburn, executive secretary of the Canadian Permanent Committee on Geographical Names, believes that "by the end of the century there may be few English and French names of native communities left in Canada."

Although these changes signify recognition of native place-names by government mapmakers, phonetic transfer alone does not meet the cartographic needs and ambitions of the Inuit. Inuktitut has its own alphabet, and geographic names transcribed in Roman letters are meaningless if not insulting to a people eager to preserve a language that is written as well as spoken. To create their own cartographic text, the Inuit recently launched a series of trilingual topographic maps, produced by integrating Inuktitut words and writing with 1:50,000-scale bilingual Canadian maps. In the collar surrounding the map, two alphabets display the sheet name and interpretative information in English, French, and Inuktitut. Within the map, prominent numeric codes link individual lakes, islands, hills, camps, and other features with an extensive trilingual key that identifies them by name. An Inuit-owned corporation used land-claim payments from the Canadian government to underwrite the map compilation and publication, and the Avataq Cultural Institute, which Inuit leaders established in 1981, compiled the native toponyms in cooperation with McGill University's Indigenous Names Surveys project. Separate copyright notices by the institute and Canada's federal mapping establishment reflect a cartographic dialogue between indigenous map users and the European Canadians who had mapped and appropriated their lands.

Historians and educators have been no less biased than government cartographers in ignoring indigenous populations: maps in most textbooks on the history of North America grossly misstate the presence of native peoples in relation to the frontier of European settlement. Figure 2.5, which appeared in *Studies in American History* above the title "Reference Map for Periods of Settlement and Revolution, Middle Atlantic States," is a good example. Omission of native settlements, place-names, routeways, and territorial boundaries implies vast empty areas with no one to conquer, swindle, decimate with disease, or displace onto so-called reservations. Published in 1898, *Studies in American History* is similar to dozens of modern high school history texts in its disrespect of North America's first nations: although an earlier map or two might use broadly curving labels such as "Algonquin" or "Iroquois" to show the approximate territories of aboriginal tribes, these maps typically represent the pre-European population as an obstacle to overcome, much like the Appalachian Mountains and their formidable Allegheny Front. A fairer, more informative view might juxtapose a map of European ports, forts, and frontier settlements with a contemporaneous view of native settlements, important trails, trading centers, hunting grounds, and farmland. Especially useful for explaining conflict and exploitation are maps that overlay the indigenous and European landscapes in contrasting hues.

In the late 1980s, multicultural critics of public education occasionally indicted textbook and wall maps for ignoring or misrepresenting native peoples and minority cultures. The media's focus on shrill charges of institutional racism and indignant countercharges of "rewriting history" quickly shifted attention from the adequacy of educational materials and course content to ideological skirmishes between the word police of the right and the left regarding whether slaves should become "enslaved persons" and the Middle East called "Southwest Asia and North Africa." This lexicographic nit-picking is unfortunate: if the multicultural movement becomes distracted with a discourse that inflames rather than enlightens, its cartographic legacy might be little more than a smattering of cosmetic revisions or the ideologically motivated replacement of one narrow-minded, single-map view with another.

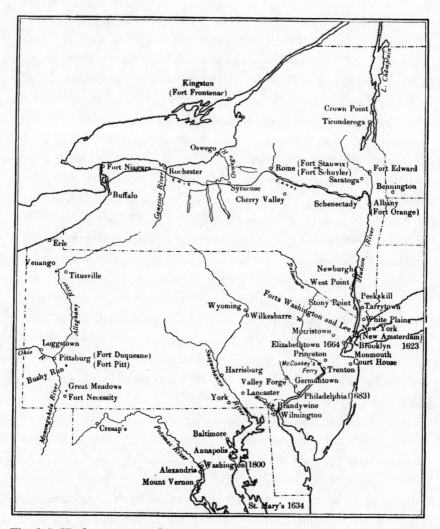

Fig. 2.5. "Reference Map for Periods of Settlement and Revolution, Middle Atlantic States" suggests that areas without European settlements were largely uninhabited. (From Mary Sheldon Barnes and Earl Barnes, *Studies in American History* [Boston: D. C. Heath and Company, 1898], map facing p. 120.)

. . .

The diverse examples examined in this chapter demonstrate that controversial geographic names can make maps controversial, and vice versa. By advertising ethnically and racially derogatory place-names beyond localities that once used and perhaps still tolerate them, maps incite controversy and stimulate change. And by transforming prejudice toward native language and culture into a tangible, written form, maps encourage protest from indigenous peoples and culturally sensitive supporters eager to integrate American Indian, Inuit, and other minority toponymies into the official cartographic record. Controversies over place-names reveal the map as a double-edged sword: an instrument of arrogance in times of oppression and exploitation, and a tool for empowerment in an era of tolerance. Maps are seldom value-neutral, but by exposing the past values that created them, maps can trigger their own evolution.

· 3 ·

The Vinland Map, Columbus, and Italian-American Pride

One of the great lies of cartography is the Vinland map. A world map with a crude representation of eastern North America, the Vinland map's fraud was its alleged age. Purportedly drawn around 1440, the map showed portions of the North American continent "discovered" by Columbus a half-century later. As a mid-fifteenth-century copy of an old Viking map, the document was enormously significant because historians had little other evidence that Europeans knew so much about the New World before 1492. Italian Americans, however, considered the Vinland map an attack on their hero, whom they had long honored each year on October 12. Even before Congress redefined Columbus Day as the second Monday in

October and made it a national holiday, many citizens shared the celebration with parades, speeches, markdown sales, and in some cases a day off. But in 1965 the announcement of this mysterious medieval map cast a shadow over the festivities and seriously threatened the gains of Columbus Day advocates still lobbying for nationwide recognition of the day. By depicting a tenth-century Viking view of the New World, the map even raised the possibility of moving the mid-October celebration ahead three days to October 9, Leif Ericson Day. That a prestigious Ivy League university chose October 11 to launch its carefully orchestrated publicity campaign for the map was taken as further insult, and because of the success of that campaign, Italian Americans had to endure nearly a decade of slurs against Columbus and the significance of his landing in 1492. Much to their relief, in 1974, scientists confirmed the suspicions of many historians and cartographic scholars by declaring the map a fraud. The Vinland map's hyped introduction to a gullible public, its enthusiastic reception by scholars and politicians, its shady sale to Yale University, and its exposure as a clever forgery make a fascinating tale of carto-controversy.

When it's not on exhibit, the Vinland map lies in repose in New Haven, Connecticut, in the Beinecke Rare Book and Manuscript Library at Yale University. Figure 3.1 is a photographic reduction of the much larger original, which measures approximately 11 by 16 inches. Brown ink with black specks records a simple, hand-drawn map of the world on a thin sheet of vellum, or parchment, with a vertical fold through the center and four pairs of worm holes. Ten square vellum patches on the back cover these eight holes, as well as two others apparently caused by the papermaker's knife, and a vertical vellum patch covers the back of the fold where the sheet had cracked. The map's outline is roughly a horizontal ellipse with the principal axes rotated slightly clockwise. Readily evident at the center is the Mediterranean Sea, visually reinforced by the well-known coastal signatures of France, Spain, Italy, and Greece. Europe's image is more accurate than those of Africa and Asia. The left third of the map describes the Atlantic Ocean, which includes two large islands west of Spain and North Africa; known to historians of cartography as the Fortunate Isles, these gross exaggerations of the Canary Islands are common on medieval world maps. At the upper left, progressively farther from the Scandinavian coast, are Iceland, Greenland, and an islandlike part of North America. Inscriptions include sixty-two geographic names and seven longer passages, or legends, describing in Latin the character or significance of specific features. The label to its upper right identifies this leftmost feature as the "Island of Vinland, discovered by Bjarni and Leif in company."

The comparatively lengthy legend above Vinland and Greenland is the reason for the island's—and the map's—significance. Cartographic historian Peter Skelton, perhaps the foremost student of the Vinland map, provided a translation:

> By God's will, after a long voyage from the island of Greenland to the south toward the most distant remaining parts of

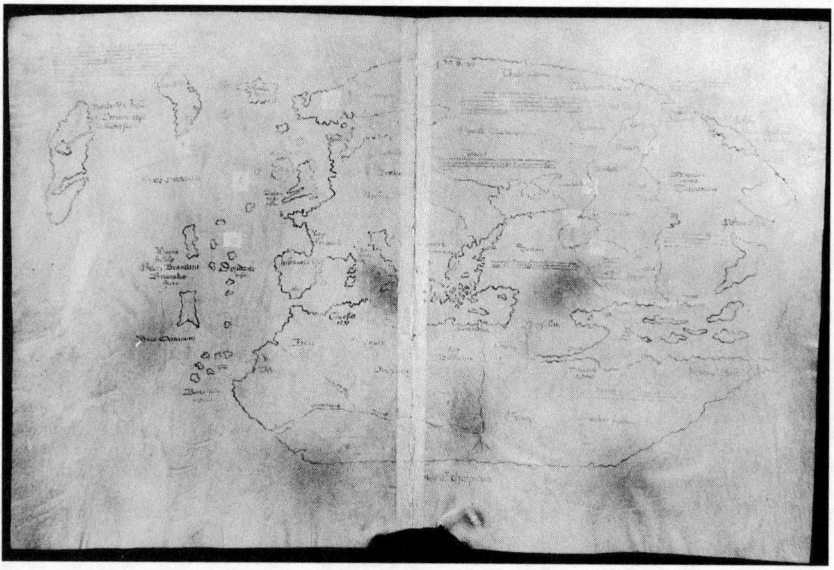

Fig. 3.1. The Vinland map. (The General Collection of Early Books and Manuscripts, Beinecke Rare Book and Manuscript Library, Yale University)

the western ocean sea, sailing southward amidst the ice, the companions Bjarni and Leif Eiriksson [*sic*] discovered a new land, extremely fertile and even having vines, the which island they named Vinland. Eric, legate of the Apostolic See and bishop of Greenland and the neighboring regions, arrived in this truly vast and very rich land, in the name of Almighty God, in the last year of our most blessed father Pascal, remained a long time in both summer and winter, and later returned northeastward toward Greenland and then proceeded in most humble obedience to the will of his superiors.

Tales and sagas of Norse voyages suggest that Bjarni Herjolfsson accidentally discovered Vinland perhaps as early as 985 A.D., that Leif Ericson explored Vinland around 1000, and that Bishop Eirik Gnupsson visited the region in 1121, more than three and a half centuries before Christopher Columbus sighted Watling Island, in the Bahamas, on October 12, 1492.

There is little doubt the Vikings explored North America nearly five hundred years before Columbus, but because of hostile natives, their brief encampments are recorded largely in quaint sagas, whereas the Columbian encounter of 1492 initiated widespread European colonization. Although the educated public was well aware of Icelandic tales of Leif Ericson's adventures, and anthropologists had even found artifacts suggesting a Viking settlement in Newfoundland, the Vinland map gave these intermittent Viking settlements renewed prominence. Moreover, scholars hailed the Vinland map as the first pre-Columbian European map of North America, and in 1965 *Time* magazine called it "by far the most important cartographic discovery of this century."

Figure 3.2, a more detailed view of the upper left portion of the map, shows Vinland as an elongated island with a north-south orientation. It is larger and farther south than Greenland, with two deep bays or inlets dividing its east coast into three peninsulas. The upper indentation, ending at a large inland lake, vaguely reflects Hudson's

Bay connected to the Atlantic by the Hudson Strait, whereas the lower penetration, markedly wider and inclined toward the southwest, resembles the Gulf of St. Lawrence. Although *Time* suggested this interpretation as "unmistakable," Skelton and other scholars recognized two other plausible correlations with modern features. In addition to a northern hypothesis associating the three peninsulas with Baffin Land in the north, Labrador in the middle, and the Canadian Maritimes at the southern end of Vinland, a southern hypothesis linked Vinland's upper peninsula to Cape Cod. Although geometric similarity clearly favors the northern hypothesis, some accounts of Norse and Icelandic voyages tie Vinland to southern New England. The British Museum's George Painter summarized the uncertainty when he called the Vinland map "a generalized and degenerate simplification of saga narratives [that can be] wrestled to support any possible theory [although] it gives evidence for none." Because the wavy appearance of the comparatively smooth western coastline reflects a style of illustration common throughout the map, scholars made no attempt to match these minor indentations to real geographic features. By comparison, the map is extraordinarily accurate in its representations of Greenland and Iceland.

When it first came to scholars' attention, the Vinland map was creased and bound at the front of a manuscript called the *Tartar Relation*, a twenty-one-page, double-column account of Franciscan friar John de Plano Carpini's expedition to central Asia in 1245–47. In this manuscript, friar C. de Bridia described the history, social customs, and religious beliefs of the Mongols, and recorded the observations of friar Benedict the Pole, who had accompanied Carpini. The link between the Vinland map and the *Tartar Relation* was puzzling, though, because the latter makes no mention of Vinland, Greenland, or the Viking exploration of North America. Moreover, because none of the eight worm holes in the map align with worm holes in the *Tartar Relation,* the map and the manuscript obviously were not originally bound together. Yet the map's text for parts of Asia suggests its compiler had used the *Tartar Relation* as a principal source.

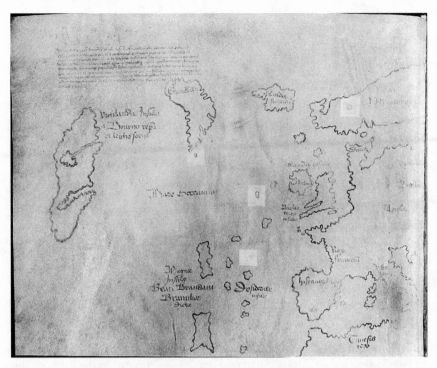

Fig. 3.2. Enlargement showing the map's treatment of Vinland, Greenland, Iceland, and western Europe. (The General Collection of Early Books and Manuscripts, Beinecke Rare Book and Manuscript Library, Yale University)

A lone inscription on the back of the map—"Delineation of the first, second and third parts of the Speculum"—suggests a link among the map, de Bridia's manuscript, and the *Speculum Historiale* (Mirror of History). Written in the mid–thirteenth century by Dominican friar Vincent de Beauvais, a prominent historian and encyclopedist, the *Speculum* is a comparatively lengthy account of world history, of which a number of copies survived. Strong evidence suggested that a copy obtained by Yale was once bound together with the map at the front and the *Tartar Relation* at the back: not only are the leaves and script similar but the worm holes at the front of the *Speculum* match those in the map, whereas the holes at the back match those in the *Tartar Relation*. As scholars interpreted the evidence, the three manuscripts had once been bound together but were separated after the original binding deteriorated; the Vinland map and *Tartar Relation* were then rebound together as a shorter volume, whereas the *Speculum* was set aside and repaired only recently. How they made this connection seems more fortuitous luck than shrewd detective work, however, and their explanation has numerous gaps.

We can trace the shorter volume containing the Vinland map and the *Tartar Relation* back to the summer of 1957, when Enzo Ferrajoli, a bookseller in Barcelona, showed it to rare-book dealers in Geneva, London, and Paris. That summer, Ferrajoli and London bookseller Joseph Irving Davis brought it to the British Museum for a brief examination by Peter Skelton, who was superintendent of the Map Room, and several experts on medieval manuscripts. Because of errors in the map's Latin inscriptions, the medievalists were skeptical. But in Geneva that September, Laurence Witten, a New Haven, Connecticut, dealer in rare manuscripts, bought it from the owner for $3,500, plus a commission for Ferrajoli. The original owner was never identified. Witten returned to New Haven and showed the manuscript to Thomas Marston, Yale's Curator of Medieval and Renaissance Literature, and Alexander Vietor, Curator of Maps. Although Vietor secured Yale's right of first refusal in the event of a later sale, Witten gave the manuscript to his wife.

In April 1958, Marston received a catalog from Davis and Orioli, a London antiquaries firm run by Joseph Irving Davis, who had accompanied Ferrajoli to the British Museum. In the catalog, he noticed a portion of the *Speculum Historiale* for sale at a modest £75. Marston had found de Beauvais's writings useful in previous research and ordered it, together with another manuscript. When his purchases arrived three weeks later, Marston was pleased with their "very unusual contemporary bindings" and invited his friend Witten to inspect them. According to Marston, "Mr. Witten came to my office late that afternoon, looked at the manuscripts, and asked if he could borrow the Vincent for a few days. I readily acceded. That evening I did not return home until after ten o'clock. I had hardly entered my house when the telephone rang. It was Mr. Witten, very excited. The Vincent manuscript was the key to the puzzle of the map and the *Tartar Relation*. The hand was the same, the watermarks of the paper were the same; and the worm holes showed the map had been at the front of the volume and the *Tartar Relation* at the back."

Discovery of this physical match posed a dilemma for Marston. By authenticating the Vinland map, his £75 purchase was now itself worth much more, and the value of Mrs. Witten's volume, which Yale's map curator had wanted for the university's collection, had increased enormously. Oddly, Marston gave the *Speculum* to Mrs. Witten. "This was not a wholly Quixotic gesture on my part," Marston later wrote, "for I hoped, although I never expressed the hope verbally, that this generosity would give the Yale Library some element of control over the disposition of the map."

In the spring of 1959, Witten offered both volumes for sale. Yale could not afford his asking price, but Marston and Vietor found an anonymous donor who purchased them for the library for nearly $1 million. Although the apparent significance of Yale's acquisition might have called for an immediate announcement, Marston and Vietor shunned publicity while seeking additional evidence to substantiate their claims. They also arranged for Yale University Press to publish a large book integrating their evidence with facsimiles of the map and manuscript. To supplement his own expertise as a medievalist, Marston consulted Skelton for his authority on the cartography of the Middle Ages and George Painter for his expertise in early books. Together they studied the map's geographic content, its ink and paper, the calligraphy and linguistic properties of its text, and the bibliographic history of the original complete manuscript. Painter focused on the *Speculum* and the *Tartar Relation*, provided a translation and interpretation of the latter, and traced their relationship to the Vinland map. Skelton described the map in detail and related it to varied historical sources, including Icelandic sagas, medieval cartography, and the mapping of Greenland and Iceland. Marston himself wrote only the relatively brief introduction to the large-format, 303-page volume, and Vietor wrote the foreword. First in his list of acknowledgments is

Laurence Witten, whom he credits with "bringing together the separate parts of the Vincent manuscript," as well as recognizing the significance of the Vinland map and the *Tartar Relation*.

From a meticulous examination of the map and his extensive knowledge of medieval cartography, Skelton concluded that an unknown compiler based the Vinland map on two prototypes. The rightmost two-thirds, especially the three-part representation of Europe, Africa, and Asia, reflect the 1436 world map of Venetian cartographer Andrea Bianco. Similarities in design, delineation, and place-names are too numerous to be mere coincidence. It was equally obvious to Skelton that this unknown compiler had copied the northwest part of the map from a different prototype, strikingly more accurate in its delineations of Iceland and Greenland than other world maps of the early fifteenth century. Links to the Viking sagas and other navigation records suggested this second prototype was originally plotted in the thirteenth or early fourteenth century. Later Scandinavian maps also reflect this earlier prototype, which has not survived. The Vinland map, which amalgamated this Scandinavian geographic tradition with the fifteenth-century European cartography, most likely had been compiled between 1430 and 1450.

Skelton's colleague George Painter offered additional insight and further narrowed the time frame by relating the map and associated manuscripts to the pre-Gutenberg publishing milieu of monastic scriptoria, where copyists (scribes) reproduced manuscripts individually by hand. Painter concluded that the Vinland map was originally compiled to provide a cartographic illustration for the *Speculum Historiale*, and that the surviving version held by Yale was itself a copy, "perhaps at several removes, from the lost original." The Yale map was copied around 1440, probably in a Franciscan monastery in western Germany or Switzerland, and most certainly by a copyist who was not the map's compiler.

More eager to tie the map to a specific site, Marston concluded his introduction to the Yale University Press volume by pinpointing the Swiss city of Basel, where an important church council held between

1431 and 1449 attracted intellectuals from throughout Europe. "Where else could such a product as this be prepared, combining an East European account of a mission to the Mongols with a medieval historical text and a map of Northern European origin?" Although largely speculative, his logic intrigued the news media, which mentioned Basel in several stories about Yale's unique discovery.

Yale promoted the book with a boldness and enthusiasm rarely observed among university presses. In addition to selecting October 11, the day before Columbus Day, 1965, as the launch date, the press arranged for *American Heritage*'s October issue to feature the book in a lead article, titled "Vinland the Good Emerges from the Mist." An attractive two-page title layout showing a photograph of a ninth-century Viking ship announced that "an astonishing discovery at Yale . . . authenticated by painstaking scholarly detective work at Yale and the British Museum . . . [would open] the door to tantalizing historical speculations." Another double-page spread presented a facsimile of the map, with short articles by Painter and Skelton under the eye-catching headlines "Was There a Lasting Colony?" and "Did Columbus or Cabot See the Map?" Yale sent bookstores an enlarged, 3-foot-square reproduction of the map, several mid-Manhattan bookstores designed special window displays, and Scandinavian Airlines System (SAS) set up exhibits at ticket offices in large U.S. and Canadian cities. As *Publishers Weekly* quoted an SAS representative, "Nobody has more experience on our trans-Atlantic route than we do, and this map proves it." The same unnamed official gloated that New York's annual Columbus Day parade would pass the prominent display in the window of their Fifth Avenue ticket office. Yale, which had approached SAS two weeks before, also supplied names of local experts, whom SAS invited to press conferences in Chicago, Los Angeles, Montreal, Philadelphia, San Francisco, Seattle, Toronto, and Washington, D.C.

Yale's planning was superb. The media could not resist the ironic revelation just before Columbus Day of a long-lost map that indicated the Vikings had beaten Columbus to America by several centuries. The *New York Times* ran the story on its front page, together with a picture of the map and the three-column headline "1440 Map Depicts the New World." In a special dispatch from New Haven, the newspaper's correspondent described the map, its significance, and the story of its discovery; it quoted Vietor and Yale's librarian, James Tannis, at length and linked the map to Viking artifacts recently discovered in Minnesota and Rhode Island. The *Washington Post* also gave the story front-page treatment, with the headline "America of Vikings Shown on Pre-Columbian Map" below the smaller kicker, "Leif Ericson Hailed as Discoverer." Staff writer Howard Simons, who recounted the map's "fascinating, now-it-can-be-told story," concluded that "Columbus Day may never be the same." Although press kits cautioned newspapers not to run the story before October 11, Yale allowed radio and television stations to announce the discovery the evening before.

Journalists who rewrote Yale's press releases and highlighted enticing quotes from the book often disregarded the scholarly restraint of Painter and Skelton for the less reticent jubilation of Marston and Vietor. A few blithely ignored the wide acceptance of Viking explorations of eastern North America centuries before Columbus. *Science News Letter,* for instance, summarized its naïvely embellished account with the subheadline "An authenticated parchment map drawn in 1440 and rediscovered by scholars in 1957 finally proves that the Norsemen reached North America before Columbus did." *Science Digest'*s correspondent Daniel Cohen, another enthusiast, wrote, "This time the proof is so solid that it will probably end the argument." In contrast, historian Samuel Eliot Morison's evaluation in the *New York Times Book Review* was more appropriately titled "It All Boils Down to What We Knew Before." Although impressed by the discovery of a pre-Columbian map, Morison eloquently chided those who had "forgotten that the real discoverers of America were the Indians, who, tens of thousands of years before the Norsemen paid their brief visits, had spread throughout the continent and established at least three great civilizations."

Numerous national magazines reviewed the book in the months following Yale's announcement, and their reviewers generally were highly impressed, if not downright enthusiastic. *Library Journal*'s consultant hailed *The Vinland Map and the Tartar Relation* as "probably one of the most important books ever published concerning the history of discovery and exploration relating to the New World." Gwyn Jones, an authority on Norse history and literature, enthusiastically told readers of the *New York Review of Books* that "what we have here is not only a first-class contribution to historical and cartographical knowledge, but a thriller not to be missed by anyone with fifteen dollars, an awareness of the past, and a belief that 'books too have their fortunes.' " Some reviewers saw the Vinland map as a phenomenon in itself. In the *New Statesman,* for instance, David Quinn linked the map's appeal to "the desire to lengthen the time-span of White America [and] the move to canonise Leif Eiriksson as the 'real' founder of the United States." But the "brilliantly expounded" scholarship of Skelton and Painter led Quinn to discount suggestions that the map might be a fake. Only a few reviewers saw nothing much to get worked up about. The *Atlantic Monthly*'s book columnist Phoebe Adams, for example, remarked that although the Yale Press book had "attracted an inordinate amount of attention from people whose only interest in the matter is a sentimental attachment to the phrase Columbus discovered America, [it] has no direct bearing on the discovery of anything [and was] full of information, bizarre or learned or both, on medieval bookmaking, papermaking, handwriting, mapmaking, the habits of bookworms, and Tartary under the khans." Yet scholarly arcana did not stop the Book-of-the-Month Club from offering *The Vinland Map and the Tartar Relation* as a dividend selection. By university press standards, the book was a best-seller: in October and early November, Yale had sold over five thousand copies of the book and printed another five thousand to meet new orders.

Newspaper stories quickly shifted to the angry Italian-American reaction to Yale's provocatively timed assault on Columbus. On

October 12, Columbus Day, the Vinland map made the front page of the *New York Times* for a second straight day. This time, however, the headline read "Columbus's Crew Won't Switch," and the lead quote came not from a Yale press release but from John La Corte, president of New York's Italian Historical Society, who threatened "to put Yale University against the wall." The story also quoted New York City mayor John Lindsay, who told a Columbus Day rally in Brooklyn that "saying Columbus didn't discover America is as silly as saying DiMaggio doesn't know anything about baseball, or that Toscanini and Caruso were not great musicians." On October 13, the *New York Times* reported similar sentiments from other politicians. Some, like New York governor Nelson Rockefeller, avowed enduring faith: "As far as the impact of Columbus's voyage is concerned, he discovered America." Others, like New Jersey senator Clifford Case, denounced the competition: "Leif Ericson is just an upstart, as far as I'm concerned." Some dismissed the new evidence as unimportant, as did congressman Richard McCarthy: "All the maps in the world will never diminish [Columbus's] fame." Mayor Lindsay even risked reminding voters of his patrician association with the offending institution: "I majored in history at Yale and I know right from the horse's mouth, Columbus discovered America."

Italian-American rage was not unique to New York. In Boston, graffiti on a wall in an Italian neighborhood declared, "Leif Ericson is a fink." At a naturalization ceremony in Washington, D.C., congressman Robert Leggett, a first-generation American with Italian parents, said, "The Nordics might have discovered America first, but they didn't know what they found, and once they'd found it, they didn't know what to do with it." In Chicago, Victor Arrigo, chairman of the Columbus Day parade, dismissed the Vinland map with the comment, "This Nordic myth makes its appearance around every Columbus Day," and called the map "a Communist plot." In Cambridge, Massachusetts, city councilman Alfred Vellucci demanded that Harvard boycott the traditional Harvard-Yale football game until Yale apologized for its effort to "disgrace the Italian race of America." And in

Philadelphia, Michael Musmanno, a justice of the Pennsylvania Supreme Court, charged that Marston and other Yale officials "have gone into the moss-covered kitchen of rumor and, on the broken-down stove of wild speculation, fueled by ethnic prejudices, have warmed over the stale cabbage of Leif's discovery of America."

Yale's book caused a stir outside the United States as well. In Spain, where Columbus Day was a national holiday, an editorial in the newspaper *A.B.C.* called Yale's announcement "an incredibly belligerent plan, prepared carefully for some time, to pulverize the glory of Spain in the discovery of the New World by Columbus." In Italy, Genoa's mayor Augusto Pedulla belittled the Viking explorations as "nothing for human history." And Italy's foreign minister, Amintore Fanfani, who visited New York in December, compared Columbus's discovery of America with Isaac Newton's discovery of the law of gravity. Like Newton, he observed, Columbus had less noteworthy forerunners. Willing to accept the map as a mildly interesting pre-Columbian curiosity, Fanfani denounced Yale's "very timely publicity" campaign as an "ink storm."

Norwegian Americans disgusted with the public Leif-bashing struck back, while other ethnic groups asserted their own claims of primacy. In Seattle, Ted Nakkerud, local president of the Leif Ericson League, challenged that "Leif knew where he was going, [whereas] Columbus was headed for India." In San Francisco, another local president, Dr. S. O. Thorlaksson, noted that although "Columbus never did set foot on American soil, Ericson explored the New World for a year." And in Oslo, Norwegian explorer Helge Ingstad argued that Christopher Columbus left a more lasting historical imprint than Leif Ericson only because Columbus had firearms whereas Leif just had spears and arrows. Spain rejoined the melee with the claim that Alfonso Sanchez de Huevla discovered America in 1484, eight years ahead of Columbus, while Wales claimed Madoc ab Owain Gwynedd established a colony along the Alabama coast around 1170, and Ireland argued Saint Brendan the Navigator landed in America around 850. But Italian Americans remained staunchly loyal to Columbus, despite a new Italian entry in the primacy sweepstakes—by linguistics professor Mario Gattoni Celli, who cited similarities in religious symbols used by South American Indians and the Etruscans as evidence that pre-Roman Italians had visited America around 1100 B.C. Meanwhile, Native American groups either remained silently smug or were characteristically ignored.

Although who discovered America remained a sensitive issue throughout 1966, Yale's Alexander Vietor dismissed the uproar in his lecture opening a seven-week exhibition of the Vinland map at New York's Pierpont Morgan Library. He explained that the Yale Press had wanted to introduce its book on October 9, proclaimed by President Lyndon Johnson as Leif Ericson Day, but had moved the book's launch to the 11th because the 9th was a Saturday. As the *New York Times* reported, the library's fellows laughed when Vietor then deadpanned, "It just happened that the next day was Columbus Day."

Mild humor at the Morgan Library contrasted with pointed sarcasm from Yale's co-conspirators at *American Heritage,* who reacted offensively as well as defensively in an editorial titled "Pride Followeth a Landfall." "Astonished [by] the roar of outrage that went up from partisans of Columbus," the magazine's editors argued that the Vinland map detracted very little from Columbus's "rediscovery of America." In addition to reporting that "all over the country, Italian-Americans hit the ceiling, shedding a tutti-frutti of charges against the map, Yale, and the Norsemen themselves," they ridiculed New Yorker John La Corte's threat to "put Yale University against the wall" and Chicagoan Victor Arrigo's denunciation of the map as a Communist plot. In the transition to Arrigo's charges, the editorial's tone shifted from sarcasm to innuendo—"In Chicago, where putting people against the wall is more or less traditional . . ."—and galled the magazine's Italian-American readers, who had little recourse beyond writing letters of protest and canceling their subscriptions.

By comparison, politicians were extraordinarily tactful when debate turned to which day schoolchildren and government workers should have off. Italians were a large, identifiable segment of the electorate in many states, and politicians were reluctant to offend them. State legislatures had long recognized Italian Americans' special attachment to Christopher Columbus; and since 1909, when New York made Columbus Day a legal holiday, more than thirty states had followed suit. At the federal level, though, Columbus Day was still merely designated by a congressionally authorized presidential proclamation having no more significance than similar proclamations for General Pulaski Memorial Day or Eating Disorders Awareness Week. Norwegian Americans, who by the 1960s had convinced several state legislatures to designate Leif Ericson Day, also won recognition at the national level in 1964, when Congress authorized the president to declare October 9 as Leif Ericson Day. Thus, in the year after the Vinland map controversy, President Lyndon Johnson continued to court Norwegian-American voters with the appropriate proclamation, but he carefully avoided commenting on whose discovery was first or more significant.

Italian Americans had also lobbied at the federal level to elevate Columbus Day from an honorary day to a legal holiday. To appreciate their zeal, one need only recall the more recent struggle of African Americans to make Martin Luther King Jr.'s birthday a federal holiday. By the 1980s, blacks, like Italians in the 1960s, had achieved significant political clout despite decades of ethnic prejudice. As the Reagan administration's cuts in social programs and affirmative action must have motivated African Americans to concentrate on establishing the King holiday, the Vinland map controversy and unfair associations in the media with organized crime motivated Italian Americans to pursue special recognition for Columbus. These efforts paid off in 1968, when Congress made Columbus Day the ninth federal holiday, to be celebrated on the second Monday in October. Although the official celebration would fall on Leif Ericson Day (October 9) one year in seven, most Italian Americans were not only satisfied but thrilled by a Columbus Day Weekend.

• • •

The most articulate and influential of Columbus's defenders was Michael Angelo Musmanno, the Pennsylvania Supreme Court justice who was quoted widely in wire-service reports following Yale's announcement. A lawyer who had worked as a coal loader in his teens, Musmanno had been a committed champion of Italian-American causes since the 1920s, when he had helped defend Italian immigrants Sacco and Vanzetti, who were executed for murder after a six-year series of trials reflecting more prejudice than evidence. Intensely patriotic, Musmanno had served in Pennsylvania as a legislator, county judge, and State Supreme Court justice, and during World War II as a rear admiral, the military governor of Sorrento, and a judge at Nuremberg. Feisty and bright, Musmanno had written more dissenting opinions than any other judge on Pennsylvania's highest court. He had also written ten books, including *Columbus Was First,* published in 1966. Columbus was his hero and was the subject of several of his books, and had he not suffered a fatal stroke on the morning of October 12, 1968, the justice would have been grand marshal of Pittsburgh's Columbus Day parade.

In *Columbus Was First,* Musmanno challenged the credibility of the Vinland map with relentless enthusiasm, and like many effective legal advocates, he not only triggered reasonable doubt but entertained his readers. He described his visits to New Haven and attacked the map's provenance in the first eleven chapters, whose titles ranged from "Holiday for Worms" to " 'History' Is Made at Yale." In the next four chapters he related the map to the Norse sagas—and again demonstrated a gift for embedding puns in cleverly crafted chapter titles: "Source for the Goose," "A Paragraph Here, a Continent There," and my favorite, "Sour Grapes." The book's final section focused on Columbus and his accomplishments, and in the last paragraph, Musmanno reverently described a visit to Columbus's tomb in the cathedral at Santo Domingo.

Musmanno was anything but reverent in chapter 3, "Out of the Blue," as he investigated Laurence Witten's curiously fortuitous purchase of the map from a mysterious, yet-unnamed seller. Particularly troubling is the short transcript of a telephone conversation in which Musmanno cross-examined Witten.

> He [Witten] expressed regret that he had not been able to confer with me while I was in his shop. I thanked him for his courtesies and said: "I'm sorry we didn't take the opportunity to talk last Friday. I wouldn't have taken much of your time because I really wanted to ask only one question."
> "And what was that question?"
> "I wanted to ask you where you got the Vinland map."
> "Well, in that case it wouldn't have done you any good to wait because I wouldn't have answered your question."
> "Why?"
> "Because I wouldn't."
> "Will you answer it now?"
> "No, I will not answer it now."
> "I'm sorry, Mr. Witten, and I don't want to seem to importune you, but it appears to me that it should be easy to say in a very few words just where you got the Vinland map."
> "No, I gave an oath never to disclose where I got the map."
> "To whom did you give such an oath?"
> "To the person from whom I bought the map."

Musmanno visited Witten's shop a few weeks later, and pursued his questioning.

> I thought he might bridle at my next question: "Well, if you won't tell us how and where you secured the map, perhaps you will tell us how and when the man you secured the map from, secured the map." Witten seemed quite undisturbed as he answered:

"Well, I will tell you frankly that I asked the owner of the map that very question, and he replied that he didn't know where the map came from."

During our conversation, Mr. Witten remarked that at the time he obtained the map he had wondered why the owner sold it, when it appeared, to Witten, such an exceedingly valuable property. "It could be that the owner thought the map was a fake, and, therefore, was willing to get rid of it." He quickly added "Of course I know it isn't a fake."

Musmanno speculated about what additional information Witten might have given. If Yale had more information, he concluded, "then the moral obligation to provide the map with its pedigree falls squarely on the shoulders of these officials."

Musmanno's book received little attention from either the media or cartographic scholars. Aside from *Library Journal,* magazines noted for reviewing books ignored it, as did scholarly journals, which traditionally shun popular literature. Wilcomb Washburn, who edited the proceedings of the Smithsonian Institution's 1966 Vinland Map Conference, included it in the annotated bibliography as "an example of the popular reaction" to the Yale book. Noting that Musmanno took "the greatest pleasure in ridiculing [Marston's, Skelton's, and Painter's] interpretation of the evidence created by the bookworms which pierced the various portions of the manuscript," Washburn concluded that the justice's "argument does little to advance knowledge of the subject [even though] it does salve the wounded feelings of the partisans of Columbus."

A number of vocal scholars shared Justice Musmanno's doubts about the map's authenticity. Most prolific among these skeptics was Gerald Crone, librarian and map curator at the Royal Geographical Society. Crone, whose prime scholarly interest was the maps of explorers, concluded that the Vinland map was probably drawn after Columbus's voyage of 1492. As Skelton had noted, except for Vinland

and other features in its northwest portion, the map was markedly similar to Andrea Bianco's 1436 world map. "It is unnecessary to search for possible alternative sources or models," he wrote. "The man who produced it was a copyist, not a cartographer." But because the copyist also reproduced a horizontal fold across the lower part of Bianco's map, Crone reasoned that at least fifty years, and probably many more, should be added to the map's age for the fold line to have developed. In this event, he concluded, the map's "representation of Vinland is not of exceptional significance, and possibly derives in part from a reading of the (Norse) Sagas." Like Musmanno and others, Crone was also skeptical about "how such a document could have escaped notice for so long."

Crone's interpretation obviously aggravated Peter Skelton, who initiated a yearlong exchange of letters in both the London *Times* and the Royal Society's *Geographical Journal.* Rich in the unctuous sarcasm common to public debate between British academics, the dialogue quickly turned to whether the Vinland map might be a fake. Skelton, who vigorously challenged Crone's interpretation that the map was copied in the sixteenth or seventeenth century, concluded one of his letters with a prophetic observation: "We have, in my view, a choice between no more than two alternatives: either the Vinland Map is of the fifteenth century, or it is a modern fabrication, of a perfection and accomplishment not conceivable before the twentieth century." This frankness is not surprising, though, for a close reading of the Yale Press book reveals an agonizing skepticism in Skelton's chapter. He recognized, for instance, the intellectual hopelessness of validating a cartographic artifact that "stands in isolation outside the main stream of cartographic evolution." Troubled in particular by "the still unidentified source" used by the map's "anonymous author" to delineate Greenland and Vinland, Skelton noted that "any attempt to trace its origin is the more speculative since we have, as point of departure, one document—the map itself."

Another vocal critic was Eva Taylor, professor emeritus of geography at the University of London and a prominent expert in medieval cartography. A feisty octogenarian, Taylor thought the Vinland map's geography was too new for the map to be anything but an elaborate fraud. Because the Vikings had not sailed north of the 76th parallel, she maintained Greenland's coastline was unbelievably realistic for a fifteenth-century map. Moreover, some perplexing similarities with other maps suggested the Vinland map might have been copied from relatively modern sources, including the U.S. Hydrographic Office's handbook on map projections. The map also exactly reflects the inaccurate relative locations of Iceland, Ireland, and the Faroe and Shetland islands on Mercator's 1569 world map, which a hoaxer could have traced from a modern facsimile reproduction. In addition, Taylor detected a number of errors not typical of mid-fifteenth-century maps and observed that "a forger often betrays himself by not knowing quite enough about the field in which he is working." Unfortunately, Taylor's health deteriorated as the controversy grew, and she died on July 5, 1966, at the age of eighty-seven. Even though the media widely publicized Taylor's allegations, Britain's Institute of Navigation did not publish her manuscript until 1974, after chemical analysis had already verified that the map was a relatively recent forgery.

Talk of fakes and forgers permeated the two-day Vinland Map Conference held at the Smithsonian's Museum of History and Technology in mid-November 1966. The forty-three participants, invited from five countries, included experts in medieval manuscripts; geography; the history of cartography; the history of exploration and navigation; and early American, early Scandinavian, and medieval European history. Peter Skelton attended, and a contingent from New Haven included Thomas Marston, Alexander Vietor, and Laurence Witten. Wilcomb Washburn, chairman of the Smithsonian's American Studies section, convened the conference and edited the proceedings, published in 1971. Although most of the participants addressed the hand copying of medieval manuscripts or the history and geography of exploration and discovery, a few focused quite pointedly on whether the Vinland map was a hoax.

In the first talk at the conference, Laurence Witten offered numerous details but few new insights about the discovery of the map and its acquisition by Yale. Although no one challenged his probity directly, Witten's stubborn refusal to identify the source from whom he purchased the map was troubling. Northwestern University history professor Franklin Scott summarized this concern: "I trust Mr. Witten. On the other hand, this kind of question will bother scholars from now until it is cleared up; not because anybody distrusts anybody, but simply because any fact that is relevant and unknown remains a question." Apparently the anonymous donor who bought the map for Yale was troubled, too. Witten triggered further skepticism by revealing he had given the donor "one bit of information [to] enable him independently to check if he so desired but which would not compromise my promise."

Armando Cortesão, a prominent cartographic historian from Portugal, followed Witten on the program. In a talk titled "Is the Vinland Map Genuine?" Cortesão observed that without further corroborating evidence scholars could neither authenticate nor discredit the map. Thus the possibility that the map was a carefully contrived conspiracy weakened the explanations and conjectures of Yale's experts and other scholars. Cortesão speculated:

> It would be quite possible for an unscrupulous scholar with the help of an expert in the falsification of old documents to have concocted the VM and to have attached it to the codex of the Tartar Relation (TR) after having separated the latter from the *Speculum Historiale* of Vincent of Beauvais, in the hope that it might be sold for a great sum of money, in the United States, of course. A blank sheet of parchment, which might have been found in the codex itself, could have been utilized for the drawing of the map. The "wormholes" could easily have been made while the manuscripts of the TR and the *Speculum* were still together in the same volume, as they certainly were, and after the forged map had been attached to

the top. Then, after the *Speculum* had been detached from the original volume, the TR with the map could have been bound anew ("a slim volume, bound in recent calf," as Mr. Marston says) and sold to some wealthy American, for nobody would be more interested in and pay a better price for such a wonderful piece of Americana. As anticipated by the forgers, the noncoincidence of the wormholes would be noted, and most appropriately the other part of the original text, that is, the *Speculum*, could then be put on the market. If Mr. Marston had not noticed it, his attention or that of somebody else in a suitable position could easily have been drawn to the discovery of the manuscript of the *Speculum*, when the coincidence of the wormholes and all the rest could be shown.

Forged manuscripts were nothing new, after all, and the plausibility of Cortesão's hypothesis troubled many scholars.

John Parker, an archivist at the University of Minnesota, also raised the possibility of forgery and conspiracy. According to Parker, such a plot would most certainly have involved a number of people:

This much-needed knowledge of binders, booksellers, collectors, or others who might have been involved in the sequence of events which saw the map and Tartar Relation separated from the *Speculum*, rebound together, sent out into the rare-book trade, and reunited, has but one purpose: to lead us to the previous owners. We need to know what sort of libraries they had, how they were assembled, who had access to them, and how they have been disposed of. We need an explanation of why this map remained unknown in the hands of one who thought enough of it to have it rebound with an important manuscript. I find it hard to believe that such treatment would have been given to a suspected forgery. I also find it hard to believe that a collector of fifteenth-century manuscripts would be ignorant of the interest the world would attach to a map of that period showing land in the western ocean.

Like Cortesão, Parker thought that Witten's refusal to identify the former owner for a fuller investigation of the map's origins tarnished its authenticity.

Next on the program was Robert Lopez, a medievalist from Yale and a blunt skeptic, who accepted the central part of the map as genuine but judged the northeastern and northwestern portions "a most clever counterfeit." Lopez cited a number of recently discovered forgeries and asserted the medievalist's responsibility to treat documents as "counterfeit until proven genuine beyond reasonable doubt." Inconsistencies and unexplainable "accidents" suggested the Vinland map was not only a forgery but a relatively recent one. "Were it possible to question the unidentified seller," he proposed, "we might perhaps bring back the counterfeiter, alive or dead — but more probably alive."

The discussion that followed focused on physical testing of the map's parchment and inks. Marston, Vietor, and Witten had examined the ink visually and photographically using a low-powered microscope and infrared and ultraviolet light. Although the ink was notably more diluted on the map than on either the *Tartar Relation* or the *Speculum*, the five or six batches of ink used by the copyist seemed chemically similar. Responding to skeptics, Vietor revealed Yale's intention "to proceed a little further than we have gone in actual physical testing." He authorized the Smithsonian's laboratory to conduct a more rigorous analysis of ink samples from the two manuscripts but not the map — at least not initially. Because spectroscopic and analytical testing of rare manuscripts was relatively new and often inconclusive, Vietor's concern that the map not be damaged was understandable. Nonetheless, he hinted, "if the results seem to work out well, we at least will have established a standard against which a sample of ink from the Vinland Map can be taken."

Mounting curiosity about anomalies in the Vinland map's ink coincided with timely advances in analytical chemistry. In 1968, when officials of the Beinecke Library discussed chemical testing with Walter McCrone Associates, a private research laboratory in Chicago, microanalysis could not yet extract meaningful results from samples small enough to spare damage to rare manuscripts. But microanalytic technology was improving rapidly, and in 1972, a Yale representative carried the Vinland map, the *Tartar Relation,* and the *Speculum Historiale* to Chicago. McCrone scientists took fifty-four small samples, the majority (twenty-nine) from the map. None of these samples weighed more than a millionth of a gram and many were closer to a billionth of a gram. After extensive testing, a number of results suggested the map was much more recent than the two manuscripts, but the detection of anatase in the ink of the Vinland map offered the most damning and convincing evidence that the map was a forgery. Transmission-electron microscopy and energy-dispersive X-ray analysis revealed spheroidal particles of anatase identical with those in commercial pigments produced in the 1920s. A form of titanium dioxide, anatase was the product of a long, costly effort by several industrial research laboratories and could not have been used in inks manufactured before 1917. In January 1974 McCrone advised Yale that the map was a fake, probably drawn around 1920.

Yale promptly advised the press of McCrone's conclusions, and the Vinland map was back in the news. On January 26, 1974, the *New York Times* reported Yale's announcement on its front page. The story included pictures of the map and Laurence Witten, and it provided a lengthy account of Witten's discovery of the map, its acquisition by the Beinecke Library, and the controversy following publication of the Yale Press book just before Columbus Day 1965. Although embarrassed by the revelation, Yale librarian Rutherford Rogers asserted philosophically, "We're in the business of trying to find out what the truth is about a lot of things." Rogers attributed the decision to pursue chemical testing to improved microanalytic techniques, not criticism from scholars. Although Witten accepted McCrone's conclusion, he still hoped additional evidence might someday prove the map genuine. Neither Yale nor the still-anonymous donor had requested a police investigation, the *New York Times* reported, and the identity of the forger remained unknown. Witten still refused to identify the previous owner. But even as a forgery, the Vinland map was a cartographic celebrity: when the Beinecke Library reopened the following Monday, the map was back on display for the first time in two years.

The Vinland map never again made the front page of the *New York Times*, but stories about the forgery and experts' reactions to it appeared off and on for the next month. The British Museum's George Painter, who had authenticated the map with Peter Skelton in the early 1960s, remained convinced the map was "a product of medieval minds and hands." (Skelton had died in 1970.) Yale medievalists Robert Lopez and Konstantin Reichardt attributed the forgery to Luka Jelic, a Yugoslavian professor of ecclesiastical law who died in 1922. Jelic, they maintained, was obsessed with demonstrating the existence of a Roman Catholic hierarchy in Vinland. In papers published in 1891 and 1894, he inaccurately called Eirik Gnupsson the "bishop of Greenland and the neighboring regions"— exactly as on the Vinland map. But no church documents ever supported this description of Eirik's diocese, and Jelic is the only one to have made such an error. In contrast, Professor Francesco Guinta, an Italian expert on the history of discovery, blamed the forgery on a team of unnamed Europeans conspiring to obliterate the significance of Columbus. Still angry over the Leif vs. Columbus flap eight years earlier, Guinta added that "the Americans apparently would be happier if their continent had been discovered by a Nordic race."

Whoever the culprit, the Italian Historical Society of America was pleased that Yale had courageously revealed the forgery. Nonetheless, the society demanded that the university "take the moral and legal steps to rectify the harm and injustice that was inflicted upon Columbus." *American Heritage*, which had eagerly promoted the map in its October 1965 issue, commiserated with Yale over the "sad discovery" after university officials had "gallantly submitted" the map for chemical analysis. In expressing disappointment at the "regretful announcement" that the Vinland map was an "apparent fraud," editor Oliver Jensen used waffling phases such as "if authentic" and "perhaps too amazing" to recapitulate the map's discovery and the controversy that followed. In contrast to the vitriolic sarcasm of the magazine's February 1966 issue, Jensen's revisionist description of the Italian-American protest that followed Yale's 1965 October surprise — "Supporters of the great navigator, incensed by the timing as well as the content of the book, dove eagerly into the controversy" — seemed even conciliatory.

Yale was equally skilled at saving face. In addition to proudly putting the discredited map back on display as a clever hoax, Beinecke officials included the Vinland map among the ninety of its "more outstanding" manuscripts described in *Medieval and Renaissance Manuscripts at Yale: A Selection,* published in 1978 as an issue of the *Yale University Library Gazette.* This abbreviated catalog's final entry is the "VINLAND MAP. Europe, 20th century. Vellum; 278 × 410 mm." The short history that follows is rich in hyperbole. Skelton's "endorsement" of the map appeared in a "most erudite cartographic essay." The forgery "misled a number of distinguished scholars all over the world, while leaving other scholars at Yale and abroad unconvinced." The handwriting was "as perfectly imitated as only a man familiar with fifteenth-century script could make it. . . . Only certain discolorations, a blunder in one of the legends, and, above all, the outline of the map invited skepticism." But chemical tests of the ink, "initiated by Yale," eventually detected this "skillful" forgery.

In June 1992, I wrote Yale asking to see the Vinland map, talk with any staff members who were involved, and examine relevant correspondence and other records in the Beinecke's own archives. The reply was disappointing: the map had been set aside for an exhibition in the fall and was unavailable, all personnel associated with the map in the 1960s and early 1970s had died, and "needless to say, the library's administrative files are not open to consultation." But the letter did mention an article by Laurence Witten in a 1989 issue of the *Yale University Library Gazette.* Witten's article, my correspondent suggested, "offers what may be presumed the final word on the subject."

"Vinland's Saga Recalled" by Laurence C. Witten II seems an honest and plausible account of what happened — or at least of as much of what happened as the bookseller really knows. It's chatty and personal, rich in the wisdom of hindsight, and full of little details missing from press accounts of the map's discovery and ultimate defamation. For instance, after McCrone reported his findings, Yale's librarian Rutherford Rogers invited Witten to a conference room, where senior staff and curators, "looking very like inquisitors," pressed for more information about the map's origin. Witten responded with a minor yet discouraging revelation: "I could only say that I did not know for sure from which library the manuscripts came, and that contrary to what I had said earlier I had never visited the library." Rogers also asked Witten's wife Cora to repay the anonymous donor. The donor never demanded a refund, however, and much of the profit had been paid to the Internal Revenue Service and to book dealers Nicholas Rauch and Enzo Ferrajoli, who in 1957 had introduced Witten to the map.

Witten's revelation must have astonished Rogers and his colleagues. Direct quotes in news stories and even his own words at the Smithsonian's Vinland Conference imply that Witten was a confidant of the original owner, a wealthy yet eccentric collector of medieval manuscripts. Ferrajoli, he told the conference, had introduced him to the man in September 1957: "I saw his library, saw the Vinland Map and the Tartar Relation, and thought the map was a genuine fifteenth-century product." He described the owner's "private library of fairly large dimensions" as containing "a large number of fifteenth- and sixteenth-century printed books and a rather large number of manuscripts." The following year, Witten questioned his anonymous supplier a second time to verify the link between the Vinland map and Marston's copy of the *Speculum*: "The owner of the library in which both volumes still resided in 1957 says he does not know the origin of the volumes but thinks they had been in his library for two or more generations." Witten also implied having discussed the man's need for anonymity: "He did not wish to pay tax and he did not wish to have export problems. He did not wish it known that he had such valuable things as his library in his possession because he would be taxed on them."

In his article for the *Yale University Library Gazette*, Witten dropped what fifteen years earlier would have been another bombshell: in addition to identifying his friend Enzo Ferrajoli as the person from whom he had purchased the map, he named Don Luís Fortuný as the likely original owner. Witten had not met Fortuný until 1963, when Ferrajoli introduced him to a number of collectors in Barcelona:

> A distinguished man of about fifty, Don Luís Fortuný was particularly interested in manuscripts and had quite an unusual collection. . . . Enzo behaved a little strangely as we went to see this man, who received us in a very courtly and flattering way and also seemed to have a clear idea of who I was and what I did, which was surprising in that apart from Enzo I had no contacts in Spain. I suddenly felt that this man Fortuný was the source of the Vinland Map, and on the way home in the taxi I challenged Enzo about it. He grinned a bit but did not say that it was from Fortuný that he got the map and the Tartar Relation; he also did not deny it outright, and his manner implied to me that there was a connection. I assumed that Fortuný was the previous owner of the manuscripts or had somehow spirited them out of a library or private collection, probably in Spain.

Of course, if Fortuný were the map's original owner, Witten had just visited his library.

When back in Spain the following year, Witten again pressed Ferrajoli for information about the map. He appealed on two points:

> I emphasized that in the absence of a reliable history for the map Enzo and I might both find ourselves embroiled in controversy, and that he could not expect a really significant bonus or share of the proceeds of sale unless he at least gave

me the basic information, even if I were not allowed to reveal specifics to others. But Enzo adamantly said that he was honor-bound not to reveal what I wanted him to tell me, and he refused to budge.

Witten attributed Ferrajoli's caginess to the typical reluctance of booksellers to reveal privileged information about wealthy suppliers eager to avoid taxes.

And so, with no firm link to the map's past, Witten adopted Ferrajoli's stance of protecting an anonymous source:

> It is no pleasure to relate that the bookseller [that is, Witten himself], back to the wall, believing passionately that the Vinland Map was not a fake but unable to provide the proof demanded, fell back on the only crutch he could think of: his source's story. Yes, I knew from which library the Vinland Map and the Tartar Relation had come, but I had promised not to reveal it to any third party. Enzo Ferrajoli's story to me became my story to the conferees, and as a result to Yale University and the donor who had purchased the manuscripts from my wife.

Any hope of prying further details from his friend vanished when Enzo Ferrajoli committed suicide shortly after the Vinland Conference.

Witten's article contains one further revelation: after Ferrajoli's death, he had given Fortuný's name to Marston and Vietor, and had urged them to contact the suspected former owner. Although Witten believed he personally would have little influence on Fortuný, he thought perhaps the distinguished curators of a prestigious university might persuade the Spanish collector to clarify the Vinland map's origin. But Witten knew of no attempt by Yale officials to pursue this lead. Vietor had died in 1981 and Marston in 1984.

Reluctant to concede once and for all that the Vinland map was a hoax, Witten ended his memoir with new evidence challenging Walter McCrone's conclusion that the map was a forgery. In 1987, scientists at the Crocker Nuclear Laboratory of the University of California at Davis examined the map and two manuscripts using their proton milliprobe, a particle-beam instrument unavailable in the early 1970s. The Crocker Historical and Archaeological Projects team reported finding no titanium in some ink lines and only trace amounts in others. Since Walter McCrone had based his assessment on having detected significant amounts of anatase, a form of titanium dioxide, the Crocker researchers concluded that "the prior interpretation that the Map has been shown to be a 20th-century forgery must be reevaluated." "Has the Vinland Map been completely rehabilitated by the Davis Cyclotron?" Witten wondered. "Or will it always be under a cloud, destined to remain controversial, as I predicted in 1957 that it would?"

The Crocker study failed, in fact, to restore the map's stature, and the cloud remained. In a careful and thorough rejoinder, McCrone attributed the conflicting results to different measuring techniques: his laboratory had used ultramicroanalysis to study very small, very pure samples whereas the Crocker researchers had employed trace-analysis techniques more suited to detecting trace concentrations in large samples. New tools are not always more appropriate, he argued in his 1988 rebuttal: "Confidence in analytical data derived from PIXE (particle-induced X-ray emission) is not necessarily based on understanding but an attitude that Hi-Tech instruments are more likely to be correct."

Kenneth Towe, a Smithsonian scientist who reviewed both studies, sided with McCrone. Moreover, Towe observed, the Crocker study "further supports [McCrone's] original interpretation of the document as a modern forgery." In their search for new evidence, the Crocker team compared the Vinland map's ink with inks from a variety of demonstrably authentic medieval manuscripts. Detailed microscopic examination revealed the Vinland map's ink was much better preserved and, by extension, more recent. "Unless plausible answers to these questions can be given," Towe argued, "the microscopial evidence . . . from other undoubtedly genuine documents, rather than detracting from the McCrone interpretation, must add additional weight to the conclusion that the Vinland Map is a painstakingly clever forgery."

So far, at least, the chemists have had the last word, although further advances in instrumentation technology and Yale's curiosity about the Vinland map's origin could again resurrect the issue of authenticity. But even full vindication of the map, however unlikely, would add nothing to historians' knowledge that the Vikings had indeed explored North America, nor should it detract at all from the significance of Columbus's "discovery." Although cartographic historians might have something to celebrate, Italian Americans ought have nothing to fear.

What, then, was the "great map flap" all about? The answer, I think, lies in the alluring believability of cartographic images. Maps have an authority that even scholars are reluctant to question. Because maps as generalizations are inherently imperfect, persons trained to be cautious willingly overlook cartographic inconsistencies. Motives for belief are varied: some maps can inspire a satisfying sense of serendipity as well as promise fame and fortune to early converts. Larry Witten, Tom Marston, and Alex Vietor quickly succumbed to the spell of the Vinland map, and George Painter too became a firm believer, committed to the end. Peter Skelton, a bit more skeptical, nonetheless cooperated enthusiastically and, until his death in 1970, was clearly the map's most effective defender. Persuaded by the map and these five apostles, Yale's administrators and a major benefactor made their own acts of faith in pledging the university's reputation and a large amount of money. The editors of *American Heritage* eagerly sought the distinction of early discipleship, and the mass media also bought in, some pushing the bandwagon and others merely along for the ride.

If this sounds like a cult or religious movement, it is perhaps because for many people, even educated ones, the nature of maps can be almost as mysterious and seductive as the nature of God. Like the Peters projection, the Vinland map had become a quasi-religious icon, complete with apostles, faithful followers, and even heretics. But unlike that of the Peters map, the Vinland map's mystique rested not in the ignorance of a cartographically untutored public but in an intriguing yet unknowable provenance.

· 4 ·

Boundary
Litigation
and the Map
as Evidence

Maps make good witnesses. In lawsuits over boundaries and land ownership, maps can testify with authority and conviction. Some might describe how the land looked decades or centuries ago, when a sovereign, government department, or land company established the original boundaries. Others narrate the history of subdivision and transfer, depict a resurvey, or explain the movement of monuments, walls, or rivers. Maps can portray what the seller purported to sell and what the buyer presumed to buy. The authority of maps in boundary litigation is long-standing and deep-rooted: maps are so closely entwined with Western civilization's concept of real estate that the owning, selling, and buying of land would be

105

impossible without them. More than signatures and deeds, surveys and property maps make real estate a reality.

Some maps are better witnesses than others. Like human witnesses, maps do not always speak coherently and reinforce one another. Surveyors' representations of a land grant or subdivision can vary as widely as the testimonies of eyewitnesses frightened during a holdup. Careless or willful omissions, faulty measurements, sloppy drafting, imprecise labels, vanished landmarks, and inconsistent use of symbols present legal adversaries with convenient opportunities for undermining the credibility of each other's cartographic evidence. Thus lawyers who use maps as proof often need to introduce experts to interpret them in their favor. In conjunction with expert testimony, however, maps become powerful courtroom propaganda.

This chapter examines the role of maps in boundary disputes between nations, neighbors, and other contestants. It begins with a concise look at the roles of maps and surveys in the European conquest of North America and the partition of the continent into colonies, land grants, farms, and building lots. Surveys that helped kings and presidents replace indigenous peoples with Euro-Caucasian settlers sometimes left a legacy of imprecision, confusion, and costly litigation between states. Early land surveys based on a first-come, first-served pattern of settlement produced equally contentious boundaries between private holdings. The next section examines the use of maps as evidence in state and federal courts and the role of surveyors, cartographers, and historians in corroborating their silent testimony. The third section looks at the law of the sea and the role of maps in extending international boundaries into the oceans. Applying modern mapping technology to the sea floor and the continental shelf through Exclusive Economic Zones perpetuates the fifteenth-century notion that maps make territory real. Examples throughout the chapter illustrate how maps can precipitate boundary disputes as well as help decide them.

Subjugation, Subdivision, and Surveying in North America

Cartographic historians have long ignored the maps of North America's pre-Columbian inhabitants. Although careful examination of American Indian rock art and folk narratives reveals they indeed had what cartographic historian J. B. Harley called a "mapping impulse," the cartography of these first Americans was markedly different from the highly practical and technical mapmaking of fifteenth-century Europe. Indigenous Americans communicated information about space and places through folktales, gestures, dances, and ephemeral drawings, but theirs was not the cartography of commerce, navigation, and warfare. Land ownership in the profane European sense of buying, selling, inheriting, recording, and taxing was an alien concept. American Indians, who considered land sacred and not "ownable," never developed a formal cartography focused on boundaries and surveys. This lack of maps—really a lack of what the European invaders recognized as maps—was one of many technological disadvantages that made the conquest of the New World not only quick and easy but also morally right in the minds of the colonists and their priests.

When Columbus returned from his first voyage in 1493, Europe lost no time asserting its own mapping impulse. Within two months Queen Isabella had persuaded Pope Alexander VI to draw an imaginary north-south "line of demarcation" awarding all non-Christian territory on its west to Spain and all heathen lands and oceans on its east to Portugal. Anchored at a point 100 leagues (roughly 340 miles) west of Portugal's colony in the Azores, the Pope's original line—although no one knew it—barely touched the easternmost tip of South America. But sensing a faulty papal estimate of the distance between Columbus's landing point and the Azores, Portugal convinced Spain to sign the Treaty of Tordesillas in 1494, which moved the boundary westward to a point 370 leagues (approximately 1,250 miles) west of the Cape Verde Islands. It mattered little that the line's anchor point was difficult to locate, much less to relate to the yet uncharted eastern coast of the yet unnamed American continent. Easily plotted at the left edge of world maps, in perfect alignment with the navigator's grid of latitude and longitude, the Tordesillas line established Spain's claim to the New World. British, Dutch, and French mapmakers defiantly ignored it, but the Portuguese cartographers who drafted the Cantino world map of 1502 boldly extended the line of demarcation through Newfoundland, which John Cabot had claimed for England in 1497. Although the Portuguese never occupied eastern Newfoundland, they used the line to claim and colonize the huge territory that is now Brazil.

Cartography helped Europe's eager empires divvy up the New World and get down to the profitable business of settlement and exploitation. As exploration progressed, maps became more geographically detailed. Cartographers made their coastlines more precise with additional rivers, bays, inlets, islands, and place-names. They added little flags and coats of arms to keep track of each country's claims and used pictures of saints and other religious symbols to signify the blessing of the Church and the approval of the Almighty. England and Spain carried divide-and-conquer even further through land grants to local governors responsible directly to the Crown. Many of the British colonies that rebelled against King George in 1776 were originally plotted by mapmakers with only a vague knowledge of inland geography. Boundaries easily established on paper by meridians, parallels, and the mouths of the Potomac, the Delaware, and other significant rivers could always be stamped on the landscape by surveyors once the king's administrators and colonial subjects were in place. In the meantime, paper maps with inked boundaries helped Lord Baltimore, John Endecott, William Penn, and other colonial governors market their unique brands of Utopia to oppressed Catholics, Puritans, Quakers, and other recruits eager for religious freedom and a piece of real estate.

. . .

Faulty knowledge of overseas territory occasionally resulted in overlapping land grants. Perhaps the best-known example is the dispute
between Maryland and Pennsylvania, ultimately resolved by the
famous Mason-Dixon line. Like most boundary controversies, its history is complex. In 1632, King Charles I of England awarded Cecil
Calvert, the second Lord Baltimore, a patent for all land "hitherto
unsettled" from the Potomac River north to the 40th parallel and
running westward from the Atlantic Ocean to a meridian through the
"first fountain" of the Potomac. A half-century later, in 1681, Charles
II granted William Penn title to adjoining territory bounded on the
south by the 40th parallel. Controversy arose when surveyors discovered that early seventeenth-century maps had placed the 40th parallel too far south. If the boundary were fixed at 40 degrees north, as
called for by both Calvert's patent and Penn's charter, Maryland
would include a nineteen-mile-wide strip of what the king's administrators had assumed would be Pennsylvania.

At the root of the controversy was a map included with a 1608
report by Captain John Smith, one of the founders of the Virginia
colony and the first European to explore Chesapeake Bay. In the
early seventeenth century most of what the English knew—or
thought they knew—about the region was based on Smith's map. As
the facsimile in Figure 4.1 illustrates, Smith oriented his map with
west at the top and north at the right so that opposing tick marks
along the top and bottom edges identified the 40th parallel. A vertical
line connecting these two 40-degree tick marks would pass somewhat
north of where the Susquehanna River enters the Chesapeake. Is the
40th parallel really where Charles I wanted the boundary? Or was
he, like many people lacking the means or rationale for greater precision, merely asserting a cavalier preference for round numbers?

Unfortunately for Maryland, Lord Baltimore, who in 1635 drew
up his own map to guide the colony's development, apparently not
only accepted Smith's placement of the 40th parallel but also located

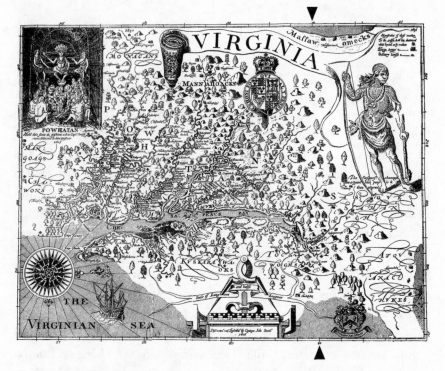

Fig. 4.1. John Smith's 1608 map of Virginia. (From J. Thomas Scharf, *History of Maryland from the Earliest Period to the Present Day* [Baltimore, 1879], Vol. 1, plate facing p. 6.)

his own northern boundary noticeably below the upper end of the Chesapeake. Had the Calverts acted quickly to survey the boundary and settle the northernmost parts of their patent, present-day Maryland might well include Gettysburg, York, Chester, and most of Philadelphia.

Disagreement between the Calverts and the Penns also included the border between Maryland and Delaware. In 1682 Penn acquired additional territory on the west shore of Delaware Bay, in an area settled by the Dutch in 1631 and by the Swedes several years later. Charles I, who concluded that the earlier presence of these Christian settlers excluded the Delaware counties from Lord Baltimore's original Maryland patent, retroactively decreed that the land between the Chesapeake and Delaware bays "be divided into equal parts by a line from the latitude of Cape Henlopen to the fortieth degree of north latitude." Unfortunately for Maryland, the "Cape Henlopen" portrayed on the map as Delaware's southern boundary was roughly twenty-five miles south of the real Cape Henlopen. Maryland historians later claimed this map was a forgery and a fraud, rather than a mistake.

Eighty years of intermittent yet costly litigation began when Charles Calvert, the third Lord Baltimore, led a small contingent of armed officials up the Delaware, measured latitude at various places with a sextant, and warned the citizens of Newcastle, Marcus Hook, and Chester to pay taxes to him, not Penn. Years later, Pennsylvania constables thwarted Thomas Cresap's attempt to survey the border after Cresap called Philadelphia "the finest city in Maryland." Border skirmishes occurred periodically until the Mason-Dixon line ended the controversy.

England's chief judge finally resolved the conflict—on paper, at least—with a decision announced in 1750 and accepted by the parties in 1760. His decree established Fenwick Island as the official "Cape Henlopen" and authorized a northern boundary for Maryland running along a parallel of latitude fifteen miles south of Philadelphia. The Lord Chancellor also fixed Maryland's eastern boundary with the Delaware counties along a line running north from the midpoint of the peninsula opposite Fenwick Island and tangent to a circle with a twelve-mile radius centered in the middle of Newcastle. (Another point of controversy was whether the "twelve-mile circle" referred to in the charter described a circumference or a radius.) The boundary now had exact specifications based on unambiguous landmarks.

Resolving a boundary in a London court proved far simpler than following straight lines, measuring distances, and placing boundary markers in a distant land often thick with trees and punctuated by swamps. Requiring precise measurements of latitude for much of its length, the Maryland-Pennsylvania border called for considerable mathematical skill, a knowledge of astronomy, and experience with precision surveying instruments. Dissatisfied with the work of local surveyors who had tried to locate the tangent line down the middle of the Delmarva Peninsula, the colonial proprietors in 1763 hired Charles Mason and Jeremiah Dixon, English astronomers trained in mathematics and surveying.

Mason and Dixon worked on the boundary from 1764 through 1767, starting with the north-south tangent line and then working westward along the southern border of Pennsylvania. They described the land surrounding the boundary on a map, and every mile they set in place a limestone monument imported from England. Figure 4.2, a portion of their 1768 map, shows the northeast corner of Maryland and the tangent point where the north-south boundary with Delaware touches the twelve-mile circle centered on Newcastle. Uncertain about Maryland's western boundary, the surveyors extended their line westward until turned back by Indians from the Five Nations, who objected to European penetration west of the Allegheny Front. A resurvey commissioned by Maryland and Pennsylvania in 1900 to check the boundary's accuracy and reset missing or vandalized markers found no need to adjust or correct the work of Mason and Dixon. The east-west portion of their border, known to historians and journalists as the Mason-Dixon line, has endured as a symbolic sociocultural divide between North and South.

Real estate boundaries in the United States commonly reflect a "chain of title" extending back to the American Revolution or predating the formation of the state in which the land is located. The War for Independence left a loosely organized confederation of states holding huge amounts of land confiscated from the British Crown, assorted English lords, and vanquished loyalists who had fled to Canada. After resolving boundary disputes with their neighbors — more quickly than did Maryland in most cases — the states partitioned those forfeited lands within their borders into large tracts, with some reserved for soldiers paid in scrip to fight in the Revolution and others sold to private land companies and speculators. Land titles in the western portions of the Atlantic states often reflect these two types of transfer. Westward from the thirteen colonies to the Mississippi

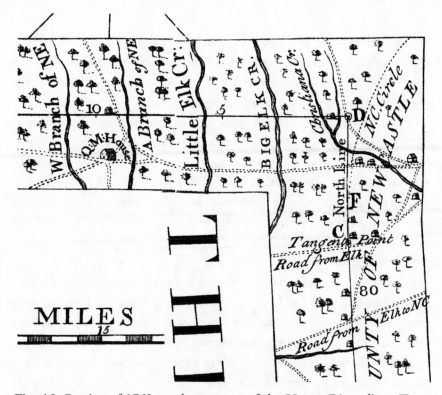

Fig. 4.2. Portion of 1768 parchment map of the Mason-Dixon line. (From *Maryland Geological Survey*, Vol. 7 [Baltimore, Md.: Maryland Geological Survey, 1908], plate 83.)

River, the federal government assumed title to a vast "public domain" larger in area than the original states. To promote settlement and raise revenues, the central government sold most of these congressional lands in rectangular blocks, large and small, and made additional grants to reward military service, support education, and promote the building of canals and railroads.

The United States has two basic types of land survey: systematic and unsystematic. Systematic land surveys, which cover most of the country west of the Appalachians, are based generally on a grid consisting of rows and columns of square *townships*, each six miles on a side and subdivided into a six-by-six array of square-mile *sections*, numbered one to thirty-six. Intersecting horizontal and vertical lines

divide each section into 160-acre quarter-sections. To adjust for measurement error and earth curvature, the U.S. National Land System
consists of thirty-five separate survey zones, each with its own system
of grid coordinates based upon an east-west *base line* (horizontal axis)
and a north-south *principal meridian* (vertical axis). Row and column
numbers identifying individual townships combine with section numbers to provide a straightforward, unambiguous land-identification
scheme in which, for instance, "NE ¼, Sec. 17, T 3 N, R 5 E" refers to
the northeast quarter of section seventeen of the township in the third
row north of the base line and the fifth range (column) east of the
principal meridian.

Comparatively easy to set up and maintain, rectangular surveys
proved ideal for a new nation in need of an efficient method of land
allocation. Government land agents readily partitioned huge regions
into 160-acre farms for sale or homesteading. Boundary conflicts
were rare because all properties were linked to a common set of maps,
monuments, and survey lines. Easy to survey and fence, straight lines
became recognized boundaries. The result was an "authored landscape" in which the survey grid had a marked effect on settlement
patterns and the shapes of counties and smaller political units. In the
typical Midwestern county, roads commonly follow section lines, the
rural population is dispersed rather than clustered, and the landscape
has a pronounced checkerboard appearance. Similar regularities
occur in western New York State and other parts of the original thirteen states where land companies used their own rectangular grid
systems to expedite subdivision and settlement.

Contentious litigation between neighboring landowners is far
more likely where land surveys are unsystematic and independent of
one another. Common in the Atlantic states, Kentucky, Ohio, Tennessee, Texas, Vermont, and West Virginia, unsystematic surveys
often reflect a first-come, first-served form of land allocation in which
the first settler claimed and marked off the most fertile and attractive
land and those who followed took what they liked of what was left.
Each claimant surveyed and sought title to a parcel described by its

metes and bounds—that is, by the lengths and directions of boundary lines that formed a closed perimeter around the property. The result often was a crazy quilt of jerky boundaries and oddly shaped land parcels.

Metes-and-bounds surveys commonly have a point of beginning identified by a monument, either natural or artificial, and consist of courses running between consecutive monuments, or *corners*. The five-sided property in the upper-left part of Figure 4.3 begins at point 1 with a straight-line course 2.70 chains long heading in a direction thirty-five degrees east of north. Field records of the distance and direction for each course allow the surveyor to draw a map, or *plat*, using a measuring scale and protractor, as in the upper-right part of Figure 4.3. If the distances and directions are accurate, course 5 will end exactly, or "close," at the point of beginning. If not—exact closure is rare—the surveyor adjusts the distances and angles by a graphic technique depicted in the lower-right part of Figure 4.3. The illustration at the lower left shows how offsets from a straight line can be used to describe winding courses that follow creeks, riverbanks, roads, or ridge tops. Metes-and-bounds is widely used throughout the country in surveys of mineral claims, Indian reservations, and irregularly shaped lots.

Retracing an old metes-and-bounds survey usually challenges the surveyor's knowledge of both local history and eighteenth-century surveying. Old surveys show the combined effects of imprecise instruments, personal idiosyncrasy, and expedient techniques for measuring directions and distances. Old bearings measured with a magnetic compass do not accord with modern bearings measured to "true" north, as defined by the polar star and the earth's axis. This effect is apparent in the street grids of some older cities such as Baltimore, where north-south streets are tilted about three degrees east from true north. Often based on obsolete units, old distances can be equally troublesome. Although surveyors usually measured length with chains consisting of one hundred long, thin links, different types of surveys could be conducted with a variety of longer and shorter

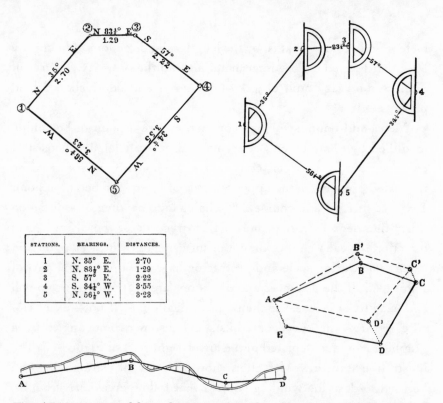

STATIONS.	BEARINGS.	DISTANCES.
1	N. 35° E.	2·70
2	N. 83½° E.	1·29
3	S. 57° E.	2·22
4	S. 34¼° W.	3·55
5	N. 56¼° W.	3·23

Fig. 4.3. A metes-and-bounds survey describes a land parcel with a closed traverse of courses. Graphic adjustment (lower right) can make the traverse close if the last course does not return to the point of beginning. Offsets measured from a straight line (lower left) can be useful in describing a winding course. (From William M. Gillespie, *A Treatise on Surveying Comprising the Theory and the Practice,* revised and enlarged by Cady Staley [New York: D. Appleton and Co., 1888], pp. 121, 125, 129, 131.)

chains. Moreover, because length itself was not standardized, the foot or mile mentioned in the old survey might well differ from the modern foot. In the early 1960s, for instance, a precise resurvey of the Delaware-Maryland boundary found that Mason and Dixon had consistently underestimated distances by about ten to twelve feet per mile, suggesting that a foot in the mid–eighteenth century was as much as a quarter percent longer than the present-day foot.

A resurvey is necessary when property is sold and the seller cannot provide a current map. A licensed surveyor can usually verify and validate recent surveys with little effort, but if the last sale occurred in the mid–nineteenth century or earlier, finding the point of beginning and other markers described in the title can be difficult at best. Trees and fence posts decay, hill slopes and stream banks erode, and "corners" and other monuments move or vanish. The surveyor must attempt to reconstruct and reestablish the original survey, mark the boundary with new, durable markers, and describe it with new bearings estimated to the nearest second ($\frac{1}{3600}$ of a degree) and new distances measured to the nearest hundredth of a foot.

A surveyor starts to reconstruct a boundary by looking for and correcting systematic errors and gross mistakes in the latest map. Because the earth's magnetic field changes slowly but systematically, the surveyor can consult a map showing rates of change for different areas, use the local rate to compensate for dated compass bearings, and then attempt to retrace the original surveyor's footsteps. The grantor's or claimant's original instructions to the surveyor as well as the surveyor's notes, if either can be found, are useful in following the original traverse. Relating the old survey to the current landscape is akin to solving a puzzle or a murder mystery. When one original corner or boundary line is evident, the surveyor can often reassemble the entire boundary bit by bit, with each new piece suggesting where to look for the next. Carefully planned digging might reveal systematically spaced remains of fence posts, for instance. A discoloration in the soil might mark the buried remains of an important tree described in the survey, and differences in vegetation can reflect field boundaries and roads. Historical records and longtime residents can yield additional clues. Careful records of field observations and archival research are important because the surveyor might later need to testify in court if an adjoining landowner files a legal objection to the reconstructed boundary.

Early in the twentieth century, many cities in the metes-and-bounds states established relatively precise control surveys to which property and other surveys could be tied. These municipal surveys consist of strategically located *street lines* set back from the curb line, anchored by carefully placed monuments, and described on maps readily available at the city hall and the county courthouse. Developed in part to avoid troublesome litigation among landowners, they not only provide a convenient reference grid for fixing and describing property boundaries but also support construction, utilities, and maintenance surveys as well as other engineering and mapping activities, such as land-use maps and zoning maps.

Modern property maps reflect a variety of necessary limitations on the use of real estate. Foremost among these restrictions are easements for sewer lines, gas lines, and other underground facilities. Acquired through the government's power of eminent domain, an easement allows a municipality or public utility to dig a trench and install pipes or cables. Although the utility must restore the surface and compensate the owner for damage, the easement becomes a permanent encumbrance on the property—that is, the landowner cannot readily build atop a gas or sewer line, and the utility can reenter the property to make repairs or improvements. Because a buyer must be made aware of these legally binding restrictions on how the property can be used or modified, survey drawings must note and describe all easements.

Surveyors must be wary of the landowner who allows a neighbor easier access to a highway or who fails to contest a misplaced fence. Many owners fail to realize that the neighbor who uses a driveway, parking lot, fence, porch, or other "encroachment" for seven to twenty years—the period varies from state to state—acquires a permanent right to continued use. To warn a prospective buyer of so-called prescriptive easements based upon the legal principle of adverse possession, the conscientious surveyor observes and plots on the survey map all visible and potentially prescriptive encroachments—even wires connecting a neighbor's house to utility poles.

Few maps of real estate generate as much anguish, anger, and litigation as zoning maps. A consequence of the "City Beautiful" movement of the early twentieth century, zoning ordinances are essential to effective urban planning and empower local zoning boards to divide a city into districts similar in activity and appearance. Each district is assigned a category that regulates the use of land and buildings, limits the size and architectural design of buildings, and controls population density and the amount of open space. Zoning prohibits retailing and manufacturing in residential districts, for instance, and the various residential categories usually specify a minimum area and width for each building lot as well as a maximum number of households for each structure. Civic leaders generally support zoning ordinances that discourage the spillover of commercial districts and the intrusion of apartments into expensive, low-density residential areas. Zoning maps thus become bitter battlegrounds between developers who want larger profits and residents who want secure investments in pleasant neighborhoods.

Zoning is an intensely political process in which board members appointed by elected officials hold open hearings on classification changes and exceptions requested by property owners. Most applicants apply for a variance, which leaves the parcel's classification intact but approves a specific "nonconforming use." Individuals, corporations, and citizens' groups can challenge these requests at the hearing, and the loser can appeal the board's decision in court. The petitioner typically describes an adverse decision as "arbitrary and capricious," and the board must then demonstrate in court that it employed procedural due process and accurately interpreted state and local zoning regulations. Lawyers specializing in real estate law often spend much of their time representing clients at zoning hearings and litigating the decisions of zoning boards.

Zoning hearings are good places to find propaganda maps, and most hearing rooms have an easel or tackboard for exhibiting them. Because location is of paramount concern in almost all zoning cases, maps help petitioners explain their plans and let residents describe their objections. Words alone are notoriously inefficient in describing spatial relationships; speakers and board members need maps for identifying features, raising questions, and illustrating arguments. In many instances, a map presents the case's most compelling argument. In straightforward contests, maps may simply reinforce the obvious. For example, a zoning board is as likely to approve extending a commercial district one lot farther into a seedy residential area as it is to reject placing a laundromat in the middle of a solidly noncommercial medium-density residential neighborhood. When the debate is less clear-cut, however, both parties might rely on conveniently selective cartographic generalizations, which exaggerate positive features and downplay negative ones. In turn, each side challenges the opposition's maps by exposing attempted distortions, ridiculing flaws, and suggesting less favorable interpretations.

Maps in Court

Trials are more formal than zoning hearings, and established principles of evidence and judicial procedure require a more rigorous use of maps as exhibits in both courtrooms and legal briefs. When arguing a boundary dispute in court, a lawyer must first determine whether a relevant map is readily admissible as evidence. In most instances, the survey's measurements and monuments and the associated deed are higher forms of evidence than the survey drawing. Moreover, a map that is not an official government survey map usually requires the testimony of the surveyor who prepared it. The lawyer who wants the court to admit the map as evidence must ask the surveyor to describe his qualifications, how he conducted the survey, and his interpretation of the deed or other relevant documents. Opposing counsel may then cross-examine the surveyor and attempt to raise doubts or have the map ruled inadmissible. Although the "ancient document" exception to the hearsay rule allows the court to accept an old map that predates the controversy, expert testimony can be useful in both authenticating the map and demonstrating a relevant link to the case.

When a surveyor testifies as an expert witness, the court not only examines the map and its author but watches the map evolve under direct and cross-examination. The surveyor typically prepares his own maps and diagrams. In an essay on effective expert testimony, California attorney Breckinridge Thomas advised surveyors, "Do not attempt to testify from someone else's work. Even if it is admitted into evidence, it detracts from your effectiveness." The surveyor must also follow the court's instructions about drawing and marking features on a map or diagram. To integrate graphic evidence with verbal testimony in the trial record, the judge typically assigns each feature pointed out by the witness a unique letter-number identifier. If the surveyor's name is Brown, for instance, the features are marked B-1, B-2, and so forth. Thomas cautioned his readers never to draw lines or mark features until instructed: "I know of one judge who would be delighted to get down there and help draw the lines. It is their prerogative to control what happens to the evidence. . . . So wait until the judge tells you to mark it before you do. Otherwise, he will think you officious, and the jury might think you are being a little overly eager too."

Maps can sometimes explain why a modern resurvey does not accord with the legal description of the property in an old deed. Most discrepancies reflect either imprecisely measured angles and distances in the original survey or an improperly copied description. Gordon Ainsworth, a Massachusetts surveyor familiar with the pitfalls of metes-and-bounds description, blamed some disputes on sloppy work by attorneys who insist upon rewriting the surveyor's description. "If a technical description by metes-and-bounds of a tract of land were submitted to a dozen different law offices," he alleged, "it would eventually wind up with the same number of interpretations on a legal form of conveyance." In one egregious case, an attorney preparing deeds for a subdivision described a building lot 180 feet wide as only 60 feet in width — apparently he was thinking in yards (180 feet = 60 yards). In another case, a lawyer who rewrote the description of a traverse inadvertently enclosed much of an adjoining tract. But Ainsworth was equally critical of surveyors with obsolete equipment who "fudge notes and fail to close traverses." A map presenting a logical, legally cogent argument for the parcel's intended boundaries can be a particularly useful prop for the expert witness whose resurvey contradicts a faulty deed or flawed survey.

As in any trial, the witness who radiates confidence is preferable to one who exudes doubt. Legal textbooks on boundary disputes underscore the importance of selecting as an expert witness a surveyor who is not only experienced and technically competent but also confident, poised, and well-spoken. The expert must carefully and courteously explain technical issues to jurors, not talk down to them. Moreover, the lawyer trying the case must not only orchestrate the expert's testimony but also make the jury comfortable with the witness. Equally important is how well the expert's testimony can survive cross-examination. If the surveyor seems even slightly hesitant about the case, or if the lawyer feels uneasy about the surveyor, the safest strategy is to find another expert witness. As the fifth edition of *Clark on Surveying and Boundaries* points out, "The use of a qualified, experienced, well-rounded expert can bring to the client hope, but the use of the wrong individual for an expert will only bring disaster."

A competent real estate lawyer will protect the client by carefully reviewing the survey and related documents before the sale. To show lawyers how to conduct a proper review, attorneys Richard White and Harlan Onsrud prepared a "survey checklist" containing 142 specific items, any of which might later become a point of controversy for the unwary buyer. Careful analysis of the survey might reveal potentially troublesome secondary boundaries that extend within or beyond the property itself. Because a landlocked lot could be largely useless, for instance, the survey drawing should show access to a public street either directly or via a right-of-way included in the sale. The map should also alert the buyer to rights-of-way and easements crossing the land, to encroachments from and protrusions onto neighboring properties, and to municipal restrictions, such as setback lines prohibiting new buildings within a specified distance of the street or adjoining properties. Since flooding and standing water also restrict use of the property, the survey should identify all water bodies on the property and delineate floodplains. In areas covered by the National Flood Insurance Program, a competent surveyor will indicate whether the property falls within the flood-hazard zone.

Water bodies make especially troublesome boundaries. Not only can boundaries shift as streams meander, lake levels fluctuate, and coasts erode, but conflicts can arise over the use of inland waters for recreation, navigation, or irrigation. Particularly important to landowners and potential buyers is whether the title explicitly mentions the water body as part of the boundary. If so, the owner may claim what are called riparian rights, which vary somewhat from state to state but usually include use of the lake or river for boating, dockage, fishing, and swimming.

Since recreational use of the water is often an owner's prime concern, the doctrine of riparian rights conveniently avoids having to find precise water boundaries. In states where the lots surrounding a pond or small lake are assumed to extend to the center, survey maps often plot straight-line boundaries from the points where property lines intersect the shore to a common intersection assumed to be the deepest point. Unless the lake level were to drop drastically, these lines are of little significance, because riparian rights provide recreational access to all shoreline residents and environmental laws prevent an individual landowner from filling in or mining part of the lake bed. River boundaries can be similarly vague. Along a nonnavigable stream, the center line or the line of fastest current marks the official, theoretical boundary, but unless the stream suddenly shifts its course, no one needs to determine the exact center. Although survey maps might show a similar division separating properties on opposite banks of a navigable waterway, each owner typically only controls the land above the high-water mark.

Water bodies occasionally shift position—slowly in most cases, rapidly in others. When erosion claims land, the owner can do little but assess the damage, file an insurance claim, and take a tax write-off. But when alluvial deposition produces new land, the owners of adjoining property can usually claim the "accretion" as a riparian right. Figure 4.4 describes the riparian allocation of new land along a portion of the south bank of the Ohio River, which shifted northward between 1830 and 1867. As a Kentucky court ruled in 1871, landowners could not merely project existing straight-line boundaries northwestward across the new land, as one landowner wanted to do. Instead, because rights to the alluvial accretion arose from titles granted when the land was originally surveyed and subdivided in 1830, surveyors first had to establish "division points" for allocating riparian rights along the old riverbank. Located where the original boundaries intersected the 1830 riverbank, these division points anchored a new set of straight-line boundaries drawn perpendicular to the "thread," or deepest part, of the river.

Riparian principles sometimes seem grossly unfair, especially when a river's arbitrary wandering creates big winners and big losers. Figure 4.5 illustrates the basic facts in a 1900 Kansas lawsuit that hinged on cartographic evidence. As the multiple channels and dates indicate, this portion of the Missouri River meandered first toward the northeast between 1855 and 1870, and then toward the southeast between 1870 and 1900. In 1855 subdivision of the large square created Lots 1, 2, 3, and 4 bordering the river and Lot A well back from the bank. By 1870 the river had moved eastward, placing Lot 1 entirely on its west bank and cutting through Lots 2 and A, among others. But by 1900 the river had receded westward, placing Lot A much farther from the bank. Nonetheless, the plaintiff, who owned Lot A, claimed a riparian right established by the easternmost, 1870 position of the river. This riparian right, the plaintiff argued, included all alluvium accreted to Lot A as the riverbank receded westward, even though some of the new alluvium lay within the original boundaries of Lots 1 and 2, owned by the defendant. In its decision, the court affirmed the plaintiff's riparian right to accreted alluvium, as established by Lot A's 1870 riverbank, labeled ∂e in Figure 4.5. In apportioning the new lands, the court recognized each lot's portion ($a\partial$, ∂e, and eh) of the west bank of the 1870 riverbank. The decision extended new straight-line bound-

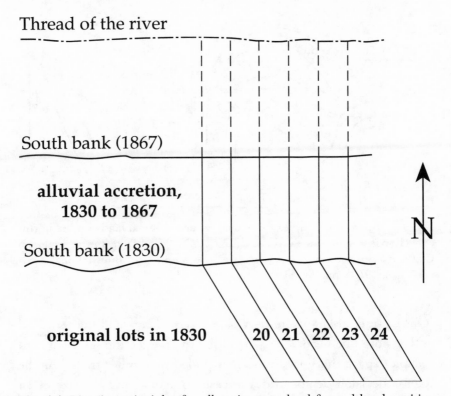

Thread of the river

South bank (1867)

**alluvial accretion,
1830 to 1867**

South bank (1830)

N

original lots in 1830   20 21 22 23 24

Fig. 4.4. Riparian principles for allocating new land formed by deposition along the south bank of the Ohio River. (Redrawn from Frank Emerson Clark, *Fundamentals of Law for Surveyors* [Scranton, Pa.: International Textbook Co., 1939], p. 33.)

aries westward toward the river as shown, thereby awarding *adcb* to Lot 1, *defc* to Lot A, and *ehgf* to Lot 2.

The Kansas court did not act arbitrarily either in making these awards or in failing to consider the owners of lots that once existed between the 1870 and 1900 positions of the river. When erosion consumed each lot's last, southwesternmost portion of dry land, these owners' legal rights disappeared forever. In contrast, the owner of Lot A was fortunate the river had meandered exactly that far east — far enough to confer riparian rights yet not sufficiently far to obliterate the lot as a legal entity. According to Frank Emerson Clark, who described these two cases in his *Fundamentals of Law for Surveyors*,

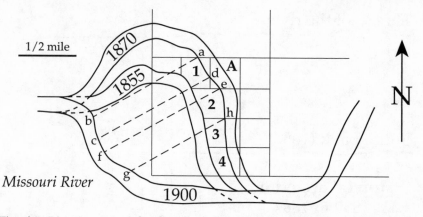

Fig. 4.5. Riparian principles for allocating new alluvial lands on the north-east (Kansas) side of the Missouri River. (Compiled from Frank Emerson Clark, *Fundamentals of Law for Surveyors* [Scranton, Pa.: International Text-book Co., 1939], pp. 34–35.)

"Once a tract of land acquires riparian rights, any land that is added to the shore line becomes a part of the tract, even though such land occupies space that was formerly included in another tract." As the Kansas decision demonstrates, a case's outcome and the size of an award depend on surveys and maps showing the location of the river when the tract acquired its ultimate riparian rights.

The Mississippi, a large river with a huge floodplain, requires a markedly different treatment of riparian rights. Draining an area of roughly 1.15 million square miles, the lower Mississippi meanders sluggishly along a broad floodplain more than a hundred miles wide in places. Topographic maps of Louisiana and Mississippi show huge, horseshoelike meanders and a course that twists and turns, some-times even flowing due north, away from the Gulf of Mexico. Although these meanders usually migrate slowly, with erosion along the outside bank balancing deposition along the inside bank a mile or so downstream, during severe floods the Mississippi can defy the best efforts of the Army Corps of Engineers by breaching its levee and gouging a shorter, radically different path through its floodplain's malleable alluvial soils. Because the old riverbed might be miles away from the new channel, farms and towns that were west of the Mississippi before the flood might be east of the river when the waters subside. As Figure 4.6 illustrates, part of the old channel typically survives as an oxbow lake, such as Lake Providence, an abandoned meander loop in northeast Louisiana. In this example, the present-day river runs roughly perpendicular to its former channel, and deposition accompanying later floods filled in part of the lake and helped create Bayou Providence. The abundance of oxbow lakes on the Mississippi floodplain testifies to the river's instability as a boundary.

To avoid widespread confusion about ownership, taxes, law enforcement, and government services in general, the courts call such radical shifts *avulsion,* and leave property and political boundaries

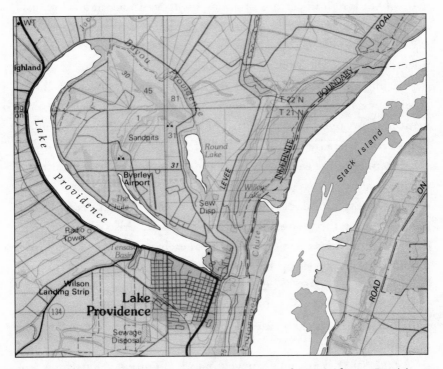

Fig. 4.6. 1982 map of the Mississippi River near Lake Providence, Louisiana. (Portion of the U.S. Geological Survey's Bastrop, Louisiana–Mississippi 1:100,000-scale metric topographic map, 1982.)

intact wherever possible. State boundaries thus run through some of the more recent oxbow lakes and occasionally place portions of the state of Mississippi uncharacteristically on the west side of the river. Because the boundary between Louisiana and Missouri ignores Lake Providence, this oxbow lake most certainly predates Mississippi's admission to the union.

Based on maps from the early nineteenth century, the dashed line labeled "Indefinite Boundary" on the 1982 map in Figure 4.6 represents the approximate thread of the river in 1817, when Mississippi became a state. Establishing the precise location of the original water boundary was difficult—flooding destroys landmarks and survey monuments, and old maps are often unreliable. The U.S. Geological Survey, which produces most of the nation's detailed topographic maps, handles an uncertain boundary with cautious neutrality, estimating its plausible location based on available evidence, representing it with a line symbol half the normal width, and announcing this uncertainty with the adjective *indefinite*.

Despite the historic roots of the approximate Louisiana-Mississippi boundary shown on modern maps, in the late 1980s both states went to court over a narrow strip of land separated from the city of Lake Providence by Cottonwood Chute (the small channel along the state boundary due east of the city) and extending six miles south (off the map) along the west side of the river. Backed by the state of Mississippi, the plaintiff claimed a riparian right to an "island" that merely migrated over to the river's west bank. Also citing riparian doctrine, Louisiana claimed the land as newly accreted sediment from unknown origins well upstream. Complicating the case was the confusing use of the name "Stack Island" for the land in question as well as for a large island several miles upstream on modern maps (Figure 4.6). Furthermore, the contested "island" is seldom completely surrounded by water.

As the 1909 Geological Survey map in Figure 4.7 illustrates, the city of Lake Providence, Louisiana, did not always have the state of Mississippi in its front yard. The clearly marked boundary between

East Carroll Parish and Issaquena County indicates that the state boundary accepted early in the century followed the thread of navigation and passed east of a feature labeled "Stack Island." Mississippi argued that this earlier Stack Island gradually moved westward to the Louisiana side of the river, whereas Louisiana challenged that erosion had slowly destroyed the original Stack Island (as shown on the 1909 map) while accretion of alluvium from upstream gradually added new land along the west bank.

Mississippi v. *Louisiana* evolved from a private lawsuit in which the plaintiff not only sought clear title but petitioned to have Stack Island declared part of Mississippi. Louisiana objected and joined the law-

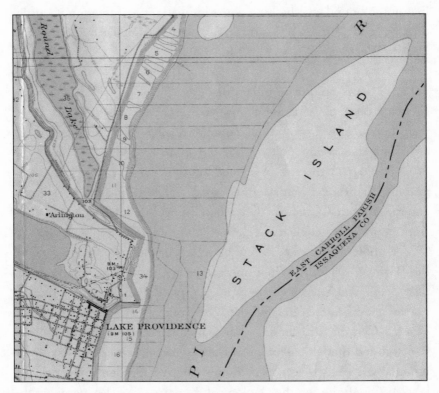

Fig. 4.7. 1909 map of the Mississippi River near Lake Providence, Louisiana. (Photoreduction of portion of the U.S. Geological Survey's 1:24,000-scale topographic map of the Lake Providence, Louisiana, quadrangle, 1909.)

suit, but a federal district court awarded the land to the plaintiff and jurisdiction to Mississippi. Louisiana took the dispute to the U.S. Court of Appeals, which reversed the district court. Mississippi then appealed to the Supreme Court, which distinguished between two very different issues: title to the land and the state boundary. In a unanimous decision, the high court agreed with the district court's decision on land ownership but held that the lower court had no jurisdiction over state boundaries. Noting that federal law gives the Supreme Court "original and exclusive jurisdiction of all controversies between two or more states," the justices asked the court of appeals to decide whether further proceedings were needed to resolve the ownership claim. If Mississippi or Louisiana wants a ruling on the boundary, the high court implied, one of them must apply directly to the Supreme Court. So far, neither state has taken the case back to Washington.

Maps can precipitate boundary disputes as well as help decide them. *Georgia* v. *South Carolina,* which was in litigation from 1977 until 1990, went to trial after negotiations failed to resolve the interstate boundary along the Savannah River. The dispute arose in 1971, after the U.S. Geological Survey published revised versions of three topographic quadrangle maps for the area between Savannah, Georgia, and the Atlantic Ocean. As the map in Figure 4.8 illustrates, the interstate boundary, which had followed the middle of the river on the 1955 edition of this map, now took an abrupt jog northward around the Barnwell Islands. When questioned by South Carolina authorities, Geological Survey cartographers replied that the new boundary line more accurately reflected the 1787 Treaty of Beaufort, which had granted Georgia all islands in the Savannah River. South Carolina, which noted that the Geological Survey had no authority to resolve boundaries, saw no reason to honor the new border, appropriately labeled "indefinite" on the Geological Survey's maps. Recognizing the potential of the Barnwell Islands for industrial development and eager to secure water rights farther downriver for its shrimp fisher-

men, Georgia attempted to work out a compromise with South Carolina. In 1977 the states seemed close to signing an agreement, when Georgia conservation rangers tried unsuccessfully to arrest a South Carolina fishing captain for illegal commercial fishing in Georgia waters. After South Carolina rejected a request to extradite the fisherman, Georgia filed suit before the Supreme Court and attached the Geological Survey's 1955 and 1971 maps as exhibits A and B.

As with many interstate boundary disputes, the Supreme Court moved slowly. Its first action was to appoint federal district court judge Walter Hoffman as a "special master" to conduct hearings, examine the evidence, and propose a solution. To allow the parties time to prepare their arguments, Hoffman delayed the hearing until 1981. Several prominent historical geographers and cartographic historians testified as expert witnesses. Georgia called Louis De Vorsey, a geographer at the University of Georgia, who later published a book about historic maps relevant to the case. South Carolina called William P. Cumming, a historian and author of *The Southeast in Early Maps*, and Arthur Robinson, a professor of geography at the University of Wisconsin widely respected for his research on both map design and the history of car-

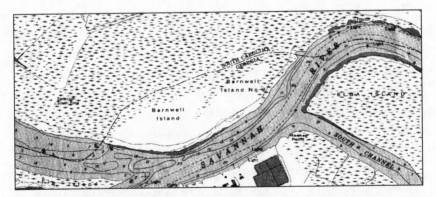

Fig. 4.8. Northward diversion of the Georgia–South Carolina border around the Barnwell Islands on 1971 U.S. Geological Survey topographic map. (Photoreduced from the Savannah, Ga.–S.C. 7.5-minute topographic quadrangle map, 1:24,000, photorevised 1971 edition.)

tography. These and other experts produced thousands of pages of transcribed testimony about the authenticity and relevance of nearly two hundred eighteenth-, nineteenth- and twentieth-century maps introduced as evidence.

Because of the need to mark exhibits as well as the reluctance of archives to loan valuable documents for an indefinite period, the expert witnesses examined photocopies of rare maps, not originals. In an article for *The Map Collector,* De Vorsey described with regret an incident during his testimony on a detailed German map showing the Savannah area during the 1730s. His research indicated that James Oglethorpe, founder of the Georgia colony, was the mapmaker's principal source. Because the map carefully described the islands in the Savannah River and identified most of them by name, De Vorsey concluded that the mapmaker considered these islands part of Georgia. In contrast, the map labeled the sparsely detailed area north of the river "Part of Carolina." When Judge Hoffman asked whether the "white shading" for a large, triangular island in the lower part of the river would suggest the island belonged to Carolina, De Vorsey agreed. He later regretted this "hasty response" when examination of the original map revealed the photostat had failed to capture the mapmaker's delicately detailed symbols for marsh grass.

In another anecdote, De Vorsey recalled with relish the South Carolina attorney's failed attempt to discredit Georgia's chief witness. Earlier in the hearing, De Vorsey had analyzed a 1748 map of Georgia and adjacent territory on which a thin dotted line parallel to the north bank of the Savannah River represented the Georgia–South Carolina boundary. During cross-examination, the South Carolina lawyer first induced De Vorsey to endorse the map's exceptional accuracy. He then took out a large magnifying lens, and asked the witness to again identify for Judge Hoffman the line of fine dots. When De Vorsey complied, the lawyer approached the witness in classic Perry Mason style and challenged dramatically, "Don't you know that those dots are Indian traders' trails?" De Vorsey looked at the map, paused, and retorted with authority, "They are not!" Apparently the South Carolina counsel, in an eager effort to undermine De Vorsey's credibility, had confused the mapmaker's double-dotted-line symbol for trails with the single-dotted-line symbol for boundaries.

However hasty the witnesses and attorneys might have been in the courtroom, the judge was in no hurry. Weeks of hearings had produced boxes of evidence and testimony. Judge Hoffman, who had other responsibilities, examined the material for several years and issued two sets of rulings. His 1986 decision addressed the boundary in the Savannah River, and his 1989 report resolved the seaward boundary, which included offshore fishing rights. As is common in cases requiring a special master, the parties then filed "exceptions" to Hoffman's rulings, and the Supreme Court announced its final decision on June 25, 1990, nearly thirteen years after Georgia had initiated the suit.

With unanimity marred only by four partial dissents on comparatively minor points, the Court supported almost all of the special master's recommendations and rejected most of Georgia's claims. In one of the complex decision's seven rulings, the Court agreed that South Carolina had acquired sovereignty over the Barnwell Islands "by prescription and acquiescence, as evidenced by . . . its taxation, policing, and patrolling of the property." According to Justice Harry Blackmun, who wrote the decision, "Inaction alone may constitute acquiescence when it continues for a sufficiently long period . . . and there has been more than inaction on Georgia's part." In another ruling, the justices endorsed a 1922 Supreme Court decision fixing the boundary in the middle of the river in stretches without islands and midway between the northernmost islands and South Carolina's bank in stretches with islands. Eager to avoid "a regime of continually shifting jurisdiction," the 1990 Court ruled that the 1787 treaty did not entitle Georgia to islands that emerged after 1787. Characteristically wary of contradicting an earlier Supreme Court decision, Justice Blackmun noted that "the Court, in its 1922 decision, did not expressly determine the treatment to be given islands that emerged after the Treaty of Beaufort."

Although maps were hardly mentioned, the mass of cartographic evidence presented at the hearings influenced the Court's decision both directly and indirectly. Directly, maps documented South Carolina's de facto control of the Barnwell Islands as well as helped the special master decide which islands were extant in 1787, and thus subject to Georgia's sovereignty. Indirectly, maps demonstrated the need for a more stable interstate boundary, by showing the effects of a whimsical river in which new islands, however small, "would alter the boundary lines to a degree that could be dramatically out of proportion to the physical change brought about by the formation of the island itself." Moreover, cartographic principles established in the special master's report will help officials fix the boundary elsewhere and grapple with future changes in the river, both natural and artificial. As it often does in boundary cases, the Supreme Court responded with a pragmatic, arguably fair-minded solution intended to avoid further litigation.

Boundaries at Sea

Georgia v. *South Carolina* went a bit further than most interstate boundary decisions. In determining who controlled Oyster Bed Island (Figure 4.9), the Supreme Court had to establish the mouth of the Savannah River, and in settling this issue the justices extended the boundary seaward onto the continental shelf. Figure 4.10 summarizes the arguments of the litigants and the special master's decision, which the Court endorsed.

Georgia had argued for a seaward boundary through A, the midpoint of the headland-to-headland "closing line" separating internal from oceanic waters. This was the usual method of defining the mouth of a river, and seemed appropriate for Oyster Bed Island, which had emerged after 1787. In an exception to the special master's report, Georgia also noted that dredging by the Corps of Engineers in the 1870s had diverted the main navigation channel from the north to

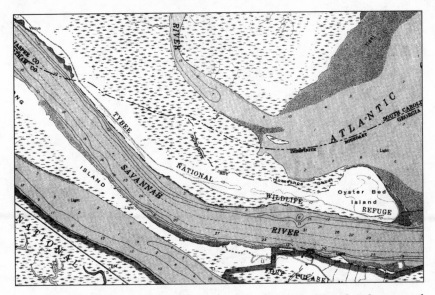

Fig. 4.9. Northward diversion of the Georgia–South Carolina border around Oyster Bed Island on 1971 U.S. Geological Survey topographic map. (Photoreduced from Fort Pulaski, Ga.–S.C. 7.5-minute topographic quadrangle map, 1:24,000, photorevised 1971 edition.)

the south side of the island. In contrast, South Carolina had proposed a division point farther south, at B, where the closing line linking Hilton Head and Tybee islands intersected the middle of the main navigation channel. A submerged shoal closer to Georgia seemed the natural northern boundary of the navigation channel and the counterpart to Tybee Island, the southern headland at the river's mouth. The Supreme Court agreed with this functional delineation of the river's mouth, awarded Oyster Bed Island to South Carolina, and anchored the state boundary at point B.

Judge Blackmun revealed the Court's willingness to accept unconventional solutions in order to avoid appearing unfair. "Given this somewhat uncommon type of river mouth," he wrote, "the Special Master's conclusion that the northern side of the Savannah's mouth is the underwater shoal is not unreasonable. To accept Georgia's proposition here would result in having Georgia waters lie directly seaward of South Carolina's coast and waters."

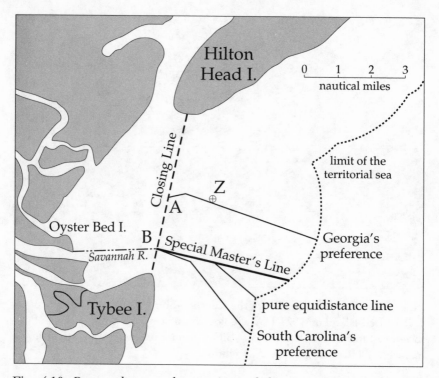

Fig. 4.10. Proposed seaward extensions of the Georgia–South Carolina boundary. (Compiled largely from Map 1 in David J. Bederman, "*Georgia* v. *South Carolina*, 110 S.Ct. 2903," *American Journal of International Law* 84 [1990], p. 911.)

The various lines seaward from points A and B in Figure 4.10 (and roughly perpendicular to their respective coastlines) reflect the marked difference between Georgia's comparatively vertical coast and South Carolina's more northeasterly shoreline. The closing line connecting the headlands provided a compromise direction that seemed equitable to Judge Hoffman, who proposed a perpendicular boundary outward from point B. In approving the special master's solution, the Court recognized the "equitable balance" between the widely different boundaries favored by Georgia and South Carolina.

The majority of the justices rejected South Carolina's proposed "coastal front" model, which bent the state boundary toward the southeast by projecting seaward two overlapping sets of straight-line coastal "fronts," one for each state. (The model generated successive images of these straight-line coasts at equal distances offshore and then connected their intersection points to describe the boundary.) Justice Stevens, in a dissent joined by Justice Scalia, asserted that a boundary "drawn in reference to the full coastlines of the respective States, rather than one drawn perpendicular to the line connecting Hilton Head and Tybee Islands, is more equitable and consistent with the equidistance principle of *Texas* v. *Louisiana*." But as the majority opinion noted, equidistance is merely a principle, not a rule of law. In searching for a legally sound, geographically equitable compromise that fit the special circumstances of the case, the Court also rejected the line equidistant (Figure 4.10) from both the Tybee Island headland and the low-tide elevation (point Z) used to delineate the territorial sea's three-mile limit. (The dotted line representing the three-mile limit bulges outward along the arc of a three-mile circle centered at point Z.)

As a state boundary extending only three miles seaward from the coast, the special master's perpendicular line seems little different from the "equidistance line" in Figure 4.10. But as Justice Stevens observed in his dissent, the Georgia–South Carolina boundary might eventually be projected much farther. On December 27, 1988, in fact, President Reagan expanded the territorial sea to twelve nautical miles, and the special master's line could be projected that far as well if the federal government awards states specific rights in the region. Future legislation or litigation could even project interstate boundaries into the 200-mile-wide Exclusive Economic Zone, which also remains federal territory. As Justice Stevens argued, a map extending the special master's line 200 nautical miles into the Atlantic would appear grossly unfair to South Carolina.

Technological advances since 1950 in deep-sea mining, submarine-launched weapons, electronic communications, and sonar mapping have encouraged nations to claim and defend ever wider strips of adjacent coastal waters. Although countries have contested marine navigation and fishing rights for centuries, new prospects for seabed mining and offshore broadcasting demanded a more consistently applied hierarchy of maritime boundaries defining a broad range of rights and jurisdictions. Decades of work by international lawyers and political geographers, and numerous widely attended Law of the Sea conferences sponsored by the United Nations, led to detailed standards in 1982 for redrawing the world's maritime boundaries. Despite the reluctance of the United States to join the more than fifty-six nations that have ratified the pact, several provisions of the UN Convention on the Law of the Sea (CLOS) have already affected America's offshore territory.

The convention recognizes five national maritime zones between the land itself, where aliens have no rights, and the high seas, where all countries have equal rights and responsibilities in fishing, mining, navigation, overflight, and scientific research. Within *internal waters*, including bays and the mouths of rivers, aliens have no rights except for the "innocent passage" of ships in designated sea-lanes. Within a nation's *territorial sea*, usually extending twelve nautical miles from the coastal baseline, aliens have few rights beyond controlled use for air and sea navigation. In the *contiguous zone*, which normally extends twenty-four nautical miles from the coastal baseline, aliens have full rights for navigation and overflight, some rights for scientific research and laying submarine cables, and whatever fishing rights apply in the *Exclusive Economic Zone* (EEZ). Extending 200 nautical miles out from the coast and overlapping the territorial sea and contiguous zone, the EEZ grants the coastal state full control over mining and substantial control over fishing. Within the *continental margin*—a geologically determined zone extending sixty nautical miles beyond the foot of the continental slope—aliens may fish and conduct scientific research in the "water column" but cannot mine or catch "sedentary" (nonmigratory) species. Only coastal nations with a broad continental shelf enjoy the rights of a continental margin extending beyond the EEZ. Additional rules apply to *archipelagic waters* surrounding "wholly archipelagic states," which enjoy less control than continental nations over navigation and overflight.

The Convention on the Law of the Sea calls for standardized zone boundaries at 12, 24, and 200 nautical miles. Prior to 1982 many nations had much narrower zones, whereas several Latin American and African countries claimed a 200-mile territorial sea—on paper at least. Most nations eagerly adopted the convention, which generally sanctioned their current claims or conferred new ones, but the United States clung to its traditional three-mile territorial sea until 1988. Seeing others' gains as its own losses, the U.S. government had long and adamantly opposed expanded maritime territories. In 1969, for example, after Peru detained several American tuna boats and fined their captains about $2,000 apiece for invading its recently proclaimed 200-mile exclusive fishing zone, Washington suspended arms sales to the highly nationalist yet fiercely anti-Communist Peruvian regime. Although declining to sign the convention, the United States nonetheless established its own 200-mile EEZ in 1983 and enthusiastically started to map it with sonar sensors. Almost all coastal states now have a 200-mile EEZ, and over sixty countries have formally signed the agreement.

Because most nations never had 200-mile economic zones before, the EEZs often created new maritime neighbors. The United States, for instance, has had to work out maritime boundaries in the Atlantic with the Bahamas, Canada, Cuba, and Mexico for the EEZ adjacent to its mainland and with the Dominican Republic, the Netherlands, the United Kingdom, and Venezuela for the EEZs centered on Puerto Rico and the Virgin Islands. In addition to extended boundaries in the Pacific and Arctic Oceans with Canada, Mexico, and the Soviet Union, the United States also found itself a maritime neighbor of Japan, Kiribati, New Zealand, and Samoa. The UN encourages nations with overlapping EEZs to negotiate whatever adjustments might be needed, draw up a map delimiting the agreed-upon legal definition, and sign a treaty renouncing other claims that might undermine the boundary's permanence.

To minimize conflict, the UN Convention on the Law of the Sea defines the baseline for a territorial sea as the tide's low-water line delineated on the coastal nation's existing navigation charts. Use of these charts was not only efficient and expedient but also discouraged countries from concocting new maps solely to exaggerate their maritime territory. The convention also permitted the territorial sea to bulge outward in twelve-mile arcs around "low-tide elevations"—offshore patches surrounded by water at low tide but inundated at high tide. But this provision applied only to low-tide elevations otherwise within the territorial sea defined by the mainland. Precise language also allowed a baseline along the seaward side of the fringing reef surrounding an atoll or other island, and allowed straight closing lines across bays and the mouths of rivers as well as along coasts that are deeply indented or fringed with islands. Although closing lines and other straight-line portions of the baseline generally should not exceed twenty-four nautical miles in length, the convention pragmatically permits exceptions. As guidelines published by the UN Office for Ocean Affairs and the Law of the Sea urge, "By judicious selection of a system of straight baselines it may be possible to eliminate potentially troublesome enclaves and deep pockets of non-territorial seas without significantly pushing the seaward limits of the territorial seas away from the coast."

Where two nations' territorial seas, continental margins, or EEZs overlap, the convention calls for median boundary lines based on the well-established equidistance principle. Grounded in a rigidly mathematical notion of fairness, the equidistance principle raises the question "Equidistant from what?" As Figure 4.11 suggests, protruding headlands can exert enormous territorial leverage and offshore islands act as nuclei for their own territorial seas and EEZs. But how much land must not be inundated at high tide for a "rock" to qualify as an island? Size, name, geology, and maximum elevation above the high-tide line are not consistently reliable criteria for distinguishing a rock from an island, and the UN requirement that the island be capable of "sustaining habitation or economic life" begs equally vexing questions. Although controversial rocks could be obstacles to resolving maritime boundaries, these flukes of marine geology can also be useful bargaining chips.

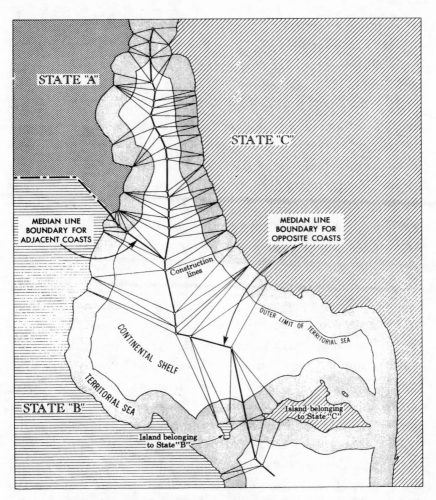

Fig. 4.11. Median boundaries between hypothetical sovereign states with adjacent and opposite coasts. (From U.S. Department of State, Office of the Geographer, "Sovereignty of the Sea," *Geographic Bulletin* no. 3 [October 1969], p. 33.)

Maritime neighbors who cannot negotiate a permanent seaward boundary have three options: leave the issue unresolved, go to war, or go to court. In international boundary disputes, going to court means appealing to the International Court of Justice at The Hague in the Netherlands. As for other disputes between countries, the International Court usually acts as an arbitrator: the parties agree to accept the court's jurisdiction; and the court listens to their arguments, evaluates the evidence, applies relevant principles of international law, and devises a solution deemed fair, legal, and useful as a model for resolving similar conflicts. In 1981, after Canada and the United States were unable to negotiate the boundary in the Gulf of Maine between their 200-mile exclusive fishing zones, they sought the wisdom of the International Court. As Figure 4.12 describes, the nations had at least agreed that their seaward boundary would begin at a common point in the mouth of the Bay of Fundy, but they disagreed over the boundary's path between the opposing coasts of Nova Scotia and Massachusetts. Canada wanted a strict-equidistance boundary reflecting the general trend of the coast and ignoring Cape Cod, Martha's Vineyard, and Nantucket, whereas the United States favored a more "functional" boundary following a natural division of the continental shelf along the Northeast Channel and treating the Georges Bank fishery as a single, indivisible ecological unit. The decision handed down in 1984 struck a balance between equidistance and these "special circumstances." As the map indicates, the court's boundary recognizes a need for both compromise and simplicity.

Like the International Court's two-segment solution to the Gulf of Maine dispute, maritime boundaries filed with the United Nations are surprisingly simple. Despite heavy use of maps, navigation charts, and tidal records during negotiations, an international accord between maritime neighbors represents the negotiated boundary as a straightforward list of points, described by their latitude and longitude. *Loxodromes*, or lines of constant geographic direction, connect successive points so that each boundary segment has a constant bearing. As with the deed to a farm or building lot, the boundary's legal

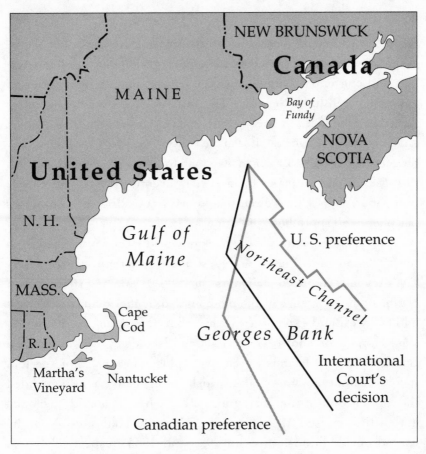

Fig. 4.12. Disputed boundaries in the Gulf of Maine.

description is more important than its map. When Britain and France used a mere six points to divide the English Channel, their agreement included a map but noted quite pointedly, "The boundary line defined [by the list of coordinates] has been drawn solely by way of illustration on the chart annexed to this Agreement." As centuries of international boundary disputes demonstrate, a map on equal footing with the list of coordinates only encourages future conflict.

Older maps of Antarctica reflect perhaps the most flagrant use of maps and the equidistance principle to grab vast overseas territories.

As the early colonial period in North America demonstrated, cartographic white space is easy to claim, and a grid of converging meridians and concentric parallels made it easy to impose boundaries on what appeared to be the relatively featureless, not readily habitable Antarctic land mass. Indeed, a map centered on the South Pole must have suggested a huge territorial pie ready to be sliced into sectors by neighboring powers. Easier to map than to occupy or defend, Antarctica was divided on paper for over half a century despite little more occupation than a few, often temporary stations on the continent's periphery and sporadic explorations of its interior.

Typical of maps toward the back of world atlases published from the 1920s onward, Figure 4.13 describes the territories of Antarctica's seven recognized claimants. In 1908 Britain asserted the first claim, a one-sixth slice of the pie bounded by meridians at 20° west and 80° west and extending northward to the parallel at 50° south. Based largely on control of the Falkland Islands to the west and the South Sandwich Islands to the east—plus generous rounding to a favorable multiple of ten—the British sector inadvertently included the tip of South America until 1917, when England's colonial authorities revised the boundary. New Zealand and Australia carved out additional slices in 1923 and 1933, respectively. Australia's enormous sector, reflecting the broad longitudinal range of its mainland and island territories in the southern Indian Ocean, overlapped France's Adélie Land, near the magnetic South Pole, which Dumont d'Urville had explored in the late 1830s. Leisurely diplomatic correspondence between 1911 and 1938 eventually fixed the French claim as a thin wedge, from 136° east to 142° east, which divided the Australian zone into two parts. In the late 1930s, Norway claimed the coast between the Australian and British territories. Lacking territory directly north, the Norwegians based their claim on exploration but acknowledged the sector boundaries of their Antarctic neighbors. In the early 1940s, Argentina and Chile proclaimed sectors that overlapped each other's claims as well as the "British Antarctic Territory." The former's broad sector reflected not only the Argentinean mainland but its historical, often frustrated claims to the Falklands and South Georgia. No nation claimed the remainder of the continent, well south of any inhabited territory, but research stations were established there and elsewhere in Antarctica by Germany, India, Japan, South Africa, the Soviet Union, and the United States.

Current maps of Antarctica are generally free of boundaries and seem likely to remain so. The 1959 Antarctic Treaty signed by the seven claimant states and numerous other "consultative parties," including the United States and the Soviet Union, shielded the continent from the cold war by forbidding new claims and freezing existing ones and by fostering cooperation in geophysical exploration, mapping, and other scientific research. Although none of the

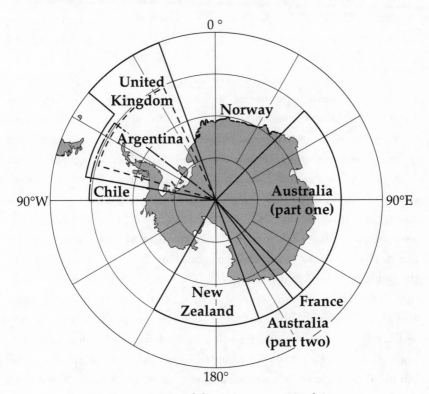

Fig. 4.13. Territories of the seven Antarctic claimants.

original seven claimants has yet renounced its claims, three decades of scientific study by others have considerably weakened their stature. In his book *Antarctica: Private Property or Public Heritage?* Keith Suter observes, "Ironically, in terms of establishing a 'presence' the US and the USSR each 'occupies' Antarctica more effectively than any of the seven claimant nations." In 1987 a *National Geographic* map supplement on Antarctica honored the national claims only on a small, marginal inset map—little more than an anecdotal fragment of geographic trivia; and in 1993, the completely new *Hammond Atlas of the World* left no hint of territorial boundaries on its one-page treatment of Antarctica. In contrast Argentinean, Chilean, and even British world atlases show several different political geographies. While these maps make individual Antarctic territories a reality for the claimant states, a conspicuous absence of boundary lines on the maps of the United States and other non-claimants declares the entire continent a scientific commons, owned collectively by all nations.

Although countless governments have used mapped boundaries to subdue, displace, or annihilate native peoples, maps can also help redress past wrongs. My final boundary story concerns upstate New York's postage-stamp Indian reservations, which enjoy some curious exemptions from state sovereignty. The region has a number of small reservations, including the Onondaga Nation Territory in my own county and the Oneida Nation Territory in the next county to the east. Maps delimit these unique "nations" with symbols more akin to the narrow dashed lines that separate towns and villages than to the bolder symbols that mark the U.S.-Canadian border. Yet reservation boundaries confer several rights denied other municipalities and ethnic groups: although they restrict where the Onondagas may sell tax-free cigarettes and gasoline and where the Oneidas can run high-stakes bingo games—and more recently, casino gambling—the boundaries do permit these concessions. These exemptions from state power apply to specific territories, which could not exist within contemporary America without maps and boundaries. As with territories defined by zoning maps, maps of EEZs, and other cartopolitical instruments, the Onondaga and the Oneida nations have a status and legitimacy that maps represent, reinforce, and defend. In the eyes of the state, maps make them real.

• 5 •

Continental Drift and Geopolitics: Ideas and Evidence

Scholarly arguments involving maps demonstrate those maps' propagandistic power in establishing the legitimacy of new scientific and political theories. This chapter examines two cartographically fueled debates, one in geology and the other in that curious intersection of geography and history called geopolitics. Although the actors and audiences are different, both controversies involved theories of global scope requiring whole-world map projections. Moreover, each theory had advocates who regarded certain maps as key evidence and detractors who viewed the same maps with scorn or amusement.

Most people have at least heard of continental drift, and most recent high school graduates consider wandering continents as much a truth as revolving planets. But how many realize the importance of maps not only in the theory's discovery but in its long fight for acceptance? Although a German meteorologist's insightful reconstruction of an ancient supercontinent was intriguing, this discovery alone did not validate the idea. After all, a geological theory invites systematic observation, detailed analysis, and conscientious refinement; and it must survive rigorous scientific scrutiny to either accommodate a growing body of evidence or yield to another model. The first part of this chapter examines how maps helped continental drift meet these requirements.

Theory is treated differently in the social sciences: vulnerable to political revolution and technological change, a paradigm that no longer fits the times is typically abandoned and forgotten. For this reason, few high school students have heard of the concepts of "Heartland" and "World Island," although in the 1940s many geographers, historians, diplomats, and military strategists enthusiastically accepted this cartographic model of the political world in which the Soviet Union occupied the pivotal position. Developed by a British geographer and statesman who wanted to warn his country about the changing tide of global power, the idea ironically became a self-fulfilling prophecy by motivating Germans' belief in their own cause and potential power. Even though enormous improvements in transportation, communications, and weaponry have undermined the "Heartland's" strategic advantage, geopolitics is alive and well in the 1990s, thriving on new, comparatively complex theories that rely on strikingly similar cartographic rhetoric.

From Continental Drift to Plate Tectonics

I first heard of continental drift in 1961, during my freshman year at Johns Hopkins, in a course on landforms and physical geography. Our text was the first edition of George Dury's *The Face of the Earth;* in those days, Dury was not a big fan of continental drift, which he defined simply as "the horizontal movement of entire continental blocks." His assessment was both skeptical and optimistic.

> Continental drift is itself a highly controversial topic, capable of raising alarm in the northern hemisphere, where the idea has fallen out of favour. It will be highly ironical if current geological research into the magnetism of rocks supplies powerful evidence in favour of drift, as it seems to promise.

Dury used the map shown in Figure 5.1 to illustrate the schematic map presented decades earlier by meteorologist Alfred Wegener, who argued that the continents could be "fitted together like a jigsaw puzzle" and had created a single continental platform on which shallow seas separated several large land masses. Our professor, M. Gordon "Reds" Wolman, was less optimistic than Dury: to a process-oriented geomorphologist firmly committed to field measurement and laboratory experimentation, Wegener's inductive leap from a modern globe or world map to this simple, nonrigorous graphic model was merely an amusing cartoon and hardly relevant to how the earth was formed.

Dury's hesitant optimism and Wolman's dismissive ridicule reflected prevailing attitudes toward continental drift. Although the theory was intriguing, the evidence was tenuous. As Figure 5.1 illus-

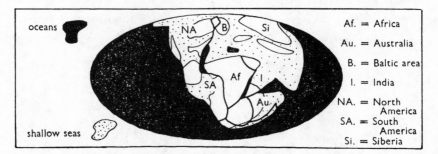

Fig. 5.1. Wegener's concept of a single continent 250 million years ago. (From G. H. Dury, *The Face of the Earth* [Baltimore: Penguin Books, 1959], p. 129. Used with permission.)

trates, the most impressive argument was geometric. If the Atlantic Ocean were removed from the globe, South America's eastward bulge could nestle snugly in Africa's southwest cavity, and removing the Indian Ocean would allow Australia and the Indian subcontinent to join Africa and South America in a huge continent that advocates of continental drift called Gondwanaland. Across the shallow sea to the north, the Baltic region of Europe meshed with eastern North America and together with adjoining parts of Asia formed another supercontinent called Laurasia.

But the evidence was not solely interlocking shapes: similar types of ancient rocks lined up perfectly when South America and Africa were fitted together, as did the occurrences of several rare, distinctive fossils. Geologists also found a significant correspondence across several other boundaries, or "contacts," in Figure 5.1. Without this lithological and paleontological verification—which skeptics dismissed as mere coincidence—continental drift would have merited little more than a smirk or a raised eyebrow. Moreover, the theory helped explain a number of mountain ranges that piled up along the leading edges of continental plates as they drifted apart. The Andes chain along the west coast of South America, advocates of continental drift argued, formed as the South American plate drifted westward after the breakup of Gondwanaland, and the Himalayas formed when the Indian plate drifted northeastward and collided with the Eurasian plate. Although lithological correspondence and the distribution of mountain chains supported the theory, the mechanism for splitting the original land masses and moving the continental fragments to their present positions was still unclear. Because geology has a rich history of plausible, attractive theories demolished by new evidence, Wolman, Dury, and thousands of other geologists and physical geographers felt a professional duty to doubt continental drift.

Alfred Lothar Wegener was the theory's most persistent advocate. Born in Berlin in 1880, Wegener studied astronomy, meteorology, and physics at the University of Berlin. After earning a Ph.D. in astronomy in 1905, Wegener worked as a meteorologist for two years in Greenland and then returned to Germany to teach meteorology at the University of Marburg and publish a textbook on atmospheric circulation and thermodynamics. After military service in World War I, he directed the Meteorological Research Department of the Marine Observatory at Hamburg for several years before moving to Austria in 1924 to accept a chair in meteorology and geophysics at the University of Graz. An enthusiastic hiker and skier, Wegener gained recognition as an Arctic explorer by participating in a 1912 expedition across the Greenland ice cap and writing several monographs on glaciology and polar meteorology. He died of a heart attack in 1930 while leading his own expedition to Greenland.

Wegener first presented his hypothesis of "horizontal displacement" in January 1912, at a meeting in Frankfurt am Main of the German Geological Association. Several months later he published an article "Die Entstehung der Kontinente" (The Origin of the Continents) in *Petermanns Geographische Mitteilungen,* and in 1915 he published a book *Die Entstehung der Kontinente und Ozeane (The Origin of the Continents and Oceans).* In his book Wegener presented a wide range of scientific arguments for the existence approximately 300 million years ago of a single continent of comparatively light rock floating atop a denser layer of crustal material. In addition to accounting for the congruence of Africa and South America, the single-continent hypothesis provided land bridges that explained worldwide similarities among many species of plants and animals. The theory was also attractive because floating continents could account for substantial differences between present and past climates, and thus explain the fossils of tropical plants Wegener had observed in Greenland as well as ancient glacial deposits in Africa and South America.

Wegener was an accomplished cartographic propagandist. Foremost among the *Origin's* numerous maps and diagrams is the dramatic demonstration in Figure 5.2, which he titled the "reconstruction of the map of the world for three [geologic] epochs." Most serious treatises on continental drift include this three-part graphic narrative.

With Africa as a stationary reference point, this temporal sequence of three maps accelerates hundreds of millions of years of geologic history to portray the breakup of the supercontinent and the horizontal displacement of the individual continental blocks to their present positions. The upper map shows the vast extent of Pangaea, the single continent of the Upper Carboniferous epoch, when many of the world's coal beds originated in enormous swamps. Concerned more with continental platforms than dry land, Wegener portrayed shallow seas with dotted area symbols and marked the present and past positions of the edge of the continental shelf with heavy, eye-catching lines. The middle map, for the Eocene epoch, indicates marked lateral displacement along several rifts; and the bottom map shows the continents and continental platforms roughly in their present positions by the Lower Quaternary. Grid lines, which the caption admits are arbitrary, and an oval whole-world map projection help make the reconstruction intriguingly realistic and believable. A similar three-part graphic narrative based on pairs of globelike hemispherical views provided visual reinforcement "in a different projection."

Casual observation of a world map triggered Wegener's serendipitous discovery of the drift hypothesis. As he recalled the incident and its aftermath:

The first notion of the displacement of continents came to me [in 1910] . . . when, on studying the map of the world, I was impressed by the congruency of both sides of the Atlantic coasts, but I disregarded it at the time because I did not consider it probable. In the autumn of 1911, I became acquainted (through a collection of references which came into my hands by accident) with the paleontological evidence of the former land connection between Brazil and Africa, of which I had not previously known. This induced me to undertake a hasty analysis of the results of research in this direction in the spheres of geology and paleontology, whereby such important confirmations were yielded that I was convinced of the

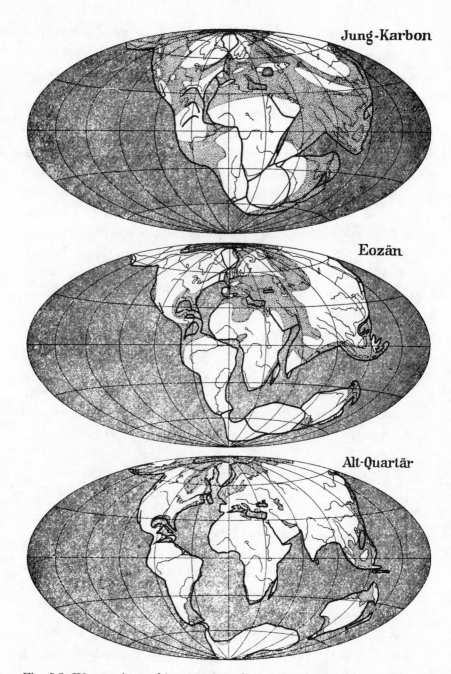

Jung-Karbon

Eozän

Alt-Quartär

Fig. 5.2. Wegener's graphic narrative of continental displacement showing the world's continents about 250, 50, and 1 million years ago during the Upper Carboniferous (top), Eocene (center), and Lower Quaternary (bottom) epochs. (From Alfred Wegener, *Die Entstehung der Kontinente und Ozeane* [Braunschweig: Samml. Vieweg & Sohn, 1915], p. 4.)

correctness of my idea. . . . Afterwards, the participation in the traverse of Greenland under J. P. Koch of 1912/13 and, later, war-service hindered me from further elaboration of the theory. In 1915, however, I was able to use a long sick-leave to give a somewhat detailed description of the theory. . . .

Wegener refined his ideas and incorporated additional evidence in second, third, and fourth editions of his book, published in 1920, 1922, and 1929. Publication of English, French, Russian, and Spanish translations in 1924 and 1925 reflected increased interest worldwide in the drift hypothesis.

Although Wegener assembled considerable evidence for continental drift and persistently promoted the theory in numerous lectures and writings, he was neither the originator of the drift hypothesis nor the first observer to report the jigsaw fit of Africa and South America. English philosopher Francis Bacon seemed close to pointing out the corresponding coastlines as early as 1620, in his *Novum Organum*. In discussing how "physical resemblances" can sometimes lead to absurd conclusions, Bacon argued that some natural phenomena cannot easily be dismissed as mere coincidence.

> The very configuration of the world itself in its greater parts presents Conformable Instances which are not to be neglected. Take for example Africa and the region of Peru with the continent stretching to the Straits of Magellan, in each of which tracts there are similar isthmuses and similar promontories; which can hardly be by accident.
>
> Again, there is the Old and New World; both of which are broad and extended toward the north, narrow and pointed towards the south.

But Bacon noted only the similarity of the west coasts of South America and Africa, not of their opposing Atlantic coasts, and he offered no suggestion that these continents were ever united.

Equally oblique was Father François Placet, a French abbot who wrote in 1668 that "America, as many islands and empires which occur today throughout the ocean, were not at that time separated from the mainland." Although Placet implied the former existence of a single continent, he ascribed its breakup to the Flood and the sinking of "the Atlantic Island." Comte de Buffon, Alexander von Humboldt, and several other seventeenth- and eighteenth-century natural philosophers noted the similarity of the Atlantic coastlines, but none offered an explanation similar to continental drift.

Growing acceptance of the theory in the 1960s raised the issue of a pre-Wegenerian originator of the displacement hypothesis. Wegener, who thought the congruent coasts a bit obvious and was too absorbed with refining his theory to claim priority, mentioned several writers who had suggested a single combined land mass before 1912. In the fourth edition of the *Origin*, for instance, he noted that as early as 1857, W. L. Green had proposed a lighter crust floating on a liquid core. A modified version of continental blocks floating in a heavier medium later became an important element in continental drift. Wegener also acknowledged the contribution of American geologist Frank Bursley Taylor, who in a presentation to the Geological Society of America in 1908 proposed the horizontal drift of continents to account for the positions of several major mountain chains.

Wegener was apparently unaware of Antonio Snider-Pellegrini, whose 1859 book on the origin of America's indigenous peoples invoked a single-continent link to Asia in support of his anthropological theories. An American geographer and natural philosopher living in Paris, Snider-Pellegrini proposed a supercontinent as a more plausible alternative to the hypothetical land bridge connecting North America with Asia. *La création et ses mystères dévoilés* (*The Creation and Its Mysteries Unveiled*) has strong biblical overtones that present-day educators would call creationist, and much of its evidence is scriptural. In the framework of a Heaven and Earth created in six days, Snider-Pellegrini was concerned more with proving the American Indian a direct descendant of Adam than with promulgating a grand geologic theory. Yet he cited the similarity of the African and South American coasts as proof that volcanic activity and crustal movement had produced a north-south crack in the earth's crust. Land east of this rift escaped the Flood, but the weight of floodwaters made the crack much wider and initiated the westward drift of the island of Atlantis, which eventually became America. For writers addressing continental drift during the 1970s and 1980s, this suggestion of horizontal displacement earned Snider-Pellegrini the dubious distinction of having stumbled upon a promising geologic discovery but doing little to promote it. Nonetheless, the continental reconstruction in Figure 5.3, which he used to account for similar fossils in coal measures in Europe and North America, demonstrates that Snider-Pellegrini was almost as effective in manipulating cartography as Wegener was. Although geologists largely ignored his cataclysmic theory of the origin of continents, at least one popular introduction to geology reproduced his map and publicized the possibility of a single continent that broke up and drifted apart.

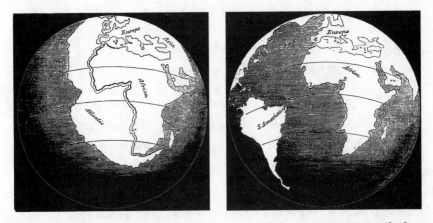

Fig. 5.3. Snider-Pellegrini's representation of the supercontinent before (left) and after (right) separation. (From John Henry Pepper, *Playbook of Metals* [London: George Routledge and Sons, 1866], pp. 8 and 10. Pepper reproduced illustrations originally printed in Antonio Snider-Pellegrini, *La création et ses mystères dévoilés* [Paris: A. Franck, 1859].)

Today, though, the recognized originator of continental drift is Abraham Ortelius, a sixteenth-century Dutch geographer and classical scholar who viewed the similarity of coastlines as evidence for a catastrophic rupture of the earth. Ortelius's explanation first appeared in 1596, in *Thesaurus Geographicus*, a revised edition of his famous dictionary of classical place-names *Synonomia*. In describing Plato's account of the mythical continent of Atlantis, Ortelius argued that the island of Gadir (present-day Cadiz)

> will be the remaining part of the island of Atlantis or America—which was not sunk (as Plato reports in the *Timaeus*) so much as torn away from Europe and Africa, by earthquakes and flood—and accordingly will seem to be elongated toward the West. But if someone refers to this as making good one fiction by telling another, he has a right to do so, as far as I'm concerned. But the vestiges of the rupture reveal themselves, if someone brings forward a map of the world and considers carefully the coasts of the three aforementioned parts of the earth, where they face each other—I mean the projecting parts of Europe and Africa, of course, along with the recesses of America. The case is such that one might say, along with Book 2 of Strabo, that what Plato related concerning Atlantis, on the authority of Solon, was not an invention.

Oddly, this passage remained unnoticed until 1994, when the prestigious British scientific weekly *Nature* published a short article by classics expert James Romm, who credited Ortelius with having first mentioned the possibility of the continents' horizontal displacement.

Shortly before Wegener's first lecture on continental drift in 1912, American geologists Frank Taylor and Howard Baker independently presented continental-displacement theories at least vaguely similar to Wegener's. Although some geologists called continental drift the "Wegener-Taylor" or "Taylor-Wegener" hypothesis, or even awarded

Taylor priority of discovery, neither Taylor nor Baker matched Wegener's energy and persistence in building and refining a comprehensive theory. Nonetheless, the coincidence of multiple drift hypotheses in the early 1910s is significant and reflects a pent-up need to explain the similarity of continental outlines and other compelling transoceanic geologic correlations.

Taylor's theory differed from Wegener's in several important ways. The centerpiece of his model was not a single continent but the gradual movement of land masses, which plowed up mountain ranges along their forward edge or formed them when two continents collided. By focusing on mountain chains, Taylor largely ignored not only the jigsaw similarity of the transatlantic continents but also an enormous body of paleontological and older stratigraphic evidence. A self-financed, Harvard-educated, amateur geologist who occasionally mapped and interpreted glaciated landscapes for the U.S. Geological Survey, Taylor first outlined his "crustal creep" hypothesis in December 1908, in a lecture presented in Baltimore at the annual meeting of the Geological Society of America and published in 1910 in the Society's *Bulletin* as "Bearing of the Tertiary Mountain Belt on the Origin of the Earth's Plan." To account for the vast belt of mountains extending from the Alps through the Himalayas and across the northern Pacific to southern Alaska and down the west coast of the Americas, Taylor called for hundreds of miles of horizontal continental movement. Convinced of the need for continental drift, Taylor was less certain of its mechanism and concluded that the most probable explanation was very slow deformation of the crust through "tidal action" that flattened the spinning earth at the poles and created a bulge around the equator.

Taylor did little to bolster his hypothesis for the next fifteen years, although in the late 1920s and early 1930s, following retirement, he published several papers that refined his earlier ideas, criticized Wegener's concept of drift, and speculated adventurously about the driving force. In his 1928 paper "Sliding Continents and Tidal and Rotational Forces," for instance, Taylor not only reiterated the need for gradual continental creep toward the equator but added the requirement of a catastrophic kickoff that started the continents drifting and explained their not having drifted about earlier. Not finding a convenient mechanism in either geology or geophysics, he turned to "astronomy and cosmogony," which could provide "a relatively sudden increase in the power of the tidal force . . . if we are permitted to believe that the moon was acquired by the earth suddenly by direct capture out of space." In other words, sudden formation of the moon during the Cretaceous epoch would in one swift stroke speed up the earth's rotation and add a new gravitational force able to pull continents toward the equator. Because the moon's initial orbit would be highly eccentric, he argued, the earth's newly captured satellite might approach as close as 24,000 miles and give the continents their required shove with a tidal force "*one thousand* times as powerful as it is now." Despite his obvious enthusiasm, Taylor also recognized this cosmological explanation might be a bit much for most geologists — and he was right.

Although more imaginative and eager to speculate beyond the evidence than Wegener, Taylor not only promoted his ideas less vigorously but used proportionately fewer graphics. Figure 5.4, a photographic reduction of his most famous, most frequently reproduced illustration, shows oceanic trenches in the southern Pacific, the comparatively stable Indo-African plateau, and the directions and amount of crustal movement relative to the world's major mountain ranges. Crammed with information inappropriately plotted on a Mercator projection, the map is more confusing than convincing and far less dramatic than Wegener's three-part graphic narrative. As drift supporter Alexander Du Toit wryly observed in *Our Wandering Continents*, Taylor's "graphic picture of the Tertiary mountain deformation rather obscures . . . the magnitude of the postulated continental displacements."

Wegener's other American rival held ideas even more bizarre than Taylor's. Howard Bigelow Baker, an otherwise obscure amateur geologist from Detroit, first presented his version of continental drift

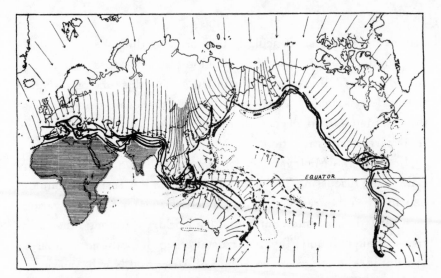

Fig. 5.4. Taylor's map relating the world's principal mountain chains to the presumed directions of crustal creep. (From Frank Bursley Taylor, "Bearing of the Tertiary Mountain Belt on the Origin of the Earth's Plan," *Bulletin of the Geological Society of America*, 1910, p. 211.)

in March 1911 at the annual meeting of the Michigan Academy of Science. Although the society's *Annual Report* for 1911 listed his paper only by title, rather than publishing its abstract or text, Baker nonetheless attracted the attention of the *Detroit Free Press*, which ran an article several weeks later in its Sunday magazine section on the "interesting theory advanced by Dr. Howard B. Baker, a Detroit student of the planetary system." Cosmology was an important part of Baker's theory, which held that the earth and the moon were once a single body. Baker attributed this idea to Reverend Osmund Fisher, a British clergyman who in 1882 extended Charles Darwin's notion that huge solar tides on a more rapidly rotating earth split the crust and propelled a huge fragment into lunar orbit. The resulting void in the crust formed the Pacific Ocean basin, a gigantic "scar" toward which the remaining continental fragments drifted. "This would make the Atlantic a great rent," Fisher concluded, "and explain the rude parallelism which exists between the contours of America and the Old World." Baker extended this hypothesis by focusing on the continental outlines, particularly the corresponding shapes of Africa and South America. In the introduction to a subsequent paper, presented to the Michigan Academy in 1912—and this time published in the academy's *Annual Report*—he attributed the location and shape of the continents to "the separation of mass from the earth" caused by "extraterrestrial gravitation."

In 1913, the Michigan Academy published two other papers by Baker, one discussing Taylor's 1908 paper on the Tertiary mountain belt and the other reconciling his single-continent model with various reconstructions of intercontinental land bridges. Among other criticisms, Baker found Taylor's crustal-creep hypothesis unable to account for the congruent coastlines of Africa and North America.

> As I have demonstrated, and it is easy to verify, the convex northwestern coast of Africa fits into the concave southeastern coast of North America in a manner not less significant than the matching together that is possible between North America and Greenland on the one hand and Africa and South America on the other.

Baker reconstructed this fit in his second paper by moving the continents around on a globe, photographing the result, and tracing a globelike map from the photograph; although, as drift advocate Alexander Du Toit observed after experimenting with his own globe, Baker apparently obtained an impressively close fit by fudging the continental outlines. In Figure 5.5, the interlocking continental outlines are visible beneath a parallel-line pattern representing dry land. In this and three other maps, Baker used his "reconstruction globe" as a base for plotting the land connections advanced by various naturalists to explain worldwide similarities in plants and animals.

First presented in his unpublished 1911 paper, Baker's reconstruction globe continued to be a persistent icon over the next four decades in the intermittent series of articles and one privately published book through which he publicized his drift hypothesis. Although he had far less impact than Wegener or Taylor, Baker's

Fig. 5.5. Baker's reconstruction globe, with parallel-line shading added to show dry land late in the Cretaceous era. (From Howard B. Baker, "Origin of Continental Forms, III," *Michigan Academy of Science Annual Report,* 1913.)

intriguing map still appears in discussions of continental drift, and seems to have secured his place in the history of geology.

Continental drift attracted little attention until the mid-1920s. The Geological Society of America's bibliography on the subject lists no entries between 1915, when Wegener's *The Origin of the Continents and Oceans* first appeared, and 1922, when Vienna paleontologist Carl Diener attacked Wegener's paleogeographic evidence yet attributed the Atlantic Ocean to the separation of Europe and America. The comparatively sudden appearance of ten essays on drift in 1923 and another twenty-three books and articles by the end of the decade reflects an increased, almost startled awareness of Wegener's hypothesis among orthodox earth scientists.

Not all of this attention was favorable. Many authors attacked Wegener's evidence, and most were openly skeptical. Some critics questioned the vague geophysical forces invoked to split Pangaea and move the huge fragments thousands of miles apart. Cambridge University geophysicist Harold Jeffreys, for instance, called Wegener's theory an "impossible hypothesis," and asserted that "the assumption that the earth can be deformed indefinitely by small forces, provided only that they act long enough, is . . . a very dangerous one, and liable to lead to serious errors." Others argued that Wegener had not only misinterpreted data but selectively ignored facts that failed to fit his hypothesis. Several critics dismissed the similarity of opposing coasts as meaningless, while others contended that forces able to split the crust and move continents would surely have destroyed the original matching shapes along the break. A few even challenged the impressive fit of Wegener's reconstructed supercontinent. "In following the edge of the continental shelf," British geologist Philip Lake charged, "[Wegener] has allowed himself a very considerable amount of latitude, and he has not hesitated to distort the shapes of the masses." Like Howard Baker, Wegener had apparently succumbed to the temptations of map authorship and indulged in some advantageously creative cartographic generalizations. As historian of science Homer Le Grand later remarked, "Wegener got his fit, as represented in his famous maps, by massaging the continents."

Some mainstream geologists were unwilling to dismiss the drift hypothesis, however, and a few strongly supported the idea after introducing minor modifications. Perhaps the most significant dialogue between its supporters and opponents occurred in 1926, at a symposium sponsored by the American Association of Petroleum Geologists. The keynote speaker, and author of the introduction and conclusion to the collection of papers published two years later, was the vice president of the Marland Oil Company, a Tulsa, Oklahoma, geologist named W. A. J. M. van Waterschoot van der Gracht. Citing "an accumulating amount of evidence," he announced his support early in his essay, and his comprehensive, critical review of this evidence included facsimiles of twelve of Wegener's maps. Anticipating criticism of Wegener's cartographic synthesis, he asserted that pro-drift "structural arguments work on a globe, not a mercator map" and argued that in comparing opposing coastlines "we should not look for a perfect fit." That aside, van der Gracht said little about congruent coasts, and went on to warn against ignoring the growing body of evidence for the drift theory:

> These hypotheses, though revolutionary, are serious and are far from having to be summarily dismissed. They are not wild dreams, but are based on very serious thought; they are sponsored by many scientists to whom we owe respect, while others at least agree that they must be seriously entertained as a possibility. Moreover, the drift idea has the advantage that it is the first that opens a vista for acceptable explanation of a series of geological problems, which so far had received no adequate answer.

But Wegener's critics at the AAPG symposium were eager to attack both the theory and the evidence, and Wegener's jigsaw puzzle was a favorite target. Retired Yale geologist Charles Schuchert reviewed much of the same evidence that intrigued van der Gracht and found it lacking. To sharpen his attack on Wegener's coastline analysis, Schuchert squeezed a quarter-inch-thick layer of transparent plasteline (a flexible substance that could be molded, cut, and hardened) onto an eight-inch globe, marked the outer edge of the continental shelf, and cut the plasteline into curved sections he could move around the globe and photograph. With three such photographs he identified several inconsistencies in Wegener's reconstructed supercontinent, including a 1,200-mile gap between Central America and Africa and a 600-mile separation between Siberia and Alaska. "It is evident," Schuchert charged, "that Wegener has taken extraordinary liberties with the earth's rigid crust, making it pliable so as to stretch the Americas from north to south about 1,500 miles, [especially] in Central America." Perhaps the intriguing similarity of the Brazilian and African coasts was, he suggested tongue-in-cheek, nothing more than Satan's way of vexing geologists and geographers.

Stanford University's Bailey Willis was one geologist who dismissed the significance of opposing Atlantic coasts. "Similarity of form . . . constitutes no demonstration," he maintained, because the horizontal displacement described by Wegener would most certainly have initiated faulting, which in turn would have destroyed the original similarity. Willis rejected Wegener's other evidence as either flawed or inadequate. Equally hostile was Johns Hopkins paleontologist Edward Berry, who not only dismissed geographic pattern as evidence for displacement but refused even to consider continental drift a worthy scientific theory. "It is inconceivable," Berry argued, "that masses of continental size should move over such large arcs and preserve their outlines of either coast or continental margin intact."

Despite his objections to Wegener's unabashed advocacy, Yale University geologist Chester Longwell was unwilling to dismiss continental drift altogether. Like others, he pointed out the mesmerizing effect of Wegener's jigsaw puzzle:

> I am convinced that this sort of argument still has very great weight with advocates of the idea. In fact, it is not improbable that gazing at the map of South America and Africa has the effect of hypnotizing the student. The coast lines appear to be such exact counterparts, even in detail—Wegener must be right!

However seductive, geographic pattern was a weak argument, Longwell maintained, yet he could not ignore growing evidence that crustal displacement could account for the Alps and the Himalayas. New data and interpretations called for open-minded research "not blinded either with the zeal of the advocate or with the prejudice of the unbeliever." Other critics shared Longwell's hunch that the drift hypothesis might have a far more solid foundation than Wegener's biased and somewhat vulnerable propaganda.

Although no one had demonstrated convincingly how and why land masses might separate from Pangaea, collide, and produce mountain chains, Wegener's most adverse critics eagerly accepted that continents could indeed float. Eduard Suess, an important predecessor of both Wegener and Taylor, had proposed as early as 1888 that the earth's outer crust consists of two kinds of material, a comparatively light layer of *sial* (largely *s*ilicates of *al*uminum, sodium, and potassium) floating on a layer of *sima* (largely *s*ilicates of *m*agnesium and calcium). According to Suess, the continents are relatively thick blocks of sial floating like ice on water atop an underlying layer of sima, and rising buoyantly above ocean floors coated with a comparatively thin layer of sial.

Geologists and geophysicists used this principle, called *isostasy*, to account for faulting and other vertical movements. In the same way that a tall cork extends farther below the water surface than a short cork, mountains had deeper "roots" of sial than plains. But just as the vertical center of a cork floats higher if its top is cut off, a mountain subject to prolonged erosion tends slowly to rise or "rebound," buoyed up by an influx of sima underneath. Similarly, basins and deltas receiving large amounts of sediment maintain isostatic equilibrium by sinking, thereby displacing horizontally some of the underlying sima, and glaciated areas tend to sink as the ice sheet grows and to rebound after it melts.

Isostasy supported continental drift because, as van der Gracht noted, "if blocks of the sial crust can be displaced vertically, as required for isostatic adjustment, and if consequently the underlying sima can be displaced horizontally, we may also accept the theory that the sial might itself be displaced horizontally." But as several participants in the AAPG symposium pointed out, proving that continents could float did not explain how and why they drifted to their present positions. Lack of a convincing geophysical mechanism for moving continents in various directions was the principal deficiency of the drift hypothesis.

How continents drifted remained a mystery until the 1960s, when plate tectonics became the prevailing paradigm for interpreting earthquakes, faulting, and volcanic eruptions. Remarkably consistent with continental drift, this new theory hypothesized crustal plates moving across the earth's surface well before the formation of Wegener's Pangaea; and like the drift hypothesis, plate tectonics relied heavily on cartographic evidence. During the 1950s geophysicists reconstructed the earth's magnetic field for widely separated points in geologic history and demonstrated the independent movement, merging, and separation of crustal blocks. In the 1960s sonar maps of the sea floor revealed a pattern of midoceanic ridges—actually parallel ridges separated by a deep trench or valley. Crustal plates diverging from these twin ridges act like giant conveyor belts, set in motion by convection currents in the earth's mantle. Molten magma rises to the surface along the midoceanic ridge, cools, expands, and wedges the plates apart. This expansion propels the plates outward, and the cooled magma welds itself to the young edge of the outwardly spreading submarine plate. At the other end of this giant conveyor system, the plate sinks into the mantle in a process called *subduction*. Lighter land masses floating atop these moving plates migrate slowly across the surface of the earth and occasionally collide to form continents or build mountains. The U.S. Geological Survey, which published the generalized map of crustal plates shown in Figure 5.6, uses plate tectonics to explain the pattern of earthquakes along the San Andreas fault, where the Pacific and North American plates are slowly and abrasively sliding past each other. The remarkable congruence of the opposing Atlantic coastlines with a plate boundary along the mid-Atlantic ridge illustrates how well plate tectonics matches earlier evidence for continental drift.

Three decades witnessed a major change in attitude toward Wegener's hypothesis. Dury's *The Face of the Earth*—the landforms text from my undergraduate days—reflects this shift in its four revi-

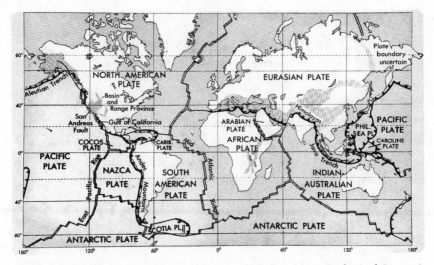

Fig. 5.6. Simplified map of the earth's crustal plates. (U.S. Geological Survey)

sions. Although all five editions include Wegener's map of Pangaea, Dury's assessment of continental drift evolved from optimistic skepticism to casual certitude. In 1986 the most recent revision credited Wegener with recognizing a "fact [that had] remained merely an oddity, or at most a topic of minor speculation" since the first global maps in the early sixteenth century. Like other geologists, Dury praised Wegener for having the vision not only to speculate about continental drift but to develop and promote it to the level of a grand theory.

Wegener was an advocate as well as a scientist, and in his hands the map became an instrument of both research and propaganda. First prompted by a map to explore similarities in the African and South American coasts, and later aided by them in collecting data, organizing facts, and devising a theory of continental drift, Wegener not only studied maps but authored them. Although zeal no doubt led him to tamper with the continental margins, this fudging was not sufficiently great or obvious to undermine the credibility of his cartographic reconstructions. Wegener's skilled and relentless use of cartography was a significant force in the history of geology, and the success of his efforts attests to the map's rhetorical power to explain and persuade.

Pivot Areas, Propaganda, and Geopolitics

Equally persuasive, for a time at least, were Halford Mackinder's geopolitical maps. Mackinder was a true geographer, and maps to him were as necessary a research tool as travel and reading. Like continental drift, geopolitical theory is fundamentally cartographic, with maps playing important roles in its discovery, synthesis, and promotion. But instead of showing how the land masses once constituted a supercontinent, Mackinder's geopolitics explained how the present configuration of continents and oceans made some places more accessible (and possibly less defensible) and how location ultimately controlled resources and won wars. Mackinder saw an inherent, preordained concentration of world power in the Eurasian land mass—the "World Island," of which Africa was merely a large peninsula. The World Island was the most strategic location on the planet, and control of its interior—the "Heartland"—was the key to world power.

Mackinder's evidence was the historical record: thousands of years of invasion and conquest. "As we consider [these] broader currents of history," he asked, "does not a certain persistence of geographical relationship become evident? Is not the pivot region of the world's politics that vast area of Euro-Asia which is inaccessible to ships, but in antiquity lay open to the horse-riding nomads, and is today about to be covered by a network of railways?" This quote from Mackinder's seminal paper "The Geographical Pivot of History" (1904) captures not only the essence of the theory but also the fervor of the theorist. Although his words now seem antiquated or even silly, in the decades to follow, some geographers, diplomats, and military strategists took them quite seriously.

Too seriously, some might say. In the 1920s and 1930s German geographer Karl Haushofer borrowed heavily from Mackinder in developing the *Geopolitik* that guided the Third Reich's pursuit of *Lebensraum* (living space). Nazi propagandists found Mackinder's concepts and images of World Island and Heartland useful as well as stimulating and developed geopolitics into an elaborate pseudo-science. Ironically, the British geographer informed—and bolstered—Hitler far more than he influenced Chamberlain and Roosevelt. Until the late 1930s, British and American foreign policy experts paid scant attention to Mackinder's theories. Only when Germany formed an alliance with Russia did they start listening.

Sir Halford John Mackinder was born to upper-middle-class parents in Lincolnshire, England, in 1861. The "Sir" reflects a knighthood conferred in 1920 after less than a year of temporary service as British High Commissioner for South Russia. Although remembered largely for his writings on geopolitics, Mackinder achieved distinction in several arenas: geographic education and writing, university administration, government service, and mountain climbing—in 1899, while leading an expedition to East Africa, he made the first ascent to the 17,040-foot summit of Mount Kenya, a feat not repeated for three decades.

Mackinder was equally adept in ascending academic ladders. After completing programs in natural science and modern history at Oxford in 1883 and 1884, and in international law in 1886, he developed a strong interest in adult education, traveled around Britain on an Oxford-sponsored lecture circuit, and delighted audiences with a series of engaging lectures on "the new geography." Mackinder's campaign for integrated study of physical and human geography led to several simultaneous academic appointments—at Oxford starting in 1887, at Reading in 1892, and at the London School of Economics in 1895. In recognition of his efforts to promote the subject among British universities, the Royal Geographical Society elected him a Fellow in 1886 and financially supported the establishment at Oxford in 1899 of Britain's first School of Geography. In addition, Mackinder served from 1892 to 1903 as principal of University College, Reading, and from 1903 to 1908 as director of the London School of Economics.

Interested in politics as well as geography, Mackinder moved from academia to public service around 1910, when he won a seat in the House of Commons. During World War I he recruited military volunteers in Scotland, was a member of the National War Savings Committee, and developed saving stamps as a means of supporting the war effort. Denied an opportunity at the war's end to participate in redrafting the boundaries of Eastern Europe—some thought him too imperialist and hawkish—Mackinder was appointed British High Commissioner to South Russia and given the futile task of helping poorly equipped anti-Bolshevik forces resist the Red Army. After several frustrating months, he returned to England to receive the knighthood announced in his absence. In addition to serving in Parliament from 1910 to 1922, Mackinder was a member or head of numerous parliamentary committees and royal commissions, and graduated to the inner circle of the British government as a member of the Privy Council in 1926. Two years before his death in 1947, the Royal Geographic Society awarded him the Patron's Medal, its highest honor, for having demonstrated geography's "fundamental role in national life, and its value as a bridge between science and philosophy."

Mackinder's writings reflect both his skill as a lecturer and his deep fascination with geography and military power. In his first book *Britain and the British Seas* (1902), he demonstrated the value of integrating physical and human geography in a comprehensive, highly readable textbook. A biographer called it "one of the few classics of modern geographical literature," and a contemporary reviewer called it "new, fresh, and forcible." Mackinder not only described the country's climate and geology but also interpreted their relevance to European history and British economic development. Geography had been important in Britain's national defense as well as in the development of its cities and industrial power. In a chapter on strategic geography Mackinder outlined important links among sea power, economic resources, and military defense; in a chapter titled "Imperial Britain" he showed that an empire divided by oceans was costly but promoted the home island's defense by "hold[ing] wide apart the

[land] masses of the ruling and of the subject peoples." Despite the British Empire's success with oceanic mobility, however, he also recognized that improved land mobility was increasing the power of continental nations. Thus while geopolitics was not an overt theme in Mackinder's book, he clearly acknowledged the strategic advantages and disadvantages of distance, accessibility, the technology of transport, and the shape and separation of land masses.

Mackinder also understood cartography's pivotal role in geography, and maps were a central element in his writings and lectures. His most famous expository map is the "Natural Seats of Power" (Figure 5.7). Mackinder used this highly generalized world map to summarize his 1904 paper "The Geographical Pivot of History," and essays on his contributions to geopolitics invariably include a facsimile. The map is seductive in the simplicity and clarity of its message. Its oval shape suggests a pseudocylindrical world map similar to those mimicked by Alfred Wegener in Figure 5.2, and close inspection indicates that Mackinder was an equally adept cartographic propagandist. Whereas Wegener used a fictitious set of meridians and parallels to create a believable map, Mackinder highlighted his theory of the Heartland by eliminating the grid and centering the map not on Greenwich but on central Asia. The strong oval frame obscures his use of a Mercator projection—which conveniently inflates the size of the Heartland—and also partly hides and renders less significant the similarly exaggerated areas of Canada and Greenland. Yet the perspective is not excessively radical; an invisible equator running horizontally through the center of the map divides the flattened globe into familiar Northern and Southern Hemispheres. In retaining the traditional equatorial axis, he conveniently separated the World Island at the upper center of his map from North and South America and the other "Lands of Outer or Insular Crescent" around the periphery. Euro-Asia not only occupied the center of this global stage but was insulated by a vast ocean from the world's less powerful outer continents. To further enhance his view of Eurasian centrality, Mackinder then duplicated the American continents along the map's left and right margins—an ingenious visual ploy dangerously close to the ethical limit of cartographic license.

Mackinder reinforced his message with symbols and text. Central Asia's "Pivot Area" has the map's only horizontal, easily read label, and a light stippled pattern highlights this text and the Heartland itself. At the top of the map, the label "Icy Sea," surrounded by a choppy, fragmented area symbol, emphasizes the inaccessibility of Asia's northern coast. To the left of the map's center, a dense symbolic barrier identifies the huge "Desert" that segregates most of Africa from the World Island. Other labels use size, shape, and orientation to underscore the subordinate locations of lesser areas. The label "Inner or Marginal Crescent" describes in small type a broad arc along the lower edge of the Heartland, and two equally small "Outer" labels accent the peripheral positions of Britain and Japan. Still farther out, the text "Lands of Outer or Insular Crescent" stretches across the map and curls at its ends like a banner proclaiming the World Island's geopolitical counterpoint.

Fig. 5.7. "Natural Seats of Power." (From H. J. Mackinder, "The Geographical Pivot of History," *Geographical Journal* 23 [1904], p. 435.)

Mackinder's cartographic message was radical and disturbing: sea power was losing its strategic importance because a single country or alliance using integrated railways to control a vast Heartland empire could dominate the world. This message not only warned of more difficult times for Britain and the empire but contradicted the prevailing view among military strategists that sea power was more crucial than land power. In 1890, in a book titled *The Influence of Sea Power Upon History, 1660–1783,* American military historian Admiral Alfred Thayer Mahan had used history, geography, and maps to make a convincing case for the effectiveness of a strong navy and merchant marine. Steam navigation and bigger, more precise gunnery reinforced Mahan's argument, as did the recent successes of the British and American navies. Yet Mackinder did not dismiss the value of sea power—indeed, he suggested, a land power controlling the rich resources of the Heartland could both defend itself against invasion by land or sea and also support a massive shipbuilding enterprise.

Mackinder's model was less an explanation than a forecast. In 1904, railway integration of Eurasia had yet to occur, and the pivot state, Russia, was still no match for the navies and armies of the inner and outer crescents. But he saw clearly the inherent strength of the Heartland:

> The oversetting of the balance of power in favour of the pivot state, resulting in its expansion over the marginal lands of Euro-Asia, would permit the use of vast continental resources for fleet-building, and the empire of the world would then be in sight. This might happen if Germany were to ally herself with Russia.

Despite little attention at the time in either Britain or America, Mackinder had accurately predicted—and as later evidence suggests, also influenced—the threat of the Nazi-Soviet nonaggression pact of August 1939.

In 1919 Mackinder repeated the warning in *Democratic Ideals and Reality.* The title reflects his belief that despite the ideals of democracy, the reality is that "the rule of the world still rests upon force." Mackinder interpreted the recent world war as the tragic consequence of an aggressive alliance of Germany, Austria-Hungary, and the Ottoman Turks. The war not only verified the strategic importance of the Heartland but confirmed the possibility that an Eastern European power might exploit its strategic location to leverage the resources of the Heartland against Western Europe and the rest of the world. Indeed, Eastern Europe was now a more potent geographical pivot than Central Eurasia. Four brutal years of worldwide conflict had demonstrated the importance of geopolitics and promised an even greater threat if the political landscape were not carefully reconstructed. As he commented in the preface, "I feel that the war has established, and not shaken, my former points of view."

Democratic Ideals and Reality elevated the World Island and the Heartland to proper nouns. In his 1904 "Geographical Pivot" paper, Mackinder never juxtaposed *world* and *island* and used *heart-land* but twice, in each instance hyphenated and without a capital, but his book's thirty-one maps make these concepts real. Several times he mentioned them in map titles, and in other cases he used symbols and labels to show their relationships to other, better-known geographic entities.

Although Mackinder relied primarily on conventional small-scale thematic maps, he included two comparatively abstract statistical graphs that academic cartographers call *area cartograms*. One portrays the relative land areas of the World Island and its geopolitical competitors, while the other compares populations. Figure 5.8, a facsimile of the population cartogram, shows how he used the relative areas of circles to show that the World Island had many more people than Britain, Japan, Australia, Malaya, and the Americas combined. (The circle for Malaya presumably represents the entire East Indies.) As with his "Geographical Pivot" map (Figure 5.7), the World Island occupies the center. The halo of markedly smaller circles emphasizes not only the World Island's majority population but its strategically dominant position—when I first saw it, I thought immediately of an astronomical diagram of Jupiter and its moons. Like other effective

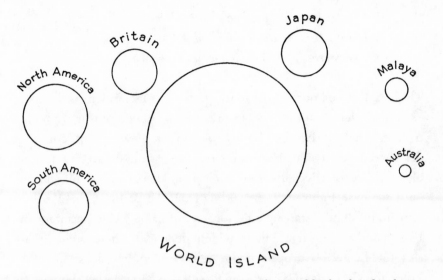

Fig. 5.8. Mackinder used circles representing the World Island and its larger satellites to show their relative areas. (From Halford J. Mackinder, *Democratic Ideals and Reality* [New York: Henry Holt, 1919, 1942], p. 69. Used with permission of Constable & Co.)

cartographic propagandists, Mackinder reinforced his cartographic symbols with seductively worded interpretative text: the map's caption refers explicitly to "the World Island and its satellites."

Concerned largely with territory, disarmament, and reparations, delegates to the Paris Peace Conference had little use for geopolitical theory. Distressed by their collective myopia, Mackinder summarized his theory with a parable that reminded readers of the Roman Empire's fatal complacency:

> A victorious Roman general, when he entered the city, amid all the head-turning splendor of a "Triumph," had behind him on the chariot a slave who whispered in his ear that he was mortal. When our statesmen are in conversation with the defeated enemy, some airy cherub should whisper to them from time to time this saying:

Who rules East Europe commands the Heartland:
Who rules the Heartland commands the World-Island:
Who rules the World-Island commands the World.

These last three lines are the most quoted of Mackinder's writings.

Mackinder's solution was to prevent any single power from controlling Eastern Europe by establishing a new set of countries designed to resist a takeover of the region yet unlikely to threaten stability with a cohesive alliance of their own. He presented this solution on the map in Figure 5.9. Extending from the Baltic to the Mediterranean, these new states would form a buffer between Germany and Russia and at the same time provide independent, defensible territories for Eastern Europe's major non-German ethnic groups: the Poles, the Czechs and Slovaks (in Greater Bohemia), the Magyars (in Hungary), the southern Slavs (the Serbs, the Croats, and the Slovenes in Greater Serbia), the Romanians, the Bulgarians, and the Greeks. Well aware that implementation would require considerable refinement, Mackinder apologized for his crude cartographic model by noting in the caption that "many boundary questions have still to be determined."

Although willing and often eager to generalize in words or maps, Mackinder was well aware of the difficulty of getting the Czechs and Slovaks to work together—not to mention the historically hostile Serbs, Croats, and other southern Slavs, for whom even peaceful noncooperation would prove a significant achievement. To minimize ethnic conflict and make his proposal work required a massive "exchange of peoples," but in the interest of world peace, neither logistics nor individual resistance was an impassable obstacle. "During the war we have undertaken much vaster things, both in the way of mere transport and also of organization," he wrote. "Would it not pay humanity to bear the cost of a radical remedy in this case, a remedy made just and even generous towards individuals in every respect?" Because the need for buffer states was obvious, the Peace Conference delegates designed a map of the Balkans not radically dif-

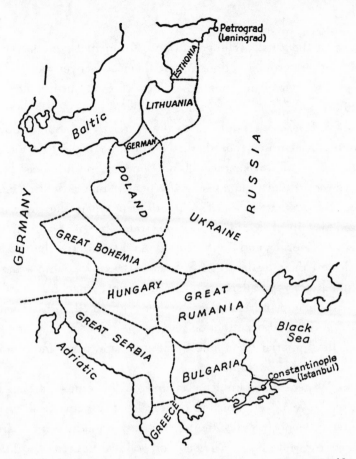

Fig. 5.9. Mackinder's plan for Eastern Europe. (From Halford J. Mackinder, *Democratic Ideals and Reality* [New York: Henry Holt, 1919, 1942], p. 161. Used with permission of Constable & Co.)

ferent from Mackinder's. But unconvinced of the strategic importance of Eastern Europe in controlling the Heartland, they merely drew an imperfect set of boundaries intended to reflect the existing distributions of the region's intermingled and potentially fractious ethnic groups. Because the massive relocations of population required for ethnically cohesive nations never occurred, Mackinder's buffer states were largely a cartographic fantasy.

Defeat of Germany and her allies in 1918 did not bring peace to Eastern Europe. Like the fall of communism seventy years later, the collapse of the Austro-Hungarian and Ottoman Empires freed the region's ethnic groups to fight among themselves over national territory—"ethnic cleansing" might be a new phrase, but it is not a new idea. Farther east, in the Ukraine and the Caucacus, the armies of Bolshevik Russia were adding to the roster of the Soviet Socialist Republics. Mackinder toured much of this area in late 1919 and early 1920, while on his mission to still-independent South Russia. He recognized the difficulty of imposing a settlement on the fighting factions in the Balkans, and he was discouraged when the government ignored his appeals to aid autonomous states in Armenia, Azerbaijan, Dagestan, Georgia, and South Russia. As they entered the 1920s, the British and their leaders had little enthusiasm for supporting self-determination, particularly through any effort smacking of military intervention; and so Mackinder, by then a knighted, upper-level civil servant, redirected his attention toward the Imperial Shipping Committee, the Imperial Economic Committee, and other important yet less controversial pursuits.

Like Wegener, Mackinder was ahead of his time—perhaps more so. But however ominous the forecast, his maps and writings attracted little more media attention than other mildly original articles or books by well-known academic or political figures. On January 26, 1904, for instance, the London *Times* devoted a full column to Mackinder's famous "Geographical Pivot" lecture to the Royal Geographical Society. But after describing Mackinder's argument, the *Times* reporter turned to a dissenter's opinion—roundly applauded by those assembled—that industrial development by "the periphery" could offset Russia's strategic advantage. Later that year across the Atlantic, *National Geographic Magazine* carried a similar report that diffused Mackinder's warning with another source's platitude that "people who have the industrial power and the power of invention and of science will be able to defeat all others."

Mackinder's academic peers seemed more willing to listen. In 1919, in a *Geographical Review* article titled "Geography as an Aid to Statecraft," University of California history professor Frederick

Teggart praised Mackinder's "new method for approaching political questions," acknowledged the "essential correctness" of his arguments, and concluded pessimistically that "the patchwork of diplomatic compromises that we call peace is no sufficient guarantee of the future." The Heartland also won a place in several leading textbooks, including Teggart's own *Processes of History* and British geographer James Fairgrieve's *Geography and World Power.* But recognition by textbook authors was far from universal, and Mackinder was also subjected to his share of academic nit-picking, as in 1920, when in a burst of Yankee chauvinism, Charles Dryer, the president of the Association of American Geographers, chided Mackinder for "missing the real geographical significance of America" and revised the geographical-pivot model by dubbing the peripheral slivers of North and South America the "World Ring."

While Mackinder's geopolitical principles found a few followers in Britain and the United States, his German converts seemed particularly receptive to his medium. In the 1920s, in reaction to the Paris Peace Conference's new political map of Europe, a group of geographers and military historians led by retired army general Karl Haushofer developed a school of thought called *Geopolitik.* In 1924, Haushofer and his followers established the journal *Zeitschrift für Geopolitik* to expound and expand the notion that Germany had been treated unfairly. To sharpen their protests as well as to support German claims to neighboring territory, practitioners of *Geopolitik* devised a richly suggestive, intentionally persuasive cartography based on selective generalization and expressive graphics. In the 1930s, the National Socialists appropriated the cartographic style of *Geopolitik,* and Nazi propaganda periodicals such as *Facts in Review,* a magazine published in New York City, began to use numerous maps—some subtle, others blatantly biased—to win sympathy and justify Hitler's actions.

One of my favorites is "Spheres of Influence," included here as Figure 5.10. Boldly outlined lobes on an interrupted world map present the case for four distinct spheres of influence: one for the United

States, one for Germany and its European competitors, one for Joe Stalin and his Soviet Union, and one for Hitler's Pacific partner, Japan. You take care of your region, it tells its American viewers, and we'll tend to ours. Although this fragmented view of the world contradicts the single bold ellipse centered on Mackinder's World Island, the British geographer's innovative approach to selecting politically convincing cartographic perspectives clearly impressed Haushofer and his followers.

Mackinder's influence on *Geopolitik* and Hitler's foreign policy remains a topic of debate among political geographers. William Parker, who wrote a biography of the British geographer, quoted a variety of sources, including Haushofer himself, to argue that Mackinder "became one of the most influential thinkers of modern times, helping to determine the course of history through his impact upon the external policies of Germany." But another biographer, Brian Blouet, saw this impact as little more than an opportune endorsement and described Mackinder as merely a "prestigious out-

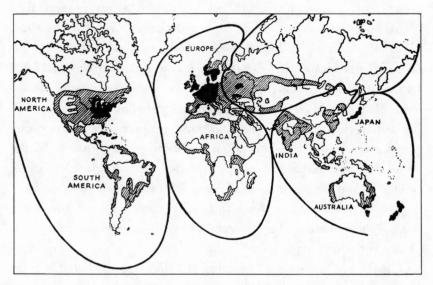

Fig. 5.10. "Spheres of Influence." (From *Facts in Review* 3, no. 13 [April 10, 1941], p. 182.)

sider quoted to buttress established orthodoxy." Nonetheless, respected geopolitical theorist Saul Cohen identified Mackinder's Heartland and World Island as key elements in Haushofer's strategy for German hegemony, which sought to control Eurasia by dominating Russia and to control the World Island by destroying the British navy. Although Blouet might be right that Germany's geopoliticians would have reached similar conclusions on their own, Mackinder's "Geographical Pivot" paper clearly found an appreciative audience in Haushofer, who once wrote, "Never have I seen anything greater than these few pages of a geopolitical masterwork!"

This last quotation appears in Hans Weigert's *Generals and Geographers*, a widely read lay introduction to geopolitics published in 1942. A distinguished German political geographer who fled to America with his family in 1939, Weigert provided not only an exposé of Nazi geographic strategy but also a convincing illustration of Mackinder's wisdom and foresight. To demonstrate that "an understanding of Mackinder's basic idea is essential to the student of Haushofer's geopolitics," Weigert reproduced and discussed Mackinder's "Natural Seats of Power" map (Figure 5.7) and reminded readers that "Haushofer refers to Mackinder time and again." Despite this intellectual debt to the British geographer, however, "to Haushofer, Mackinder was and remained the enemy." Indeed, in a 1925 review of *Democratic Ideals and Reality*, Haushofer had written, "This book should not be translated into German unless the German people are willing to lose all their self-respect, confronted with such a hateful enemy."

Weigert's own praise as well as his account of Haushofer's begrudging endorsements brought Mackinder overdue recognition in both the United States and Britain. In 1944, at a ceremony in London at the American Embassy, the American Geographical Society awarded Mackinder its Charles P. Daly Medal for, among other accomplishments, having "previsioned the true basis of what to-day has won fashion as geo-politics" and for being "the first to provide us with a global concept of the world and its affairs." Not to be outdone, the Royal Geographical Society awarded him its Patron's Medal the following year.

However belated, Western respect for Mackinder was not just ceremonial: in the years following World War II, his work had a notable impact on U.S. foreign policy. Although nearly a half century old, his ominous view of the World Island, with the Soviet Union at its heart, dramatized Russia's threat to world peace and led to the policy of containment that the United States and its allies adopted around 1947 and pursued intensively for two decades. Viewed differently by the U.S.S.R. and the Communist bloc as a policy of "encirclement," the West sought to deny Russia the geopolitical potential of the Heartland through military alliances such as NATO, CENTO, and SEATO as well as by direct assistance to Greece, Turkey, Iran, Malaya, South Korea, and other countries along Mackinder's Inner or Marginal Crescent. Although Mackinder's influence on the cold war is easily exaggerated—particularly after the intense critical review of his theory during the 1950s—as Saul Cohen observed in 1963, "when all is said and done, most Western strategists continue to view the world as initially described by Mackinder."

Critical examination revealed numerous defects in Mackinder's geopolitical model. Some critics pointed out that the Eurasian heartland played a pivotal role only rarely in world history, while others noted that America had evolved from a mere satellite of the World Island to a significant secondary heartland. Best known among Mackinder's critics was Yale political science professor Nicholas John Spykman, who in 1942 inverted Mackinder's model to focus not on the Heartland but on the surrounding "Rimland." Spykman maintained that the Rimland was the geopolitically superior part of Eurasia—clearly much more than a Marginal Crescent—and because of its better climate, agriculturally superior as well. Like others, Spykman considered Mackinder's land power–sea power dichotomy misleading: after all, they argued, a world power would have both a strong army and a strong navy. Other critics, led by aviation innova-

tor Alexander de Seversky, pointed out the increased strategic importance of air power. As demonstrated by the atomic bomb that precipitated Japan's surrender in World War II, geopolitical theories tied to
railroads, tanks, and battleships were technologically obsolete in an era
of satellites, strategic bombers, and intercontinental ballistic missiles.

While the cold war fostered the dullest of debates over Heartland
versus Rimland, advocates of air power demanded fresh cartographic
perspectives reflecting distance and paths of attack more realistically
than the staid Mercator projection. Maps configured like Mackinder's
along a horizontal equator ignored the frightening over-the-pole proximity of the United States and the U.S.S.R. According to de Seversky
and other air-age globalists, an azimuthal equidistant projection centered on the North Pole offered a more meaningful view. On this map
straight lines approximating the routes of "strategic bombers" and
ICBMs demonstrated the need for a network of Distant Early Warning (DEW line) radar stations strung across Alaska and northern
Canada. Eliminating areas south of the equator reduced the projection's more severe areal distortions and focused attention on the world
"power hemisphere," containing the two superpowers and most of the
world's people and resources—although it myopically ignored the
growing fraternity of nations able to build an atomic bomb.

Although a useful reminder of the conventional world map's egregious distortion of the Arctic, de Seversky's polar perspective was
inadequate for addressing most of the pressing international issues of
the 1960s: neocolonialism and self-determination, guerrilla warfare
and counterinsurgency, regional development and resurgent nationalism, military aid and energy dependency, the Population Bomb and
the Green Revolution. Polar-centric maps typically played only a
supporting role in international affairs—in the study of Antarctic
land claims and the hole in the ozone layer, for instance. Because most
of the world's trouble spots—geopoliticians call them *shatterbelts* or
crush zones—were in the largely equatorial Third World, Mackinder's
worldview provided a more suitable framework for discussing foreign policy.

· · ·

Although many political scientists think geopolitics died with Mackinder in 1947—or should have—the geographic study of power and strategic advantage is neither dying nor brain-dead. Mackinder's view of the world may be too simplistic for the late twentieth century, but persistent regional disparities in population pressure, ethnic hostility, accessibility, defensible terrain, mineral wealth, and food resources continue to drive global politics. These disparities create the tension and disorder that geographer Saul Cohen addresses in his theory of global geopolitical equilibrium. Unlike Mackinder's "closed," deterministic model, Cohen's "open," developmental schema does not forecast a single inevitable result. Instead, Cohen views the world as a "global system" evolving in dynamic equilibrium and resistant to an imposed Pax Americana or New World Order that would substitute force for leadership.

I have tried to capture the key elements of Cohen's complex model in Figure 5.11. The map's thickest boundary lines and most distinct area symbols illustrate a primary division of the world into two "geostrategic realms": the trade-oriented Maritime realm (shown in gray), which includes Western Europe as well as nations on both sides of the Pacific; and the more inner-directed Eurasian Continental realm, which includes East Asia and much of Mackinder's Heartland. Shatterbelts (represented by the map's horizontal-line pattern) occur in Eastern Europe and the Middle East, where the realms interact on accessible terrain, without a water barrier to dissipate tension. Neutral zones surround both poles, and South Asia exists as an independent geopolitical region, insulated on the north by the rugged Himalayas and on the south by the Indian Ocean. Thinner boundaries partition each realm into several other geopolitical regions, within which national states (omitted here) form the hierarchy's third, lowest level. Cohen then defined a "Quarter-Sphere of Marginality" to point out the comparatively minor strategic and commercial importance of two relatively remote and less developed regions, sub-Saharan Africa and South America.

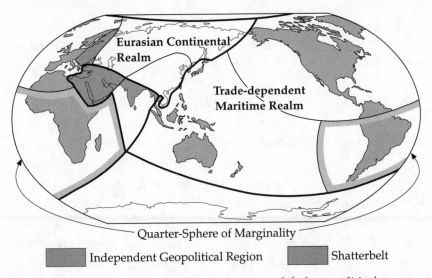

Fig. 5.11. Elements of Saul Cohen's dynamic global geopolitical system. (Compiled from Saul B. Cohen, "Global Geopolitical Change in the Post–Cold War Era," *Annals of the Association of American Geographers* 81 [1991], p. 553.)

Despite liberties taken in condensing Cohen's complex symbols onto a smaller map, Figure 5.11 faithfully reproduces his choice of projection. Accepting cartographic wisdom that an uninterrupted general-purpose, small-scale, whole-world map should *minimize* areal distortion, not avoid it altogether, Cohen chose the Robinson projection, which balances the distortions of area, shape, and distance. Like Mackinder, he omitted meridians and parallels; a grid would not only interfere visually with the hierarchy of boundary lines but imply greater precision than intended by the sweeping, necessarily vague arcs of his maritime boundaries. Most distinctively, Cohen provides a balanced view of the Maritime and Eurasian Continental realms by placing the central meridian at 154° east to keep Eurasia, the Americas, and the Pacific Ocean intact. The result is a broadly inclusive picture of an uncertain geopolitical system that must adjust to recurrent imbalances caused by technology, interethnic conflicts, and alliances among regionally important military and economic powers. Just as Mackinder provided a geographic perspective arguably appropriate for the beginning of the twentieth century, Cohen's world map informatively reflects the diplomatic challenges of the twenty-first century.

The controversial maps of Alfred Wegener and Halford Mackinder illustrate revealing cartographic similarities between the world of physical science and the world of social science. For both scholars, the map was initially important as a vehicle of discovery. Although Wegener's discovery that continental outlines fit together seems more serendipitously sudden, Mackinder most likely grasped the significance of a "geographical pivot" in a similar flash of insight. Wegener and Mackinder used numerous maps to develop and expand their theories, and both relied heavily on maps to promote their ideas in lectures, articles, and books. Their success demonstrates the power of cartographic generalizations as scientific propaganda: both men chose map projections that favored their purposes, and while Wegener fudged coastlines to fit his concept of continental drift, Mackinder conveniently selected features and labels that highlighted the importance of a Heartland and a World Island. Maps account for much of the early interest in these theories, and maps were important in their revival, modification, and wider dissemination. Because much of the evidence and logic was cartographic, opponents and skeptics of the theories often attacked the maps: eminent geologists at the AAPG symposium ridiculed Wegener's maps, for instance, whereas Spykman inverted Mackinder's cartographic evidence to assert the strategic dominance of the Rimland. Although new theories eventually replaced the simplistic models of Wegener and Mackinder, plate tectonics and Cohen's global geopolitical equilibrium depend as heavily on both cartographic evidence and cartographic rhetoric. In the end the rhetoric proved more important than the evidence: physical scientists stimulated by Wegener's writings have demonstrated that continents do indeed drift, and social scientists who have read Mackinder appreciate location's crucial role in diplomacy and world trade.

• 6 •

Maps,
Votes,
and Power

Viewed with suspicion, if not fear, by most politicians, political
reapportionment brings us to the local side of the map story.
Election district lines have become the subject of much controversy,
as they can substantially increase the electoral strength of a party or
candidate and can directly affect the power of a citizen's vote. With
such basic rights at stake, redistricting is an inherently emotional process and frequently sparks bitter disputes. Instinctively wary of oddly
shaped legislative districts, people are quick to question districts that
deviate from traditionally compact shapes, and anything perceived as
gerrymandering, regardless of the social cause, can further heighten
the pitch of the debate. However, issues of racial justice and equal

representation raise interesting questions about mapping modern political priorities, and as the saga of electoral remapping in the state of New York shows, implementing these priorities cartographically is not easy. Faced with the possibility of orchestrating the success of various constituencies and special interests by the maps they draw, legislators and voters have had to reevaluate the objectivity of straightforward district shapes.

I'll begin the reapportionment story close to my home on a cold, snowy, Saturday afternoon in Syracuse. A new show had come to town, and it was here for just one day, for a single performance. It's not a play, a concert, or a movie, I explained to Jo, my thirteen-year-old daughter, but you might find it interesting. "I've never attended one before," I told her, "and it concerns something I'll be writing about. Besides, it'll be better than sitting around the house, and we can stop at the library on the way back. Or perhaps the mall." That was how I talked Jo into attending a reapportionment hearing.

An hour later we entered the spartan lower lobby of the Onondaga County Court House. The elderly guard sitting behind a beat-up desk looked up from his newspaper and told us to go up to the fourth floor. He was alone: the lawyers, defendants, witnesses, bailiffs, judges, gawkers, and passport seekers were enjoying the weekend as best they could, at home, on the ski slopes, or in jail. We nodded to the guard, climbed three wide marble staircases, and walked down a long hallway past oversize doors with cut-glass panes stenciled in large black letters. The door to Room 407, the Legislative Chamber, was open. At a table just inside the anteroom sat three clerks, and one of them asked if I wanted to speak, but I declined. Jo, who wasn't invited, declined too. We signed in, picked up a booklet containing maps and data, and walked to the spectators' benches in the back of the room, where nearly a hundred people were seated.

On the raised platform at the front of the room sat a half-dozen men and one woman, all in weekday business dress. In front of the dais, two easels holding large maps faced the spectators. A balding, slightly overweight man with a raspy voice faced the people on the platform and spoke into a microphone. He was upset about both the map and the state legislature, which had appointed the people on the dais—the people who had drawn the map. The young woman who testified after him was no less indignant. We had arrived around 1 P.M., and before we left an hour later, several other people spoke. A prominent article in the morning newspaper had discussed the hearing's purpose, and notices posted several weeks earlier at local libraries described its format, which included "pre-notified oral testimony" by anyone who signed up in advance. The bulletin also encouraged attendees and others to file written comments about the map.

If this event had been a movie, we would have missed the beginning and much of the plot. But although a dozen people had spoken since 11 A.M., what they said was probably no different from what we heard later: everyone denounced a small part of the map, some particular boundary. Anyone who might have been pleased with the map and its boundary lines either kept silent or stayed home. According to a press release, the "open-ended" hearing would end only "when all testimony has been presented," and a follow-up story in the local Sunday newspaper reported that altogether twenty-eight people testified.

Listening to speakers' complaints, taking notes, and trying to appear attentive were the members and staff of New York State's Legislative Task Force on Demographic Research and Reapportionment. Their "task" was to draw maps of congressional and legislative boundaries that satisfied the "one person, one vote" principle in Article I, Section 2 of the U.S. Constitution as well as several more specific requirements added by the Federal Voting Rights Act passed in 1965 and strengthened in 1982. The Constitution requires a population census every ten years to determine whether each state should have more or fewer members of the House of Representatives and to provide data for dividing states into House districts nearly equal in population. The 1990 census had not been kind to New York, which grew more slowly during the 1980s than the nation as a whole, and consequently would lose three of its thirty-four congressional seats. Migration to the outer suburbs and other population shifts required substantial readjustment of district boundaries, especially in New York City, where some ethnic neighborhoods grew, others withered, and a few new ones appeared. Like other states, New York faced rejection by the courts or the U.S. Department of Justice if its reapportionment plan diluted the political strength of minorities. Because the state's minority populations had increased both numerically and proportionately during the 1980s — the non-Hispanic white population had actually decreased — New York was obligated to reconfigure several districts to increase the likelihood of electing more African Americans and Hispanics to the House of Representatives. Even more complex was the required reapportionment of the 150-member State Assembly and the 61-member State Senate.

Many of the speakers were politicians, and a few held seats threatened by the new redistricting plan. One was a school-board member who knew and liked a particular incumbent and resented reassignment to someone else's district — it was much easier to lobby a friend or neighbor than a stranger from the next county or beyond. Several speakers came from Cortland County, just south of Onondaga County. These Cortlanders pointed with rage at the thick dashed-and-dotted line that jogged east, then south, and then west just above the county name "Cortland" in Figure 6.1 — the task force had drawn a line that divided their county between two Assembly districts, neither with county residents in the majority. "Our county has always been 'kept whole' in the Assembly," one argued, "and to make us deal with two Assemblymen, neither of whom particularly cares about our problems, diminishes our clout in Albany!" "Why pick on us?" another asked. "I'll tell you why," one Cortlander answered, "the new boundaries are Albany's retribution for Cortland's organized local opposition to a proposed state-run landfill for low-level radioactive waste." Homemade signs saying "Dump the dump!" and "Waste Cuomo!" had screamed from lawns and windows throughout the county for the past half-dozen years. "This is just one more example of Governor Cuomo's well-known vindictive streak!" But a few politically savvy Cortlanders saw a purely partisan motive: a plot to strengthen the newly pro-Democratic 125th District in neighboring Tompkins County by tacking on the city of Cortland and the town of Cortlandville. (In 1988, for the first time in seventy-five years, Tompkins County, which includes Ithaca and Cornell University, had sent a Democrat to the Assembly.) "How can the New York City Democrats who control the Assembly get away with such blatant gerrymandering?" they demanded.

Cortlanders were not the only speakers charging partisan bias: Republicans from Oneida County, fifty miles east of Syracuse, objected to the innermost thick boundary near the center of Figure 6.2—this new 116th Assembly District, they charged, increased the chances of electing a Democrat by combining the cities of Rome and Utica. (Although the two cities combined had more Democrats than Republicans, separate districts centered around Rome and Utica included sufficient rural residents to elect Republicans to the Assembly.) "Rome and Utica are different," said speakers from both cities, "and each needs its own Assembly seat." "Once again the Democrats from downstate are having their way with upstate New York," another irate Upstater protested. "Reapportionment is nothing more than political rape!" Rate the show PG-13, for adult situations.

As in most states, politicians in New York's legislature control redistricting. And as elsewhere in the country, the process of redrawing New York's congressional and legislative boundaries has been contentious: some groups must lose prestige or power, and some winners are not appeased by small gains. Reflecting most, if not all, of the redistricting controversies found in other states, New York affords a case study perhaps more colorful than most because of its pronounced ethnic and cultural diversity, its large population with enormous expectations paying high taxes to a big bureaucracy with numerous programs and regulations, the conflicting goals of "upstate" and "downstate" as well as several big cities and their suburbs, and a delicate balance between a Democratic party that controls the State Assembly and the governorship and a Republican party that controls the State Senate.

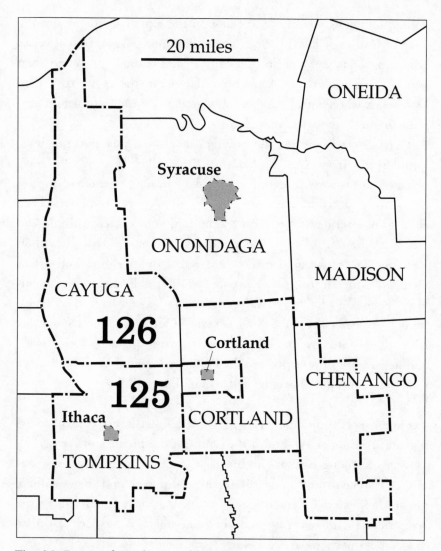

Fig. 6.1. Proposed 125th Assembly District groups all of Tompkins County with a portion of Cortland County, to the east. The remainder of Cortland County shares another district with Cayuga County and a substantial portion of Chenango County, farther east.

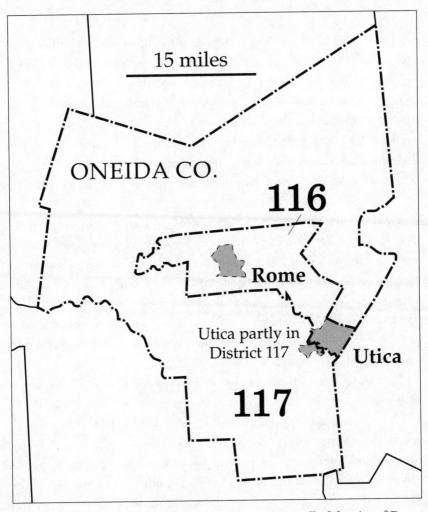

Fig. 6.2. Proposed 116th Assembly District combines all of the city of Rome with most of the city of Utica, about twelve miles east.

Redistricting takes years of preparation, and is more a process than a task or project. And as a process, redistricting is fundamentally cartographic: census enumeration depends on detailed maps, and the same maps provide a basis for drawing new district boundaries, adding up residents, and nudging the boundaries this way and that to make the counts satisfy federal guidelines as well as accommodate incumbents, power elites, and at least a few minorities and special interests. Maps are important, too, in presenting and debating proposed redistricting plans. I say "proposed" because, as New York's experience demonstrates, a reapportionment plan typically is modified in the later stages of the process, when the public, the governor, the Justice Department, the state courts, the federal courts, and perhaps even the U.S. Supreme Court have a say. However authoritative and final the maps might look, both the hearing process itself and the perceived effects of the new districts invite challenge and controversy.

New election districts can generate heated quarrels, and the pejorative of choice is "gerrymander." Both a noun and a verb, *gerrymander* draws attention to noncompact shapes, calls to mind the prototypical caricature of an outrageously curved district with vicious clawlike appendages, and suggests malevolent manipulation by ruthless political hacks in smoke-filled rooms. Yet as I will show, the gerrymander got a bum rap—a compact shape need not be as important as many politicians and political scientists might think.

Elbridge Gerry, after whom the term was coined, is also unfairly maligned. Gerry, a Republican, was governor of Massachusetts in 1812, two years after population shifts revealed by the census of 1810 called for reapportioning the Bay State's senatorial districts. Gerry's party, which controlled the state legislature, used its power to pass a redistricting bill that diminished the strength of its opponents, the Federalists. After looking carefully at voting and registration data, the party's political cartographers created two kinds of districts— those in which Federalists were very strong and those in which Republicans held a comparatively weak majority. If this sounds like political suicide, think again: however weak, a majority of only one vote was sufficient to elect a candidate. Even though more Massachusetts voters marked their ballots for a Federalist candidate in the next election, the Republicans captured three-quarters of the seats in the new State Senate—packing lots of Federalist voters into relatively few districts proved an effective way to "waste" thousands of Federalist votes.

Creating a statewide minority of Federalist strongholds sometimes required highly contorted boundaries, such as those in Figure 6.3, a map of Essex County, north of Boston, printed in the March 6, 1812, issue of the Boston *Weekly Messenger*. Rather than split the county into two relatively compact districts with a comparatively straight east-west boundary, the Republicans drew a boundary that began with the north-south line separating Beverly from Danvers, ran to the north-west as far as the Merrimack River, and then meandered eastward along the river to the coast. The result was an upper Essex district that stretched from Salsbury (spelled back then without the *i*) to Chelsea by linking a string of four towns north of the river with eight other towns along the county's western edge. The Federalists quickly spotted the ploy, and when a reporter at the Boston *Gazette* noted the similarity between the upper Essex district and a salamander, his editor exclaimed scornfully, "Salamander! Call it a Gerrymander!" This clever outburst inspired political cartoonist Elkanah Tisdale to draw the well-known part map in Figure 6.4. On March 26 the *Gazette* published Tisdale's Gerrymander, richly embellished with claws, teeth, and wings; several weeks thereafter other pro-Federalist Massachusetts newspapers borrowed the printing plate, reprinted the cartoon-map, and stirred up sufficient anti-Republican sentiment to defeat the governor. Although Gerry himself did little more than sign the bill, most historians remember him not for signing the Declaration of Independence or helping found the Library of Congress but for endorsing his party's self-serving carto-chicanery.

Despite losing the governorship, the Republicans did extraordinarily well in the statehouse sweepstakes: their party captured twenty-nine seats with a total of only 50,164 votes statewide, whereas their Federalist opponents collected only eleven seats with a total vote of 51,766. Anyone skeptical about how gerrymandering can be so effective need only consider the hypothetical region in Figure 6.5. The area's 150,000 voters are scattered among twenty-four townships, which must be divided into two legislative districts, each with 75,000 voters. Two of the townships, marked A and B, are largely

ESSEX COUNTY.

Fig. 6.3. The news map that inspired the Gerrymander cartoon, which appeared in the Boston *Weekly Messenger*, March 6, 1812. (From Elmer C. Griffith, *The Rise and Development of the Gerrymander* [Chicago: Scott, Foresman and Company, 1907], p. 69.)

urban, and each has 20,000 voters. All of the other twenty-two townships are rural, and each has only 5,000 voters. As shown in the lefthand map, a simple solution is to draw a vertical boundary through the middle of the region, thereby creating two districts identical in area, shape, and voting population. In addition to meeting the one person, one vote requirement, each district is as geometrically compact as the region's overall shape allows.

Now let's add two political parties—call them the Extremists and the Mugwumps—and some typical rural-urban differences in political

Fig. 6.4. Elkanah Tisdale's original Gerrymander, as it appeared in the Boston *Gazette,* March 26, 1812. (From James Parton, *Caricature and Other Comic Art* [New York: Harper and Brothers, 1877], p. 316.)

preference. Based on party registration data and results from the last election, each rural township has 3,000 pro-Extremist voters and only 2,000 pro-Mugwump voters. By contrast, each urban township has 18,000 pro-Mugwump voters and only 2,000 pro-Extremist voters — town dwellers solidly support Mugwump principles while country people favor the Extremist agenda by a more modest margin. In an election based on the two compact districts, the Mugwumps would win both seats. Each district gives the Mugwump candidate 40,000 votes:

18,000 *votes in the single urban township*

+**22,000** *votes in the 11 rural townships (11 ×2,000)*

40,000 *votes for each Mugwump candidate*

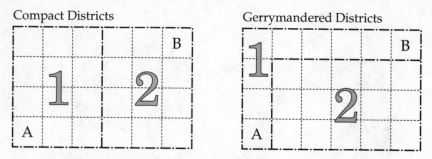

Fig. 6.5. Alternative reapportionment plans for a hypothetical area with twenty-two rural townships and two urban townships, A and B.

This is hardly a landslide because each Extremist candidate captures a respectable 35,000 votes:

> **2,000** *votes in the single urban township*
> +**33,000** *votes in the 11 rural townships (11 ×3,000)*
> _____
> **35,000** *votes for each Extremist candidate*

But despite these close votes, the majority rule of "winner take all" gives the Mugwumps both seats.

What the Extremists need is a political map that puts the two urban townships in the same legislative district, leaving the other district with an Extremist majority. Because Townships A and B are at opposite corners of the region, a district that wastes Mugwump votes is necessarily long and thin, like the inverted L-shape of District 1 in the right-hand part of Figure 6.5. Containing the two urban townships and only seven rural townships, this District 1 will give its Mugwump candidate a solid 50,000 votes:

> **36,000** *votes in the 2 urban townships (2 ×18,000)*
> +**14,000** *votes in the 7 rural townships (7 ×2,000)*
> _____
> **50,000** *votes for District 1's Mugwump candidate*

In comparison, the Extremist opponent will do poorly, capturing a mere 25,000 votes:

4,000 *votes in the 2 urban townships (2 × 2,000)*

+21,000 *votes in the 7 rural townships (7 × 3,000)*

25,000 *votes for District 1's Extremist candidate*

Despite the likelihood of losing the District 1 seat for the next ten years, the Extremist political cartographers know what they are doing: the fifteen rural townships that constitute District 2 afford an almost equally safe seat for the Extremist candidate, who might expect 45,000 votes (15 × 3,000), in contrast to only 30,000 ballots (15 × 2,000) favoring the opposing Mugwump. A paradox of the reapportionment game is that one party can benefit by designing a district that guarantees landslide victories for the other party—redistricting experts call this practice *packing.*

As with other games, strategies vary with strength and need. If the Mugwumps were not quite so strong in the two urban townships, gerrymandering District 1 to contain both City A and City B might help them rather than the Extremists. Were each urban center to yield only 15,000 Mugwump votes, for example, with the remaining 5,000 enriching the Extremist opponent's tally, the Mugwump candidate would lose in each of the two compact districts by a mere 1,000 votes! Yet a Mugwump could win handily in the thin, L-shaped district with a healthy 13,000-vote lead. Work it out if you don't believe me. Reapportionment is perhaps the main reason parties and politicians keep meticulous records of election results and party affiliation. Such records no doubt told the Democrats in the New York Assembly that stretching—or *shoestringing,* as the experts say—a district to include strongholds in Rome and Utica, about fifteen miles apart, was an easy way to pick up an additional seat.

Politicians and political parties are not the only players in reapportionment. The Voting Rights Act of 1965, which outlawed the poll tax and other racially motivated impediments to voting, required that the Justice Department and the courts monitor and enforce state and local compliance with the constitutional guarantee of equal representation. The act forbade wide variations in population among congressional and legislative districts, and allowed the courts to redraw district boundaries if states and localities resisted. Moreover, the law required "preclearance" by the Justice Department for changes to election laws, districts, and other procedures in states with historically significant pockets of low voter participation. Because of low voter registration and turnout among blacks and Hispanics in three of New York City's five counties, New York State could not implement a redistricting plan without federal preapproval.

In 1982 Congress strengthened the Voting Rights Act—and the role of the courts—by prohibiting reapportionment officials from splitting a natural black or Hispanic constituency among two or more congressional or legislative districts. By ignoring the issue of *intent* and focusing on the *effects* of redistricting, the new law not only made judicial intervention easier but forced officials to look carefully at the geographic distribution of minorities and to seek their participation. Although requirements were frustratingly imprecise, decisions by both the Supreme Court and the Justice Department encouraged states to empower minorities through remedial racial gerrymandering.

To understand how redistricting can promote ethnic empowerment, we need only substitute Hispanics and non-Hispanics for the Mugwumps and Extremists in our hypothetical example in Figure 6.5. Whether a district actually sends a Hispanic to the legislature would depend, of course, on voter turnout, ethnic solidarity, and whether one of the parties nominates a Hispanic candidate. In practice, a "winnable" minority-majority district generally requires a minority population of about 65 percent to compensate for a smaller proportion of eligible adults, lower registration rate, and lower turnout. Whatever the outcome of the election, though, the shoestring district's Latino population would clearly enjoy greater clout than if the region were divided into two comparatively compact districts, each only 49 percent Hispanic.

Which plan is fair? To answer such questions political scientists and federal judges often look at a redistricting plan's net effect statewide. An election in which a party's candidates collect only 50 percent of the votes cast yet win 80 percent of the seats, for example, would raise serious doubts about the plan's fairness—whether districts looked long, thin, and gerrymandered would be irrelevant. But extreme disparities between total votes and seats won almost always requires some conspicuous manipulation of the state's electoral geography. In questions of minority representation, the Justice Department and the courts look especially closely at boundaries within metropolitan areas. For example, if bloc voting by 50,000 residents were sufficient to win a seat by a slim majority in a city with 100,000 Hispanics, a fair plan would give the city's Hispanics at least one clearly winnable district. Thus a district containing approximately 65,000 Latinos should meet both the spirit and the letter of the law as long as its shape were not too outrageous and its configuration did not infringe the rights of another minority. Because of the contiguity requirement, an interlocking pattern of ethnic enclaves could make the empowerment of one group difficult without the disenfranchisement of another. Yet one or two highly contorted—and quite obviously gerrymandered—districts are unlikely to disqualify a reapportionment plan that redresses a historical pattern of egregious underrepresentation.

Losers in the redistricting game are seldom happy, of course, and if the political map affords an opportunity to cry foul, they will. As the sinuous shape of North Carolina's 12th Congressional District in Figure 6.6 demonstrates, the most powerful and dramatic evidence of what white conservative columnist James J. Kilpatrick denounced as "gerrymandering madness" is typically cartographic. In 1992 five white North Carolina residents appealed to the U.S. Supreme Court that District 12's highly unusual shape was not only "ugly" and "uncouth" but an outrageously unconstitutional response to the 1982 amendments to the Voting Rights Act. "The Constitution is color-blind," the plaintiff's attorney Robinson Everett told the *New York Times*, but "this redistricting is just the opposite and amounts to the use of a racial quota system." Yet defenders contended that despite its serpentine shape, District 12 was both fair and functional. In rejecting the contention that congressional districts should be cohesive and compact, American Civil Liberties Union attorney Kathleen Wilde argued that "you should be able to make it ugly for racial reasons in order to comply with the Voting Rights Act."

As the map testifies, District 12, if not ugly, is anything but compact. Following Interstate 85 for most of its 150-mile length, from Durham through High Point to Charlotte, with offshoots to Winston-Salem and Gastonia, the district tenaciously maintains contiguity while linking concentrations of black voters in six widely separated cities—*Newsweek* described it as "a skinny dragon with Interstate 85 as its spinal cord." "At some points," attorney Everett told *Congressional Quarterly*, "it's no wider than two lanes of I-85." This unusual

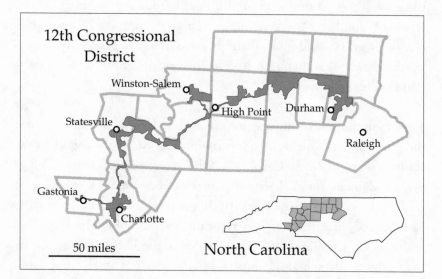

Fig. 6.6. The contorted shape of North Carolina's 12th Congressional District invites attack as a "racial gerrymander."

shape reflects a Justice Department effort to force the North Carolina legislature to create two districts in which African Americans form a 57 percent majority; since blacks constitute 22 percent of the state's population, attorneys in the Civil Rights Division reasoned that blacks should have the opportunity to elect two (17 percent) of the state's twelve congressmen. After the Justice Department rejected a plan creating only one winnable black district, the legislature's Democratic-controlled reapportionment committee drew new boundaries that established two "minority-majority" districts while still protecting the party's white incumbents in other districts. Although the revised plan achieved a closer racial balance — North Carolina sent two black representatives to Washington for the first time in this century — the plaintiffs in *Shaw* v. *Reno* (all white Democrats) argued that District 12 unfairly discriminated against white voters.

In June 1993, by a close 5-to-4 vote, the Supreme Court ruled that District 12 might violate the Constitution and remanded the case back to a lower court. Writing for the majority, Justice Sandra Day O'Connor observed, "It is unsettling how closely the North Carolina plan resembles the most egregious racial gerrymanders of the past." But not all the justices were as outraged by the district's shape. In one dissenting opinion, Justice David Souter noted, "The Court offers no adequate justification for treating the narrow category of bizarrely shaped district claims differently from other districting claims." In another dissent, Justice John Paul Stevens conceded that "the existence of bizarre and uncouth district boundaries is powerful evidence of an ulterior purpose behind the shaping of those boundaries," but argued that "in this case, however, we know what the legislators' purpose was: The North Carolina Legislature drew District 12 to include a majority of African-American voters." Moreover, Justice Stevens asserted, "There is no independent constitutional requirement of compactness or contiguity. . . ." Although many press accounts portrayed the high court's wishy-washy decision as a setback for civil rights, I am not so sure it is. After all, by ruling that a contorted shape merely requires further study, the Court implied a willingness to value motives and effects more highly than shape. Legislatures, courts, and voters can get used to bizarre and possibly even fragmented shapes as long as these districts contribute to fairness and equity.

Racial balance is not the only explanation for District 12's serpentine shape: like its counterparts elsewhere, the Democratic majority in the North Carolina legislature is committed to protecting incumbents. Because clout is linked to seniority, few legislators see any conflict between protecting their own seats and looking out for their constituents. (In the next reapportionment, after the year 2000 census, the state's new African-American Democratic incumbents will no doubt receive similar protection when partisan gerrymandering steps in again.)

A legislature typically protects incumbents by not placing two members in the same district, which would force one of them to lose, move, or step down. Reapportionment committees prefer to give an incumbent a district where he or she is well known, has won the last election, and is likely to win again. If the incumbent is well liked by constituents, the ideal new district is very similar to the old district, but where voters supporting an incumbent have moved out of a city into the suburbs, both the election district and its legislator sometimes move with them. By manipulating boundaries, creative political cartographers can sometimes shelter a vulnerable incumbent by jockeying a strong potential opponent into another district, and party bosses can enforce discipline by threatening a disloyal incumbent with a district in which the opponent is certain to win. When legislative rules or a nearly equal balance of power requires bipartisan cooperation, parties will strike deals in order to avoid sacrificing their strongholds — yet maintaining the political status quo is seldom easy when some areas have grown while others have stagnated or declined. Because both parties need to protect incumbents, each legislator who declines to run adds flexibility to the reapportionment plan while each member who seeks reelection makes the resulting boundaries more irregular.

$\bullet \quad \bullet \quad \bullet$

Although New York's Legislative Task Force admits a conscious effort to help incumbents, it chooses its words carefully. The book of maps Jo and I picked up at our public hearing lists three principles that guided reapportionment of the State Assembly: the equal-population requirement, the federal mandate to create new minority-majority districts, and the need to protect incumbents. In describing this third principle, however, the preface cautiously avoided the word *incumbent*:

3) When not in conflict with Federal equal population and minority representation requirements, efforts were made to preserve communities and continuity of constituencies by preserving county borders and retaining existing district boundaries when possible.

Highlights listed for the task force's senatorial plan described similar goals, again without direct reference to incumbents:

the recommendation seeks to preserve town and county borders as well as maintain existing district boundaries except for Voting Rights Act and equal population concerns.

Constituency preservation is politically understandable. The most irate speakers at the Syracuse hearing were concerned about maintaining their communities' unique identities and denounced boundaries that split Cortland County or merged communities, like Rome and Utica. Even if an incumbent were not threatened, local politicians were certain to complain; political parties are organized by city and county, and the local committees of both parties want to keep their domains intact. Because of wide variation in county populations, however, no reapportionment plan could preserve all constituencies and still satisfy the equal-population requirement. To demonstrate awareness of these concerns—better to acknowledge unmet needs than to ignore them—the task force proudly noted that by splitting only twenty-three counties, the proposed reapportionment plan for the

Assembly was a marginal improvement over the current districts, which split twenty-seven of the state's sixty-two counties.

Although unhappy county committees could be troublesome, the task force was far more wary of how the Justice Department and the courts would view its compliance with the Voting Rights Act. Four bulleted items listed in the book of maps under "Assembly Highlights" summarized the proposed plan's "strict adherence" to the federal mandate:

- Thirty-one Assembly Districts have been proposed within which the minority community will have the ability to elect a candidate of their choice.
- Five new minority Assembly Districts are proposed with no sitting incumbent. . . .
- Twenty-six additional Assembly Districts have substantial minority populations and will make it possible for the minority community to elect candidates of their choice. . . . Three more proposed districts have populations which have a majority of its residents who are members of a minority community.
- Even when it was not possible to create minority seats, minority communities throughout the state were "kept whole."

Highlights listed for the proposed Senate districts pointed to the separate empowerment of African-American and Hispanic communities in New York City:

This new State Senate proposal will create eleven Senate Districts of which seven have a majority non-Hispanic Black voting population and four have a majority Hispanic voting age population. . . .

Additionally, under this plan, there are four districts with a Hispanic majority voting population. . . .

Because the new districts would exist through the year 2002, the latter claim implied future gains for the city's growing Hispanic population.

• • •

Congressional remapping proved more troublesome than the state government's legislative reapportionment. Unlike the compromise that perpetuated Democratic control of the Assembly and Republican control of the State Senate, no mutually convenient deals could protect incumbents and both parties at the federal level. Moreover, because New York had lost three seats in the House of Representatives, at least three incumbents would have to go. Yet by early March, only one—Long Island Democratic congressman Robert Mrazek— seemed willing to step down. By late March no one else had announced plans to retire, so the task force needed to target at least two additional incumbents. Because each party controlled half of the state legislature, the six-person task force was deadlocked. (That twenty-one of New York's thirty-four members of Congress were Democrats gave their brethren on the task force no additional clout.) Although the Democrats proposed eliminating Mrazek's seat on Long Island, a Republican seat somewhere in western New York, and an evenly matched "fair-fight" seat somewhere else, the Republicans preferred eliminating three "fair-fight" districts. The Republicans also saw an advantage in the stalemate: if the task force and state legislature didn't act, the courts would redraw congressional boundaries themselves, and a map drawn by judges would most likely jeopardize more Democratic incumbents than a map drawn by the task force.

Meanwhile, time became tighter still, as incumbents and eager challengers demanded to know where they might run and which residents could legally sign their nominating petitions. Statewide primary elections were scheduled for the end of the summer, and to give candidates sufficient time to circulate petitions, state law required full approval of all three new electoral maps by June 9. Since review by the Justice Department could take up to two months, the congressional redistricting plan should have been presented, discussed at public hearings, revised, approved by the legislature, and sent to

Washington no later than April 9. But as of March 25, after missing and extending its own deadline four times, the task force was unable to resolve the impasse, and its Republican co-chairman Dean Skelos openly conceded that a court-appointed "special master" might be asked to remap the state's congressional districts. Although an at-large election might be all right for Rhode Island or Delaware, for a state as large and diverse as New York the courts would most certainly impose their own redistricting plan rather than sanction a preposterously complex choose-any-thirty-one-candidates political slugfest.

Demographic shifts and the need to establish additional minority-majority districts created further pressure, further wrangling, and further delays. New York City's Latinos, who constituted about a quarter of the population, were understandably concerned that only one of the city's fourteen members of Congress was Hispanic. Not content with protests, lobbying, and threats of lawsuits, the Puerto Rican Legal Defense and Education Fund drew its own map, propos-ing a 60 percent Hispanic district reaching into portions of three of the city's five boroughs. Showing how Hispanic representation could easily be enhanced was a powerful cartographic gauntlet tossed at the feet of the task-force members and any other politicians who might need reminding about the intent and requirements of the Voting Rights Act. That the suggested boundaries might have made Elbridge Gerry's colleagues sigh with envy made little difference. *New York Times* political correspondent Sam Roberts, who described the pro-posed district as "an asymmetrical lobster [with] one claw in Brook-lyn, another in Queens and with a disembodied head across the East River in lower Manhattan," noted similarities to other geographically droll yet apparently acceptable congressional districts such as North Carolina's "Interstate 85" district (shown in Figure 6.6), a district in Texas that resembled "four spiders having an orgy," and an Illinois district likened to "a pair of earmuffs." As Roberts wryly observed, to previously disenfranchised minorities, gerrymander "is no longer a pejorative word, it's a goal."

On April 7, the courts intervened and ruled that unless the state legislature reached a consensus by April 27, the Democrats, the Republicans, the Hispanics, and any other interested party could submit their own congressional redistricting plans by May 4. Then, starting May 5, either the judges themselves or a special master yet to be appointed would review all proposals submitted and either choose one or develop his or her own. So on April 27 both the Assembly and the State Senate submitted separate redistricting plans, which African-American and Hispanic groups immediately denounced as diluting minority voting strength in violation of the Voting Rights Act. A week and a half later, on May 5, the court tried again to force an agreement by giving Governor Mario Cuomo and the legislature another week to reach a compromise. Speaking for a Federal District Court panel, Judge George Pratt noted that in 1982 appointing a special master to resolve a legislative impasse over congressional redistricting had cost taxpayers a million dollars—an enormous waste because legislative leaders ultimately worked out a last-minute compromise. Nonetheless, on May 12, after officials still could not agree, the court appointed seventy-one-year-old former U.S. district judge Frederick Lacey and ordered him to draw up by May 26 a redistricting plan meeting both the equal-population and civil-rights requirements. The court also placed the task force's data, software, computers, staff, and facilities at Lacey's disposal.

On May 26 Lacey presented his plan. The *New York Times*, which compared the districts Lacey created with those presented separately by the Assembly Democrats, the Senate Republicans, the Republican minority in the Assembly, and the group of congressional Democratic incumbents, noted that Lacey's map showed markedly less regard for incumbents and other political concerns yet created not one but two new Hispanic-majority districts. Not surprisingly, incumbents bitterly attacked Lacey's proposal, which radically modified many districts and called for senior congressmen to oppose each other for four seats—putting added pressure on legislative leaders to reach a more politically palatable compromise.

To further complicate the picture, experts previously appointed by the State Supreme Court presented yet another congressional redistricting plan on June 1. Unlike the federal court's plan, the state court's offering was kinder to incumbents and their traditional districts but created only one new Hispanic-majority district. Both proposals provided four black-majority districts, all with an African-American incumbent.

Although unwilling to fully concede its redistricting prerogative to the courts, the state legislature recognized that the state court plan was far less disruptive than Judge Lacey's proposal. During a June 3 federal court hearing to examine the special master's plan, legislative leaders informed Judge Pratt that they had nearly compromised on the state court's proposal. With patience that even Job might envy, the federal judges gave the state yet another week to enact a final redistricting plan. Despite threats by the Puerto Rican Legal Defense and Education Fund to appeal to the Justice Department for a third Hispanic seat, the majority leaders in both houses dared not risk their fragile truce. On June 8 the Senate approved the compromise plan, on June 9 the Assembly added its endorsement, and on June 11 the governor signed the bill into law, barely meeting the federal court's final deadline. Although he supported the Hispanic position, Governor Cuomo signed the measure, confident that the Justice Department would insist on modification when it saw from Judge Lacey's map how a second new Hispanic district could be configured. When instead the Justice Department certified compliance with the Voting Rights Act on July 2, Cuomo realized that rejecting the compromise and letting the courts—probably the Supreme Court—settle the final reapportionment might have been the better stratagem.

If anything, controversies arising during New York's latest decennial redistricting effort suggest that election district boundaries can be drawn in a surprising number of perfectly legal ways. Without well-defined statutory constraints on the shape of districts or a clear mandate to protect political parties in the same way that the Voting Rights

Act now protects African Americans and Latinos, the courts willingly tolerate collusive bipartisan gerrymanders that protect incumbents and favor the majority party. As long as politicians responsible for reapportionment meet equal-population requirements and respect geographically compact minority constituencies, nothing is inherently wrong with geometrically irregular districts designed to protect incumbents. Because the Justice Department and the federal courts are willing to approve remedial gerrymanders yet unwilling to require them, two radically different redistricting plans can be equally legitimate — as in 1992, when New York could have added a third Hispanic-majority congressional district but wasn't required to do so.

Is there a more efficient, less contentious way to choose among the many possibilities? Some geographers and political scientists think there is. Why not, they say, let a computer draw the boundaries. Give the machine specific goals for promoting minority representation and geographic compactness, keeping counties whole, requiring contiguity, minimizing variation in population — and even helping incumbents, if you like — and let it apply these criteria automatically to a large, highly detailed electronic geographic database with information about population, race and ethnicity, the street network, and town and county boundaries. Surely an objective, nonpartisan computer can draw an optimal map that reflects precisely drafted goals, trade-offs, and constraints worked out through public discussion and legislative debate.

An attractive idea, perhaps, but hardly an objective, unanimously acceptable solution because the redistricting problem has no completely unbiased, inherently amicable solutions. Computerized legislative districting is an enormously complex computational task. There are many ways of setting up the "objective function" that tells the computer a boundary drawn here is better than a boundary drawn there and many ways of telling the machine that some solutions, however meritorious, are unacceptable. How much weight should be given a map that reduces the range of district populations to less than one percent, in contrast, say, to a map with measurably less irregular shapes?

And that's not all. Even when decision makers agree on specifications, no one can be certain the computer will find the best possible solution. Because of an indefinitely large number of possible boundary configurations, not even a supercomputer can test every possible redistricting plan. The best a computer can do is find a stable, "locally optimal" map better than any comparatively minor rearrangement of its boundaries and better than other local optima it encounters. There is no guarantee, though, that a slightly better "global optimum" does not exist. Since "best" often is not much better than "very good," anyone can argue the merits of a very good solution that also helps the worthy Extremists, the erstwhile Mugwumps, or the especially brilliant and deserving incumbent Smith. Moreover, because legislators, ethnic leaders, potential challengers, and other interested parties would need to see the likely effects of various objective measures and constraints in order to debate their merits intelligently, can we be certain that politicians aware of the effects on their own ambitions can be trusted to select objective functions, weights, and constraints that serve the public at large? Don't bet on it.

Although computers programmed to enforce compactness might be an attractive way to prevent gerrymanders, in legal circles compactness is not only fuzzy but controversial. As a partly intuitive geometric concept, compactness can be measured using a variety of formulas, each perhaps yielding a somewhat different electoral map. Since there's no legal way to hide the outcome from the contestants, everyone can argue for the measure most helpful to a candidate or political party. Yet, some legal scholars contend that compactness is more than a geometric or geographic issue: whether or not a district is a gerrymander depends, they say, not on visual assessments of shape but on its political or racial consequences. The courts reflect this tolerance in their ambivalence toward partisan gerrymandering; although willing to entertain lawsuits, most judges are reluctant to intervene. Perhaps this is best. According to Bernard Grofman, an expert witness in many key judicial rulings, "only the most egregious types of partisan gerrymanders should be overturned by the courts." And you shouldn't need a computer to recognize one.

Even so, using computers has merit, if for no other reason than to enforce consistency and gauge the extent to which political bias distorts legislators' abilities to know and serve their districts. Richard Morrill, a geographer at the University of Washington, used an automated redistricting model to examine the objectivity of manually produced reconfigurations of his state's legislative and congressional districts. Morrill had drawn these plans himself in the early 1970s when a federal court appointed him special master, but because of time pressures he was unable to use a computer. Impressed with the way the computer-generated districts not only reflected principles of compactness and geographic efficiency but produced districts arguably superior to his own, Morrill concluded that "either we were unable to discover the most efficient patterns manually or we intentionally or unintentionally emphasized other criteria such as the preservation of incumbents by trying not to change districts too much."

By considering the highway network and travel time, Morrill's computer model offered a more functional approach to compactness. Because travel time is more relevant to human interaction than straight-line distance—especially in the rugged terrain of western North Carolina—districts based on a high-speed road such as I-85 might well be more functionally compact than their irregular shapes suggest. Nevertheless, disadvantaged parties and the media will eagerly seize on the map as evidence of gerrymandering.

Although the New York legislature is unlikely to surrender reapportionment to a computer, automated districting worked well in Iowa in both 1981 and 1991. The computer's instructions are straightforward: make the districts nearly uniform in population yet avoid irregular shapes, divided counties, and illegally diluted minority communities. The machine knows nothing about where incumbents live or which neighborhoods favor a particular party. Iowa's review process is also simple: the nonpartisan Legislative Services Agency pushes a button, generates a plan, and submits it to the legislature, which must vote without recommending changes. If legislators reject the first plan, the agency adjusts the model and generates another map. Only if the legislature rejects the second plan may individual members recommend changes. To avoid the bickering and lobbying likely at this stage, most legislators approve the first or second plan. In 1991, when the legislature approved an initial plan that placed 40 percent of them in a district with another incumbent, most of the 60 percent who were protected supported the new map, and public pressure encouraged others to go along. Despite the expertise sacrificed when a respected incumbent loses to a fellow legislator, most Iowans have come to welcome a periodic shake-up.

However skeptical about automated redistricting, New York's Legislative Task Force on Demographic Research and Reapportionment relies heavily on computers for handling census and voting data, tabulating population counts for trial configurations, plotting maps, and converting cartographic representations into the verbal boundary descriptions on which lawmakers actually vote. Like political cartographers elsewhere, New York officials readily recognize that interactive computer graphics is ideal for tinkering with boundaries. State and local governments can take advantage of electronic street maps embedded in the Census Bureau's TIGER (Topologically Integrated Geographically Encoded Referencing) database, and choose among dozens of software systems and consulting firms. Redistricting software that accepts addresses of incumbents warns users when two legislators are in the same district. The Justice Department's Civil Rights Division also uses interactive computers to verify each voting district's compliance with the Voting Rights Act; by overlaying a map of district boundaries on a detailed map showing a region's percentages of blacks or Hispanics, an analyst can immediately spot any line that splits a minority community. Using similar hardware and software, New York's Task Force expedited the Department of Justice review by submitting the state's final redistricting plans in electronic form. State taxpayers made a big investment in the post-1990 redistricting exercise, and task-force personnel promoted a truly bipartisan effort by supporting five separate interactive systems: a multiple-workstation system for themselves, separate systems for the majority and minority "conferences" of the State Senate, and separate systems for their counterparts in the Assembly.

Recent court opinions about equal protection and public access assure a more open redistricting process than before, although the system is far from ideal in practice. Technically, access now includes both open hearings and access to the data. For a fee, task-force staff members provide paper or electronic copies of relevant population and recent voting data as well as maps showing census blocks and other small areal units to be used in redistricting. Although the full statewide set might overwhelm the average citizen in both cost and bulk—$900 for 137 large, 36-by-48-inch map sheets and 313 personal computer diskettes—for individual counties the information is affordable. A bulletin issued in October 1991 described how residents or organizations might submit their own statewide or local redistricting plans for legislative or congressional districts. Unfortunately, few people actually took advantage of the opportunity. Although the task force advertised these products and offered assistance in Spanish and Chinese, as well as English, the information was of little use to most citizens. Wisconsin, by comparison, was markedly more helpful: its Legislative Reference Bureau published a list of access dates for ten sites throughout the state at which staff would be available with a computer terminal. The bureau not only invited residents to make appointments for drawing legislative and congressional district boundaries but prepared its amateur political cartographers by sending ward maps and statistical data.

The task force's public hearings also fell short of effectively enabling citizens to examine its legislative redistricting plan as some residents had difficulty finding or interpreting the maps. Figure 6.7, a full-size facsimile of a portion of a map in the booklet distributed at the Syracuse hearing, reveals a dearth of well-known streets and landmarks. Many nonlabeled parts of district lines could be identified — or guessed — only by overlaying the map on a more detailed cartographic reference. Had local newspapers not interpreted the task force's maps, urban residents would have been poorly informed about the new district lines. Moreover, sets of "high-resolution" maps shipped to reference departments of central public libraries in sixteen regionally important cities were both too bulky and too poorly detailed for easy use. Uncertain about what to do with map sheets larger than the average library table, reference librarians typically rolled them into a bundle and placed them atop a row of filing cabinets. As I observed at libraries in both Syracuse and Utica, the heap grew more unsightly and disordered after each use. Despite invitations to participate, the process was hardly inviting.

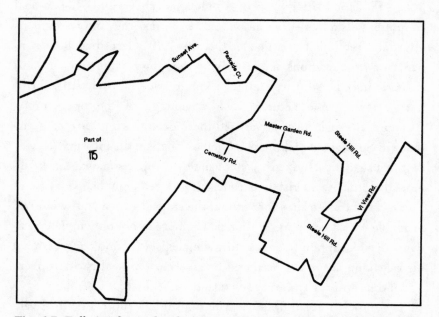

Fig. 6.7. Full-size facsimile of portion of the task force's map showing the boundary between the 115th and 116th Assembly Districts within the city of Utica. (From New York State Legislative Task Force on Demographic Research and Reapportionment, "Co-Chairmen's Proposed 1992 Assembly and State Senate District Boundaries," 21 January 1992, p. 38.)

• • •

Can the redistricting process be a cartographic smoke screen that diverts attention from other, more effective approaches to sharing power? Most controversies over political cartography imply that geography-based representation is the best way, if not the only way, to promote democracy. But there are alternatives. For example, cumulative voting schemes similar to those in Italy, Israel, and other non-English-speaking democracies could empower geographically dispersed "communities of interest" whose votes are diluted by distance and winner-take-all elections. Although potentially divisive for national elections, cumulative voting within the legislature would be a fairer mechanism for allocating committee chairmanships. Less radical measures include election reform and initiative and referendum—all of which would strengthen our present system. Mario Cuomo is right: true empowerment of minorities and everyone else includes making it easier for residents to register to vote, easier for candidates to get on the ballot, and more difficult for incumbents to amass huge campaign war chests. Equally important are mechanisms that let citizens initiate legislation and enact laws directly, bypassing—if they choose—a legislative bureaucracy encumbered by committee fiefdoms and seniority. Because legislative committees control much of the power in state government, any map suggesting that all senate districts are equally influential is disingenuous. Without broader reforms, redrawing boundaries every ten years is little more than an amusing ritual staged by professional legislators more for their own benefit than for ours.

Siting, Cartographic Power, and Public Access

Although politicians traditionally dread reapportionment, their constituents are far more apprehensive when planners use maps to site an incinerator, landfill, or nuclear waste dump. When the proposal triggers a NIMBY (not in my backyard) reaction, maps are pulled into the fray to convince local people, legislators, and judges that the chosen location is efficient and safe. We have to put the damned thing somewhere, the planners argue, and a formal system of map analysis offers an "objective," logical method for evaluating plausible locations. When persons living near the proposed site perceive threats to health, happiness, or property values, maps help officials play down the risk to bodies or wallets. Very conveniently, the tool of analysis doubles as an instrument of propaganda, useful in refuting residents' objections, whether based on myth, uncertainty, experience, or fact. Trust us, say the bureaucrats: it's not a hazard; it won't be disruptive; we know about these things. Backed by a technologically superior arsenal of impressive and complex maps that few dare challenge, the planners enjoy an enormous advantage over residents ill-prepared to question geographic data or abstract graphics. But as this chapter will show, maps in locational disputes also can be turned against a bureaucratic aggressor like a double-edged sword.

In 1986 New York State set just this type of cartographic controversy in motion when it sought a disposal site for low-level radioactive waste, such as booties and gloves worn by workers in nuclear power plants, syringes used in hospitals to inject scintillation fluids for diagnostic testing, and contaminated filters and petri dishes from university research laboratories. Experts on waste disposal consider such materials markedly less hazardous than such "high-level" goodies as spent fuel rods from nuclear power plants and warheads from decommissioned nuclear weapons. Lacking efficient methods of treatment, nuclear experts prefer to pack radioactive waste into fifty-five-gallon drums or protective containers and bury them in a landfill or mine shaft, covering the low-level stuff with a thick layer of soil and putting the more potent atomic garbage as far underground as they can. But the public is skeptical of all things radioactive, its fears reinforced by the Three Mile Island incident in 1979; the Chernobyl disaster in 1986; and medical evidence that radon, an invisible, odorless radioactive gas seeping into some houses from the soil, can be as lethal as tobacco smoke. Assurances that low-level radioactive waste is more akin to industrial rubbish than to atomic weapons are seldom reassuring. Despite its name, low-level radioactive waste is a broad category that includes comparatively hazardous, "greater than class C" material—"low-level" really means "not high-level."

New York acted cautiously in deciding where to dispose of its low-level radioactive trash, and ultimately, selecting the site was as much a political process as it was a research problem in environmental science. That my neighbors forty miles south in Cortland County reacted angrily is not surprising: the announcement that their county contained two of the five potential sites was both sudden and frightening. That the state siting commission resisted local efforts to examine and challenge its maps, locked away in a geographic information system—a moral if not illegal affront to New York's Freedom of Information Act—was also not surprising: much of the power of cartographic weapons depends upon keeping information out of the opponent's hands. Whenever government seeks sites for hazardous waste dumps, atomic or otherwise, local residents and siting authorities almost always become adversaries.

Cortland County's story began in the late 1970s, when the nation's radioactive waste generators were sending all their low-level waste to licensed landfills in just three states: South Carolina, Nevada, and Washington. A decade earlier there had been six regional burial sites, established in the 1960s as a low-cost alternative to dumping fifty-five-gallon drums of waste at sea. One site was in West Valley, New York, in the western part of the state, about thirty-five miles southeast of Buffalo. Landfill operators dug shallow trenches in the clay soil, distributed the waste throughout the trench, and added a layer of compacted clay on top. Because clay is nearly impermeable, so the theory goes, the clay soil below isolates the radioactive waste from groundwater below the trench, while the clay cap on top prevents the infiltration of rainwater. Like many engineering principles, this one doesn't always work, especially in humid climates, where migration of radioactive materials beyond a disposal site is often evident within a decade. At West Valley, the clay caps failed and the trenches filled up like bathtubs and overflowed at the surface. According to Sierra Club activist Marvin Resnikoff, "The landfill acts a lot like tea bags: the water goes in, the flavor goes out." Migrating radioactivity forced the closing of the West Valley site in 1975, and similarly contaminated sites in Kentucky and Illinois shut down in 1977 and 1978, respectively.

Growing concern about radioactive mismanagement prompted action by state officials in Washington, Nevada, and South Carolina. Because of continued violations of poorly enforced federal regulations, Governor Dixie Lee Ray temporarily closed the low-level waste dump at Richland, Washington, on October 4, 1979. And less than three weeks later, on October 23, Governor Robert List temporarily closed the disposal site at Beatty, Nevada, after inspectors discovered radioactive waste buried outside the fence. Fearing that Barnwell, South Carolina, might become the nation's sole graveyard for low-level radioactive waste, state authorities quickly announced new restrictions, including a 50 percent reduction by October 1981 in the volume of waste the site could accept. Worried that his state was receiving an increased share of the nation's atomic rubbish, South Carolina governor Richard Riley called for federal and state governments to distribute the burden more equitably by establishing regional disposal sites. In 1979 Riley and the governors of Nevada and Washington pressed the issue at the National Governors' Conference and threatened to cut off access to disposal sites within their states unless Congress acted the following year.

Congress narrowly averted a crisis by passing the Low-Level Radioactive Waste Policy Act on December 22, 1980. The new law bought some extra time for the other forty-seven states by encouraging interstate compacts that would establish regional disposal sites for low-level radioactive waste. After January 1, 1986, compact states could exclude waste from nonmember states. Modeled on the governors' proposal, the statute required Congress to approve every compact but contained no direct penalties for states that failed to join one.

As 1986 approached, governors and lawmakers realized the policy was not working. Several interstate compacts had formed, but only three had an operating disposal site—the existing dumps in Nevada, South Carolina, and Washington. On January 1, these three compacts could begin to exclude low-level radioactive waste from the remaining thirty-one states. The other regional compacts were years behind schedule, and several states, including New York, had neither joined a compact nor developed a suitable plan for dealing with their low-level waste.

Efforts to form a northeast compact were frustrated by contrasts between Massachusetts, New York, and Pennsylvania, which were among the five states that generated the most low-level radioactive waste, and Maine, New Hampshire, and Vermont, which were among those that generated the least. None of the big generators was eager to host a regional disposal site, and none of the smaller generators was willing to help out its neighbors. Enthusiasm for a three-state northern New England compact waned when Maine had second thoughts about volunteering an aboveground site near the Maine Yankee nuclear power plant. In Massachusetts a 1982 state law requiring direct voter approval in a public referendum effectively blocked an in-state site. Although Connecticut, Delaware, Maryland, and New Jersey had submitted a compact for congressional approval, the region's other seven states had missed the proposal's June 30, 1984, deadline. As it often does, New York commissioned a study. In 1984 a report by the State Energy Office recommended a carefully monitored aboveground storage site for very-low-level, "Class A" radioactive waste and a more thorough study to find a suitable location within the state for a full-range, permanent disposal site. The governor and the legislature read the report and waited to see what would happen; perhaps the generators themselves could sit on the waste, through on-site storage, until the federal government solved the problem.

Alarmed by the approaching December 31 deadline and threats by Nevada, South Carolina, and Washington to exclude outside waste, on December 19, 1985, Congress enacted a new law with penalties for noncompliance. Its first deadline was July 1, 1986, when each state was to either join a regional compact or begin looking for an acceptable disposal site within its own borders. By January 1, 1988, each interstate compact was to identify a specific site within a designated "host state." Each noncompact state had to select an in-state disposal site as well, all sites were to be operating by January 1, 1993, and each compact could exclude waste from outside its region. Unlike the 1980 legislation, the new statute also had teeth. Existing landfills could increase charges for waste from noncompact states without a site, and after January 1, 1996, any generator in a noncomplying state could ask the state to "take title" to its low-level radioactive waste. If the state refused, its taxpayers would be liable for all damages, both direct and indirect. Most states, the bill's authors reasoned, would comply rather than risk having to reimburse hospitals, research laboratories, and industrial firms for expenses incurred in closures, accidents, or lawsuits resulting from low-level radioactive waste. The new law also listed the seven approved interstate compacts and their thirty-seven member states.

New York responded in July 1986 by establishing the Low-Level Radioactive Waste Siting Commission, charged with selecting a site and choosing an appropriate method of disposal. Aware of the West Valley incident and leakage problems in other radioactive-waste landfills, the legislature specifically prohibited conventional shallow-land burial. To demonstrate a sense of urgency to Congress and South Carolina, the bill's authors emphasized "the need to expedite the completion of these facilities and to minimize or eliminate any delay in beginning facility operation." Like their federal counterparts who had precipitated the search, state lawmakers set several deadlines: the Department of Environmental Conservation (DEC) was to publish a draft of site-selection criteria by July 31, 1987, and issue final guidelines by December 31, 1987; and the Siting Commission was to select the site and disposal method by December 1, 1988—allowing slightly more than four years to meet the January 1, 1993, deadline for an operating disposal site. Appointed by the governor, the five-member commission consisted of a geologist, a health physicist, a medical doctor, a professional engineer, and a private citizen to serve as chair. The legislature authorized a first-year budget of $1,630,000, with additional funds to support siting work and other related activities in three state agencies (the Energy Research and Development Authority and the Departments of Health and Environmental Conservation). Money was never a problem. Because the state's six nuclear power plants generated most of the low-level waste, the law that established the siting commission also required the plant operators to pay its bills. And like regulated utilities everywhere, the power companies passed these costs along to consumers.

By the end of 1987, the siting commission was up and running. The first issue of its prophetically named newsletter, *LLRW Frontline*, reported on the appointments of commission members, an executive director, and an advisory committee of representatives from state agencies, environmental groups, radiation health professionals, and radioactive waste generators. The commission had hired its staff, advertised for an experienced environmental consulting firm, launched a public outreach program, and set up a library for staff and the public.

As instructed by the legislature, the DEC developed a list of criteria for site selection. An astute mixture of common sense, sound environmental science, and interest-group politics, the DEC siting regulations eventually eliminated about 30 percent of the state from consideration. In addition to excluding floodplains, areas with highly permeable soils, vulnerable aquifers, fault zones, and other geologically unsuitable features, the DEC regulations exempted all state, federal, and municipal parkland; protected wetlands, wildlife habitats and management areas, and the "critical habitat" of any endangered or threatened species; military reservations; Indian reservations; cities, incorporated villages, and other places with a population density greater than 1,000 persons per square mile; the Western New York Nuclear Service Center at West Valley, where the state had closed a low-level radioactive waste dump in 1975; and any area where existing radioactive material might interfere with monitoring. (It was more than mere coincidence, though, that West Valley was in the district represented by the majority leader of the State Assembly.) Anticipating objections from farmers, the DEC insisted that the site "minimize adverse impacts on agricultural operations." And to counter objections based on safety and aesthetics, the regulations called for a buffer zone sufficient to maintain security and "contribute to a desirable land disposal facility appearance"—however obnoxious, the site need not look ugly.

To cope with the DEC's complex set of explicit regulations as well as to provide the required public involvement, the siting commission devised the hierarchical, multistage siting strategy described in Figure 7.1. Relying on readily available maps and other data, Phase 1 consisted of four screening steps, each using different criteria to narrow the search. Starting with the entire state, thirty candidate areas were selected; then this was narrowed to ten areas; and the third and fourth steps screened the candidate areas for eight and then four candidate sites for a more thorough examination in Phase 2. Boxes representing public hearings asserted a concern for citizen participation, while rectangles labeled "Evaluate 'Volunteers'" reflected the naïve hope that a friendly landowner would step forward with an unimpeachable site.

To carry out this plan, the siting commission adopted a method of map overlays managed electronically with a geographic information system (GIS). Figure 7.2, an illustration used in several commission reports, explained overlay analysis to the public with a hypothetical screening based on population density, major transportation routes, and primary aquifers. The top row represents source maps from the U.S. Bureau of the Census, the New York State Department of Transportation (DOT), and the DEC, and the second row reflects the generalization of these maps to a huge grid of square cells aligned in rows running east-west and columns aligned north-south. For the initial step of statewide exclusionary screening, these cells were one mile on a side. GIS technicians "digitized" their data by placing each source map at its appropriate position within the statewide grid and using numbers to rate each cell according to the relevant mapped features. Each criterion became an overlay (or layer) with a grid identical to all other layers. With a common grid of cells, the GIS was able to store numerous overlay maps, add weighted scores from selected overlays, and create a map of composite favorability (bottom of Figure 7.2).

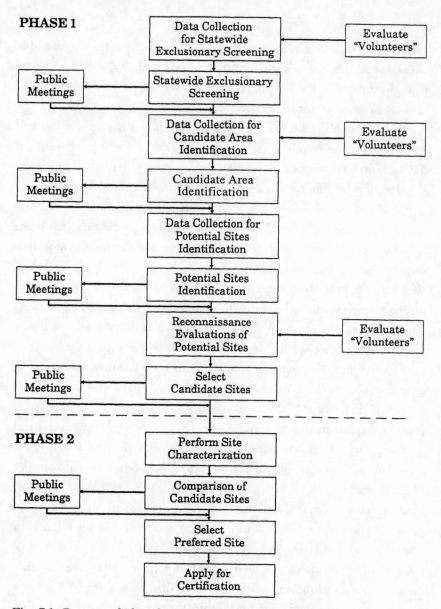

Fig. 7.1. Process of identifying sites for low-level radioactive waste siting. (From New York State Low-Level Radioactive Waste Siting Commission, *Candidate Area Identification Report,* Albany, N.Y., December 1988, p. S-8.)

Source Maps

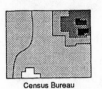

Census Bureau · DOT · DEC

Digitized Maps with Grid Overlay

Highly Populated Areas · Major Transportation Routes · Aquifers

Cell Assignments

Scoring (buffers for example only)

2	1	1	0	0
2	1	1	0	0
2	2	1	1	1
2	2	2	1	1

1: ≤ 2 mi, least favorable
2: 2–4 mi, less favorable

1	0	1	2	3
1	0	1	2	3
0	0	1	2	3
0	1	2	3	4

1: ≤ 1 mi, least favorable
2: 1–2 mi, less favorable
3: 2–3 mi, favorable
4: 3–4 mi, more favorable

1	1	3	5	5
0	0	1	3	5
0	0	1	3	5
1	1	3	5	5

1: ≤ 1 mi, least favorable
3: 1–2 mi, favorable
5: >2 mi, most favorable

Weighted Scores

40	20	20	E	E
40	20	20	E	E
40	40	20	20	20
40	40	40	20	20

Factor = 20

30	E	30	60	90
30	E	30	60	90
E	E	30	60	90
E	30	60	90	120

Factor = 30

50	50	150	250	250
E	E	50	150	250
E	E	50	150	250
50	50	150	250	250

Factor = 50

Composite Scores

120	E	200	E	E
E	E	100	E	E
E	E	100	230	360
E	120	250	360	390

Composite Favorability Map

Excluded by law

100–199, Low favorability

200–349, Middle favorability

350–500, High favorability

Fig. 7.2. Conceptual application of GIS in site screening. (From New York State Low-Level Radioactive Waste Siting Commission, *Candidate Area Identification Report*, Albany, N.Y., December 1988, p. 3-16.)

Cartographers call this system of gridded overlays a *raster model*. The other principal type of GIS is the *vector model*, which represents boundaries, routes, and other map features more precisely as lists of points and coordinates. A vector GIS can provide a more realistic description of irregularly shaped geographic features, but the massive vector database required to encode many features for a large area is costly to develop and difficult to process, even by modern high-speed computers. In contrast, a raster GIS is less demanding, especially if the cells are large. But because a square-mile area contains 640 acres, describing its condition or favorability with a single number yields crudely imprecise overlays, especially for siting criteria that can vary greatly over short distances. Although a raster GIS can be an appropriate tool for environmental analysis, accurate results depend on geographically detailed overlays based on much smaller cells, which can require a staggering amount of data. For example, the minimum area the U.S. Geological Survey uses for its nationwide land-use and land-cover database would require nearly sixty-five times as many cells as the square-mile grid the siting commission adopted for exclusionary screening. Because the magnitude and cost of a raster database depends on number of layers as well as cell size, the Geological Survey compensated for its finer mesh by including comparatively few overlays, with categories too vague for exclusionary screening. Faced with a tight deadline, the commission sacrificed precision in its early stages of screening. Only after the coarse, square-mile mesh reduced the search to ten candidate areas, with a combined area of 1,125 square miles, or about 2 percent of the state, did siting officials adopt a forty-acre grid of squares a quarter-mile (1,320 feet) on a side—still quite coarse for environmental analysis.

Although intended principally as an overview of the screening process, the siting commission diagram in Figure 7.2 reveals the coarse grid's tendency to exclude too much territory. As examples, the "highly populated area" (in the left column) covers only one of the four cells marked "E" (for "excluded") on the population overlay, and the aquifer (in the right column) covers only about half (or less) of the four cells excluded in the hydrologic overlay. Despite the appearance of a soundly conservative practice for eliminating areas close to residential neighborhoods and vulnerable aquifers, buffers of "least favorable" cells surrounding the excluded features (in the fourth row of Figure 7.2) propagate these crudely jagged boundaries outward across the grid. More troublesome is the chain of cells describing a major transportation route (in the second column). Any cell the highway penetrates, however briefly, is excluded in toto, whereas sites within adjacent cells a mere hundred feet away are merely downrated as "least favorable."

Equally suspicious are the numerical scores describing relative favorability and the weighting factors specifying each overlay's relative clout. Although treating a cell within two miles of a populated area as less favorable than a cell two to four miles away is intuitively appropriate, assigning favorability scores of 1 to the closer cell and 2 to its more distant neighbor seems unduly arbitrary. Why not, for instance, make the weights 1 and 3? More puzzling are the relative weights of 20 for the overlay representing highly populated areas, in which many people live, sleep, and breathe day after day, and a weight of 30 for transportation corridors, which most people inhabit only fleetingly. Would the average citizen (or siting official) consider living 1.5 miles away (weighted favorability = 20) from a radioactive dump as only three times more risky than driving along a road 1.5 miles away (weighted favorability = 60)? Is GIS screening a reliably objective process, as touted in the siting commission's reports, or an expedient but arbitrary exercise in pseudoscientific numerology?

Several of the scoring systems reported with technical precision in the siting commission's *Candidate Area Identification Report* are more egregious than the hypothetical illustration in Figure 7.2. A particularly revealing example is criterion 23, the commission's interpretation of the DEC's concern with meteorology and climatology. For a nationwide search, rainfall is a relevant siting factor because heavy precipitation would cause water infiltration and the migration of radioactivity beyond the site through the "teabag effect." New York lacks deserts and monsoon belts, however, and its precipitation varies more from year to year than from place to place. Nonetheless, the commission eagerly devised an overlay of cells rated as either 1 (least favorable), 3 (somewhat favorable), or 5 (most favorable) according to whether the average annual precipitation was greater than fifty inches, between forty and fifty inches, or less than forty inches. Based on the highly generalized map in Figure 7.3, the precipitation overlay identified only three "least favorable" areas, in the Adirondacks and the Catskills. Yet, 75 of the 105 weather stations used to make the map had recorded more than fifty inches of precipitation at least once in thirty years. Had siting officials not heard of the statistician who drowned trying to wade across a river with an average depth of two feet?

Even if annual averages were reliable, how relevant are the forty- and fifty-inch rainfall contours in Figure 7.3? Experience suggests

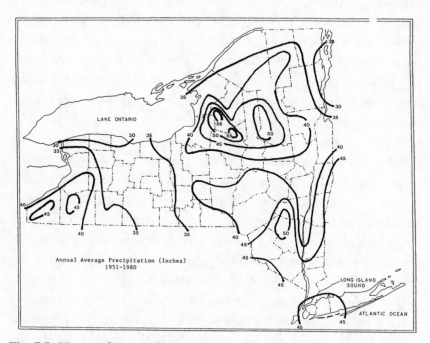

Fig. 7.3. Map used to rate favorability for the precipitation criterion. (From New York State Low-Level Radioactive Waste Siting Commission, *Candidate Area Identification Report,* Albany, N.Y., December 1988, Figure 4-22, n.p.)

rain infiltration and radioactive migration can be troublesome wherever annual rainfall exceeds thirty inches. (Only 3 of the 105 weather stations used to make the map had annual averages less than thirty inches, but even there the annual maximum was at least thirty-seven inches.) Is a precipitation overlay at all necessary in a generally humid state with only modest variation in average rainfall? More to the point, is a precipitation overlay at all relevant to waste stored in a proper vault on a site not subject to flooding, a seasonably high water table, or other factors addressed more directly by other siting factors? After all, a legislature wary of seepage at West Valley had wisely ruled out traditional shallow-land burial.

Another example of spurious objectivity in screening is criterion 51, "Proximity to Generators." Concerned with the risk and cost of shipping hazardous materials over great distances, the siting commission used data on the volume of waste shipped by individual waste generators to compute, for each county, the "weighted average distance of sending counties to receiving county." To compute a crude, unweighted mean distance, a computer would merely add up distances to a particular county from each of the other sixty-one counties and then divide by 61. To compute a *weighted* mean distance, the computer multiplies each distance by the volume of waste generated in the county and divides the resulting sum by the total amount of waste generated in the state. (Really just a simple weighted average.) Hypothetically, if all waste generators were on one side of the state, weighting would give these counties much lower weighted mean distances than counties at the opposite end of the state, to which lots of waste would have to be transported over much greater distances. Location analysts use a similar approach to minimize the total cost of transporting goods from a regional warehouse to stores in various cities: a high-volume retail outlet thus exerts more pull on the location of a new warehouse than a store with mediocre sales. Siting officials used these weighted distances to rate each county as either 1 (least favorable), 3 (somewhat favorable), or 5 (most favorable). A map with symbols too illegible to reproduce here identified half of western New York and half of northern New York as "least favorable." Although this map assigned Cortland and twenty-six other counties to the "most favorable" category, its patterns were flawed because the raw, intercounty distances used were straight-line, "air" distances between county centers, not more realistic over-the-road highway distances. The commission completely ignored rough terrain, the Finger Lakes, and the highway network in general! A day spent counting distances between road intersections in the *Rand McNally Road Atlas* could have produced more accurate unweighted distances, and consequently, better proximity ratings. Ironically, the analysts processing the radioactive-waste data seemed to ignore the adage "Garbage in, garbage out."

Not all siting criteria were as outlandish as these examples. Geologic criteria addressing groundwater and seismic hazards were comparatively reasonable, and only the coarse square-mile grid diminished the reliability of exclusionary data on American Indian reservations and other protected lands. But how did the commission integrate the overlays for its various preference criteria? How did it choose the relative weights used in computing a composite map of overall favorability? In August 1988 the commission held a workshop "that sought input from the Low-Level Radioactive Waste Advisory Committee and other representatives of State and local government, industry, interest groups, and others." As I interpret the *Candidate Area Identification Report,* attendees at the workshop discussed the criteria, voted on weights, and tweaked the numbers here and there until most people close to the process were satisfied with the results. The commission then compared these findings with "the results of similar exercises performed independently by the Commission staff and its contractor," considered relevant state regulations and additional comments from the public, and made a few more, comparatively minor adjustments. Not an unreasonable strategy, I suppose, but hardly objective.

Weights for the fourteen preference criteria used to select candidate areas, listed here by name and number in order of decreasing weighting factor, reveal equal ratings for several of the criteria:

Aquifers (criterion no. 12)	147
Population Density (45)	120
Groundwater Discharge (13)	107
Retardation of Radionuclides (14)	107
Thickness/Areal Extent (5)	93
Surface Water Quality (22)	80
Chronic Severe Weather (24)	53
Earthquake Acceleration (2)	53
Oil and Gas Fields (10)	53
Precipitation (23)	53
Proximity to Generators (51)	53
Federal Protected Lands (37)	27
Indian Lands (42)	27
State Protected Lands (39)	27

Perhaps the equally weighted criteria reflect a politically savvy reluctance to treat state and reservation land differently as well as the difficulty of comparing climatic, seismic, and transportation factors with the risk of contaminating fossil fuels. Although skeptical about weights assigned individual factors, I cannot quibble about the relative order. Because of demonstrated threats to groundwater, distance from aquifers seems the most significant preference criterion, with proximity to population concentrations a reasonable choice for second place. Yet equally apparent is the virtual certainty that somewhat different weights would alter the list of candidate areas.

What would I have done differently? Aside from insisting upon a finer grid, throwing out precipitation altogether, and figuring distance along highways (not as the crow flies), probably very little—until I computed my first map of composite favorability. At this stage the issues of data quality and method-produced error are crucial if officials are to convince skeptics that the siting process is not an environmental gerrymander, driven by expediency if not malice. I would take advantage of the faster processing and display of raster GIS and tinker obsessively with the weights in hope of identifying highly stable, absolutely unshakable candidate areas. GIS experts call this sensitivity analysis, and the better ones use it to compensate for uncertainty about data and requirements. I also would have let the public use the GIS; citizen participation should mean more than public hearings and newsletters. To their credit, the siting commission ran a few sensitivity analyses, but they buried the results in a largely inaccessible technical report. So without access to the GIS, citizens of Cortland County and other targeted areas never had a fair opportunity to challenge the screening.

Until the siting commission announced its ten candidate areas on December 20, 1988, the news media paid little attention to its activities. The day after the announcement, the *New York Times* ran a medium-length Associated Press report noting the exclusion of New York City, Long Island, and the Adirondacks. An accompanying map identified the affected counties and named all thirty-two rural towns composing the ten candidate areas, which ranged in size from 50 to 150 square miles. The article went on to report that within these ten areas the commission would, by fall, identify a few smaller, more specific sites for a facility described as both a "dump" and a "plant" and expected to cost as much as $25 million and employ thirty people. Hearings scheduled during January in the ten areas were expected to draw protests. In mid-January, a longer article titled "Residents Assail Plan for A-Waste Dump" described the first of these hearings, in a high school auditorium in Hudson Falls, about fifty miles north of Albany. Five hundred people attended, and no one was pleased that one of the ten candidates was in Washington County. Housing the waste in a huge concrete vault was foolish, a local chemical engineer charged: concrete is too porous to isolate the waste for 500 years. To dramatize the wider effect of the dump, a town supervisor held up a bottle of milk from a dairy three miles from the candidate area. "I'm using this as Exhibit A," he said. David Maillie, the commission's radiologist member, anticipated similar objections at the other hearings. "Mention radioactivity and people panic," Maillie said. The other hearings must have been too similar and predictable to interest the *New York Times* editors, who then ignored the story until September 1989, when the siting commission had narrowed its search to five specific sites.

Newspapers in the affected areas were far more apprehensive. Even though southeastern Cortland County was but one of ten candidate areas, its six towns were a big part of the *Cortland Standard*'s backyard. A small, independently owned afternoon newspaper published in the city of Cortland, the *Cortland Standard* circulates throughout the county. Like most small-city dailies, it mixes local news and sports with state, national, and foreign news from the Associated Press, but local reporting and an image of commitment to the community are its prime defense against *USA Today* and the regional editions of Syracuse's newspapers. Ordinarily the paper would have welcomed thirty new jobs to the local economy, but not when radioactivity, however low-level, might threaten the air, the water, and heaven knows what else.

Five days before Christmas 1988, the *Cortland Standard* broke the story with the front-page headline "County on Nuke Waste Dump List." Details were still sketchy. The day before, an unnamed member of the siting commission had told the chairman of the county legislature that six towns in southeastern Cortland County were one of ten areas under consideration. By late summer, the search was to narrow to four specific sites. Public hearings would begin in January, and the county health director predicted intense local opposition. The following day the paper outlined the six-town candidate area on a county map and interviewed several local officials for a story headlined "Dump Listing Raises Anxiety." Some were adamantly opposed to a dump site anywhere in the county, others were confident their town would not be chosen. An editorial titled "Not at All Reassuring" defended local fears about groundwater contamination and transportation accidents, and scolded the state for not joining a regional compact.

Opposition developed rapidly over the five weeks preceding the commission's January 24 local hearing. The county legislature, the

Cortland City Council, and town boards throughout the county passed resolutions opposing the dump. Two citizens' groups, Concerned Citizens of Cortland County and Citizens Against Radioactive Dumping (CARD), formed to distribute literature, lobby legislators, and organize protests. School districts provided buses to transport residents to the hearing, and bus drivers volunteered their time. A *Cortland Standard* editorial urging residents to attend exclaimed defiantly, "Not here, not now, not ever."

Residents packed the hearing, and jeered and booed as commission members and staff struggled for more than four and a half hours to satisfy doubts about safety and responsibility. An estimated 1,300 persons attended, while thousands more watched the boisterous meeting on television. Local officials and other residents raised dozens of issues, including the failure to consider Cortland County's aquifers during the exclusionary screening, the state's cart-before-the-horse strategy of picking a dump site before choosing a disposal method, the legislature's taboo against siting the dump on state-owned land, the admission that no commission member lived near a candidate area, the hazards of transporting nuclear waste over winding snow-covered highways, and the paradoxical need to site an allegedly safe dump in a "sparsely populated area." Hardly reassuring was commission counsel Douglas Eldridge's comment, "There are no guarantees in this business." Anticipating a long fight against the dump, the county legislature voted the next day to hire a researcher to coordinate data collection and monitor commission activities.

A second wave of bad news arrived on Saturday, September 9, 1989. Dave Fuller, a Cincinnatus resident who owned a small, three-acre parcel in the Cortland County town of Taylor, received an early morning call from his local post office, which had an Express Mail package for him. The nine-by-twelve-inch envelope he opened later that morning contained a cover letter from the siting commission and maps of five potential dump sites, one of which included his land. Although the official announcement was scheduled for Monday, local residents were shocked by the bold headline of their Saturday after-noon *Cortland Standard*: "2 N-dump Sites Here." Two of the commission's five "recommended potential sites" were in Cortland County.

Both sites were in the town of Taylor, at the northern end of the candidate area, about twelve miles due east of the city of Cortland (Figure 7.4). The other three sites were in northern Allegany County in western New York, roughly seventy miles south-southeast of Rochester and about thirty miles north of the Pennsylvania border. A report released in Albany that Monday described how the commission had applied additional exclusionary and preference criteria to the ten candidate areas. GIS screening identified ninety-six initial sites, and "qualitative" analysis of topographic maps trimmed the list to fifty-one sites. A team consisting of a geologist, a biologist, and a planner conducted "windshield surveys"—driving around and look-ing without getting out of the car—at these fifty-one locations, plus another four "volunteer sites," and after examining air photos and consulting with local geologists, commission staff had pared the list to nineteen sites. Relatively thorough, map-based "comparative assess-ments" then narrowed the menu to the five sites described in the commission's anticlimactic September 11 announcement.

Enlarged excerpts from maps in the commission's report described the narrowing of potential sites in Cortland County (Figure 7.5). Small triangles and dots in the top frame show the seventeen local sites identified by GIS screening. Further map analysis eliminated all but the seven sites represented by triangles. The three triangle symbols in the middle frame indicate that drive-by inspection and other informa-tion eliminated four more sites, and the bottom frame shows the two sites that survived the more intensive "comparative assessment."

Notice a discrepancy between the middle and lower frames in Fig-ure 7.5? Both of the triangles in the bottom excerpt are within a sin-gle town (Taylor) yet the middle map depicts but one site in the same town. As local officials were quick to point out, GIS screening had not endorsed the Taylor North site, number 13-18. How, they asked, had an area rejected by the commission's own procedures become one

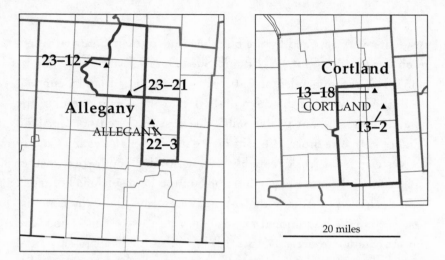

Fig. 7.4. Recommended potential sites in Allegany (left) and Cortland (right) Counties. (From New York State Low-Level Radioactive Waste Siting Commission, *Report on Potential Sites Identification*, September 1989, Figure 9-1, n.p.)

of the five "recommended potential sites"? The conclusion seemed obvious: as a "volunteer site" owned by a single individual, Taylor North had received special treatment in the commission's eagerness to put the dump somewhere — anywhere.

Local antidump activists were not surprised. They had already read in the September 1 *Cortland Standard* a story headlined "Taylor Residents Fear Their Town Tops State's N-dump List," which reported that Arthur Allen, a farmer in the town of Taylor, had offered to sell his 730-acre property to the state. Two days earlier, at a meeting in the Taylor Community Center, the fifty-two-year-old Allen defended his offer and complained of high taxes and low income. His eagerness to sell was understandable, and many area farmers must have dreamed of selling out; dairy farming has always been a hard way to make a living, and falling prices were driving more and more milk producers out of business. Even so, Allen's neighbors felt angry and betrayed. At least one threatened to kill him.

Residents vowed to resist on-site inspection and testing. On Saturday, September 9, the same day the mailing error prematurely

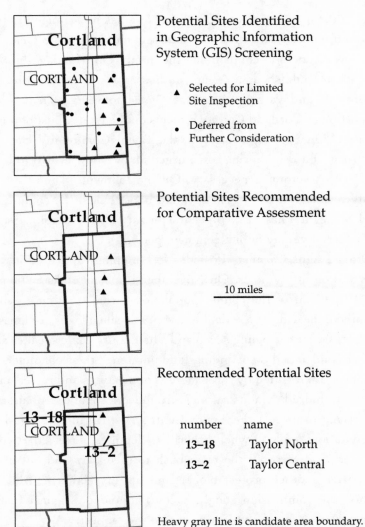

Potential Sites Identified in Geographic Information System (GIS) Screening

▲ Selected for Limited Site Inspection

• Deferred from Further Consideration

Potential Sites Recommended for Comparative Assessment

10 miles

Recommended Potential Sites

number	name
13–18	Taylor North
13–2	Taylor Central

Heavy gray line is candidate area boundary.

Fig. 7.5. Identification and progressive narrowing of potential sites in Cortland County. (From New York State Low-Level Radioactive Waste Siting Commission, *Report on Potential Sites Identification*, September 1989, Figures 4-1, 5-1, and 9-1, respectively.)

revealed the list of finalists, 250 people at a hastily arranged rally in Courthouse Park heard the area's state senator and assemblyman join other local leaders in condemning the dump. On Monday, the *Cortland Standard* published names and addresses of elected state and federal officials, and warned readers they "must not remain silent as long as a portion of Cortland County remains on the [state's] list of potential sites." On Tuesday evening at the Taylor Community Hall, the supervisor of a neighboring town urged an overflow crowd to block roads with tractors, if necessary. On the following Thursday, 250 county residents drove to Albany, and booed when the commission voted to begin Phase 2's "precharacterization" site reconnaissance after a forty-five-day public-comment period.

Orange signs warning "Posted! No Dumping" appeared in windows and on utility poles. On November 15, as many as 5,000 people—10 percent of the county population—tried to attend a siting commission hearing at the field house of a local college. Commission members sat for a grueling seven and a half hours while speaker after speaker condemned their findings and impugned their character. On December 13, fortified by three months of meetings and rallies, more than one hundred demonstrators surrounded a small reconnaissance team trying to inspect the Taylor Central site. An intense two-hour confrontation ended when sheriff's deputies arrested twenty protesters and escorted the team back to their cars and out of the area. After another protest blocked on-site inspection in Allegany County, the commission sought a court injunction against further demonstrations, but a state judge denied the request.

Despite threats from Albany, local prosecutors and judges defiantly dropped charges or levied token fines. On January 19, 1990, protesters set up roadblocks on all routes leading to the two Cortland County sites, and turned away another reconnaissance team. On several occasions in early March, police arrested protesters for blocking access to a local information office the siting commission had recently opened in Cincinnatus, near the two potential sites. Town building inspectors registered numerous code violations by the office, and farmers left animal carcasses on its doorstep at night. On April 12, the county sheriff asked the commission to temporarily close the office. A week earlier, near the other three finalist sites in Allegany County, a violent protest had injured several people. Finally, on April 19, in response to repeated requests from citizens, legislators, and other local officials, Governor Cuomo halted on-site inspections and ordered the Cincinnatus office closed. Unable to carry out inspections, the commission suspended the siting process indefinitely.

Embarrassed by charges of slipshod work and political ineptitude, the legislature rewrote the siting commission's duties and responsibilities. The new law, passed in July 1990, directed the commission to select a disposal method before, not after, choosing a site and to demonstrate that its staff and consultants had acted fairly and consistently in excluding sites. Without saying so, the act implied the commission might have to reconsider its five finalist sites. The legislature also added a social scientist and an environmentalist to the commission, changed the advisory committee into a *"citizen* advisory committee" independent of the commission, and called for the Department of Health to arrange a thorough review of the commission's work by an independent panel of technical and scientific experts.

With siting far behind schedule, New York sought to overturn the portentous "take title" provision of the federal Low-Level Radioactive Waste Policy Act of 1985. If South Carolina cut off access to the Barnwell dump in 1996, New York's generators could make state taxpayers responsible for their low-level radioactive waste and any related damages. Fearing this worst-case scenario, Governor Cuomo and Attorney General Robert Abrams filed a lawsuit in federal court in early 1990. Allegany and Cortland Counties joined the state in challenging Congress's right to compel, rather than merely encourage, New York to provide for disposal of waste generated within its borders. The plaintiffs based their argument on the Tenth Amendment — "The powers not delegated to the United States by the Constitution, nor prohibited to it by the States, are reserved to the States respectively, or to the people." After the federal district court and the court of appeals dismissed the case, New York appealed to the U.S. Supreme Court, which heard oral arguments on March 30, 1992. Three months later, in a 6-to-3 decision, the Supreme Court struck down the take-title provision but upheld the law's monetary and access incentives. In writing for the majority, Justice Sandra Day O'Connor concluded, "While there may be many constitutional methods of achieving regional self-sufficiency in radioactive waste disposal, the method Congress has chosen is not one of them." South Carolina could close its dump to New York, but the generators, not the taxpayers, would be responsible for their state's radioactive waste.

In early 1991 siting commission members acknowledged publicly the futility of locating a low-level radioactive waste dump in an area with strong citizen opposition. Adopting a new tactic, they asked the state to offer cash and construction grants to any community willing to accept the disposal site. Fear and rage ruled out productive discussions with officials in Cortland and Allegany Counties, where residents denounced the move as blatant bribery. "It could be a million dollars, and we'd say no," John Smith, a member of Taylor's town board, told the *New York Times.* "What's a million dollars compared to your health?" By then, though, state officials realized any serious offer would have to be much more generous. When the proposed "benefits package" became public in June 1991, the offer was $4.2 million in the first year, for road improvements, a town park, library books, a fire engine, college scholarships, and other niceties, with annual payments thereafter of $1.5 million. Negotiated secretly by local business leaders and a group of waste generators, the package was announced on June 12 at a public meeting in Ashford, a Cattaraugus County town that included the village of West Valley—site of the defunct Western New York Nuclear Service Center and the only place in New York State specifically excluded by the 1986 law that established the siting commission.

To the town's 2,200 residents, $4.2 million was a lot of money. On the northern fringe of Appalachia, Ashford was poor. Its fire engine was twenty-four years old, and $1.5 million was about what the town collected each year in property taxes. And radioactivity was old hat. Despite leaking radioactive waste at the low-level dump that closed in 1975 and more serious atomic pollution in another part of the West Valley plant—an expensive, $1.4 billion federally financed cleanup begun in 1982 was still under way—local people were less apprehensive about radioactivity than most rural New Yorkers. Or at least their leaders were. In a nonbinding referendum on July 9, with a record 80 percent participation, Ashford voters rejected the dump 702 to 533. Nonetheless, on the following evening the five-member town board unanimously approved the proposal. But the town's majority had its day in court three months later, when a State Supreme Court judge ruled that the town board's resolution violated the 1986 radioactive waste siting law as well as state regulations requiring an environmental impact statement.

Although the state had halted site selection, Cortland County felt very much threatened by the continued existence of the siting commission and its list of finalist sites. Local officials and environmental activists relentlessly studied documents and technical reports, attended commission meetings, asked embarrassing questions, and demonstrated with increasing confidence that the selection process had, at times, been inconsistent and untrustworthy. Some of their best weapons in this administrative war of attrition were the siting commission's maps.

Convinced the commission had misrepresented the Taylor North site, residents familiar with local soils, aquifers, springs, and farmland compiled their own maps and detected several striking contradictions. On May 3, 1991, county low-level radioactive waste coordinator Cindy Monaco presented their cartographic evidence to New York's junior senator, Alfonse D'Amato. Intrigued by the apparent contradictions and eager to help constituents, D'Amato requested a

General Accounting Office (GAO) audit of the commission's decision to include Taylor North among the five finalists.

Fifteen months later, in August 1992, the GAO announced its findings in a report with the telling title *Nuclear Waste: New York's Adherence to Site Selection Procedures Is Unclear.* After examining commission records and quizzing staff about the land volunteered by Arthur Allen, the auditors concluded:

> For two reasons, this decision was inconsistent with the commission's requirement that . . . an offered site must be "at least as good as" other sites under consideration at that time. First, Taylor North contained more than 5 acres of agricultural land in active production, contrary to state regulations that prohibited a low-level radioactive waste disposal site from containing more than 5 acres of such land. Second, the site scored below the minimum score that the commission's staff had established to identify promising sites for further consideration.

Taylor North had made the short list principally because the owner was willing to sell.

To explain the commission's misuse of GIS scores, the GAO report included a map (Figure 7.6) of the forty-acre squares wholly or partly within the Taylor North site. Diamonds mark excluded cells partly within a state reforestation area, and numbers in the nonexcluded cells represent favorability scores for a near-surface disposal site. In selecting other sites, the commission had required at least five contiguous cells scoring 3900 or higher (out of a possible 5000), but the shaded squares indicate Taylor North had only two pairs of contiguous cells above the cutoff. The site should have flunked, but the commission picked a new cutoff and passed it anyway.

Confusion about boundaries and ownership raised questions about the site's size and shape. Confirming local suspicions, the federal auditors reported that Allen did not own all the land he had offered: the site consisted of several different parcels, and Allen's parents owned about 20 percent. Moreover, he had based the offered

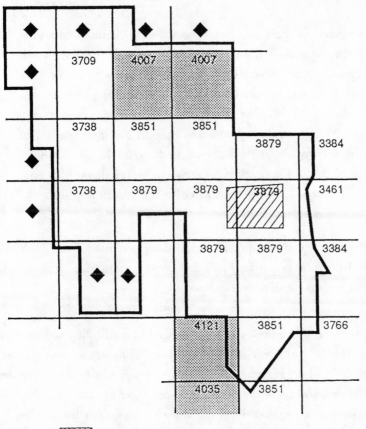

 Excluded the 28-acre parcel that was not offered.

◆ Excluded because the 40-acre cell contained state reforestation land.

Fig. 7.6. Taylor North near-surface grid cell scores. (From U.S. General Accounting Office, *Nuclear Waste: New York's Adherence to Site Selection Procedures Is Unclear*, report no. GAO/RCED-92-172, August 1992, p. 19.)

site's boundaries on an imprecise conservation plan map. Aside from checking local tax maps, the commission had not verified Allen's ownership or his authority to represent his parents. Furthermore, as the diagonally ruled area in Figure 7.6 shows, the Taylor North site completely surrounds a 28-acre parcel not offered for sale.

In summarizing their findings, the government's auditors stopped just short of charging bureaucratic fraud:

Our review of the commission's actions leads us to conclude that the commission's consideration of the Taylor North offered site was inconsistent with some of its procedures — and may not have followed others — in evaluating and eventually selecting the Taylor North site for on-site investigation.

. . . Without thorough documentation and articulation of the siting process in the case of Taylor North, for example, the public — and perhaps ultimately the state's disposal licensing authority — could have difficulty understanding how the commission selected the site.

The inconsistency was especially troubling because of the NIMBY context. In siting sensitive facilities, local acceptance depends on trust, and trust depends on openness.

In an intense attack on the siting commission's credibility, county residents searched the *Report on Potential Sites Identification*—everyone I spoke with used its acronym ROPSI, pronounced *rop-see*—for other contradictions and inconsistencies. Aware of my interest in deceptive maps, assistant county low-level radioactive waste coordinator Denise Cote-Hopkins pointed out a flawed procedure discovered by town supervisor Kimberly Abbey, with the help of Taylor residents Mary Anne Diaz and Nancy Moon. For one of five factors used to reduce the number of potential sites from ninety-six to fifty-one, analysts had based measurements on new maps for some areas and on old maps for others. Because state regulations for hazardous waste facilities required an evaluation of "incompatible structures" near the site, the commission rated each potential site by its proximity to "residences, schools, hospitals, churches, commercial centers, nursing homes or other sensitive populated structures." Each site received one of three ratings: (+) for sites with no incompatible structures "within 0.5 miles of the site boundary"; (0) for sites with one or more structures "within 0.25 miles to 0.5 miles of the site boundary"; and (−) for sites with one incompatible structure "within 0.25 miles of the site boundary." To avoid excluding too many sites, analysts estimated distance from the center of site, not the boundary, thereby ignoring some structures just beyond the perimeter of a large site. Moreover, they considered only those structures portrayed on 1:24,000-scale topographic maps from the U.S. Geological Survey and the state's Department of Transportation (DOT). Unfortunately for Cortland County, the most recent topographic maps of the Taylor area were several years older, on average, than maps of the other sites and so failed to tally its newer structures. As Cote-Hopkins observed in a detailed presentation to the siting commission,

> Clearly, the Commission did not use maps "of the same vintage" to compare sites; in claiming it did so, the Commission misrepresented itself. Moreover, the use of the oldest maps discriminated against the Cortland County sites.

As if this revelation were not enough, her presentation compared the two siting commission maps in Figure 7.7, one describing distance estimates for the Taylor Central site and the other showing the site boundary on a 1974 topographic map. Bold arrows added to the example on the right identify two structures relatively close to the site boundary but missing from the map on the left. The question was obvious, but Cote-Hopkins asked it anyway: "Why were structures whited-out from the map which did, indeed, lie within a ½ mile radius of the digitized centroid, and which are, indeed, found on the DOT topographic maps?"

Commission reports yielded other graphic propaganda. Figure 7.8 is an especially suggestive illustration that Cindy Monaco reproduced in the Cortland County Low-Level Radioactive Waste Office's bimonthly newsletter. Obtained from the "Drift Mine Repository Design Report," prepared by one of the commission's consultants, these maps superimpose general layouts of a drift-mine disposal structure on the two Cortland County sites. How valid was the drift-mine analysis, Monaco asked, if both sites required considerably more acreage than originally stated?

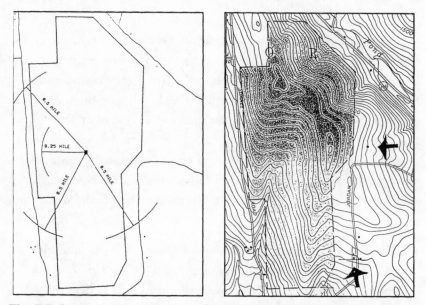

Fig. 7.7. Suggestive juxtaposition of a diagram describing distance estimation for the Taylor Central site (left) with a topographic map showing structures in the vicinity (right). (From Attachment B of Denise Cote-Hopkins, Taylor Against LLRW, *Presentation to the New York State Low-Level Radioactive Waste Siting Commission on the Issue of Incompatible Structures,* November 21, 1991, Cincinnatus, N.Y., 1991.)

Although Cortland County used maps to challenge the GIS findings, Allegany County chose a more direct attack and asked for a copy of the data. In early 1989 Concerned Citizens of Allegany County (CCAC) made several requests for information about the GIS, but the siting commission resisted, claiming the database was changing daily and not yet in a form suitable for release. CCAC took the state to court and charged the commission with violating New York's Freedom of Information Law. "The Siting Commission publicly proclaims time and again that the public must educate itself," their attorney argued, "yet it monopolizes and keeps secret the very factual and statistical data it uses for its decision-making." On July 17, State Supreme Court Justice Frederick Marshall ruled that CCAC was "immediately entitled to access to the data" and ordered the commis-

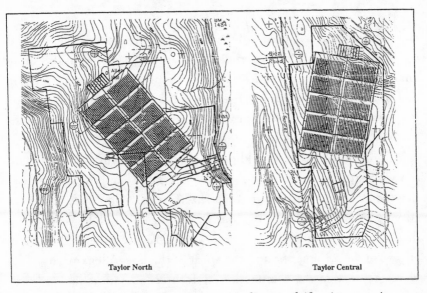

Taylor North   Taylor Central

Fig. 7.8. Overlays in a consultant's report indicate a drift-mine repository at either of the two Cortland County sites would require additional land. (From *Cortland County Low-Level Radioactive Waste Office Newsletter* 1, nos. 5 and 6 [September/October 1991], p. 5.)

sion to "comply fully" by September 11. As ordered, the commission released the statewide data in mid-July and the candidate-area GIS data used to select potential sites in mid-September.

Allegany County soon realized the commission's legal loss was actually a technological checkmate. The siting commission had complied with the court order by giving CCAC's attorney two sets of reel-to-reel magnetic tapes generated on a mainframe computer in Seattle, Washington, by Roy F. Weston, Inc., the commission's principal technical contractor. But the county's computers could not read the tapes, nor could computers at Alfred University, where several faculty members had become active in antidump efforts. Although the county legislature appropriated $5,000 for data translation, officials soon discovered the information was useless without the GIS software used to recommend candidate areas and potential sites. Weston's GIS was a one-of-a-kind, in-house system developed by the firm's programmers for earlier siting projects and not compatible with any commercial GIS software. Under contract to the commission, Weston had entered the data into its proprietary system and carried out the siting analysis three thousand miles away, in Seattle. New York's Freedom of Information Law might have granted access to the data, but the computer program needed to explore and critique the data was confidential, and Weston refused to provide a copy. State employees, citizens' groups, or judges could use the system directly—if they went out to Seattle. The siting commission could, I suppose, have augmented Weston's contract to include setting up a duplicate system in Albany, or even in Allegany County, but the deadline for selecting a disposal site was approaching too rapidly. Aware now of the differences between the letter and the spirit of the law, Allegany residents were skeptical of the commission's offer to run a workshop and provide the assistance of a Weston engineer.

In an interview with the *Olean Times-Herald*, CCAC's attorney David Seeger expressed the county's anger and sense of betrayal:

> The siting commission's staff told my client and others . . . that the computer program the contractor used for the GIS is a commercial program called Arc/Info . . . but I discovered from [the commission's attorney] that there is no Arc/Info component to these tapes. They run solely on that trade-secret program. [This withholding of the software is] directly contrary to the spirit of the Freedom of Information Law. It puts [Allegany County] in the position of having to . . . safe-crack the GIS data during the 45-day comment period.

The state learned from its mistakes in the Alleghany County case and true access to siting data is now possible. The county's efforts made state lawmakers and the governor aware that effective public access requires more than an unreadable reel of magnetic tape. In amending the state's low-level radioactive waste management act, the legislature not only required public access to computerized data but mandated "a format that is accessible for general use." Accordingly, the siting commission arranged with Weston to convert the GIS data to a form compatible with the Intergraph workstations used by the state transportation department. To further improve general access, the commission bought an Intergraph workstation and hired an operator to assist staff and the public. In May 1992, shortly after its system arrived, the commission announced that the GIS was available for public review, even "on nights and weekends to accommodate the schedule of interested citizens who might find weekdays inconvenient." Ironically, on a visit to the commission in June 1993, I discovered that no one from Allegany County or Cortland County had requested a work session. But four years earlier, the response would have been overwhelming.

New York's misguided use of geographic analysis reflects a heavy reliance on outside experts. In requesting proposals from experienced contractors, the siting commission encouraged use of a GIS but overlooked the cost and time involved in obtaining reliable data. Only one sentence mentioned GIS, and potential contractors were told merely to follow exclusionary screening with progressive narrowing, first to ten candidate areas and then to four potential sites. As long as a proposal addressed the legislature's timetable and the DEC's regulations about places to avoid, the contractor could design whatever data analysis scheme seemed "appropriate." Nine qualified contractors submitted bids, and by offering the lowest price, Weston got the job. (Apparently, no one questioned the square-mile cell proposed for exclusionary screening. After all, if the area of the dump site would be about a square mile, why add to the cost with a more refined GIS?) Supervision was minimal at best: the small commission staff (about twenty people) that monitored the contractor's work also dealt with the legislature, the public, and the five commissioners (none of whom had experience in GIS or locational analysis) who met monthly to review progress and offer suggestions. And because the siting strategy New York ultimately settled upon was unique, little could be learned from other states seeking low-level radioactive-waste disposal sites.

Like any weapon in incompetent hands, the GIS dealt the commission an embarrassing self-inflicted wound. But there's another interpretation, in which maps and the GIS were important props to impress state residents, the federal government, and the Southeast Compact Commission, which controls the Barnwell, South Carolina, dump. I know this sounds cynical, but hear me out. Humiliated a decade earlier by its ill-conceived, mismanaged low-level radioactive waste dump at West Valley, the state sought to redeem its tarnished reputation with a formal, technologically sophisticated search for the truly optimum disposal site. Even though the siting effort failed, the process as a whole demonstrated a valiant attempt to deal with problems and do the right thing. Siting efforts also gave lawmakers a politically useful opportunity to protect wetlands, farmland, and the Adirondack and Catskill parks, and to assist local governments; under the same law that established the siting commission, Allegany and Cortland Counties received hundreds of thousands of dollars in special grants. Moreover, for at least half a decade, this bureaucratic soap opera convinced South Carolina and Congress that New York—unlike Massachusetts, say—was really trying to dispose of its low-level radioactive waste. And besides, the siting process wasn't costing taxpayers a dime, at least not directly.

Although New York still has no place to collect its low-level radioactive waste, it has learned a lot. From the protests of citizens and the admonishments of judges, its planners have learned the importance of openness and the folly of arrogant reliance on experts with maps and computers. In favoring a negotiated site, officials have acknowledged that a hierarchical, bureaucratic siting strategy is costly and futile if the "host community" chooses to resist. New York can, I think, find a home for a monitored, retrievable storage facility designed so that low-level waste can be watched carefully, repackaged when necessary, and moved safely should technology ever discover a truly optimal solution. Areas like West Valley and the neighborhoods of existing nuclear plants are good candidates because of their residents' experience with demonstrably more hazardous facilities. The benefits package will be costly, though; tricking the rural poor into bartering their health for lower taxes and a few new school buses is not the solution. Guaranteed oversight will be as important as money, and government must accept long-term responsibility for the facility's impact on the surrounding area. If I let you put your waste in my backyard, after all, you must not only pay me but let me watch what you're doing.

• 8 •

Risk Maps and Environmental Hazards

Maps are a crucial link in the defense against environmental deterioration and other disasters. When development threatens drinking water or fragile wildlife habitats, maps tell landowners not to build, factories to shut down, or communities to find a better way to dispose of garbage. Where people have built homes on floodplains, along seacoasts, and in other hazardous locations, maps help authorities plan orderly evacuations as well as restrict further growth. For nuclear power plants and other hazardous facilities, government requires an evacuation plan as well as an environmental impact statement, and if maps suggest a difficult evacuation, authorities can deny an operating license. The families, businesses, and municipalities that oppose these restrictions sometimes attack the environmental maps that threaten their activities and ambitions. Cartographic controversies can also occur when citizens, local governments, or environmental groups charge state and federal agencies with ignoring important evidence or doing too little to avoid a natural or technological disaster. As in New York State's fretful search for a disposal site for low-level radioactive waste, these controversies occasionally involve enigmatic computer models, data of questionable quality, or maps that are more rhetorical than scientific.

Rather than focus on a specific dispute or a particular type of map, this final chapter looks broadly at the map's role in environmental protection and emergency management. The first section examines the design of environmental maps and the role of maps in risk communication. In presenting maps to local officials, the news media, and the public, an environmental agency must be especially wary of either exaggerating or understating the risk of a natural or technological disaster. The second section examines emergency-response maps, such as evacuation maps for low-lying coastal zones or areas around nuclear power plants. Evacuation maps can thwart developers' plans by dramatizing the geographic impact of a severe storm or technological accident, however unlikely. The third section addresses the conflict between environmentalists and landowners over the delineation and mapping of wetlands and other fragile wildlife habitats. The final section explores how maps promote public health and environmental protection by suggesting or demonstrating links between contamination and disease.

Maps for Risk
Communication

Several years ago a research contract with what was then called New Jersey's Department of Environmental Protection (DEP) offered some valuable insight into the use of maps in risk communication. My task was to examine the agency's use of risk maps and to prepare guidelines for their design and evaluation. DEP's managers were concerned with the effectiveness of maps used to inform local officials, the media, and the public about remediation (cleanup) plans and other geographic aspects of environmental hazards. As my study progressed, "risk map" proved a vague and potentially confusing term. Although many DEP staff had responded to my request for examples of existing risk maps, few of the maps submitted focused on risk and uncertainty. Halfway through the project we resolved this troublesome issue of definition by changing my report's title to "Design Guide for Environmental Maps." The change was appropriate because a survey of both DEP personnel and the risk communication literature indicated that nontechnical people readily interpret any map of an environmental hazard as a risk map.

Environmental maps become risk maps by showing the distance between an identifiable hazard and the viewer's residence or workplace. Simply put, if the incinerator or toxic dump is close by, the perceived risk is greater than if the site were farther away. Even better, of course, is the map that excludes the viewer's neighborhood entirely. Interviews with scientists and officials experienced in the use of environmental maps indicate that the territory shown—cartographers call this the map's "geographic scope"—is particularly important because viewers often equate the area portrayed on the map with the area at risk. Although no one has tested this principle systematically, I have observed visible expressions of either anxiety or relief when people inspecting a new environmental map find or fail to find their neighborhoods. As my design guide points out, map authors must be careful neither to create undue apprehension by showing too much of the surrounding region nor to invite mistrust or complacency by showing too little.

Interviews with DEP personnel and a careful examination of their maps revealed a tendency to confuse the roles of environmental maps and locator maps. Because of the importance of geographic scope, it is bad practice to plot earth science information describing the hazard on a map intended primarily to relate the site to a large audience's wider geographic framework of landmarks, boundaries, and numbered highways. Although a map covering one or two counties, or even the entire state, might be quite suitable for showing the hazard's location, rarely would a geographic scope this large be suitable for showing the area at risk. Use of a wide-area map merely limits the amount of detail it can convey intelligibly. Moreover, when political units define a map's geographic scope, an accident of geography can shove the hazardous site to one side of the cartographic stage. A site that is the center of interest should be near the center of the map.

Failure to separate the functions of location and site description are but one example of the single-map syndrome, whereby a well-intentioned map author crams too much information onto a single, poorly focused map that confuses rather than enlightens. (This foolish, one-map parsimony might reflect a concern with production costs, which were more burdensome with the pen-and-ink map drafting of a decade ago than with contemporary electronic cartography.) A well-focused map needs a straightforward theme, summarized with its own straightforward title, and each theme needs its own map. When presenting a hazard remediation strategy to the public, an environmental protection agency might require six or more maps to answer basic questions: Where is it? What do we know about it? What might happen if we do nothing? What are we doing? How much have we done? And, perhaps most important, What should you (local officials, newspaper readers, the public) do? These themes can vary substantially in information content, map symbols, and descriptive text, and might even require individually labeled profile diagrams to help the viewer interpret the maps.

Most maps have a rectangular format, which rarely matches the shape of the area at risk. Wherever possible, the map author should delimit irregularly shaped hazard zones, such as aquifers vulnerable to contamination from a landfill or steep slopes prone to landslides. Where part of the affected area's boundary is uncertain, a dashed or dotted line can distinguish inexact segments from sharp, comparatively reliable portions. Map readers quickly interpret a dashed line, once explained, to represent a less certain stretch of an otherwise solid boundary line. Maps intended for the public should rely on simple graphic metaphors, not complex symbols and technically challenging map keys.

Environmental maps designed for nontechnical audiences also should be heavily (and carefully) interpreted. For example, although a geologist or regional planner might reliably infer seismic risk from a set of topographic and soils maps enriched with fault lines and other geologic information, the city administrator or potential homeowner usually needs a more explicit representation. Unable in most cases to interpret technically complex cartographic descriptions, the general user requires a map that clearly identifies where he or she should not build a house, a road, or any other structure. A particularly effective graphic metaphor for maps showing restrictions on land use is the stoplight color sequence of red, yellow, and green.

Although colored maps can be highly effective, map authors must anticipate what an Environmental Protection Agency (EPA) worker I know once called "the Xerox effect." This increasingly common problem occurs when a colored map is reproduced on a black-and-white copier or transmitted by a standard black-and-white fax machine. Confusion arises because red and orange symbols emerge as black, while blue symbols typically disappear altogether. To avoid misinterpretation, a conscientious map author not only adds a cautionary note that the map was originally printed in color but also supplements color with other graphic stimuli, such as differences in texture or shape, so that users of a photocopy or fax copy—as well as the 2 percent of the population with impaired color vision—can distinguish between black and red symbols.

Although the better risk maps typically use abundant color, black-and-white examples can illustrate several useful approaches to the cartographic description of environmental risk. An especially effective example is the U.S. Geological Survey map in Figure 8.1, which describes the earthquake hazard along California's San Andreas fault.

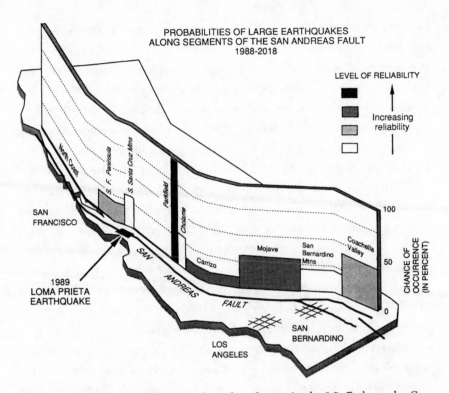

PROBABILITIES OF LARGE EARTHQUAKES
ALONG SEGMENTS OF THE SAN ANDREAS FAULT
1988-2018

LEVEL OF RELIABILITY

Increasing
reliability

SAN
FRANCISCO

North Coast

S. F. Peninsula

S. Santa Cruz Mtns

Parkfield

Cholame

100

Mojave

San
Bernardino
Mtns

Coachella
Valley

50

Carrizo

SAN

ANDREAS

FAULT

1989
LOMA PRIETA
EARTHQUAKE

CHANCE OF
OCCURRENCE
(IN PERCENT)

0

LOS
ANGELES

SAN
BERNARDINO

Fig. 8.1. Forecast map for an earthquake of magnitude 6.5–7 along the San
Andreas fault. (From Robert A. Page et al., *Goals, Opportunities, and Priorities
for the USGS Earthquake Hazards Reduction Program*, U.S. Geological Survey
circular no. 1079 [Washington, D.C.: U.S. Government Printing Office,
1992], p. 29.)

Originally prepared in 1988, this map addressed a specific risk: an
earthquake of magnitude 6.5 to 7 on the Richter scale—large but not
the most severe. Like all forecast maps, it refers to a particular time:
the twenty-year period from 1988 to 2018. In addition to showing
variation in the degree of risk along the fault zone, the map distinguishes between the likelihood of a large earthquake, portrayed with
the vertical dimension of a bar chart aligned with the fault, and the relative reliability of these estimates, represented by the gray interiors of
individual vertical bars. The Geological Survey was especially proud
of this map, which predicted the highly destructive 1989 Loma Prieta

earthquake; the low level of reliability for the portion of the fault in the southern Santa Cruz Mountains merely reflects uncertainty about the probability of occurrence, which would be higher by some criteria but lower by others. Especially ominous is the highly reliable forecast of a large earthquake along the Parkfield segment, about midway between Los Angeles and San Francisco. Although moderate earthquakes have occurred here roughly every twenty-two years since 1857, U.S. Geological Survey geologists stuck out their necks in 1985 by predicting a 5.5–6.0 quake sometime before the end of 1992. Their first comparatively precise prediction proved an embarrassment when 1993 arrived without a significant earthquake along the twenty-mile stretch. With further research, however, seismologists will not only improve the reliability of earthquake forecasts but shorten the forecast period.

Intriguing as well as ominous for many California residents, this highly generalized seismic-risk map holds little information directly useful to local officials, who need more geographic detail about specific hazards and populations at risk. Yet as Figure 8.2 demonstrates, seismological forecasting can address these local concerns as well. Encompassing a small portion of metropolitan Salt Lake City near the Wasatch Fault, this simple, three-category map relates the earthquake hazard to populations at risk by pointing out schools and residential areas with a high potential for severe damage when the soil behaves like a liquid (liquefaction) during a large earthquake. Irregular patterns show that vulnerability to shaking and collapse depends more on soil and subsurface material than on proximity to the fault (dashed and solid lines). The geographic information system (GIS) that combined these overlay maps on a single plot could easily provide an appropriate geographic framework by adding the street network, landmarks, and place-names. Although lacking numerical forecast estimates, this geographically detailed map identifies schools requiring careful inspection for structural integrity as well as residential areas requiring rigorous enforcement of building codes. The map would also be useful in siting fire stations and hospitals and for developing an emergency-response plan.

HAZARD INFORMATION PRODUCT

SCHOOLS AND RESIDENTIAL AREAS
IN HIGH LIQUEFACTION–POTENTIAL ZONES

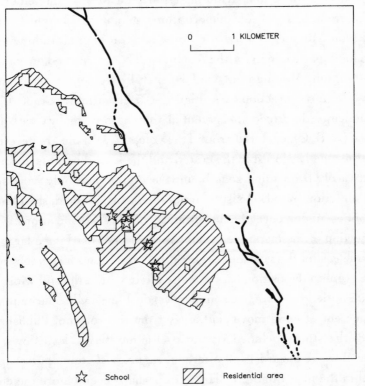

☆ School   ▨ Residential area

Fig. 8.2. Highly specific computer-generated map of earthquake hazard. Created with a geographic information system (GIS). (From Robert A. Page, et al., *Goals, Opportunities, and Priorities for the USGS Earthquake Hazards Reduction Program,* U.S. Geological Survey circular no. 1079 [Washington, D.C.: U.S. Government Printing Office, 1992], p. 37.)

For hazards such as earthquakes, avalanches, and mudslides, a scientist with good data can predict the general shape and orientation of the area at risk with reasonable certainty. But for a volcanic eruption or an accidental release of radioactive gas, the affected area depends largely on the direction the wind is blowing at the time. Forecasting the geographic impact of an atmospheric hazard thus requires data on wind direction and perhaps the local climatological effects of terrain. Because air flow at high altitudes is more important than surface winds in distributing tephra (particles of molten or solid rock produced by a volcanic eruption), geologists Dwight Crandell and Don Mullineaux used Air Force data on high-altitude winds in developing a risk map for the Mount St. Helens volcano, in southern Washington. Although the wind is generally from west to east, it can blow in any direction. As Figure 8.3 illustrates, Crandell and Mullineaux described the potential thickness of tephra with three zones (A, B, C), centered around the volcano, with the greatest thickness in zone A, closest to the crater, and the least thickness in zone C. Eastwardly protruding zone boundaries reflect not only wind speed and direction but also existing tephra deposits, while dashed boundary lines underscore the uncertainty of these delineations. Close inspection of the map reveals faint vertical lines marking a large sector due east of the volcano in which the wind blows 50 percent of the time. Tephra deposition is even more likely within the still wider sector identified by *either* horizontal or vertical lines, which accounts for 80 percent of wind movement away from the volcano. Published in 1978, the study included risk maps for mudflows, lava flows, and other flowage hazards, and offered a vague prediction of an eruption "within the next hundred years, and perhaps even before the end of this century." When a spectacularly destructive eruption of Mount St. Helens validated this U.S. Geological Survey study only two years later, in May 1980, winds blowing uncharacteristically from the south left an unlikely yet (obviously) plausible northward pattern of tephra deposition. For comparatively continuous atmospheric hazards, such as a municipal incinerator or coal-burning power plant, average conditions provide a more reliable base for assessing risk.

Controversies over risk maps have been comparatively rare, mostly because this kind of mapping is relatively new. The most common disputes involve flood mapping, as when homeowners complain about the cost of flood insurance. Yet a major scandal is likely in the next few decades as earth scientists become better at assessing hazards—or more venturesome in their claims. It seems inevitable that a

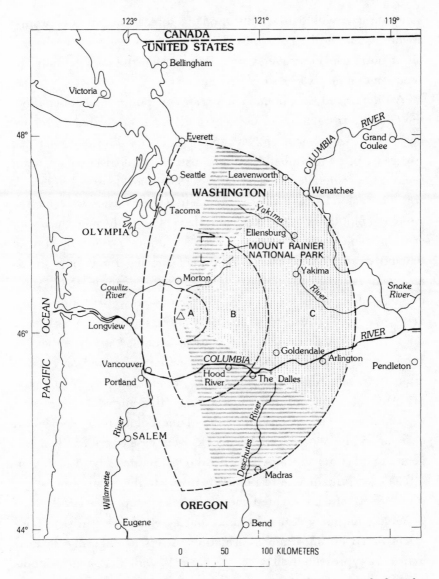

Fig. 8.3. Tephra-hazard zones on a risk map published two years before the 1980 eruption of Mount St. Helens. (From Dwight R. Crandell and Donal R. Mullineaux, *Potential Hazards from Future Eruptions of Mount St. Helens Volcano, Washington,* U.S. Geological Survey circular no. 1383-C [Washington, D.C.: U.S. Government Printing Office, 1978], p. C18.)

major disaster will follow publication of a forecast map that appropriate authorities either never heard of or chose to ignore, or perhaps a bit of both. The controversy will focus not on the map's content or accuracy but on awareness of its significance.

A 1985 tragedy in Colombia illustrates my point. In October 1985, the Colombian Institute of Geology and Mines circulated a preliminary hazard-zone map for Nevado del Ruiz, a volcano dormant for 140 years. On November 12, 1985, institute geologists completed a revised, more detailed version (Figure 8.4) that identified high-risk zones for ash falls, lava flows, pyroclastic material expelled from the vent, and mudflows. As a reduced-scale version of the map shows, an erupting volcano can trigger massive mudflows that travel well beyond areas affected by lava and pyroclastic material to bury towns and villages located along rivers. In 1595 and again in 1845, mudflows resulting from rapid melting of snow and ice had killed hundreds of people at Armero, a small city about thirty-five miles east of the volcano, but each time the town was rebuilt. On November 13, 1985, a month after release of the preliminary map and only a day after its revision, the volcano erupted again, killing over 25,000 people. Most of the victims died in a mudflow that buried Armero. Had the eruption occurred several months later, failure to implement an evacuation plan would have been a clear-cut case of criminally negligent bureaucratic incompetence. Although nature gave Colombian officials too little time to respond effectively to the hazard, investigation of how officials treated the preliminary map reveals a lack of understanding of volcanic hazards and a reluctance to risk a false alarm based on uncertain information. A less geologically detailed map that represented all volcanic hazards with the same symbols might have counteracted the popular perception of volcanic eruptions that led authorities to focus only on areas in danger of lava flows, which they considered a more serious hazard than mudflows. The Nevado del Ruiz disaster demonstrates that public officials cannot afford risk maps designed only for scientists.

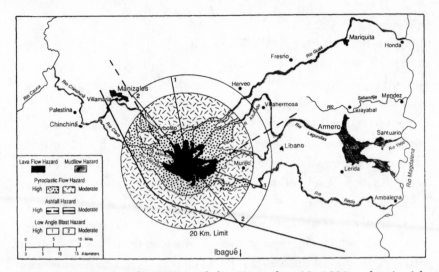

Fig. 8.4. Reduced-scale version of the November 12, 1985, volcanic-risk map completed by the Colombian Institute of Geology and Mines a day before the eruption of Nevado del Ruiz. (From Barry Voight, "The 1985 Nevado del Ruiz Volcano Catastrophe: Anatomy and Retrospection," *Journal of Volcanology and Geothermal Research* 44 [1990], p. 362. Reprinted with permission.)

Emergency Management, Evacuation Plans, and NIMBY Resistance

Although maps have an important role in emergency management, cartographic planning is not equally appropriate for all hazards. Hazard-zone maps are more suitable for reducing losses from earthquakes and volcanoes, for instance, than in coping with a hepatitis epidemic or a severe drought. Maps of seismic hazards and flood zones allow officials to safeguard life and property by discouraging development in damage-prone areas or by requiring that special precautions be taken. Where preventive restrictions are not practicable, carefully developed emergency-response maps promote efficient evacuation of areas threatened by coastal flooding, industrial accident, or fire. In such emergencies, geographic information systems can help fire and rescue officials rapidly identify nearby schools and nursing homes as well as factories or warehouses where explosives and other hazardous materials are stored. Effective emergency mapping includes not only readily accessible, on-line inventories of dangerous chemicals and vulnerable populations but also interactive electronic systems and simulation models with which local officials can assess the impacts of a variety of plausible natural and technological disasters. Lack of these valuable planning tools might soon become controversial.

Careful assessment of locally relevant hazards can help a community decide which emergency maps are most appropriate. Because of obvious geological differences, for example, California is more concerned about earthquakes than New York, and Hawaii is more concerned about volcanoes than California. Hazard perception is highly subjective, of course, as demonstrated by a ranked list of hazards that New York's State Emergency Management Office prepared in the mid-1980s. A survey asked county emergency management coordinators to rate forty-one different hazards according to their history, probability of occurrence, and local vulnerability. State officials then weighted individual ratings by county population and computed summary ratings, for which a score of 100 or more identifies a hazard requiring emergency planning. Statewide results identified twenty-seven of the forty-one hazards as important, with severe winter storms (score: 215) at the top of the list. Weighting by population diminished the importance of wave action (score: 63), which few officials in noncoastal counties considered severe. Statewide scores lower than I would have expected are also apparent for infestation, blight, forest fires, and other hazards that typically threaten only sparsely populated rural counties:

Rank	Hazard	Score
1	Severe winter storm/blizzard	215
2	Hurricane	204

31	Wave action	63
32	Blight	58
33	Levee/dam failure	52
34	Forest fire	46
35	Landslide	42
36	Radiological accident, fixed site	27
37	Mudslide	21
38	Structural collapse	3
39	Volcanic activity	1
40	Tsunami (tidal wave)	1
41	Mine cave-in	0.2

These rankings reflect the 1980s as well as the state's unique geography. County public safety officials apparently not only remembered the unexpected power blackout that shut down metropolitan New York on July 13, 1977, but seemed to fear a shooting war with the former Soviet Union. Yet the scores suggest they had largely forgotten the Three Mile Island nuclear accident of 1979 and the race riots of the late 1960s, and were less wary of terrorism than after the World Trade Center bombing of 1993. Even so, the winter of 1993 validated their top two choices with a blizzard that dumped forty-two inches of snow on Syracuse and a severe winter storm that wrecked many homes on the Long Island coast. Despite obvious historical influences and other limitations of let's-put-it-to-a-vote surveys, the twenty-seven hazards with a score above the 100 threshold clearly warrant emergency planning, as do several locally important hazards such as forest fires (score: 46). A nineteenth-place ranking for earthquakes (score: 144) is deceptively reassuring because New York has active faults and earthquakes are a significant hazard in parts of the state.

Because of the controversy over efforts to site a low-level radioactive waste dump in New York, the most perplexing entry on the list of hazards is a score of 27 for a fixed-site radiological accident. This score is puzzling because the state's six operating nuclear power plants as well as the Knolls Atomic Power Laboratory, thirteen miles northwest of the state capital, and several other research facilities contain not only low-level "radwaste" but highly radioactive fissionable material, which could be released into the atmosphere with tragic consequences in the event of an operations error, earthquake, plane crash, or terrorist attack. Although the minuscule risk associated with an earthquake or unintentional plane crash is impossible to calculate, much less worry about, both an operations accident and a terrorist attack are sufficiently plausible to warrant rigorous monitoring of plant safety as well as a recent federal directive to guard against truck bombs. As some of the public officials surveyed were surely aware, the Federal Emergency Management Agency (FEMA) requires detailed emergency-response plans, including evacuation maps, for all nuclear power stations.

Around each nuclear plant FEMA guidelines identify two Emergency Planning Zones (EPZs), with radii of ten and fifty miles. The ten-mile EPZ includes a hypothetical "plume exposure pathway" requiring rapid evacuation during a worst-case "core melt sequence" to prevent exposure of the whole body to deadly gamma radiation as well as internal exposure of the thyroid, lungs, and other organs through inhalation. FEMA requires a detailed evacuation plan describing the plume-exposure EPZ's boundaries and showing exit routes from the area. Although the immediate, short-term effects of a core-meltdown accident ought not extend beyond ten miles, the fifty-mile "ingestion exposure EPZ" recognizes the possible contamination of soil and pastures, which in turn could damage the bone marrow and internal organs of anyone consuming milk and vegetables produced there. Although an evacuation plan is not needed for this outer zone, FEMA requires careful planning to avoid the long-term effects of a contaminated food supply. In particular, farmers must immediately remove cows from pastures, and officials must carefully monitor milk-processing plants. The ten- and fifty-mile circles around the site are merely guidelines, which the plant operator should adjust to reflect terrain, transport routes, population distribution, local political jurisdictions, and other characteristics of the site.

Before the momentous Three Mile Island incident near Harrisburg, Pennsylvania, in March 1979, the Nuclear Regulatory Commission (NRC), which licenses commercial nuclear power plants, had sole authority over off-site emergency planning. The NRC considered a radiological accident most unlikely and required emergency plans only for the plant site and the two- or three-mile surrounding area. Although licensees had to prepare emergency-response plans, submission was voluntary and the NRC did not assess quality or require modifications. Because of inadequate preparations by both the plant operator and local governments, more than ten times the appropriate number of people decided to evacuate, including roughly 144,000 people within a fifteen-mile radius. Although the advisory broadcast by state officials called for the evacuation only of pregnant women and preschool-age children living within five miles of the plant, as many as 9 percent of people living twenty-five miles away evacuated. Evacuees overwhelmed hastily arranged shelters, and many traveled much farther than necessary. Harsh criticism of the NRC's lax attitude toward radiological-emergency planning led to tighter regulations, which included a mandatory quality assessment by FEMA. Under the new rules, the NRC could not license a new nuclear plant until FEMA had approved the site's evacuation plan.

These reforms are not altogether reassuring. A recent study by cartographers Ute Dymon and Nancy Winter reveals that many of the maps used by plant operators to inform the public about evacuation plans are poorly designed. Dymon and Winter obtained public information materials for thirteen U.S. nuclear power plants, mostly in the Northeast. Mailed annually to persons living within the plume-exposure EPZ, these materials vary in format from calendars (permanent but not convenient for a car's glove compartment) to brochures (easily mislaid or discarded). Most of the EPZs are circular, not adjusted to reflect terrain and population, as the FEMA guidelines suggest; and few of the publications reflect coordination or approval by state emergency-management officials. In general, evacuation maps are overly vague and lack adequate landmarks and reference features. A quarter of the sample do not show designated evacuation routes, a quarter omit the EPZ boundary, and a third neither identify relocation centers nor describe how to reach them. None of the maps identifies schools, hospitals, major employers, police stations, or decontamination centers. Moreover, the graphic hierarchy is often too weak and the symbols too complex for ready and reliable use by someone under considerable stress.

Curious about evacuation plans in my own region, I obtained a current copy of the emergency-response booklet for the three nuclear power plants at Oswego, New York, about thirty-five miles north of Syracuse. The booklet's maps are more reassuring than most of those examined by Dymon and Winter. One page contains an evacuation map showing an adjusted EPZ that extends somewhat beyond ten miles. The map (Figure 8.5) identifies evacuation routes by street name or route number and divides the EPZ into twenty-nine Emergency Response Planning Areas (ERPAs), including four on Lake Ontario and three in the usually navigable Oswego River. (Printed with black and orange ink, the full-size original is better looking and more informative than this black-and-white, photoreduced facsimile—a victim of the "Xerox effect.") For the twenty-two nonmarine ERPAs, five pages of detailed street maps show densely placed bus pickup points, from which residents can ride to the "reception center" at the New York State Fairgrounds, just west of Syracuse. Verbal directions tell evacuees how to reach the fairgrounds by car, and a generalized map describes the location. Although the plan seems straightforward on paper, I wonder how people would respond in an emergency.

As snarled traffic after a large concert or major sporting event demonstrates, evacuating tens of thousands of people is neither rapid nor straightforward. A survey of U.S. nuclear-plant managers suggests that evacuation of the ten-mile EPZ might take between 4.8 and 21 hours in good weather, and from 5.3 to 27 hours in bad weather; medians of 5.8 and 7.3 hours for fair and foul conditions are hardly reassuring. Just a few vehicles running out of gas would slow traffic

Emergency planning zones and evacuation routes

This map shows the designated evacuation routes for the entire 10-mile emergency planning zone around the Nine Mile Point and FitzPatrick plants. That zone is divided into Emergency Response Planning Areas (ERPAs). Designated evacuation routes are indicated on the map by blue arrows.

These routes have been chosen to minimize traffic congestion and provide the quickest way out of the emergency planning zone.

To find how you would evacuate, locate the designated route nearest your home. You would follow that route out of the emergency planning zone regardless of your final destination. Be sure that your route does not take you back into the emergency planning zone.

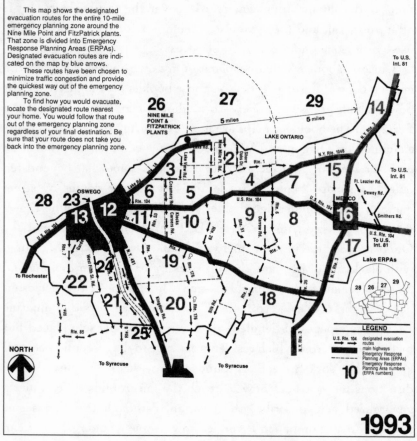

1993

Fig. 8.5. Evacuation route map for evacuation zone surrounding three nuclear power plants at Oswego, New York. (From Oswego County Legislature, Oswego County Emergency Management Office, New York State Disaster Preparedness Commission, Niagara Mohawk Power Corporation, and New York Power Authority, *Oswego County Emergency Planning and You: Radiation, Hazardous Materials, and Weather* [Oswego, N.Y., 1993], p. A1.)

and cause others to run out or overheat, and with the proliferation of personal firearms, a panicked evacuation of a heavily populated area could make the fall of Saigon look like a picnic. Likely overevacuation by residents fleeing areas beyond the suggested EPZ further threatens public safety around relocation centers and along routes leading from the EPZ. The scenario of a disastrously disorderly evacuation not only terrifies state and local officials but provokes residents fifty miles away to protest the possible takeover of their local schools and hospitals. Because of the possibility that large numbers of people might decide to get the hell out, the mere existence of an evacuation plan becomes a center of controversy.

Controversy over the evacuation plan can kill a nuclear plant. In the case of the Shoreham nuclear power station, death came quite late, after the plant operator had run the reactor for a few months in 1985. Shoreham's principal handicap was its location on the north shore of Long Island, about fifty-five miles east of midtown Manhattan (Figure 8.6). The plant's ten-mile plume-exposure EPZ extends almost completely across Long Island, and intersects principal routes to the eastern half of Suffolk County. Although only 57,000 people lived within ten miles of the plant in the early 1980s, the area's summer population is much greater. That slightly over 5 million people lived within fifty miles of the site magnified the consequences of a panic evacuation. Geographers Don Zeigler and Jim Johnson, who surveyed households in the area, predicted a behavioral reaction vastly exceeding the health risk posed by a radiological emergency. Their telephone-survey data indicated that a carefully orchestrated, limited evacuation of the ten-mile zone was highly unrealistic. Even a public announcement merely advising persons living within five miles of the plant to stay indoors would precipitate a massive exodus westward, toward New York City. Because of widespread ignorance of ionizing radiation, Zeigler and Johnson doubted even an extensive education campaign could substantially reduce public fear. A radiological emergency at the Shoreham plant would trigger a behavioral overresponse similar to the one that accompanied the 1979 Three

Mile Island accident, but with far greater repercussions because of the larger population and fewer exit routes.

Although planning for Shoreham had begun in the late 1960s, before emergency-response plans were required, local government reluctantly cooperated with the plant operator until the early 1980s, when mass evacuation became an issue. Because FEMA approval of the emergency-response plan was now a prerequisite to NRC approval, state and county leaders chose to fight the plant by blocking the evacuation plan. A U.S. General Accounting Office study requested by Senator Daniel Patrick Moynihan includes a chronological saga of seventy-eight actions and responses between 1981 and 1987 among various officials and agencies. Believing no evacuation plan could ever be suitable, state and local officials chose not to cooperate with LILCO and rejected a proposal for a plan with a twenty-mile radius—twice the federal requirement. LILCO developed an

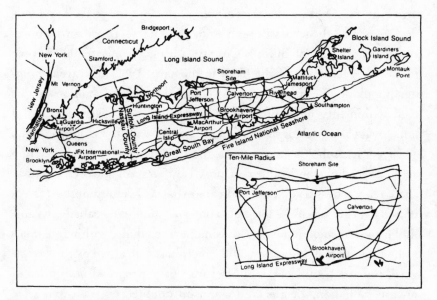

Fig. 8.6. The Shoreham site and its plume-exposure EPZ on Long Island. (From U.S. General Accounting Office, *Emergency Planning: Federal Involvement in Preparedness Exercise at Shoreham Nuclear Plant*, report no. GAO/RCED-87-45, December 1986, p. 15.)

independent plan, FEMA found several "inadequacies," LILCO revised its plan, and in 1986, at the NRC's request, FEMA conducted an "emergency response test" involving seventy-five federal evaluators and about 1,000 "participants" (mostly LILCO employees), some of whom enacted the roles of state and local officials. In a pointedly sarcastic statement to the House Subcommittee on Energy Conservation and Power, Moynihan ridiculed this crude attempt to validate the LILCO plan:

> Some looked forward to a FEMA-run drill of LILCO's plan, saying that it would tell us, once and for all, whether Long Island could be evacuated safely and quickly. It was to be an experiment, with a scientific answer to a knotty question.
>
> Rubbish! No FEMA emergency drill actually tests the number of cars that the highways can carry, but every other drill the agency has done has at least been a test of the plans that would be used by the people who would use them in an emergency. This was not a real test of a real question. . . .
>
> And what have they got? A 156-page "Post Exercise Assessment" that recounts in graphic detail exactly who watched whom doing what. The logistical planning that went into the exercise by those who watched and those who were watched was impressive. But the report tells us precisely what those of us who thought it through expected—nothing relevant to the question at hand.

Shoreham's opponents eventually won out, but at a price. FEMA's regional director Frank Petrone resigned rather than endorse the LILCO evacuation plan, the Chernobyl accident in late April 1986 reinforced fears of a nuclear-plant disaster, Congress refused to allow the pro-Shoreham NRC to change the rules, and Governor Cuomo arranged for the Long Island Power Authority to buy the Shoreham plant for a dollar and begin decommissioning and for LILCO to recover its $3.7 billion investment through increased electrical rates. In the end, thousands of simple mental maps of the plant's location had defeated an elaborately detailed plan based upon the tenuous assumption of an orderly evacuation.

Wetlands: Fragile and Flexible

If the key question in the bitter controversy over the Shoreham plant's evacuation plan was "Would it work?" the central issue in an equally contentious dispute about wetlands mapping is "Where do we draw the boundaries?" For landowners eager to build homes or extend their fields by filling marshes or draining swamps, wetlands mapping creates a conflict between property rights and government regulation. An editorial in the *Wall Street Journal* once called wetlands regulation "a monster that lurks in America's swamps and eats landowners" and "a federal zoning code for every damp piece of ground in the country." But for environmentalists frightened by the loss of hundreds of thousands of acres of wetlands every year, the issue is not only wildlife protection but flood control and the purity and abundance of drinking water. Further discord arises from the varied criteria (soil, hydrology, and vegetation) used to define wetlands, and from the diverse definitions and procedures used by the different federal agencies responsible for various aspects of wetlands protection. Moreover, although states and localities typically map their wetlands, the federal government is unwilling to accept these maps. Favoring instead a complex and often costly process for determining on an *ad hoc* basis whether an area is protected, federal agencies cannot agree on a practicable standard for nationwide general-purpose wetlands mapping.

With less money and fewer responsibilities, state and local governments adopt a more pragmatic approach to wetlands. Because a readily available wetlands map allows officials to determine whether a permit is required for land use, states and municipalities don't just regulate wetlands, they map them. Moreover, state-level regulation is easier because overall responsibility typically resides in a single department of environmental protection or natural resources, with broad responsibilities for groundwater, air quality, hazardous materials, and flood control. A single agency with limited resources and a comparatively local constituency knows that maps at hand are infinitely more useful than bureaucratic or scientific debates about proper procedure.

Although consolidating wetlands regulation in a single federal department might improve communications and lower costs, diverse federal goals would still frustrate the development of a national set of general-purpose wetlands maps. I say this because several past interagency efforts to develop standard procedures have produced complicated, allegedly objective rules subject nonetheless to substantial differences in interpretation. One collaboration culminated in 1989 with the *Federal Manual for Identifying and Delineating Jurisdictional Wetlands*, a joint publication of the Army Corps of Engineers, the Environmental Protection Agency (EPA), the Fish and Wildlife Service, and the Soil Conservation Service. In part, the 130-page manual attempts to merge different approaches described in technical manuals developed independently by the Corps of Engineers and the EPA. Its glossary offers a deceptively simple definition—wetlands are "areas that under normal circumstances have hydrophytic vegetation, hydric soils, and wetland hydrology"—which in fact reflects separate procedures for examining plants, soils, and moisture. Although none of these criteria is easy to assess, all were to be evaluated using direct or indirect evidence.

Some federal officials work closely with the manual, but the typical user is a private consultant hired by a landowner or local government to determine whether a proposed project is expressly forbidden, requires a permit, or threatens a developer with penalties or a farmer with loss of subsidies and loan guarantees. Carrying out these complex procedures has been lucrative for many consulting engineers, biologists, and testing laboratories, yet many of their clients see the manual's complex criteria as a needless expense, especially when a more straightforward examination might yield the same result.

The scientific community generally applauded the increased wetland area under federal jurisdiction—an earlier manual, published by the Corps of Engineers in 1987, was less inclusive—although some otherwise sympathetic scientists complained that the manual's authors confused policy issues with scientific issues, did not clearly state their goals, called for evidence that is technically unsound, and failed to consider the needs of users. In contrast, landowners objected strongly to the liberal interpretation of "normal circumstances." In particular, a new criterion treated as wetland any area ponded, flooded, or saturated to a depth of six to eighteen inches for one week or more during the growing season. Much of the nation's farmland, some complained, was now wetland. Although this charge is an obvious exaggeration, the 1989 manual roughly doubled the amount of federally regulated wetland from 100 million to 200 million acres. Other critics warned that increased complaints about troublesome wetlands regulations might lead to less enthusiastic enforcement.

Ironically, the 1989 manual appeared less than two months after George Bush took office. During the 1988 presidential campaign, as part of his pledge to become the "Environmental President," Bush had promised "no net loss" of protected wetlands. But he and Vice President Dan Quayle clearly sympathized with conservative landowners and oil companies, who feared interference with their activities on the Alaskan tundra. Responding to pressure from lobbyists and political kinfolk, therefore, the White House and Republican members of Congress recommended substantial modifications to agency staff already at work on a revised version of the manual. After the EPA's chief wetlands scientist resigned in disgust from the writing team, lawyers and other nonscientists from the Council on Competitiveness and the Office of Management and Budget assumed a dominant role in developing a new manual, released in July 1991. Among other changes, a revised hydrologic criterion increased the required seven-day wetness period to either fifteen consecutive days of inundation or twenty-one consecutive days of saturation. By some estimates, this new "consensus" definition would reduce the amount of protected wetlands from 200 million to 150 million acres.

Environmentalists were appalled and outraged. Field tests in selected states suggested the new criteria would wipe out protection for more than half the wetlands in Connecticut, Idaho, Maine, New York, Washington, and Wisconsin, as well as shrink the country's total federally protected acreage to around 50 million acres—roughly half the national inventory before the 1989 manual. Especially troubling was a shift of the burden of proof from the landowner, who previously had to demonstrate the land was not a wetland, to the government, which now had to prove it was. A sixty-day public comment period, extended several more months because of widespread interest, drew a staggering 80,000 formal responses in addition to countless letters to senators and representatives. In the end, Congress and the administration resolved the issue, as they often do, by doing nothing. William Reilly, the environmentalist Bush appointed to direct the EPA, adopted the older 1987 Corps of Engineers criteria and refused to implement "the Quayle manual." The 1991 revisions died when Bill Clinton took over the White House in January 1993.

Reaction to the proposed 1991 revisions reflected public alarm over both historical and recent losses of wetlands. Fueling this reaction was a report published the previous year by U.S. Fish and Wildlife Service scientist Thomas Dahl. *Wetlands Losses in the United States, 1780s to 1980s* estimated that two centuries of land conversion, largely to agriculture, had reduced the amount of wetland in the forty-eight contiguous states from 221 million to 103 million acres. To arrive at this figure, Dahl had sampled air photos and various historical sources. Among the report's illustrations, a map showing the percentage of wetlands lost since colonial times by individual states revealed enormous losses in California, the Midwest, and other significant agricultural states. *Scientific American* and various environmental preservation groups eagerly reproduced the map and data. The following fall, during the comment period for the proposed rule change, another Fish and Wildlife Service report, *Status and Trends of Wetlands in the Conterminous United States, Mid-1970s to Mid-1980s,* recycled the message with the map reproduced in Figure 8.7. As the table below the map indicates, twenty-two states had lost over half their original wetlands in the past 200 years. Although the difficulties of delineating wetlands and reconstructing past landscapes preclude precise estimates, the map makes the numbers believable, dramatizes the cumulative effects of wetlands conversion, and demands regulatory intervention.

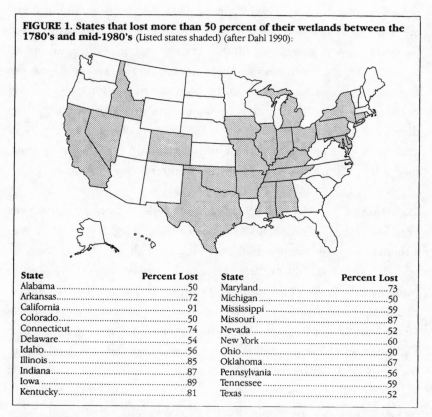

FIGURE 1. States that lost more than 50 percent of their wetlands between the 1780's and mid-1980's (Listed states shaded) (after Dahl 1990):

State	Percent Lost	State	Percent Lost
Alabama	50	Maryland	73
Arkansas	72	Michigan	50
California	91	Mississippi	59
Colorado	50	Missouri	87
Connecticut	74	Nevada	52
Delaware	54	New York	60
Idaho	56	Ohio	90
Illinois	85	Oklahoma	67
Indiana	87	Pennsylvania	56
Iowa	89	Tennessee	59
Kentucky	81	Texas	52

Fig. 8.7. Map of states that lost more than half their wetlands since colonial times. (From Thomas Dahl et al., *Status and Trends of Wetlands in the Conterminous United States, Mid-1970s to Mid-1980s,* U.S. Fish and Wildlife Service, 1991, p. 2.)

Maps, Environmental Health, and Public Policy

Mapping is an important but unheralded tool of epidemiologists, public health authorities, and other scientists and officials committed to understanding and preventing disease. Perhaps the most widely applauded environmental-health map is an 1854 plot of cholera deaths by London physician John Snow. Suspicious that drinking water was responsible for a recent epidemic in Soho, Snow mapped the homes of victims and discovered a cluster of deaths around the now-infamous Broad Street Pump—a location marked more recently by the well-known John Snow Pub. Disabling the pump by removing its handle confirmed his theory: new cases in the district plummeted when Snow's action forced residents to get water from other, noncontaminated sources. Local residents were fortunate that Snow so quickly demonstrated a significant epidemiological principle before skeptical colleagues could question the handful of mapped deaths much closer to a different pump. Although careful investigation later revealed these distant victims enjoyed the taste of the Broad Street water and were willing to walk there to fill their jugs, Snow's map never had time to become controversial.

Cartographic studies of disease are rarely so straightforward. Few infections are as rapid and indiscriminate as cholera in dispatching victims, and few diseases yield as compact and blatant a spatial pattern. People move around a lot and sample a variety of environments at home, work, school, restaurants, and playgrounds, as well as when they take vacations and change neighborhoods. Epidemics are far less common today than in the mid–nineteenth century, and in the developed world fatal communicable diseases usually reflect tainted food, unsafe sex, or dirty hypodermic needles. Most Americans rely on carefully monitored municipal water systems rather than neighborhood wells, and the rare precautionary warning to boil drinking water is generally effective in preventing waterborne infections, however mild. Moreover, despite evidence that cumulative exposure to pesticides and other airborne hazards can be lethal, mixed pollutants from multiple sources make it difficult to prove that emissions from factory X killed person Y—the rare exceptions are widespread and suddenly fatal exposures, such as the accidental release of lethal gas from a Union Carbide pesticide plant in Bhopal, India, in 1984. Even so, health officials watch for case clusters of cancers and other diseases that might be caused or triggered by polluted air, contaminated water, and other environmental toxins.

Few scientific controversies are as intriguing as the Sternglass correlation, named for Ernest Sternglass, who in the late 1960s and 1970s challenged the nuclear establishment with evidence that low-level radiation was a significant health hazard, particularly for infants exposed as an embryo or fetus. Sternglass was a radiation physicist who had received a Ph.D. from Cornell in 1953 and worked for the Westinghouse Electric Corporation for fifteen years. He was deeply interested in genetic damage caused by radiation. His son, who was born in 1947 with Tay-Sachs disease, might have inherited the condition from Sternglass's father, a dermatologist who worked closely with X rays. In 1962, an article in *Science* by respected nuclear physicist Ralph Lapp refreshed his interest in the harmful effects of low-dose radiation. Lapp reported the chance discovery in April 1953, by a radiochemistry class at the Rensselaer Polytechnic Institute, of extraordinarily high radioactivity in Troy, New York. The class determined that a cloudburst had brought down fallout from an atomic test two days earlier thousands of miles away, in Nevada. With a population of a half-million people, the Albany-Troy area provided a good opportunity, Lapp suggested, for a statistically sound study of the possible link between radiation and childhood leukemia. Starting in 1945, the Atomic Energy Commission (AEC) conducted numerous aboveground tests of nuclear weapons at test sites in New Mexico and Nevada. Hoping to reduce the risk to humans—and avoid political resistance to testing—AEC officials deferred detonation until the wind was blowing eastward, away from California and other heavily populated areas; in principle atmospheric circulation would eliminate the hazard by spreading fallout around the globe. Although southern Utah and other places with high levels of radiation had comparatively high rates of leukemia and various cancers, their small populations obviated statistically definitive conclusions.

In 1967, an appointment as professor of radiation physics at the University of Pittsburgh gave Sternglass more time to explore links between radiation and health. Two years later he reported his results in the *Bulletin of the Atomic Scientists*, a respected forum for dissident or environmentally conscious physicists, as well as in *Esquire* magazine, a decidedly unorthodox outlet. Although other scientists studied fallout and worried about its consequences, Sternglass "went public" with alarming findings even supportive colleagues considered tentative and unproven. Attempts by the AEC to discredit him and suppress his work attracted further attention from the media and made him the darling of antinuclear activists. Sternglass also collided with the nuclear-power industry by presenting evidence that accidental releases of low-level radiation from nuclear power stations, research reactors, and other atomic facilities not only harmed local residents but accounted for a national decline in Scholastic Aptitude Test (SAT) scores, a proposition difficult to demonstrate conclusively even if true.

Despite extensive use of graphical analysis and geographic data, Sternglass employed surprisingly few maps. Among those he did use, the most convincing are the pair of maps reproduced in Figure 8.8. In the *Bulletin* article, where these maps appeared, Sternglass used time-series graphs to show that atmospheric testing of nuclear weapons since 1945 appeared to reverse a long-term downward trend in fetal mortality (stillbirths), which had fallen more or less steadily since about 1930. In the Albany-Troy area, for instance, the rate was still falling after the 1953 fallout, but with a less pronounced decline. For the country as a whole, another time-series graph revealed a marked temporal correlation in which the infant mortality rate (deaths of children less than a year old) resumed its previous, steeper decline after

Britain, the Soviet Union, and the United States signed a test-ban treaty in 1963. To answer critics who considered these representations irrelevant, Sternglass compared each state's actual infant mortality rates for 1946 and 1950 with hypothetical rates estimated for those years by extrapolating the state's trend for the period 1940 to 1945. For example, the −5 value for Texas on the upper map indicates that for 1946 the Lone Star State experienced an infant death rate 5 percent lower than extrapolation of its 1940–45 trend would have predicted. That is, infant mortality had not only declined but was 5 percent better than expected. (Sorry if this sounds complex—perhaps that's why few media reports of Sternglass's findings reproduced his charts and maps.) In contrast, on the lower map a positive 9 percent value indicates that in 1950 Texas had an infant death rate 9 percent higher than predicted by its 1940–45 trend. On both maps, Sternglass shaded the states where the infant mortality rate was more than 5 percent greater than the forecast. Only the lower map shows an astonishingly coherent pattern of higher-than-forecast infant death rates for all southeastern states in the plume path of the first man-made atomic explosion, code name Trinity, detonated on July 16, 1945, at Alamogordo, New Mexico. According to Sternglass, who mapped similar calculations for other years,

> by 1949, a pattern of excess infant mortality had appeared, with significant increases occurring in a narrow band of states to the east and northeast [of the detonation], over which the high altitude portion of the fallout cloud was carried by the prevailing jet streams that blow steadily from west to east, while the mortality rates in the states to the south, the north and the west were essentially unaffected.

Few people can ignore a geographic pattern this strong. Reinforced by other data and further analyses, Sternglass's maps helped demonstrate a link between nuclear testing and infant death, significant even to colleagues who deplored his sloppy data and use of the media. In 1972, David Inglis and Allan Hoffman, former critics who

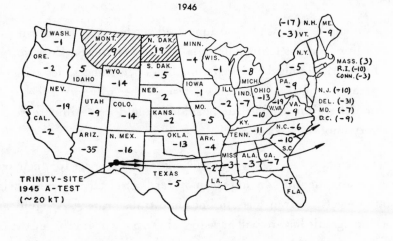

1946

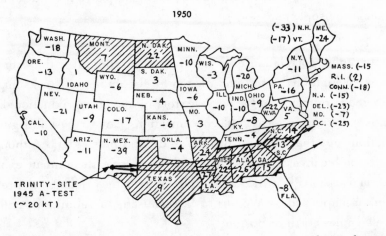

1950

Fig. 8.8. Maps showing the emergence of a strong correlation between excess infant mortality and the plume of the July 16, 1945, atomic test at Alamogordo, New Mexico. (From Ernest J. Sternglass, *Bulletin of the Atomic Scientists* 25 [October 1969], pp. 29–32. Reprinted with permission.)

reviewed his book *Low-Level Radiation,* remained skeptical yet showed appreciation for his contributions:

> Institutions have a tenacity for life. Change is resisted almost everywhere. Sternglass' early papers, presented with his enthusiastic flair, left even more room for doubt than does the more complete assembly of cases in his book. . . . Now that

even more evidence has accumulated, we have even less justification for not taking Sternglass' thesis seriously. Even so, his work should be a beginning. Independent studies with greater facilities, untainted by desire either to verify a thesis or defend an established position, should delve into these very important correlations. . . .

Today, scientists are showing similar guarded respect for their colleagues investigating the unproven hazards of electromagnetic fields around electrical transmission lines and transformers. A significant electromagnetic hazard will have to yield strong geographic evidence, but if a correlation does exist, maps may again prove the crucial tool in making people take notice.

Geographic correlation does not prove causation; if it did, similar maps for high birth rates and the world distribution of storks might be a good argument for the nursery story of human reproduction. But maps showing patterns and concentrations can suggest intriguing relationships that demand further study and better data. Although cartographic investigations are often inconclusive, the ultimate explanation occasionally proves a bit different from the original interpretation, as happened in Woburn, Massachusetts, a suburban satellite twelve miles north of Boston.

Woburn's story reached a climax, if not a conclusion, in September 1986, when the W. R. Grace Company, a large chemical and food producer, settled a lawsuit for $8 million. Four years earlier a group of Woburn families had sued Grace, Beatrice Foods, and a local dry cleaner for the leukemia deaths of as many as twelve local children. According to the plaintiffs, negligent storage and disposal of toxic chemicals by the three companies had contaminated the community's drinking water and unleashed a poison that destroyed their children's immune systems. The dry-cleaning firm settled out of court, but Grace and Beatrice Foods considered the evidence weak and flawed, and chose a vigorous defense. To simplify this complex case for jurors, the judge divided the trial into two parts, the first concerning groundwater contamination and the second addressing the link between chemical pollution and leukemia. At the end of the first phase, after weeks of conflicting testimony by expert witnesses, much of it based on maps and statistics, the jury acquitted Beatrice Foods. But because jurors seemed willing to find the other defendant guilty on legally questionable grounds, the judge ordered a new trial for W. R. Grace. Like many defendants in "toxic tort" cases, Grace maintained its innocence but settled the lawsuit to avoid further expense. Although scientists following the case continued to debate issues of cause and effect, the jurors seemed not at all reluctant to link the deaths to a specific chemical dumped at a single site.

The map in Figure 8.9 describes key facts in the case. The boundaries show how the Bureau of the Census divided Woburn into six more or less socioeconomically homogeneous census tracts, and the twelve dots represent the homes of children diagnosed with leukemia between 1969 and 1979. At the top of the map a shaded symbol labeled "Industriplex" represents an old manufacturing area redeveloped as an industrial park in the early 1970s. Since 1853, when a local entrepreneur set up a plant to supply tanneries and paper mills with acids and other caustics, the site has served numerous chemical firms, including manufacturers of insecticide, glue, grease, and explosives. In July 1979 EPA investigators discovered enormous quantities of toxic waste buried on the property, which quickly became a Superfund site targeted for environmental cleanup.

Farther south on the map, between the industrial park and Walker Pond, two small open circles labeled G and H represent wells developed in 1964 and 1967, respectively, to supplement Woburn's municipal water system. Officials originally intended to pump these wells only in the summer, during peak usage, but when consumption increased in the late 1960s because of residential growth, water authorities extended the pumping period to six months. In May 1979

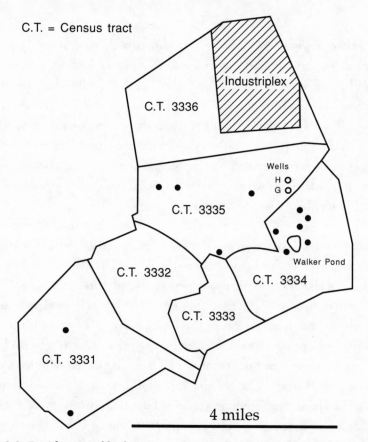

C.T. = Census tract

C.T. 3336

Industriplex

Wells
H ○
G ○

C.T. 3335

Walker Pond

C.T. 3332

C.T. 3334

C.T. 3333

C.T. 3331

4 miles

Fig. 8.9. Residences of leukemia patients at the time of diagnosis, Woburn, Massachusetts, 1969–1979. (From John J. Cutter et al., "Childhood Leukemia in Woburn, Massachusetts," *Public Health Reports* 101 [1986], pp. 201–205.)

the Massachusetts Department of Environmental Quality Engineering shut down both wells after tests revealed dangerously high concentrations of chloroform, trichloroethylene, and other carcinogens.

Anne Anderson, an eastern Woburn resident whose son Jimmy had leukemia, was the first person to notice the cluster of cases near Walker Pond. Leukemia patients require frequent treatment, and Anderson took Jimmy to a clinic in Boston several times a week. A form of cancer affecting the bone marrow and causing excessive production of white blood cells, leukemia is a rare disease with fewer than two thousand new diagnoses each year in the entire country. After recognizing in Woburn shops the faces of several other mothers who took their children to the same clinic, Anderson began to suspect the number of local cases might be abnormally high, perhaps because an unknown virus had infected the local water supply. Troubled by this theory, she sought the counsel of her pastor, the Rev. Bruce Young. Hoping to replace fear and speculation with reason and facts, Young suggested she count cases. To help her collect data, he asked other local parents of children with leukemia to meet at the church on October 4, 1979. Young and Anderson advertised the meeting in the local newspaper, bought a street map at a stationery store, and with the help of her son's physician, prepared a list of relevant questions to ask attendees. At the meeting, Anderson saw some familiar faces and several new ones, and when she plotted their addresses on the map, she discovered that six of Woburn's twelve cases of leukemia were quite close to her home. The map surprised not only Young but Jimmy's physician, who called the Centers for Disease Control (CDC), in Atlanta, to request an investigation.

Released in 1981 by Massachusetts public health officials who collaborated with the CDC, the Woburn study validated the case cluster on Anderson's map but failed to identify a probable cause. Statistics confirmed the rarity of the pattern of dots: not only was the incidence of leukemia in one Woburn census tract 7.5 times greater than expected, but the probability of finding six of the city's twelve cases in this single tract was less than 1 in 100. To explore typical causal links with genetic, dietary, environmental, medical, and other factors, the investigators administered a detailed questionnaire to the families of leukemia victims and to similar families in two carefully chosen control groups. The results revealed no significant differences between case families and control families. Moreover, no leukemia sufferer had been in contact with a hazardous waste site, and the contaminants found in wells G and H (Figure 8.9) were not known causes of leukemia. Because the researchers were unable to either reconstruct a history of past contamination or identify the sources of present contaminants, they failed to establish a link between the site and the leukemia cases. Their conclusion was hardly reassuring: "It is not possible to rule out exposure to this water as a factor, particularly in the eastern Woburn residents."

Anderson and other local parents were not impressed. They had read the chemical analyses from the wells, and they recalled the foul air around the industrial park as well as their sour summer tap-water—a consistent complaint of eastern Woburn residents mentioned in the CDC report. When the Industriplex site appeared on the Superfund list, many believed more firmly than ever that poisoned groundwater killed or was killing their sons and daughters— including Jimmy Anderson, who died on January 18, 1981, five days before health officials released the Woburn study. Eager to obtain justice, Anne Anderson, Bruce Young, and other volunteers formed Friends for a Clean Environment (FACE) and enlisted the assistance of Marvin Zelen and Steven Lagakos, biostatisticians at the Harvard School of Public Health, who designed a telephone survey to detect other health-related problems caused by contaminated drinking water. It was more than a fishing expedition. FACE hoped to corroborate the link between leukemia and Woburn's contaminated groundwater by finding similar geographic patterns for other health problems. Three hundred volunteers worked for nearly a year interviewing over 3,000 families about medical histories, personal habits such as smoking and drinking, and socioeconomic characteristics.

Because the contaminated wells supplied more water to some neighborhoods than to others, Zelen and Lagakos developed a computer model that estimated the relative importance, block by block, of various sources supplying the Woburn water system. (Among other patterns, the model demonstrated that wells G and H never supplied residents of western Woburn, which had two-thirds of the city's population but comparatively few leukemia cases.) Using data collected by FACE volunteers, the Harvard researchers then computed incidence rates for a variety of diseases and birth defects. Estimated exposure (access) to water from the contaminated wells, they found, was positively correlated not only with leukemia but also with perinatal deaths; births with eye or ear anomalies or chromosomal or oral cleft anomalies; and childhood disorders of the kidneys, urinary tract, and lungs. Released at a public meeting in February 1984, the Harvard study shocked local residents, encouraged Anderson and others in their lawsuit against W. R. Grace and Beatrice Foods, and incurred the wrath of scientists who criticized the researchers' methods and their decision to go public without peer review.

Without the perseverance of state hydrologists, Anderson might have sued the wrong firms, or not sued at all. Although state health officials doubted the exercise would prove anything, hydrologists in the Massachusetts Department of Environmental Quality Engineering (DEQE) decided to locate the source of contaminants found in wells G and H by reconstructing groundwater conditions during the 1970s. In 1981, using water samples, discharge rates, and depth measurements recorded for various wells in the area as well as geophysical data collected by seismic sounding, they developed a computer model of the aquifer. Released in March 1982, their report surprised residents and local officials by implicating not the Industriplex Superfund site but a small manufacturing plant just southeast of the industrial park. Computer models indicated that subsurface water flowed toward the contaminated wells from the Cryovac plant, where a division of W. R. Grace producing food-packaging equipment used trichloroethylene (TCE) as a degreasing agent to clean machinery. Investigators detected TCE in a dump directly behind the plant, and test wells drilled in the area found high concentrations of the chemical in groundwater "downstream" from the site, between the Cryovac property and the contaminated wells. Other sources of TCE contamination included a local tannery that Beatrice Foods had purchased in 1978. In May 1982 the DEQE groundwater study became a principal argument in the lawsuit Anderson and ten other families filed against W. R. Grace, Beatrice Foods, and other unnamed defendants.

If the trial had progressed to its second phase, dealing with the causes of leukemia, expert witnesses would have testified for the defense that TCE did not induce leukemia in humans. Such assertions were irrelevant, however, to the plaintiffs' argument that TCE exposure damaged the children's DNA, weakened their immune systems, and made them more susceptible to the disease. Whether Grace could have won the case is questionable. While the scientific community debated causal linkages and called for further research, the plaintiffs' theory, however novel, seemed far more compelling to a less patient legal system unwilling to await more evidence.

In addition to Grace's $8 million settlement, the Woburn families won an important moral victory. Publicity surrounding Anne Anderson's discovery influenced the establishment or enhancement of environmental monitoring programs, including the Massachusetts Cancer Registry, which the governor had vetoed in 1979 but approved the following year. As in most other states, Massachusetts health officials now have a database that supports a systematic search for case clusters and other geographic patterns that might otherwise go undetected.

A cancer registry is but one component of the systematic monitoring of disease and pollution essential for environmental protection and public health. As in clinical medicine, early detection can prevent the spread of infection and increase the chance of recovery. Public health agencies must be alert to case clusters, especially those suggesting heretofore unknown effects of chemicals and other toxins dispersed by water, air, or food. Mortality data alone are insufficient; epidemiological monitoring systems such as cancer registries must encourage clinics and physicians to report tumors and other suspicious symptoms.

Because effective monitoring addresses causes as well as effects, environmental health officials need a well-designed network of sampling stations to measure the levels of both known and suspected pollutants. An important consideration in developing a monitoring network is an inventory of waste dumps and other sites where toxic materials are produced, stored, used, interred, or incinerated.

Equally essential are reliable computer models of groundwater, air circulation, and municipal systems for distributing drinking water and collecting sewage. A geographic information system (GIS) model of local wells and aquifers, for example, would allow environmental analysts to forecast the short- and long-term range of a newly discovered contamination plume or assess the vulnerability of the community's groundwater reservoir to a new industry, proposed landfill, or substantially increased pumping at a particular well.

Maps of local monitoring networks are less controversial than they should be. Local libraries seldom acquire these maps, most citizens never see them, and some communities don't have them. Yet the increased threat of technological hazards to human health indicates clearly that local and regional defenses against disease, pollution, and other byproducts of economic development are as fundamental as a strong national defense against foreign powers. What you can't see because you don't map it can hurt you.

Epilogue

These tales of cartocontroversy reflect our widespread ignorance about how maps work. Most viewers willingly accept maps as factual and objective, but as we have seen, challenging only those maps that blatantly threaten your wallet or beliefs is naïve and risky. Subtle cartographic propaganda is common in many contexts and frequently misleads citizens who are unaware that maps are highly selective and necessarily biased. Moreover, a stylish or intriguing map becomes a flag to rally around and parade behind. When a cartographic banner passes by and its supporters shout, bystanders take notice. Therein lies much of the power of maps—enchanting displays divert attention from their authors' motives.

Recent developments in cartography and information networks leave little doubt that map viewing in the twenty-first century will be highly interactive, with few distinctions between mapmakers and map users. But the consequences of this greater freedom are troubling. Multimedia is a powerful technology for integrating information, and as the controversies examined here illustrate, integration of maps and words is the key to cartographic persuasion. The effect is symbiotic: the map attracts the eye and the accompanying words seduce the mind. As multimedia makes maps interactive, informed skepticism will become even more essential.

Responsibility for cautious map viewing falls on the user—the alternatives of licensing users and restricting access to data are too onerous for a free society. Viewers of interactive maps must rely on caveats embedded in cartographic data, perhaps with a pop-up warning label linked to each feature. The curious user who clicks on a symbol could then review a statement of uncertainty. But how many viewers are so compulsively diligent? An abundance of information can be confusing, after all, and the hidden "fine print" of hypertext is easily ignored. To encourage caution without obstructing use, display software must provide summary warnings (often maps themselves) to make these cautionary notes more obvious and meaningful—and map users must be sufficiently savvy to demand that software developers offer these safeguards.

Responsible map authors also need to understand that lying with maps is not something one always does intentionally. The consequences of ignorance and inattention are readily apparent in maps drawn with computers by uninformed newspaper artists and careless software users. Because unintended blunders can be as misleading as deliberate distortions, it is not enough to be skeptical only when the map author is a likely propagandist. Other dangers are cartographic databases that contain meaningless or misleading information and mapping software that encourages inappropriate displays. Naïve viewers of interactive maps who think themselves in control too easily overlook the biases and blunders of commercial firms and government agencies whose products they use.

Even with radical changes in technology, however, cartography is still encumbered by traditional practices. A noteworthy example is the topographic map sheet, which represents the infrastructure and terrain for a rectangular area called a quadrangle. Used as a starting point for both traditional and electronic maps, these base maps reflect the view of government cartographers and engineers that roads, boundaries, coastlines, and similar "cartographic objects" are the only true basic geographic features. The result is a detailed map showing drive-in movies, sports arenas, and abandoned railways while ignoring Superfund sites, flood-prone areas, active fault lines, and other human information less relevant to military defense and economic development. The classic excuse (or explanation, to be kind) for this distortion is that adding more features would clutter the paper with symbols and make the map confusing and less effective. Nonetheless, electronic base maps for interactive mapping are similar in content to their conventional counterparts even though concurrent display of all features in the database is not essential. And the files themselves, one per quadrangle, reflect an arbitrary division of territory appropriate only for sheets of paper fed through a printing press.

Informed, unbiased mapmaking is not the only concern for the future of cartography, however. In fact, laws, regulations, and policies affecting what maps are made and who can use them is a problem bigger than map content or file structure. Automated mapping can integrate topographic databases with environmental information, for instance, but this promising possibility cannot be realized unless government mapmakers collect and disseminate relevant data. Because geographic information is costly to develop, policy makers must weigh carefully the benefits of adding new features, improving resolution, and bringing old coverages up to date. But as several of my examples point out, governments do not always do the right thing, even when the law says they must. Future cartographic controversies will most certainly involve costly litigation over missing data and misinformation about data as well as errors in data.

More ominous are severe restrictions on public access to protect the government against lawsuits arising from incompatible data sets—for example, hazards data not suitable for use with topographic data. Ironically, mapmaking agencies long proud of the precision of their products need to cultivate caution (if not mistrust) among users.

Notes

1. THE PETERS PROJECTION CONTROVERSY

GENERAL SOURCES

Critiques of Peters and his work by geographers and cartographers include Iain Bain, "Will Arno Peters Take Over the World?" *Geographical Magazine* 56 (July 1984), pp. 342–343; John Loxton, "The Peters Phenomenon," *Cartographic Journal* 22 (December 1985), pp. 106–108; D. H. Maling, "A Minor Modification to the Cylindrical Equal-Area Projection," *Geographical Journal* 140 (1974), pp. 509–510; D. H. Maling, "Personal Projections," *Geographical Magazine* 46 (August 1974), pp. 599–600; D. H. Maling, "Peters' Wunderwerk," *Kartographische Nachrichten* 24 (1974), pp.

153–156; Phil Porter and Phil Voxland, "Distortion in Maps: The Peters Projection and Other Devilments," *Focus* 36 (Summer 1986), pp. 22–30; Arthur H. Robinson, "Arno Peters and His New Cartography," *The American Cartographer* 12 (1985), pp. 103–111; John P. Snyder, "Social Consciousness and World Maps," *The Christian Century* 105 (24 February 1988), pp. 190–192; "The So-called Peters Projection," *Cartographic Journal* 22 (December 1985), pp. 108–110; Andi Spicer, "Controversial Cartography," *Geographical Magazine* 61 (September 1989), pp. 42–44; Peter Vujakovic, "Arno Peters' Cult of the 'New Cartography': From Concept to World Atlas," *SUC Bulletin* (Society of University Cartographers) 22, no. 2 (1989), pp. 1–6; Peter Vujakovic, "Mapping for World Development," *Geography* 74 (1989), pp. 97–105; and Peter Vujakovic, "The Extent of Adoption of the Peters Projection by 'Third World' Organizations in the UK," *SUC Bulletin* (Society of University Cartographers) 21, no. 1 (1987), pp. 11–15. The numerous media reports examining the Peters map include "A Fairer View of the World," *Science 84* 5 (May 1984), p. 7; "A New View of the World," *Christianity Today* 28 (17 February 1984), pp. 39–40; Dana Sachs, "Cutting the Old World Down to Size," *Mother Jones* 11 (December 1986), pp. 16–17; and "The Real World," *Harpers* 268 (April 1984), p. 15.

Books and articles published by Peters or his supporters include Ward L. Kaiser, *A New View of the World: A Handbook to the World Map: Peters Projection* (New York: Friendship Press, 1987); Ward L. Kaiser, "Prejudice and the Peters Projection," *The Christian Century* 105 (13 April 1988), p. 369; Arno Peters, *Die Neue Kartographie (The New Cartography)* (New York: Friendship Press, 1983); Arno Peters, *Peters Atlas of the World* (London: Longman, 1989; New York: Harper and Row, 1990). Geographic essays on the cultural and political significance of map projections include J. B. Harley, "Can There Be a Cartographic Ethics?" *Cartographic Perspectives* 10 (Summer 1991), pp. 9–16; J. B. Harley, "Deconstructing the Map," *Cartographica* 26 (Summer 1989), pp. 1–20; John Pickles, "Texts, Hermaneutics and Propaganda Maps," in Trevor J. Barnes and James S. Duncan, eds., *Writing Worlds: Discourse, Text and Metaphor in the Representation of Landscape* (London and New York: Routledge, 1992), pp. 193–230; and Thomas F. Saarinen, "Centering of Mental Maps of the World," *National Geographic Research* 4 (1988), pp. 112–127.

A comprehensive general reference on map projections is John P. Snyder, *Map Projections —A Working Manual,* U.S. Geological Survey professional paper no. 1395 (Washington, D.C.: U.S. Government Printing Office, 1987).

NOTES

10   "land masses . . .": Robinson, "Arno Peters and His New Cartography," p. 104.

10   "West German historian . . .": *Christianity Today,* 17 February 1984, p. 40.

11   "All that I would ask . . .": Ibid., p. 123.

12   "we may obtain . . .": Ibid., p. 119.

12   "For general purposes . . .": Ibid., p. 121.

13   On Lambert's solution, see Snyder, *Map Projections — A Working Manual,* pp. 76–85.

14   On the sinusoidal projection, see ibid., p. 243.

15   On Goode's projection, see J. Paul Goode, "The Homolosine Projection: a New Device for Portraying the Earth's Entire Surface," *Annals of the Association of American Geographers* 15 (1925), pp. 119–125.

18   "In our epoch . . .": *A New View of the World,* p. 40.

20   "the Mercator world map . . .": Erwin Raisz, *General Cartography* (New York: McGraw-Hill, 1938), p. 87.

20   "one of the most . . .": Arthur H. Robinson, *Elements of Cartography* (New York: John Wiley and Sons, 1953), pp. 42–43.

21   "use of the Mercator . . .": S. W. Boggs, "Cartohypnosis," *Scientific Monthly* 64 (June 1947), pp. 469–476.

21   "has been a favorite . . .": David Greenhood, *Mapping* (Chicago: University of Chicago Press, 1964), p. 128.

21   "people's ideas . . .": G. J. Morrison, *Maps: Their Uses and Construction* (London, 1902), p. 38.

21   "as a world map . . .": Bain, "Will Arno Peters Take Over the World?" p. 342.

22   In his deconstruction . . . : Robinson, "Arno Peters and His New Cartography," p. 109.

22   In 1991 the U.S. Postal Service . . . : Jeff Stage, "Stamps Mark Olympics, WW II," *Syracuse Herald-American,* 1 September 1991, Stars section, p. 29.

22   In his book . . . Peters described ten criteria . . . : Peters, *The New Cartography,* pp. 105–118. Jointly published in Austria and the United States, the book juxtaposes German and English text on each page.

23   "direction to the north . . .": Ibid., p. 108.

23   "have fidelity of area . . .": Ibid., p. 66.

23   "by extreme distortion . . .": Ibid., p. 118.

23   In a checklist comparing eight projections . . . : Ibid., p. 114.

24 "to detach a section . . .": Ibid., p. 116.

24 "all points . . .": Ibid., p. 109.

24 "longitudinal distortion . . .": Ibid., p. 113.

24 "permits the construction . . .": Ibid., p. 113.

24 A sister organization in Britain . . . : Translation, attributed to a Mrs. Jutta Müller, was published in the British Cartographic Society's journal. See "The So-called Peters Projection."

25 *Who's Who in the World, 1980–81*, 5th ed. (Chicago: Marquis Who's Who, 1980), p. 749.

25 His academic training . . . : Loxton, "The Peters Phenomenon."

25 Several sources note a bit ominously . . . : Porter and Voxland, "Distortion in Maps: The Peters Projection and Other Devilments," and Robinson, "Arno Peters and His New Cartography."

26 His critics also mention . . . : Porter and Voxland, "Distortion in Maps," p. 28.

26 "written by a Communist . . .": "German Sells U.S. a Red Textbook," *New York Times*, 15 November 1952, p. 4.

26 Three days later, though, . . . : "Germans Defend Book," *New York Times*, 18 November, 1952, p. 5.

26 In *The New Cartography*, Peters traced . . . : Peters, *The New Cartography*, p. 146.

27 "with geographical leanings.": Ibid., p. 146.

27 I never had an opportunity . . . : Vujakovic, "Arno Peters' Cult."

27 "He claims not . . .": Ibid., p. 2.

28 The first major offensive . . . : Maling, "Peters' Wunderwerk."

28 "a remarkable example . . .": Maling, "A Minor Modification."

28 Peter Vujakovic sent questionnaires . . . : Vujakovic, "The Extent of Adoption"; and Vujakovic, "Mapping for World Development."

29 Vujakovic traced Christian Aid's decision . . . : Vujakovic, "Mapping for World Development," p. 99.

30 Its first report, . . . : Independent Commission on International Development Issues, "North–South: A Program for Survival" (Cambridge, Mass.: MIT Press, 1980).

30 In 1977 one of these missives . . . : "Peters Projection — to Each Country its Due on the World Map," *ACSM Bulletin* (American Congress on Surveying and Mapping) 59 (November 1977), pp. 13–15. Endnote states "Article from The Bulletin, Press and Information Office of the Government of the Federal Republic of Germany, Bonn, Vol. 25, No. 17, Aug. 17, 1977, pp. 126–127."

31 In its very next issue, . . . : "American Cartographers Vehemently Denounce German Historian's Projection," *ACSM Bulletin* 60 (February 1978), p. 27.

33 "Since Mercator . . .": Peters, *The New Cartography*, p. 149.

33 His opinion of world maps . . . : Ibid., p. 33.

33 "This error led . . .": Ibid., p. 75.

33 "too large and too regular . . .": Maling, "Peter's Wunderwerk," p. 156.

33 The map displaced some parallels . . . : Maling, "Personal Projections."

33 In an article in *Kartographische Nachrichten*, . . . : Maling, "Peter's Wunderwerk," p. 156.

34 In a similar vein, . . . : Porter and Voxland, "Distortion in Maps," p. 26.

34 Porter and Voxland demonstrated that . . . : Ibid., p. 28.

34 In early 1984, for instance, . . . : Patrick O'Driscoll, "The Global Perspective: Third World Gains," *USA Today*, 23 January 1984, p. 9A.

35 "makes the predominantly . . .": "Cutting the Old World Down to Size," p. 16.

35 "portrays a colonialist . . .": "A Fairer View of the World."

35 . . . and in response to cartographer John Snyder's . . . : Kaiser, "Prejudice and the Peters Projection."

35 "at least Peters's . . .": Snyder, "Social Consciousness and World Maps."

35 Snyder, who had focused . . . : Kaiser, "A New View of the World."

35 He pointed out how Kaiser . . . : Snyder, "Social Consciousness and World Maps," p. 191.

35 "Support for . . .": Kaiser, "A New View of the World," pp. 9–10.

36 "You may wish . . .": Ibid., pp. 30–31.

36 "These people . . .": Ibid., p. 24.

36 "We know that . . .": Ibid., p. 25.

36 "Dr. Peters has chosen . . .": Ibid., p. 25.

36 "The Mercator map . . .": Ibid., pp. 11–12.

37 "takes the world's . . .": Ibid., p. 18.

37 "The transformation . . .": Ibid., p. 38.

38 But they also found fault . . . : Russell King and Peter Vujakovic, "Peters Atlas: A New Era of Cartography or Publisher's Con-trick?" *Geography* 74 (1989), pp. 245–251; and Vujakovic, "Arno Peters' Cult," p. 5.

39 "intriguing and challenging . . .": Chauncy D. Harris, "Peters Atlas of the World" (review), *American Reference Books Annual* 22 (1991), p. 171.

39 "such occasional lapses . . .": Kenneth F. Kister, "Peters Atlas of the World" (review), *Library Journal* 116 (January 1991), pp. 94–95.

39 In the critical camp . . . : James Rettig, "Peters Atlas of the World" (review), *Wilson Library Bulletin* 65 (March 1991), pp. 131–132.

39 "appears to view . . .": "The World Turned Upside Down," *Economist* 310 (25 March 1989), p. 97.

39 Indeed, in the early 1980s, . . . : Michael Kidron and Ronald Segal, *The State of the World Atlas* (New York: Simon and Schuster, 1981).

40 In 1989, for instance, seven . . . : "Geographers and Cartographers Urge End to Popular Use of Rectangular Maps," *The American Cartographer* 16 (1989), pp. 222–223. The organizations are the American Cartographic Association (ACSM's cartographic affiliate), the American Geographical Society, the Association of American Geographers, the Canadian Cartographic Association, the National Council for Geographic Education, the National Geographic Society, and the Geography and Map Division of the Special Libraries Association.

41 The three ACSM projection booklets are American Cartographic Association, Committee on Map Projections, "Which Map Is Best?: Projections for World Maps" (Falls Church, Va.: American Congress on Surveying and Mapping, 1986); American Cartographic Association, Committee on Map Projections, "Choosing a World Map: Attributes, Distortions, Classes, Aspects" (Falls Church, Va.: American Congress on Surveying and Mapping, 1988); and American Cartographic Association, Committee on Map Projections, "Matching the Map Projection to the Need" (Bethesda, Md.: American Congress on Surveying and Mapping, 1988).

41 "a colonial mentality . . .": Saarinen, "Centering of Mental Maps," p. 124.

41 "modify[ing] old Eurocentric . . .": Ibid., p. 125.

41 "cartography shares . . .": Pickles, "Texts, Hermaneutics and Propaganda Maps."

42 Cartographers, Harley observed, . . . : Harley, "Deconstructing the Map."

42 "The real issue . . .": Harley, "Can There Be a Cartographic Ethics?" p. 11.

43 In a 1974 lecture, Peters proudly . . . : Loxton, "The Peters Phenomenon," p. 106.

43 During a 1989 visit to Britain . . . : Spicer, "Controversial Cartography."

2. PLACE-NAMES, ETHNIC INSULTS, AND IDEOLOGICAL RENAMING

GENERAL SOURCES

My principal geographic source for specific place-names was Frank R. Abate, ed., *Omni Gazetteer of the United States of America* (Detroit, Mich.: Omnigraphics, 1991). For information on the history, objectives, and operations of the U.S. Board of Geographic Names, I relied on Donald J. Orth, *Principles, Policies, and Procedures: Domestic Geographic Names* (Reston, Va.: U.S. Geological Survey, 1987, revised 1989); Donald J. Orth, "The Mountain Was Wronged: The Story of the Naming of Mt. Rainier and Other Domestic Names Activities of the U.S. Board on Geographic Names," *Names* 32 (1984), pp. 428–434; Donald J. Orth, "The U.S. Board on Geographic Names: An Overview," *Names* 38 (1990), pp. 165–172; Roger L. Payne, "Geographic Names Information System: Philosophy and Function," *World Cartography* 18 (1986), pp. 49–52; and Roger L. Payne, *Geographic Names Information System*, Data Users Guide no. 6 (Reston, Va.: U.S. Geological Survey, 1987).

Useful sources of information on the significance of place-names and practices in naming places included W. F. H. Nicolaisen, "Placenames and Politics," *Names* 38 (1990), pp. 193–207; Alan Rayburn, "Unfortunate Connotations Acquired by Some Canadian Toponyms," *Names* 36 (1988), pp. 187–192; Robert A. Rundstrom, "Mapping, Postmodernism, Indigenous People and the Changing Direction of North American Cartography," *Cartographica* 28 (Summer 1991), pp. 1–12; George R. Stewart, *Names on the Globe* (New York: Oxford University Press, 1975); George R. Stewart, *Names on the Land* (New York: Random House, 1945); and Yi-Fu Tuan, "Language and the Making of Place: A Narrative-Descriptive Approach," *Annals of the Association of American Geographers* 81 (1991), pp. 684–696.

NOTES

47 The upper-left panel reproduces . . . : A 7.5-minute topographic quadrangle map covers a rectangular area 7.5 minutes (or one-eighth of a degree) of latitude from north to south and 7.5 minutes of longitude from east to west. Most U.S. Geological Survey 7.5-minute topographic maps have a scale of 1:24,000, for which one inch on the map represents 2,000 feet on the ground. The lower-left panel is an excerpt enlarged from a 1:250,000-scale topographic map covering a degree of latitude and two degrees of longitude.

48 The Geological Survey began to remove . . . : Steven Lagerfeld, "Name that Dune," *Atlantic Monthly* 266 (September 1990), pp. 20–22.

48 As its preface acknowledges . . . : Payne, "Geographic Names Information System: Philosophy and Function"; and Payne, *Geographic Names Information System*.

50 Town folklore connects . . . : Marjorie Perez, Wayne County Historian, telephone interview, 30 July 1992; and Shirley Eygnor, Town Clerk, Town of Huron, telephone interview, 31 July 1992.

50 Perhaps to deal with this . . . : Shirley Eygnor, Town Clerk for the Town of Huron, searched town records unsuccessfully for a local ordinance changing the name to Graves Point. Shirley Eygnor, telephone interview, 31 July 1992.

52 "Names which may be considered . . .": Morris M. Thompson, *Maps for America: Cartographic Products of the U.S. Geological Survey and Others* (Reston, Va.: U.S. Geological Survey, 1979), p. 90.

52 In 1963, at the initiation of . . . : Roger Payne, telephone interview, 29 July 1992. The minutes of the board, available in the National Archives, describe this policy and its adoption.

53 Chinamans Spring in Yellowstone National Park . . . : Lagerfeld, "Name that Dune," p. 20.

53 For example, features may not . . . : Orth, *Principles, Policies, and Procedures: Domestic Geographic Names*, p. 26.

53 "either some direct association . . .": Ibid., p. 13.

53 Local opposition can block . . . : Orth, "The Mountain Was Wronged," pp. 429–430.

53 Even risqué and kinky names . . . : Eric Schmitt, "Ultimate Arbiter of Hill and Vale," *New York Times*, 27 November 1986, pp. C1, C6.

53 In 1990, for instance, Dorothy Igau . . . : Lisa Belkin, "On Geographic Names and Cleaning Them Up," *New York Times*, 14 February 1990, p. A16.

54 The story began when the DEC . . . : Patrick E. Brown, first assistant counsel to the governor, letter to author, 9 September 1992.

54 Investigating further with local historians, . . . : Fredric Dicker, "Racist Name for a Swamp Banned by Gov after Uproar," *New York Post*, 5 October 1988.

54 Despite these roots, this cartographic insult . . . : "Offensive Names Are Ordered Removed," *New York Times*, 1 November 1988, p. B2.

54 Executive Order no. 113, Office of the Governor, State of New York, signed 27 October 1988.

55 The historian originally in charge of the effort . . . : "Summary: Derogatory Names Initiative," prepared in early July 1992 by Philip Lord, on the State Historian's staff. Lord was appointed acting secretary of the Committee on Geographic Names in 1991. Edward Winslow, the previous secretary, died in 1990.

55 "the Department feels . . .": Marc S. Gerstman, Deputy Commissioner and General Counsel of the Department of Environmental Conservation, letter to Edmund Winslow, Education Department, 25 August 1989.

55 Even though the *Wap* . . . : Staughton Lynd, *Anti-Federalism in Dutchess County, New York* (Chicago: Loyola University Press, 1962), pp. 46–47, 86–88; and Henry Noble MacCracken, *Old Dutchess Forever: The Story of an American County* (New York: Hastings House, 1956), p. 3.

55 By November 1989, the State Committee . . . : "Policies and Procedures for the Committee on Geographic Names," draft submitted by Murray Heller on 3 November 1989, circulated to the committee with a memo from Edmund Winslow dated 14 November 1989, and (according to minutes dated 24 January 1990) approved at the 23 January 1990 meeting of the committee, footnote 2.

56 "Dingle Hole and Dingle Hole WMA do not . . .": "Policies and Procedures for the Committee on Geographic Names," footnote 3.

56 "pieces of excrement . . .": Jonathon Green, *Newspeak: A Dictionary of Jargon* (London: Routledge and Kegan Paul, 1989), pp. 69–70.

57 Meron Benvenisti, *Conflicts and Contradictions* (New York: Villard Books, 1986), pp. 191–202.

59 "We have done more than . . .": Ibid., p. 196.

59 "maps cease being geographical . . .": Ibid., p. 199.

60 As did the Indians, the Pakistanis . . . : Jonathan Addleton, "The Fate of British Place Names in Pakistan," *Asian Affairs* 18 (February 1987), pp. 37–44.

61 The Turkish Cypriots proclaimed . . . : Sarah Ladbury and Russell King, "Settlement Renaming in Turkish Cyprus," *Geography* 73 (1988), pp. 363–367.

61 Furthermore, the newly independent Soviet Socialist Republics . . . : "The Names They Are A-Changin'," *Time* 139 (27 January 1992), p. 36.

62 In the late 1950s, for instance, . . . : "What's in a Name?" *Time* 131 (18 January 1988), p. 43.

62 Political renaming accelerated . . . : Felicity Barringer, "It's a Mad, Mad World for Map Makers," *New York Times*, 27 January 1991, p. A14.

62 . . . and in 1991, after years . . . : Edmund Stevens, " 'Peter' Dislikes Place Name Changes," *The Times* (London), 29 January 1974, p. 5.

62 "every time a city . . .": "The Game of the Name," *Economist* 319 (8 June 1991), p. 50.

62 Differences run deeper . . . : Tom Post et al., "Making War on Muslims," *Newsweek* 120 (10 August 1992), p. 40.

63 As a move to appease . . . : N. L. Nicholson and L. M. Sebert, *The Maps of Canada: A Guide to Official Canadian Maps, Charts, Atlases, and Gazetteers* (Hamden, Conn.: Archon Books, 1981), p. 223.

63 Elsewhere in Canada . . . : Mordecai Richler, "Reporter at Large: Inside/Outside," *New Yorker* 67 (23 September 1991), pp. 40–92, esp. pp. 40 and 46.

63 As part of this crusade, . . . : Nicolaisen, "Placenames and Politics," pp. 197–198.

64 Québec Commission de Toponymie, *Repertoire Toponymique du Québec*, Gouvernement du Québec, 1987, p. xii.

65 Language differences can lead . . . : Nicolaisen "Placenames and Politics," p. 202.

65 Stewart, *Names on the Land*, pp. 108–110; also see Stewart, *Names on the Globe*, pp. 54–62.

66 In Michigan, for example, . . . : Ibid., p. 156.

66 In Maine, for instance, . . . : J. N. Hook, *All Those Wonderful Names: A Potpourri of People, Places, and Things* (New York: John Wiley and Sons, 1991), p. 260.

66 As further testimony to the power . . . : Tuan, "Language and the Making of Place: A Narrative-Descriptive Approach," p. 688.

66 For unnamed features off the reservation, . . . : Walter Sullivan, "Honoring, and Unearthing, Indian Place Names," *New York Times*, 9 September 1990, p. 34; and Roger Payne, telephone interview, 21 August 1992. The board had approved the native-names policy in early 1990, but as of summer 1992 had not officially submitted the policy statement to the Secretary of the Interior. Several other policy statements were undergoing revision, and the board intended to submit them as "a package."

66 Especially suitable are distinctive names . . . : Orth, *Principles, Policies, and Procedures*, p. 25.

67 Alaskans have requested . . . : Orth, "The Mountain Was Wronged," pp. 430–431.

67 Not really, say associates . . . : Lagerfeld, "Name That Dune," p. 22; and Schmitt, "Ultimate Arbiter of Hill and Vale," p. C1.

67 Because the board refuses . . . : Orth, *Principles, Policies, and Procedures*, p. 11.

67 In the 1980s Canada began . . . : Helen Kerfoot and Alan Rayburn, "The Roots and Development of the Canadian Permanent Committee on Geographical Names," *Names* 38 (1990), pp. 183–191, esp. p. 189.

68 "by the end of the century...": Alan Rayburn, "Native Names for Native Places," *Canadian Geographic* 107 (April/May 1987), pp. 88–89.

68 To create their own cartographic text,...: Rundstrom, "Mapping, Postmodernism, Indigenous People and the Changing Direction of North American Cartography," p. 10.

69 The media's focus on shrill charges...: Joseph Berger, "Arguing About America," *New York Times*, 21 June 1991, pp. A1, B4; and Sam Howe Verhovek, "Plan to Emphasize Minority Cultures Ignites a Debate," *New York Times*, 21 June 1991, pp. A1, B4.

3. THE VINLAND MAP, COLUMBUS, AND ITALIAN-AMERICAN PRIDE

GENERAL SOURCES

The Yale University Press introduced the Vinland map in R. A. Skelton, Thomas E. Marston, and George D. Painter, eds., *The Vinland Map and the Tartar Relation* (New Haven, Conn.: Yale University Press, 1965). A scathing attack on the Yale discovery by a distinguished Italian-American jurist is found in Michael A. Musmanno, *Columbus Was First* (New York: Fountainhead Publishers, 1966), and an important scholarly evaluation is Wilcomb E. Washburn, ed., *Proceedings of the Vinland Map Conference* (Chicago: University of Chicago Press, 1971).

Other critical evaluations of the Yale University Press book or the Vinland map itself include Lee Ash, "The Vinland Map and the Tartar Relation" (review), *Library Journal* 90 (1 December 1965), pp. 5275–5276; G. R. Crone, "How Authentic Is the 'Vinland Map'?" *Encounter* 26 (February 1966), pp. 75–78; G. R. Crone, "The Vinland Map Cartographically Considered," *Geographical Journal* 32 (1966), pp. 75–80; Gwyn Jones, "International Thriller," *New York Review of Books* 5 (25 November 1965), pp. 28–30; Francis Maddison, "A Skeptical View of the *Tartar Relation*," *Geographical Journal* 140 (June 1974), pp. 187–191; Samuel Eliot Morison, "It All Boils Down to What We Knew Before," *New York Times Book Review* (section VII), 7 November 1965, pp. 7, 92; Herbert Musurillo, "The Vinland Map and the Tartar Relation" (review), *America* 114 (15 January 1966), pp. 87–88; and Helen Wallis, "The Strange Case of the Vinland Map: Introduction," *Geographical Journal* 140 (June 1974), pp. 183–187.

Physical testing of the Vinland map is discussed in T. A. Cahill et al., "The Vinland Map, Revisited: New Compositional Evidence on Its Inks and Parchment," *Analytical Chemistry* 59 (15 March 1987), pp. 829–833; "Ink Study Suggests Vinland Map Fraud," *Chemical and Engineering News* 52 (11 February 1974), p. 21; Walter C. McCrone and Lucy B. McCrone, "The Vinland Map Ink," *Geographical Journal* 140 (June 1974), pp. 212–214; Walter C. McCrone, "The Vinland Map," *Analytical Chemistry* 60 (15 May 1988), pp. 1009–1018; Vera Rich, "Titanium Content Brings Vinland Map Back into Play," *Nature* 328 (July 1987), p. 195; and Kenneth M. Towe, "The Vinland Map: Still a Forgery," *Accounts of Chemical Research* 23 (March 1990), pp. 84–87.

NOTES

73 Figure 3.1: For a description of the Vinland map, see R. A. Skelton, "The Vinland Map," in *The Vinland Map and the Tartar Relation*, pp. 107–239, esp. pp. 109–110.

74 "Island of Vinland, discovered . . .": Ibid., p. 139.

74 "By God's will, after a long . . .": Ibid., p. 140.

75 Tales and sagas of Norse voyages . . . : Ibid., pp. 209–212, 223–226.

75 "by far the most important . . .": "Map of History," *Time* 86 (15 October 1965), pp. 120–123.

76 Although *Time* suggested this interpretation . . . : Skelton, "The Vinland Map," esp. pp. 208–221; and George D. Painter, "The Tartar Relation and the Vinland Map: An Interpretation," in *The Vinland Map and the Tartar Relation*, pp. 241–262, esp. pp. 251–255.

76 "a generalized and degenerate . . .": Painter, "The Tartar Relation and the Vinland Map," pp. 254–255.

76 When it first came to scholars' attention, . . . : Francis Maddison, "A Skeptical View of the *Tartar Relation*"; and George Painter, "The Tartar Relation and the Vinland Map: Edited, with Introduction, Translation, and Commentary," in *The Vinland Map and the Tartar Relation*, pp. 19–106.

77 As scholars interpreted the evidence, . . . : Thomas E. Marston, "The Manuscript: History and Description," in *The Vinland Map and the Tartar Relation*, pp. 1–16, esp. pp. 4–6; and Painter, "The Tartar Relation," pp. 21–27.

78 That summer, Ferrajoli . . . : Wallis, "The Strange Case of the Vinland Map: Introduction."

78 "Mr. Witten came to my office . . .": Marston, "The Manuscript: History and Description," p. 3.

79 "This was not a wholly . . .": Ibid., p. 4.

79 Yale could not afford . . . : "Map of History," *Time*, p. 123.

79 Britain's *Sunday Times* reported widespread rumors that the donor was millionaire Paul Mellon, but neither Yale nor Mellon confirmed these reports. See "Is the Vinland Map a Forgery?" *Sunday Times*, 6 March 1966, p. 13.

80 "bringing together the separate parts . . .": Alexander O. Vietor, Foreword, in *The Vinland Map and the Tartar Relation*, p. vii.

80 From a meticulous examination . . . : Skelton, "The Vinland Map," pp. 111–123, 141, 228–229.

80 The Vinland map, which amalgamated . . . : Ibid., pp. 228–230.

80 Painter concluded that . . . : Painter, "The Tartar Relation and the Vinland Map: An Interpretation," p. 243.

81 "Where else could such a . . .": Marston, "The Manuscript: History and Description," p. 16.

81 Yale promoted the book . . . : "Yale Press Publishes 'Map Discovery of the Century,' " *Publishers Weekly* 188 (11 October 1965), pp. 31–34.

81 In addition to selecting October 11, . . . : Oliver Jensen, "Vinland the Good Emerges from the Mist," *American Heritage* 16 (October 1965), pp. 4–8.

81 Another double-page spread . . . : R. A. Skelton and George Painter, "Was There a Lasting Colony?" *American Heritage* 16 (October 1965), pp. 100–103; and R. A. Skelton, "Did Columbus or Cabot See the Map?" *American Heritage* 16 (October 1965), pp. 103–106.

81 "Nobody has more experience . . .": "Yale Press Publishes 'Map Discovery of the Century,' " p. 34.

81 Yale, which had approached SAS . . . : Walter Carlson, "Advertising: S.A.S. Discovers New World of Promotion," *New York Times*, 14 October 1965, p. 68.

82 the three-column headline . . . : "1440 Map Depicts the New World," *New York Times*, 11 October 1965, pp. 1, 48.

82 Staff writer Howard Simons, . . . : "America of Vikings Shown on Pre-Columbian Map," *Washington Post*, 11 October 1965, pp. A1, A3.

82 Although press kits cautioned . . . : Carlson, "Advertising: S.A.S. Discovers New World of Promotion," p. 68.

82 "An authenticated parchment map . . .": "New World Mapped, 1440," *Science News Letter* 88 (23 October 1965), p. 263.

82 "This time the proof . . . ": Daniel Cohen, "Columbus vs. Ericson—What Science Says," *Science Digest* 59 (January 1966), pp. 10–15; quote on p. 10.

82 Morison's evaluation . . . : "It All Boils Down to What We Knew Before," p. 92.

83 *Library Journal*'s consultant . . . : Ash, "The Vinland Map and the Tartar Relation" (review).

83 Gwyn Jones, an authority . . . : Jones, "International Thriller," p. 28.

83 David Quinn linked . . . : "A Map of the Norse World," *New Statesman* 70 (15 October 1965), p. 568.

83 Phoebe Adams, for example, . . . : "Potpourri" (column), *Atlantic* 216 (December 1965), pp. 161–162.

83 Yet scholarly arcana did not . . . : Harry Gilroy, "Book Club Picks the Vinland Map," *New York Times*, 25 November 1965, p. 33.

84 "to put Yale University against . . .": Philip Benjamin, "Columbus's Crew Won't Switch," *New York Times*, 12 October 1965, pp. 1, 58.

84 On October 13, the *New York Times* reported . . . : Natalie Jaffe, "Did Ericson Do It? Battle Rages On," *New York Times*, 13 October 1965, p. 27.

84 In Boston, graffiti . . . : Jaffe, "Did Ericson Do It?" p. 27.

84 At a naturalization ceremony in Washington, D.C., . . . : "Fuss Over Who Got Here First Livens Naturalization Ceremony," *Washington Post*, 13 October 1965, p. A2.

84 In Chicago, Victor Arrigo, . . . : " 'Myth' Gibe in U.S.," *Times* (London), 13 October 1965, p. 10.

84 "a Communist plot": "A Windblown Leif," *Time* 86 (22 October 1965), p. 25B.

84 In Cambridge, Massachusetts, . . . : "Pride Followeth a Landfall," *American Heritage* 17 (February 1966), pp. 100–101.

84 And in Philadelphia, Michael Musmanno, . . . : "A Windblown Leif," p. 25B.

85 In Spain, where Columbus Day . . . : "Fuss Over Who Got Here First," p. A2.

85 In Italy, Genoa's mayor . . . : Ibid., p. A2.

85 Willing to accept the map . . . : Richard J. H. Johnston, "Fanfani, Supporting Columbus, Likens Explorer to Isaac Newton," *New York Times*, 11 December 1965, p. 35.

85 "Leif knew where . . .": "New Light on the Discovery of the New World," *U.S. News and World Report* 59 (25 October 1965), pp. 8–9.

85 In San Francisco, another local . . . : Jerry Buck, "Who Was Here First? Don't Count Out the Irish!" *Syracuse Herald-Journal*, 12 October 1965, p. 1.

85 And in Oslo, Norwegian explorer . . . : "Explorer Says a Lack of Arms Kept Leif Ericson from Staying," *New York Times*, 14 December 1965, p. 45.

85 Other nationalities joining the melee . . . : Buck, "Who Was Here First?" p. 1; Richard Deacon, *Madoc and the Discovery of America: Some New Light on an Old Controversy* (New York: George Braziller, 1966); "A Windblown Leif," p. 25B; and "New Light on the Discovery of the New World," p. 9.

86 But Italian Americans remained . . . : "Good-by, Vikings; Hello, Etruscans," *New York Times*, 1 November 1965, p. 3.

86 He explained that the Yale Press . . . : Harry Gilroy, "Morgan Library Girds for Critics," *New York Times*, 16 February 1966, p. 40.

86 Mild humor at the Morgan . . . : "Pride Followeth a Landfall," pp. 100–101.

87 State legislatures had long . . . : Jane M. Hatch, *The American Book of Days* (New York: H. W. Wilson, 1978), p. 920.

87 Thus, in the year after . . . : "Johnson Proclaims Oct. 9 as Leif Erikson Day," *New York Times*, 17 September 1966, p. 19.

87 To appreciate their zeal, . . . : "The Continuing Struggle for a National King Holiday," *Ebony* 43 (January 1988), pp. 27–32.

87 These efforts paid off . . . : "House Votes to Shift Holidays to Mondays," *New York Times*, 9 May 1968, p. 40; and "Senate Panel Backs Plans for Holidays on Mondays," *New York Times*, 21 June 1968, p. 22.

88 A lawyer who had worked . . . : Gary Null and Carl Stone, *The Italian-Americans* (Harrisburg, Pa.: Stackpole Books, 1976), pp. 154–156.

88 Feisty and bright, Musmanno . . . : Albin Krebs, "Long and Colorful Career [Michael A. Musmanno]," *New York Times*, 13 October 1968, p. 84.

89 "He [Witten] expressed . . .": Musmanno, *Columbus Was First*, p. 20.

89 "I thought he might . . .": Ibid., p. 22.

90 "then the moral obligation . . .": Ibid., p. 23.

90 "argument does little to advance . . .": *Proceedings of the Vinland Map Conference*, p. 165.

90 I was unaware of Musmanno's book until Frances Robotti, director of Fountainhead Publishers, sent me a copy in early 1991, shortly after the publication of *How to Lie with Maps*. Ms. Robotti, who had edited several

of Musmanno's books, observed quite accurately that "the Vinland map surely qualifies to be considered one of the big 'lies' in cartography." Frances Robotti, letter to author, 20 May 1991.

90 "It is unnecessary to search . . .": Crone, "How Authentic is the 'Vinland Map'?" p. 78.

91 "how such a document . . .": Ibid., p. 75.

91 Crone's interpretation obviously . . . : See, for example, correspondence from Skelton and Crone in the March, June, and September 1966 issues of the *Geographical Journal*.

91 "We have, in my view, . . .": R. A. Skelton, "Correspondence: The Vinland Map," *Geographical Journal* 132 (1966), pp. 336–339.

91 "any attempt to trace . . .": Skelton, "The Vinland Map," pp. 177–178.

91 A feisty octogenarian, Taylor . . . : "Is the Vinland Map a Forgery?" *Sunday Times*, 6 March 1966, p. 13; and "The Map Flap," *Newsweek* 67 (11 April 1966), p. 68.

92 "a forger often betrays . . .": "Is the Vinland Map a Forgery?" p. 13.

92 publish her manuscript . . . : E. G. R. Taylor, "The Vinland Map," *Journal of the Institute of Navigation* 27 (April 1974), pp. 195–205.

92 Wilcomb Washburn, chairman . . . : Edited by Wilcomb E. Washburn, the proceedings were published in 1971, five years after the conference, by the University of Chicago Press.

92 In the first talk at the conference . . . : Laurence Witten, "Vinland's Saga Recalled," in *Proceedings of the Vinland Map Conference*, pp. 3–10.

93 "I trust Mr. Witten . . .": *Proceedings*, p. 11.

93 "one bit of information . . .": *Proceedings*, p. 13.

93 "It would be quite possible . . .": Armando Cortesão, "Is the Vinland Map Genuine?" in *Proceedings of the Vinland Map Conference*, pp. 15–18.

94 "This much-needed knowledge . . .": John Parker, "Authenticity and Provenance," in *Proceedings of the Vinland Map Conference*, pp. 19–22.

95 "a most clever counterfeit": Robert S. Lopez, "The Case Is Not Settled," in *Proceedings of the Vinland Map Conference*, pp. 31–34.

95 "counterfeit until proven genuine . . .": Ibid., p. 31.

95 "Were it possible to question . . .": Ibid., p. 34.

95 "to proceed a little further . . .": *Proceedings*, p. 39.

96 On January 26, 1974, the *New York Times* reported . . . : Michael Knight, "Yale Says Prized 'Vinland Map' of North America Is a Forgery," *New York Times*, 26 January 1974, pp. 1, 38.

97 "a product of medieval minds . . .": Albin Krebs, "Notes on People" (column), *New York Times*, 7 February 1974, p. 46.

97   Yale medievalists Robert Lopez and . . . : " 'Vinland Map' Forgery Revives Debate on Columbus," *New York Times,* 3 February 1974, p. 42.

97   But no church documents . . . : In 1966, Lopez, Reichardt, and a few other scholars at the Vinland Map Conference also mentioned Jelic as a possible forger. See *Proceedings of the Vinland Map Conference,* pp. xii, 113–114; Konstantin Reichardt, "Linguistic Observations of the Captions of the Vinland Map," in *Proceedings of the Vinland Map Conference,* pp. 109–110; and Stephan Kuttner, "Observations on the Relationship between Church History and the Vinland Map," in *Proceedings of the Vinland Map Conference,* pp. 111–113.

97   "the Americans apparently would be . . .": " 'Vinland Map' Forgery Revives Debate on Columbus."

97   "take the moral and legal steps . . .": Ibid.

97   *American Heritage,* which had eagerly . . . : Oliver Jensen, "Letter from the Editor" (column), *American Heritage* 25 (June 1974), p. 2.

98   Yale was equally skilled . . . : Walter Cahn and James Marrow, *Medieval and Renaissance Manuscripts at Yale: A Selection* (published as vol. 52, April 1978 issue, of the *Yale University Library Gazette*), pp. 277–278.

98   "needless to say, the library's . . .": Robert Babcock, letter to author, 30 June 1992.

99   "Vinland's Saga Recalled" . . . : Laurence C. Witten II, "Vinland's Saga Recalled," *Yale University Library Gazette* 64 (October 1989), pp. 11–37. Witten used the same title for his statement in the *Proceedings of the Vinland Map Conference.*

99   "I could only say . . .": Ibid., pp. 34–35.

99   "I saw his library, saw the Vinland Map . . .": Witten, "Vinland's Saga Recalled" (Vinland Map Conference), p. 5.

99   He described the owner's "private library . . .": Ibid., p. 14.

99   "The owner of the library . . .": Ibid., p. 8.

99   "He did not wish to pay tax . . .": Ibid., p. 12.

100 "A distinguished man of about fifty . . .": Witten, "Vinland's Saga Recalled," pp. 31–32.

100 "I emphasized that in the absence . . .": Ibid., pp. 32–33.

101 "It is no pleasure to relate . . .": Ibid., p. 34.

101 Witten's article contains one further . . . : Ibid., p. 36.

102 The Crocker Historical and Archaeological . . . : Cahill et al., "The Vinland Map, Revisited"; and Rich, "Titanium Content Brings Vinland Map Back into Play."

102 "the prior interpretation that the Map . . .": Cahill, "The Vinland Map, Revisited," p. 829.
102 "Confidence in analytical data . . .": McCrone, "The Vinland Map," p. 1017.
102 "further supports . . .": Towe, "The Vinland Map: Still a Forgery."
102 "Unless plausible answers to these . . .": Ibid., pp. 86–87.

4. BOUNDARY LITIGATION AND THE MAP AS EVIDENCE

GENERAL SOURCES

Gordon E. Ainsworth, "Metes-and-Bounds Location Problems in New England," *Surveying and Mapping* 11 (1951), pp. 124–130; David J. Bederman, "Georgia v. South Carolina, 110 S.Ct. 2903," *American Journal of International Law* 84 (1990), pp. 909–914; Breckinridge Thomas, "The Surveyor in Court as Witness," *Surveying and Mapping* 23 (1963), pp. 565–573; Louis De Vorsey, Jr., "Historical Maps Before the United States Supreme Court," *Map Collector* 19 (June 1982), pp. 24–31; Carl L. Everstine, "The Potomac River and Maryland's Boundaries," *Maryland Historical Magazine* 80 (1985), pp. 355–370; Martin Ira Glassner, *Neptune's Domain: A Political Geography of the Sea* (Boston: Unwin Hyman, 1990); Edward C. Papenfuse and Joseph M. Coale, *The Hammond-Harwood House Atlas of Historical Maps of Maryland, 1608–1908* (Baltimore: Johns Hopkins University Press, 1982); J. R. V. Prescott, *The Maritime Political Boundaries of the World* (London: Methuen, 1985); Walter G. Robillard and Lane J. Bouman, *Clark's Treatise on the Law of Surveying and Boundaries*, 5th ed. (Charlottesville, Va.: Michie, 1987); Keith Suter, *Antarctica: Private Property or Public Heritage?* (Australia and London: Pluto Press, 1991); Morris M. Thompson, *Maps for America: Cartographic Products of the U.S. Geological Survey and Others* (Reston, Va.: U.S. Geological Survey, 1979); United Nations Office for Ocean Affairs and the Law of the Sea, *The Law of the Sea — Baselines: An Examination of the Relevant Provisions of the United Nations Convention on the Law of the Sea* (New York: United Nations, 1989); and William C. Wattles, "The Surveyor in Court," *Surveying and Mapping* 13 (1953), pp. 286–288.

NOTES

107 Although careful examination . . . : J. B. Harley, *Maps and the Columbian Encounter* (Milwaukee, Wis.: Golda Meir Library, University of Wisconsin, 1990), p. 5.

107 Land ownership in the profane European sense . . . : Jerry Mander, *In the Absence of the Sacred: The Failure of Technology and the Survival of the Indian Nations* (San Francisco: Sierra Club Books, 1991), pp. 215–217; see also W. Stitt Robinson, "Conflicting Views on Landholding: Lord Baltimore and the Experiences of Colonial Maryland with Native Americans," *Maryland Historical Magazine* 83 (1988), pp. 85–97.

107 Within two months . . . : J. H. Parry, *The Discovery of South America* (London: Paul Elek, 1979), pp. 68–69, 90–92.

107 Harley, *Maps and the Columbian Encounter,* pp. 63–65.

109 For a history of the Mason-Dixon line, see Edward Bennett Matthews, "History of the Boundary Dispute between the Baltimores and the Penns Resulting in the Original Mason and Dixon Line," *Maryland Geological Survey* 7 (1908), pp. 103–203; Edward B. Matthews, "The Maps and Map-Makers of Maryland," *Maryland Geological Survey* 2 (1898), pp. 337–488, esp. pp. 432–442; Papenfuse and Coale, *Hammond-Harwood House Atlas,* pp. 5–9, 37–41; and Harold E. Vokes, *Geography and Geology of Maryland* (Baltimore: Maryland Geological Survey, 1957), pp. 2–5.

109 In 1632, King Charles I . . . : Vokes, *Geography and Geology of Maryland,* p. 2.

109 Controversy arose . . . : Papenfuse and Coale, *Hammond-Harwood House Atlas,* pp. 37–38.

109 In the early seventeenth century . . . : Walter B. Scaife, "The Boundary Dispute between Maryland and Pennsylvania," *Pennsylvania Magazine of History and Biography* 9 (1885), pp. 241–271, esp. pp. 246–247.

109 Unfortunately for Maryland, Lord Baltimore, . . . : Matthews, "The Maps and Map-Makers of Maryland," pp. 360–363.

110 "be divided into equal parts . . .": Ibid., p. 433.

111 Unfortunately for Maryland, the "Cape Henlopen" . . . : Papenfuse and Coale, *Hammond-Harwood House Atlas,* pp. 38–39.

111 Maryland historians . . . : Charles C. Tansill, *The Pennsylvania-Maryland Boundary Controversy* (Washington, D.C.: National Capital Press, n.d. [ca. 1915]), pp. 52–53.

111 Eighty years of intermittent yet costly litigation . . . : Matthews, "History of the Boundary Dispute," pp. 128–130.

111 Years later, . . . : Maryland Writers' Project, *Maryland: A Guide to the Old Line State* (New York: Oxford University Press, 1940), p. 33.

111 England's chief judge . . . : Matthews, "History of the Boundary Dispute," pp. 173–179.

111 Dissatisfied with the work of local surveyors . . . : Ibid., pp. 183–191.

112 Uncertain about Maryland's western boundary, . . . : Matthews, "History of the Boundary Dispute," pp. 188–189.

112 A resurvey . . . : W. C. Hodgkins, "Report of the Engineer in Charge of the Resurvey of the Boundary between Maryland and Pennsylvania, part of the Mason and Dixon Line," *Maryland Geological Survey* 7 (1908), pp. 37–102, esp. p. 81.

112 After resolving boundary disputes with their neighbors . . . : D. W. Meinig, *The Shaping of America: A Geographical Perspective on 500 Years of History*, vol. 1: *Atlantic America, 1492–1800* (New Haven, Conn.: Yale University Press, 1986), pp. 256–261.

114 To adjust for measurement error . . . : U.S. Bureau of Land Management, *Manual of Instructions for the Survey of the Public Lands of the United States* (Washington, D.C.: U.S. Government Printing Office, 1947), pp. 164–171.

114 In the typical Midwestern county, . . . : Vernon Carstensen, "Patterns on the American Land," *Surveying and Mapping* 36 (1976), pp. 303–309; and Norman J. W. Thrower, *Original Survey and Land Subdivision: A Comparative Study of the Form and Effect of Contrasting Cadastral Surveys* (Chicago: Rand McNally, 1966).

114 Similar regularities occur . . . : William Wyckoff, *The Developer's Frontier: The Making of the Western New York Landscape* (New Haven, Conn.: Yale University Press, 1988).

114 Common in the Atlantic states . . . : Meinig, *The Shaping of America*, p. 240.

115 metes and bounds . . . : John G. McEntyre, *Land Survey Systems* (New York: John Wiley and Sons, 1978), pp. 314–364.

116 In the early 1960s, for instance, . . . : B. K. Meade, "Delaware-Maryland North-South Boundary Resurvey of 1961–62," *Surveying and Mapping* 24 (1964), pp. 33–36.

117 Relating the old survey . . . : Ibid., pp. 572–576.

117 Early in the twentieth century . . . : George D. Whitmore, *City Surveying* (Scranton, Pa.: International Textbook Co., 1942).

118 Although the utility must . . . : Daniel F. Hinkel, *Practical Real Estate Law* (St. Paul, Minn.: West Publishing Co., 1991), p. 63.

118 Many owners fail to realize . . . : Ibid., pp. 6–7, 91–92.

118 To warn a prospective buyer . . . : Susan M. Reid, "Minimum Standards for Property Boundary Surveys (1986) and Beyond the Minimum," in *Land Surveys: A Guide for Lawyers* (Chicago: American Bar Association, 1989), pp. 67–78.

118 A consequence of the "City Beautiful" movement . . . : Robert C. Klove, "City Planning in Chicago: a Review," *Geographical Review* 38 (1948), pp. 127–131; and Rutherford H. Platt, *Land Use Control: Geography, Law, and Public Policy* (Englewood Cliffs, N.J.: Prentice Hall, 1991), pp. 166–169.

119 Zoning prohibits . . . : Frederick H. Bair, Jr., *The Zoning Board Manual* (Chicago: American Planning Association [Planners Press], 1984).

119 Zoning is an intensely political process . . . : Bair, *The Zoning Board Manual*; Hinkel, *Practical Real Estate Law*, pp. 61–62; and Platt, *Land Use Control*, pp. 177–193.

119 Lawyers specializing . . . : Richard F. Babcock and Charles L. Siemon, *The Zoning Game Revisited* (Boston: Oelgeschlager, Gunn and Hain, 1985), pp. 255–265.

120 Moreover, a map that is not . . . : Robillard and Bouman, *Clark's Treatise*, pp. 590–594.

120 The lawyer who wants . . . : Ibid., pp. 382–383; and Wattles, "The Surveyor in Court."

120 Although the "ancient document" exception . . . : Robillard and Bouman, pp. 381–382.

121 "Do not attempt . . .": Thomas, "The Surveyor in Court as Witness," p. 569.

121 "I know of one judge . . .": Ibid., p. 571.

121 Most discrepancies reflect . . . : Robert B. Mathias, "The Surveyor, a Professional Witness in Court," *Surveying and Mapping* 29 (1969), pp. 65–69.

121 "If a technical description . . .": Ainsworth, "Metes-and-Bounds Location Problems in New England," p. 126.

121 In one egregious case, . . . : Ibid., pp. 126–127.

121 In another case, . . . : Ibid., p. 128.

122 Legal textbooks on boundary disputes . . . : Robillard and Bouman, *Clark's Treatise*, pp. 34–36.

122 "The use of a qualified . . .": Ibid., p. 36.

122 J. Richard White and Harlan Onsrud, "Survey Checklist: A Practical Guide to Review and Analysis of Land Surveys," in *Land Surveys: A Guide for Lawyers* (Chicago: American Bar Association, 1989), pp. 79–116.

123 If so, the owner may claim . . . : Robillard and Bouman, *Clark's Treatise*, pp. 656–673.

123 Although survey maps might . . . : Ibid., pp. 759–760.

124 As a Kentucky court ruled . . . : *Miller* v. *Hepburn*, 71 Kentucky (8 Bush) 326 (1871), cited in Robillard and Bouman, *Clark's Treatise*, pp. 705–706.

124 Nonetheless, the plaintiff, who owned Lot A, . . . : *Peuker* v. *Canter*, 628 Kan. 363, 63 P. 617 (1901), cited in Robillard and Bouman, *Clark's Treatise*, pp. 704–705.

126 "Once a tract of land . . .": Clark, *Fundamentals of Law for Surveyors*, p. 35.

126 Draining an area of . . . : *World Almanac and Book of Facts*, 1990, p. 535.

127 To avoid widespread confusion . . . : Andrew L. Harbin, *Land Surveyor Reference Manual* (San Carlos, Calif.: Professional Publications, 1985), pp. 25–15, 25–16; and Milton O. Schmidt and William Horace Rayner, *Fundamentals of Surveying*, 2nd ed. (New York: D. Van Nostrand, 1978), p. 313.

128 The U.S. Geological Survey, which . . . : Thompson, *Maps for America*, p. 78.

128 Despite the historic roots . . . : Ronald Smothers, "An Island in a Stream Is in Court," *New York Times*, 14 December 1992, p. A12.

130 In a unanimous decision, . . . : *Mississippi et al.* v. *Louisiana et al.*, *United States Law Week* 61 (1992), pp. 4025–4027.

130 The dispute arose in 1971, after . . . : Louis De Vorsey, Jr., *The Georgia–South Carolina Boundary: A Problem in Historical Geography* (Athens, Ga.: University of Georgia Press, 1982), pp. 1–20.

131 Several prominent historical geographers . . . : De Vorsey, "Historical Maps Before the United States Supreme Court."

132 Because of the need to mark . . . : Ibid., pp. 26–27.

132 In another anecdote, . . . : Ibid., p. 27.

133 His 1986 decision addressed . . . : David K. Secrest, "Judge Draws the Line on Ga.–S.C. Border Dispute," *Atlanta Constitution*, 31 March 1989, p. A17.

133 With unanimity marred . . . : 110 S. Ct. 2903 (1990), pp. 2905–2906. Also see Bob Dart, "S.C. Gets Islands Ga. Claimed," *Atlanta Constitution*, 26 June 1990, p. D2; and Stanley Rodgers, "Interstate Boundary

Dispute Between South Carolina and Georgia Resolved," *South Carolina Law Review* 43 (1991), pp. 150–156.

133 "by prescription and acquiescence . . .": 110 S. Ct., p. 2905.

133 In another ruling, . . . : *Georgia* v. *South Carolina,* 257 U.S. 516 (1922).

133 Characteristically wary of contradicting . . . : 110 S. Ct., p. 2915.

134 "would alter the boundary . . .": Ibid., p. 2916.

135 "Given this somewhat uncommon . . .": 110 S. Ct., p. 2917.

136 In approving the special master's solution, . . . : Ibid., p. 2922.

136 The majority of the justices . . . : Ibid., p. 2920.

137 Justice Stevens, in a dissent . . . : Ibid., p. 2923.

137 In searching for a legally sound . . . : Ibid., pp. 2921–2922.

137 On December 27, 1988, in fact, President Reagan . . . : Proclamation No. 5928, 27 December 1988, 54 Federal Register 777 (1989), cited in Bederman, "*Georgia* v. *South Carolina,* 110 S.Ct. 2903," p. 912.

137 Although countries have contested . . . : Lee Kimball, "Implications in the Law of the Sea as They Concern Marine Boundary Demarcation," *Surveying and Mapping* 37 (1977), pp. 39–44; and A. L. Shalowitz, "Boundary Problems Associated with the Continental Shelf," *Surveying and Mapping* 15 (1955), pp. 189–211.

138 Decades of work . . . : Clyde Sanger, *Ordering the Oceans: The Making of the Law of the Sea* (Toronto: University of Toronto Press, 1987), pp. 1–55; also see Glassner, *Neptune's Domain.*

138 Despite the reluctance of the United States . . . : UN Office for Ocean Affairs and the Law of the Sea, telephone inquiry, 10 August 1993. An additional thirty-four countries (for a total of 90) have proclaimed an Exclusive Economic Zone. Additions and other changes are noted in the *Law of the Sea Bulletin,* a UN publication. (As of April 1991, forty-seven countries had signed. See United Nations Office for Ocean Affairs and the Law of the Sea, *The Law of the Sea—National Claims to Maritime Jurisdiction: Excerpts of Legislation and Table of Claims* [New York: United Nations, 1992], p. iii.)

138 The convention recognizes . . . : Prescott, *Maritime Political Boundaries of the World,* pp. 36–46.

139 Prior to 1982 many nations . . . : Robert W. Smith, *Exclusive Economic Zone Claims: An Analysis and Preliminary Documents* (Dordrecht: Martinus Nijhoff, 1986), map facing p. 482; also see Ted L. McDorman, Kenneth P. Beauchamp, and Douglas M. Johnson, *Maritime Boundary Delimitation: An Annotated Bibliography* (Lexington, Mass.: Lexington Books, 1982); and Sanger, *Ordering the Oceans,* p. 198.

139 In 1969, for example, after Peru . . . : "Fish and Oil," *Time* 93 (30 May 1969), p. 40; and "Gunboat Diplomacy," *Newsweek* 73 (24 February 1969), p. 50.

139 Although declining to sign . . . : Patrick Huyghe, "Seafloor Mapping," *Oceans* 19 (November-December 1986), pp. 22–29; and Andrew C. Revkin, "Mapping a Wet Frontier," *Discover* 10 (September 1989), p. 30.

139 Almost all coastal states: As of August 1993, ninety countries had established an EEZ.

139 Because most nations never . . . : Harm J. de Blij, "The U.S. and its Maritime Neighbors," *Focus* 36 (Fall 1986), pp. 30–31.

139 To minimize conflict, the UN Convention . . . : United Nations Office for Ocean Affairs and the Law of the Sea, *The Law of the Sea — Baselines: An Examination of the Relevant Provisions of the United Nations Convention on the Law of the Sea*, pp. 1–5.

140 The convention also permitted . . . : Ibid., pp. 12–16.

140 Precise language . . . : Ibid., pp. 5–12.

140 and allowed straight closing lines . . . : Ibid., pp. 17–38.

140 Although closing lines and other . . . : Ibid., pp. 29–34.

140 "By judicious selection . . .": Ibid., p. 18.

140 and the UN requirement . . . : Prescott, *Maritime Boundaries*, pp. 72–74.

142 In 1981, after Canada and the United States . . . : John Cooper, "Delimitation of the Maritime Boundary in the Gulf of Maine Area," *Ocean Development and International Law* 16 (1986), pp. 59–90; Douglas M. Johnson, *The Theory and History of Ocean Boundary-Making* (Kingston and Montreal: McGill–Queen's University Press, 1988), pp. 178–191; and Anthony F. Shelley, "Law of the Sea: Delimitation of the Gulf of Maine," *Harvard International Law Journal* 26 (1985), pp. 646–654.

142 The decision handed down in 1984 . . . : Glassner, *Neptune's Domain*, pp. 40–41.

143 When Britain and France used . . . : United Nations Office for Ocean Affairs and the Law of the Sea, *The Law of the Sea — Maritime Boundary Agreements (1985–1991)* (New York: United Nations, 1992), p. 1.

144 Typical of maps toward the back . . . : F. M. Auburn, *Antarctic Law and Politics* (Bloomington, Ind.: Indiana University Press, 1982), pp. xvi–xvii, 5–31; Jack Child, *Antarctica and South American Geopolitics: Frozen Lebensraum* (New York: Praeger, 1988), pp. 15–17; and Prescott, *Maritime Boundaries*, pp. 140–144.

145 The 1959 Antarctic Treaty signed by . . . : Suter, *Antarctica: Private Property or Public Heritage?* pp. 9–27. For information on the mapping of

Antarctica, see J. Holmes Miller, "The Mapping of Antarctica," pp. 81–98, in Trevor Hatherton, ed., *Antarctica* (London: Methuen, 1965), and William A. Radlinski, "Mapping of Antarctica," *Surveying and Mapping* 21 (1961), pp. 325–331; also see "Frozen Waste," *Economist* 319 (20 April 1991), p. 46.

146 "Ironically, in terms of . . .": Suter, *Antarctica: Private Property or Public Heritage?* p. 16.

146 In 1987 a *National Geographic* . . . : For commentary accompanying the folded map, see Priit J. Vesilind, "Antarctica," *National Geographic* 171 (1987), pp. 556–560.

146 *Hammond Atlas of the World* (Maplewood, N.J.: Hammond Incorporated, 1993), p. 137.

5. CONTINENTAL DRIFT AND GEOPOLITICS: IDEAS AND EVIDENCE

GENERAL SOURCES

Wegener's most significant geophysical publication was Alfred Wegener, *Die Entstehung der Kontinente und Ozeane* (Braunschweig, Germany: Samml. Vieweg, 1915), which he revised and expanded three times. General sources for the concept of continental drift and the contributions of Alfred Wegener include Albert V. Carozzi, "New Historical Data on the Origin of the Theory of Continental Drift," *Geological Society of America Bulletin* 81 (1970), pp. 283–286; A. Hallam, "Alfred Wegener and the Hypothesis of Continental Drift," *Scientific American* 232 (February 1975), pp. 88–97; H. E. Le Grand, *Drifting Continents and Shifting Theories: The Modern Revolution in Geology and Scientific Change* (Cambridge: Cambridge University Press, 1988); John A. Stewart, *Drifting Continents and Colliding Paradigms* (Bloomington, Ind.: Indiana University Press, 1990); and H. Takeuchi, S. Uyeda, and H. Kanamori, *Debate About the Earth: Approach to Geophysics through Analysis of Continental Drift* (San Francisco: Freeman, Cooper and Co., 1967), pp. 20–21.

Mackinder's principal geopolitical works are H. J. Mackinder, "The Geographical Pivot of History," *Geographical Journal* 23 (1904), pp. 421–444; H. J. Mackinder, *Britain and the British Seas* (New York: D. Appleton and Company, 1902); and Halford J. Mackinder, *Democratic Ideals and Reality: A Study in the Politics of Reconstruction* (London: Constable, 1919).

General sources on Mackinder and the Heartland concept include Brian W. Blouet, *Halford Mackinder: A Biography* (College Station, Texas: Texas A&M University Press, 1987); E. W. Gilbert, "The Right Honourable Sir Halford J. Mackinder, P.C.," *Geographical Journal* 110 (1947), pp. 94–99; Gerry Kearns, "Halford John Mackinder, 1861–1947," *Geographers Biobibliographical Studies* 9 (1985), pp. 71–84; "Obituary: Sir H. Mackinder," *Times* (London), 8 March 1947, p. 6; J. F. Unstead, "H. J. Mackinder and the New Geography," *Geographical Journal* 113 (1949), pp. 46–57; and Hans W. Weigert, *Generals and Geographers* (New York: Oxford University Press, 1942).

NOTES

150 "the horizontal movement of . . .": G. H. Dury, *The Face of the Earth* (Baltimore, Md.: Penguin Books, 1959), p. 66.

150 "Continental drift is itself . . .": Dury, p. 129.

151 Born in Berlin in 1880, . . . : Hallam, "Alfred Wegener and the Hypothesis of Continental Drift," pp. 88–90; and Le Grand, *Drifting Continents and Shifting Theories*, pp. 37–39.

152 Wegener first presented . . . : Alfred Wegener, "Die Entstehung der Kontinente," *Geologische Rundschau* 3 (1912), pp. 276–292.

152 Several months later . . . : Alfred Wegener, "Die Entstehung der Kontinente," *Petermanns Geographische Mitteilungen* 58 (1912), pp. 185–195. Shorter comments followed in two additional notes in later issues of the same volume on pp. 253–256 and 305–309.

152 and in 1915 . . . : Alfred Wegener, *Die Entstehung der Kontinente und Ozeane* (Braunschweig, Germany: Samml. Vieweg, 1915).

153 "The first notion of the displacement . . .": Alfred Wegener, *The Origin of the Continents and Oceans*, 3d ed. trans. (London: Methuen, 1924), p. 8; also quoted in Takeuchi, Uyeda, and Kanamori, *Debate About the Earth*, pp. 20–21.

155 Wegener refined his ideas . . . : Tina Kasbeer, *Bibliography of Continental Drift and Plate Tectonics* (Special paper no. 142) (Boulder, Colo.: Geological Society of America, 1972), p. 4. Also see the short biography by his brother Kurt, in Alfred Wegener, *The Origin of the Continents and Oceans*, 4th rev. ed., trans. John Biram (New York: Dover Publications, 1966), pp. iii–v.

155 "The very configuration of the world . . .": Francis Bacon, *The Works of Francis Bacon*, vol. 8, *Novum Organum* (English-language translation), ed.

James Spedding, Robert Leslie Ellis, and Douglas Denon Heath (Boston: Taggard and Thompson, 1863), p. 235.

156 Equally oblique was . . . : François Placet, *La corruption du grand et petit monde où il est montré que toutes les Creatures qui composent l'Univers, sont corrompues par le peché d'Adam . . .* , 3d ed. (Paris: Vve G. Alliot et G. Alliot, 1668), quoted in Carozzi, "New Historical Data on the Origin of the Theory of Continental Drift," p. 284.

156 Comte de Buffon, Alexander von Humboldt, and . . . : Carozzi, "New Historical Data," and J. Tuzo Wilson, "Mobility in the Earth," in *Continents Adrift and Continents Aground: Readings from Scientific American* (San Francisco: W. H. Freeman and Co., 1976), pp. 3–7.

156 In the fourth edition of . . . : Wegener, 4th rev. ed. trans., pp. 2–4.

156 he noted that as early as 1857 . . . : W. L. Green, "The Causes of the Pyramidal Form of the Outline of the Southern Extremities of the Great Continents and Peninsulas of the Globe," *Edinburgh New Philosophical Journal,* new series 6 (1857), pp. 61–64.

156 Wegener also acknowledged . . . : Frank Bursley Taylor, "Bearing of the Tertiary Mountain Belt on the Origin of the Earth's Plan," *Geological Society of America Bulletin* 21 (1910), pp. 179–226. The text of Taylor's 1908 oral presentation was not published until 1910.

156 Wegener was apparently unaware . . . : Antonio Snider-Pellegrini, *La création et ses mystères dévoilés* (Paris: A. Franck, 1859).

156 In the framework of a Heaven and Earth . . . : Le Grand, *Drifting Continents and Shifting Theories,* p. 29.

157 Land east of this rift . . . : Carozzi, "New Historical Data," pp. 284–285.

157 For writers addressing continental drift . . . : See, for example, Stewart, *Drifting Continents and Colliding Paradigms,* p. 28; and Don Tarling and Maureen Tarling, *Continental Drift: A Study of the Earth's Moving Surface* (Garden City, N.Y.: Doubleday and Company, 1971), pp. 2–4.

158 "will be the remaining part of the island . . . ": Abraham Ortelius, *Thesaurus Geographicus* (Antwerp: Plantin, 1596), leaf Nnn verso.

158 Oddly, this passage . . . : James Romm, "A New Forerunner for Continental Drift," *Nature* 367, pp. 407–408. Ironically, Romm's finding preceded by only three years the discovery by Canadian geologist Alan Goodacre that British philosopher Thomas Dick had described continental drift in his 1838 book *Celestial Scenery; or The Wonders of the Planetary System Displayed, Illustrating the Perfections of the Deity and a Plurality of Worlds.* See Alan Goodacre, "Continental Drift," *Nature* 354, p. 261.

158 Although some geologists called . . . : See Frank Leverett, "Memorial to Frank Bursley Taylor," *Proceedings of the Geological Society of America*, 1938, pp. 191–200, esp. p. 193; and H. W. Menard, *The Ocean of Truth: A Personal History of Global Tectonics* (Princeton, N.J.: Princeton University Press, 1986), p. 22.

159 Taylor first outlined his . . . : Taylor, "Bearing of the Tertiary Mountain Belt on the Origin of the Earth's Plan."

159 Taylor did little to bolster . . . : Leverett, "Memorial."

159 following retirement, he published several papers . . . : See, for example, the abstract published as Frank Bursley Taylor, "Wegener's Theory of Continental Drifting: A Critique of Some of His Views," *Bulletin of the Geological Society of America* 43 (1932), p. 73.

160 Not finding a convenient mechanism . . . : Frank Bursley Taylor, "Sliding Continents and Tidal and Rotational Forces," in *Theory of Continental Drift: A Symposium on the Origin and Movement of Land Masses both Inter-Continental and Intra-Continental, as Proposed by Alfred Wegener* (Tulsa, Okla.: American Association of Petroleum Geologists, 1928), pp. 158–177; quote on pp. 174–175.

160 Because the moon's initial orbit . . . : Ibid., p. 176.

160 As drift supporter . . . : Alexander L. Du Toit, *Our Wandering Continents: An Hypothesis of Continental Drifting* (Edinburgh and London: Oliver and Boyd, 1937), p. 12.

161 an article several weeks later . . . : "Origin of the Moon; Interesting Theory Advanced by Dr. Howard B. Baker, a Detroit Student of the Planetary System," *Detroit Free Press*, 23 April 1911, Feature Magazine Section, p. 7.

161 Baker attributed this idea . . . : O. Fisher, "On the Physical Cause of the Ocean Basins," *Nature* 25 (1882), pp. 243–244.

162 In the introduction to a subsequent . . . : Howard B. Baker, "Origin of Continental Forms, II," *Michigan Academy of Science Annual Report* 14 (1912), pp. 116–141; quote on p. 116.

162 "As I have demonstrated . . .": Howard B. Baker, "Origin of Continental Forms, IV," *Michigan Academy of Science Annual Report* 15 (1913), pp. 26–32; quote on p. 28.

162 Baker reconstructed this fit . . . : Howard B. Baker, "Origin of Continental Forms, III," *Michigan Academy of Science Annual Report* 15 (1913), pp. 107–113; Figure 5.5 appears as Figure 4 on p. 113.

162 although, as drift advocate . . . : Du Toit, *Our Wandering Continents*, p. 15.

162 First presented in his unpublished 1911 paper, . . . : Born in 1872, Baker died in 1957. The Geological Survey's *Bibliography of Geology of North America* suggests that his last published paper appeared in 1954, when he was approximately eighty-two years old. See Howard B. Baker, "The Earth Participates in the Evolution of the Solar System," *Detroit Academy of Science Occasional Paper* no. 3 (1954). According to the National Union Catalog, Baker published his 305-page mimeographed book *The Atlantic Rift and Its Meaning* in 1932. A halftone photograph of his reconstruction globe appears as plate 1, facing p. 4 in Howard B. Baker, "Structural Features Crossing [the] Atlantic Ocean," *Pan-American Geologist* 66 (1936), pp. 1–11.

162 Although he had far less impact . . . : See, for examples, Du Toit, *Our Wandering Continents*, p. 14; and Tarling and Tarling, *Continental Drift*, p. 5.

162 The Geological Society of America's bibliography . . . : Kasbeer, *Bibliography of Continental Drift*.

163 when Vienna paleontologist . . . : Carl Diener, "Major Features of the Earth's Surface," *Pan-American Geologist* 37 (1922), pp. 177–197.

164 Cambridge University geophysicist . . . : Harold Jeffreys, *The Earth: Its Origin, History and Physical Constitution* (Cambridge: Cambridge University Press, 1924), p. 261.

164 "In following the edge . . .": Philip Lake, "Wegener's Hypothesis of Continental Drift," *Nature* 111 (1923), pp. 226–228; quote on p. 227.

164 As historian of science . . . : Le Grand, *Drifting Continents*, p. 56.

164 The keynote speaker, and author . . . : W. A. J. M. van Waterschoot van der Gracht, "The Problem of Continental Drift," in *Theory of Continental Drift: A Symposium on the Origin and Movement of Land Masses both Inter-Continental and Intra-Continental, as Proposed by Alfred Wegener* (Tulsa, Okla.: American Association of Petroleum Geologists, 1928), pp. 1–75.

165 "an accumulating amount of evidence" . . . : Ibid., p. 40.

165 "structural arguments work . . .": Ibid., p. 54.

165 "we should not look for . . .": Ibid., p. 55.

165 "These hypotheses, though revolutionary, . . .": Ibid., p. 75.

165 Retired Yale geologist . . . : Charles Schuchert, "The Hypothesis of Continental Drift," in *Theory of Continental Drift*, pp. 104–144.

165 "It is evident," Schuchert charged, . . . : Ibid., p. 111.

166 Stanford University's Bailey Willis . . . : Bailey Willis, "Continental Drift," in *Theory of Continental Drift*, pp. 76–82; quote on p. 81.

166 Equally hostile was . . . : Edward W. Berry, "Comments on the Wegener Hypothesis," in *Theory of Continental Drift*, pp. 194–196; quote on p. 195.

166 "I am convinced that . . .": Chester R. Longwell, "Some Physical Tests of the Displacement Hypothesis," in *Theory of Continental Drift*, pp. 145–157; quote on p. 152.

166 "not blinded either with the zeal . . .": Ibid., p. 157.

167 Eduard Suess, an important predecessor . . . : Eduard Suess, *Das Antlitz der Erde* (Vienna: F. Tempsky, 1888–1901).

167 Geologists and geophysicists used . . . : See, for example, Reginald Aldworth Daly, *Our Mobile Earth* (New York: Charles Scribner's Sons, 1926), pp. 90–102; Arthur Holmes, *Principles of Physical Geology*, rev. ed. (New York: Ronald Press, 1965), pp. 27–31; Takeuchi, Uyeda, and Kanamori, *Debate About the Earth*, pp. 31–41.

167 "if blocks of the sial crust . . .": van der Gracht, "The Problem of Continental Drift," pp. 33–34.

168 How continents drifted . . . : For discussion of the emergence of plate tectonics and its intellectual ties to continental drift, see R. P. Beckinsale, "Eustasy to Plate Tectonics: Unifying Ideas on the Evolution of the Major Features of the Earth's Surface," in *History of Geomorphology: From Hutton to Hack*, ed. K. J. Tinkler (Boston: Unwin Hyman, 1989), pp. 205–221; Le Grand, *Drifting Continents and Shifting Theories*; Menard, *Ocean of Truth*; Peter Molnar, "Continental Tectonics in the Aftermath of Plate Tectonics," *Nature* 335 (1988), pp. 131–137; and Stewart, *Drifting Continents and Colliding Paradigms*.

168 During the 1950s geophysicists . . . : Henry Frankel, "From Continental Drift to Plate Tectonics," *Nature* 335 (1988), pp. 27–130; F. J. Vine and D. H. Matthews, "Magnetic Anomalies over Oceanic Ridges," *Nature* 199 (1963), pp. 947–949.

168 In the 1960s sonar maps . . . : See, for example, Menard, *Ocean of Truth*, pp. 56–107.

168 Crustal plates diverging . . . : For a concise discussion of plate tectonics, see David G. Howell, *Tectonics of Suspect Terranes: Mountain Building and Continental Growth* (London: Chapman and Hall, 1989), pp. 1–51, and Lester C. King, *Wandering Continents and Spreading Sea Floors on an Expanding Earth* (Chichester: John Wiley and Sons, 1983), pp. 1–33.

168 Dury's *The Face of the Earth* . . . : The second through fifth editions appeared in 1966, 1971, 1976, and 1986, respectively.

169 Figure 5.6 appears in several Geological Survey publications and decorates the inside front cover of Robert E. Wallace, ed., *The San Andreas*

Fault System, California, U.S. Geological Survey professional paper no. 1515 (Washington, D.C.: U.S. Government Printing Office, 1990).

169 "fact [that had] remained merely an oddity, . . .": G. H. Dury, *The Face of the Earth,* 5th ed. (London: Allen and Unwin, 1986), p. 26.

170 "As we consider [these] broader currents . . .": Mackinder, "The Geographical Pivot of History," p. 434.

172 Denied an opportunity at the war's end . . . : Kearns, "Halford John Mackinder," p. 76.

172 Mackinder was appointed . . . : B. W. Blouet, "Sir Halford Mackinder as British High Commissioner to South Russia, 1919–1920," *Geographical Journal* 142 (1976), pp. 228–236.

172 Two years before his death . . . : "Annual General Meeting, 25 June 1945," *Geographical Journal* 105 (1945), pp. 230–232; quote on p. 230.

172 In his first book . . . : Mackinder, *Britain and the British Seas.*

172 A biographer called it . . . : Gilbert, "The Right Honourable Sir Halford J. Mackinder, P.C.," p. 98.

172 a contemporary reviewer called it . . . : Hugh Robert Mill, "Review: Britain and the British Seas," *Geographical Journal* 19 (1902), pp. 489–495; quote on p. 489.

172 In a chapter on strategic geography . . . : Mackinder, *Britain and the British Seas,* p. 349.

173 The strong oval frame . . . : S. W. Boggs, "Cartohypnosis," *Scientific Monthly* 6 (1947), pp. 469–476, esp. pp. 470–472.

175 In 1890, in a book titled . . . : Alfred Thayer Mahan, *The Influence of Sea Power Upon History, 1660–1783* (Boston: Little, Brown and Co., 1890).

175 "The oversetting of the balance of power . . .": Mackinder, "The Geographical Pivot of History," p. 436.

175 In 1919 Mackinder repeated . . . : Mackinder, *Democratic Ideals and Reality,* p. v.

176 "the rule of the world . . .": Ibid., p. 208.

177 "A victorious Roman general, . . .": Ibid., p. 150.

178 Well aware that implementation . . . : Ibid., p. 161.

180 He recognized the difficulty . . . : Blouet, "Sir Halford Mackinder as British High Commissioner."

180 But after describing Mackinder's argument, . . . : "The Geographical Pivot of History: Strategical Position of Russia," *Times* (London), 26 January 1904, p. 11.

180 Later that year across the Atlantic, . . . : "The Geographical Pivot of History," *National Geographic Magazine* 15 (1904), pp. 331–335; quote on p. 335.

180 In 1919, in a *Geographical Review* article . . . : Frederick J. Teggart, "Geography as an Aid to Statecraft: An Appreciation of Mackinder's 'Democratic Ideals and Reality,' " *Geographical Review* 8 (1919), pp. 227–242; quotes on pp. 227, 241, and 242.

181 The Heartland also won a place . . . : See, for example, Frederick John Teggart, *Processes of History* (New Haven: Yale University Press, 1918; Berkeley and Los Angeles: University of California Press, 1962, reprint), pp. 247–267; and James Fairgrieve, *Geography and World Power*, 8th ed. (New York: E.P. Dutton and Co., 1941), pp. 328–334.

181 But recognition by textbook authors . . . : See, for example, Samuel Van Valkenburg, *Elements of Political Geography* (New York: Prentice-Hall, 1939, 1944). The book's extensive index contains no listing of Mackinder, the Heartland, or the World Island, nor do any of Van Valkenburg's maps reflect the theory.

181 Mackinder was also subjected to . . . : Charles Redway Dryer, "Mackinder's 'World Island' and its American 'Satellite,' " *Geographical Review* 9 (1920), pp. 205–207.

181 In the 1930s, the National Socialists . . . : G. Henrik Herb, "Persuasive Cartography in *Geopolitik* and National Socialism," *Political Geography Quarterly* 8 (1989), pp. 289–303.

182 William Parker, who wrote . . . : W. H. Parker, *Mackinder: Geography as an Aid to Statecraft* (Oxford: Clarendon Press, 1982), p. 147; also see pp. 158–160.

182 But another biographer, . . . : Blouet, *Halford Mackinder*, p. 178.

183 Nonetheless, respected geopolitical theorist . . . : Saul Bernard Cohen, *Geography and Politics in a World Divided* (New York: Random House, 1963), pp. 40–41.

183 "Never have I seen anything . . .": Haushofer quoted in Weigert, *Generals and Geographers*, p. 116. Weigert's source is K. Haushofer, *Space-Conquering Powers*, 1937, p. 76.

183 A distinguished German political geographer . . . : Parker, *Mackinder*, footnote about Weigert on p. 179.

183 To demonstrate that . . . : Weigert, *Generals and Geographers*, quote on p. 117, map on p. 114, discussion on pp. 126–133.

183 "Haushofer refers to Mackinder . . .": Ibid., p. 115.
183 "to Haushofer, Mackinder . . .": Ibid., p. 116.
183 "This book should not . . .": Ibid., p. 116.
183 In 1944, at a ceremony . . . : "Presentation of the Medals Awarded by the American Geographical Society to Two British Geographers," *Geographical Journal* 103 (1944), pp. 131–134.
184 Although nearly a half-century old, . . . : Parker, *Mackinder,* pp. 192–197.
184 "when all is said and done, . . .": Cohen, *Geography and Politics in a World Divided,* p. 40.
184 Critical examination revealed . . . : For a summary, see Parker, *Mackinder,* pp. 213–247.
184 Some critics pointed out . . . : Arthur R. Hall, "Mackinder and the Course of Events," *Annals of the Association of American Geographers* 45 (1955), pp. 109–126.
184 Best known among Mackinder's critics . . . : See Nicholas John Spykman, *America's Strategy in World Politics: The United States and the Balance of Power* (New York: Harcourt, Brace and Co., 1942); and Nicholas John Spykman, *The Geography of the Peace* (New York: Harcourt, Brace and Co., 1944).
184 Spykman maintained that . . . : Donald W. Meinig, "Heartland and Rimland in Eurasian History," *Western Political Quarterly* 9 (1956), pp. 553–569.
184 Other critics, led by aviation innovator . . . : Alexander P. de Seversky, *Air Power: Key to Survival* (New York: Simon and Schuster, 1950); also see Harm J. De Blij, *Systematic Political Geography* (New York: John Wiley and Sons, 1967), pp. 132–137.
185 While the cold war fostered . . . : Alan K. Henrikson, "The Map as an 'Idea': The Role of Cartographic Imagery During the Second World War," *The American Cartographer* 2 (1975), pp. 19–53.
185 According to de Seversky . . . : Ibid., pp. 36–38; also see Alan K. Henrikson, "Maps, Globes, and the Cold War," *Special Libraries* 65 (1974), pp. 445–454.
186 These disparities create . . . : Saul B. Cohen, "Global Geopolitical Change in the Post–Cold War Era," *Annals of the Association of American Geographers* 81 (1991), pp. 551–580. For an earlier version of Cohen's concepts, see Saul B. Cohen, "A New Map of Global Geopolitical Equilibrium: A Developmental Approach," *Political Geography Quarterly* 1 (1982), pp. 223–241.

6. MAPS, VOTES, AND POWER

GENERAL SOURCES

General sources addressing the theory and practice of redistricting include Arthur J. Anderson and William S. Dahlstrom, "Technological Gerrymandering: How Computers Can Be Used in the Redistricting Process to Comply with Judicial Criteria," *Urban Lawyer* 22 (1990), pp. 59–77; Gordon E. Baker, "Representation and Apportionment," in Jack P. Greene, ed., *Encyclopedia of American Political History*, Vol. 3: *Studies of Principal Movements and Ideas* (New York: Charles Scribner's Sons, 1984), pp. 1118–1130; Michelle H. Browdy, "Computer Models and Post-Bandemer Redistricting," *Yale Law Journal* 99 (1990), pp. 1379–1398; Bernard Grofman and Lisa Handley, "Identifying and Remedying Racial Gerrymandering," *Journal of Law and Politics* 8 (1992), pp. 345–404; and Richard L. Morrill, *Political Redistricting and Geographic Theory* (Washington, D.C.: Association of American Geographers, Resource Publications in Geography, 1981).

Sources examining court decisions on redistricting cases include James U. Blacksher, "Drawing Single-member Districts to Comply with the Voting Rights Amendments of 1982," *Urban Lawyer* 17 (1985), pp. 347–367; Bernard Grofman, "An Expert Witness Perspective on Continuing and Emerging Voting Rights Controversies: From One Person, One Vote to Partisan Gerrymandering," *Stetson Law Review* 21 (1992), pp. 783–818; and Stephen J. Thomas, "The Lack of Judicial Direction in Political Gerrymandering: An Invitation to Chaos Following the 1990 Census," *Hastings Law Journal* 40 (1989), pp. 1067–1093.

NOTES

191 A prominent article in the morning newspaper . . . : Local news stories about the Syracuse hearing include Mike Fish, "Redrawn Lines Don't Draw Mob," *Syracuse Post-Standard*, 1 February 1992, p. B1; Charles Miller, "In the Valley, Politics Creates Odd Boundaries," *Syracuse Herald American*, 2 February 1992, pp. B1–B2; "Reapportionment Plans Met with Anger," *Syracuse Post-Standard*, 3 February 1992, p. B3; Robert B. Haggart, "Legislature Brings Bureaucratic Battle before Public," *Syracuse Post-Standard*, 3 February 1992, p. B1; and "Redistricting Input Needed Fertilizer," *Syracuse New Times*, 5 February 1992, p. 5.

192 Even more complex was . . . : Article III, section 4 (amended 6 November 1945) of the state constitution uses a formula to fix the size of the Senate. See Gail S. Shaffer, *Manual for the Use of the Legislature of New York*, 154th ed. (Albany: New York State Department of State, 1988), pp. 16–18.

193 As in most states, politicians . . . : Deanna Marquart and Winston Harrington, "Reapportionment Reconsidered," *Journal of Policy Analysis and Management* 9 (1990), pp. 555–560.

196 Gerry, a Republican, was . . . : Elmer C. Griffith, *The Rise and Development of the Gerrymander* (Chicago: Scott, Foresman and Company, 1907), pp. 16–19.

197 The Federalists quickly spotted . . . : John Ward Dean, "The Gerrymander," *New England Historical and Genealogical Register* 46 (1892), pp. 374–383.

201 A paradox of the reapportionment game . . . : Richard Morrill, "Gerrymandering," *Focus* 41 (Fall 1991), pp. 23–27.

201 As with other games, . . . : Morrill, *Political Redistricting and Geographic Theory*, pp. 11–16.

201 The Voting Rights Act of 1965, . . . : Baker, "Representation and Apportionment."

202 Because of low voter registration . . . : Alan Ehrenhalt, "Pulling Away from the Racial Gerrymander," *Perspectives: The Civil Rights Quarterly* 15 (Winter-Spring 1983), pp. 32–37.

202 In 1982 Congress strengthened . . . : "Voting Rights Act Extended, Strengthened," *Congressional Quarterly Weekly Report* 38 (1982), pp. 373–377.

202 By ignoring the issue of *intent* . . . : Thomas, "The Lack of Judicial Direction in Political Gerrymandering."

202 Although requirements were frustratingly imprecise . . . : Grofman, "An Expert Witness Perspective on Continuing and Emerging Voting Rights Controversies"; Pamela S. Karlan, "Minority Reps," *The Nation* 256 (8 March 1993), p. 292; and Morrill, "Gerrymandering."

202 In practice, a "winnable" . . . : Blacksher, "Drawing Single-member Districts to Comply with the Voting Rights Amendments of 1982."

203 James J. Kilpatrick, "Gerrymandering 'Madness' in North Carolina," *Syracuse Herald American*, 16 February 1992, p. B3.

203 In 1992 five white North Carolina residents . . . : Ronald Smothers, "Fair Play or Racial Gerrymandering? Justices Study a 'Serpentine' District," *New York Times*, 16 April 1993, p. B7.

204 "The Constitution is color-blind . . .": Ibid.

204 "you should be able to . . .": Ibid., p. B7.

204 Following Interstate 85 . . . : Jay Mathews, "Color-Coded Congress-men," *Newsweek* 120 (21 September 1992), pp. 67–68.

204 "At some points . . .": Holly Idelson, "Supreme Court Considers Racially Conscious Map," *Congressional Quarterly Weekly Report* 51 (24 April 1993), pp. 1034–1035.

204 This unusual shape reflects . . . : Charles Mahtesian, "Blacks' Political Hopes Boosted by Newly Redrawn Districts," *Congressional Quarterly Weekly Report* 50 (25 April 1992), pp. 1087–1090. Other sources say 53 percent, e.g., *Newsweek*, 21 September 1992.

205 After the Justice Department rejected a plan . . . : Charles Mahtesian, "North Carolina Tries Again," *Congressional Quarterly Weekly Report* 50 (25 January 1992), p. 88; and Charles Mahtesian, "Green Light to N.C. Races," *Congressional Quarterly Weekly Report* 50 (14 March 1992), p. 656.

205 Although the revised plan . . . : Holly Idelson, "Supreme Court Consid-ers Racially Conscious Map," *Congressional Quarterly Weekly Report* 51 (24 April 1993), pp. 1034–1035.

205 In June 1993, by a close . . . : "Excerpts from High Court's Opinions on Racial Gerrymandering," *New York Times*, 29 June 1993, p. A12; also see *Shaw* v. *Reno, United States Law Week*, 29 June 1993, pp. 4818–4834.

207 "3) When not in conflict with Federal . . .": New York State Legislative Task Force on Demographic Research and Reapportionment, "Co-Chairmen's Proposed 1992 Assembly and State Senate District Bound-aries," January 21, 1992, p. 2.

207 "the recommendation seeks . . .": Ibid., p. 3.

209 Although the Democrats proposed eliminating . . . : Kevin Sack, "Impasse Blocking Accord in Albany on New Districts," *New York Times*, 25 March 1992, pp. A1, B8.

210 But as of March 25, . . . : Kevin Sack, "Impasse Blocking Accord"; also see Luther F. Bliven, "Task Force Delays Redistricting Action," *Syracuse Post-Standard*, 18 March 1992, p. A7.

210 Not content with protests . . . : Kevin Sack, "Lawsuit Takes Redistrict-ing to U.S. Court," *New York Times*, 27 March 1992, pp. B1, B5; and Kevin Sack, "Ruling Gives Deadline to Albany for Redrawing Dis-tricts," *New York Times*, 8 April 1992, p. B5.

210 "an asymmetrical lobster . . .": Sam Roberts, "Will District Mappers Ever Draw the Line?" *New York Times*, 23 March 1992, p. B3.

211 So on April 27 . . . : Kevin Sack, "Two Methods to Redistrict Sent to Court," *New York Times*, 28 April 1992, p. B2.

211 A week and a half later, . . . : Todd S. Purdum, "Congressional Redistricting Deadline Given," *New York Times*, 6 May 1992, p. B4.

211 Nonetheless, on May 12, . . . : Kevin Sack, "Ex-Judge Given Albany's Task for Redistricting," *New York Times*, 13 May 1992, pp. A1, B4.

211 The *New York Times*, which compared . . . : Sam Howe Verhovek, "Congressional District Plan Unveiled," *New York Times*, 27 May 1992, pp. B1, B5.

211 Not surprisingly, incumbents . . . : Sam Howe Verhovek, "Redistricting Plan Spurs Lawmakers' Criticism," *New York Times*, 28 May 1992, pp. B1, B4.

212 To further complicate the picture, . . . : Sam Howe Verhovek, "New York Court Throws a State Plan into Redistricting Fray," *New York Times*, 2 June 1992, pp. B1, B2.

212 During a June 3 federal court hearing . . . : Kevin Sack, "Albany Legislators Agree on Plan for Revised Congressional Lines," *New York Times*, 4 June 1992, pp. A1, B8.

212 Despite threats by the Puerto Rican . . . : Bob Benenson and Ines Pinto Alicea, "State Legislators in New York Near Compromise on Map," *Congressional Quarterly Weekly Report* 50 (6 June 1992), p. 1641.

212 On June 8 the Senate . . . : Kevin Sack, "Court Backs New Plan for Districts," *New York Times*, 10 June 1992, p. B6; and Kevin Sack, "Court May Impose Redistricting Plan," *New York Times*, 12 June 1992, p. B4.

212 Although he supported the Hispanic position, . . . : Sam Howe Verhovek, "Cuomo Won't Veto Redistricting Bill," *New York Times*, 11 June 1992, pp. A1, B8.

212 When instead the Justice Department certified compliance . . . : Kevin Sack, "Redistricting Plans Approved," *New York Times*, 3 July 1992, pp. B1, B4.

212 Without well-defined statutory constraints . . . : Thomas, "The Lack of Judicial Direction in Political Gerrymandering."

213 As long as politicians responsible . . . : Sam Kaplan, "Court's Redistricting Ruling Draws Mixed Reaction," *Congressional Quarterly Weekly Report* 51 (6 March 1993), p. 538.

213 Some geographers and political scientists . . . : Anderson and Dahlstrom, "Technological Gerrymandering," pp. 74–77; Browdy, "Com-

puter Models and Post-Bandemer Redistricting"; and Morrill, *Political Redistricting and Geographic Theory,* pp. 36–44.

213 An attractive idea, perhaps, . . . : Michelle H. Browdy, "Simulated Annealing: An Improved Computer Model for Political Redistricting," *Yale Law and Policy Review* 8 (1990), pp. 163–179.

214 The best a computer can do . . . : Ibid., pp. 169–172.

214 Although computers programmed . . . : Blacksher, "Drawing Single-member Districts," pp. 363–364; and Andrew Hacker, *Congressional Districting: The Issue of Equal Representation* (Washington, D.C.: The Brookings Institution, 1962), pp. 66–70.

214 As a partly intuitive geometric concept, . . . : Daniel D. Polsby and Robert D. Popper, "The Third Criterion: Compactness as a Procedural Safeguard Against Partisan Gerrymandering," *Yale Law and Policy Review* 9 (1991), pp. 301–353; and Ernest C. Reock, Jr., "Measuring Compactness as a Requirement of Legislative Apportionment," *Midwest Journal of Political Science* 5 (1961), pp. 70–74.

214 Yet, some legal scholars contend . . . : Bernard Grofman and Howard A. Scarrow, "Current Issues in Reapportionment," *Law and Policy Quarterly* 4 (1982), pp. 435–474, esp. pp. 454–455.

214 "only the most egregious . . .": Grofman, "An Expert Witness Perspective," p. 813.

215 Richard Morrill, a geographer . . . : Richard J. Morrill, "Redistricting Revisited," *Annals of the Association of American Geographers* 66 (1976), pp. 548–556.

215 Morrill had drawn these plans . . . : Richard J. Morrill, "Ideal and Reality in Reapportionment," *Annals of the Association of American Geographers* 63 (1973), pp. 463–477.

215 "either we were unable to discover . . .": Morrill, "Redistricting Revisited," pp. 555–556.

215 Although the New York legislature . . . : Rob Gurwitt, "The Messiest Job in Politics," *Governing* 5 (November 1991), pp. 43–48; and Don Wolf, "Iowa Hailed as Model of Redistricting without Politics," *Pittsburgh Press,* 10 November 1991, pp. B1, B5.

216 Like political cartographers elsewhere, . . . : Baker, "Representation and Apportionment," p. 1129.

216 State and local governments . . . : Mark Thompson, "The Gerry-manderer's Dream Machine," *California Lawyer* 10 (January 1990), pp. 20–22.

216 The Justice Department's Civil Rights Division . . . : Gary H. Anthes, "GIS Eases Redistricting Worry," *Computerworld*, 7 October 1991, p. 65.

217 task-force personnel . . . : Interviews on June 1, 1990, with Dan Hennessey (demographic analysis expert), Lou Hoppe (co–executive director), Debra A. Levine (co–executive director), and Mark Radin (GIS manager), and on April 8, 1992, with Hennessey, Levine, Radin, and Vera Tostanoski (cartographic products coordinator).

217 Recent court opinions about equal protection . . . : Anderson and Dahlstrom, "Technological Gerrymandering," pp. 74–76.

217 For a fee, task-force staff . . . : New York State Legislative Task Force on Demographic Research and Reapportionment, "Public Access Procedures" (ca. 1991, no date), 14 pp.

217 A bulletin issued in October 1991 . . . : New York State Legislative Task Force on Demographic Research and Reapportionment, "How to Submit a Redistricting Plan," *Special Report from the Task Force*, October 1991, p. 3.

217 Wisconsin, by comparison, . . . : "Try Your Hand at Redistricting by Computer," *Wisconsin Mapping Bulletin* 18 (January 1992), p. 14.

218 Moreover, sets of "high-resolution" maps . . . : "Locations for Viewing High-Resolution Maps," *News Alert from the Legislative Task Force on Demographic Research and Reapportionment*, 1 January 1992, p. 4.

219 For example, cumulative voting . . . : Grofman and Handley, "Identifying and Remedying Racial Gerrymandering," pp. 394–400; Grofman and Scarrow, "Current Issues in Reapportionment," pp. 457–466; Marquart and Harrington, "Reapportionment Reconsidered," pp. 558–560; and Richard H. Pildes, "Gimme Five," *New Republic* 208 (1 March 1993), pp. 16–17.

219 Although potentially divisive . . . : Michael Lind, "A Radical Plan to Change American Politics," *Atlantic Monthly* 270 (August 1992), pp. 73–83.

7. SITING, CARTOGRAPHIC POWER, AND PUBLIC ACCESS

GENERAL SOURCES

General sources on disposal of low-level radioactive waste include Frans Berkhout, *Radioactive Waste: Politics and Technology* (London and New York:

Routledge, 1991); Michael E. Burns, ed., *Low-Level Radioactive Waste Regulation: Science, Politics and Fear* (Chelsea, Mich.: Lewis Publishers, 1988); Mary R. English, *Siting Low-Level Radioactive Waste Disposal Facilities: The Public Policy Dilemma* (New York: Quorum Books, 1992); Ray Kemp, *The Politics of Radioactive Waste Disposal* (Manchester, U.K.: Manchester University Press, 1992); Scott Saleska, "Low-level Radioactive Waste: Gamma Rays in the Garbage," *Bulletin of the Atomic Scientists* 46 (April 1990), pp. 18–35; Gale Warner, "Low-Level Lowdown," *Sierra* 70 (July/August 1985), pp. 19–23; and Irvin L. White and John P. Spath, "How Are the States Setting Their Sites?" *Environment* 26 (October 1984), pp. 16–20, 36–42.

Principal reports by the New York State Low-Level Radioactive Waste Siting Commission are *Candidate Area Identification Report*, Albany, N.Y., December 1988, cited below as *CAIR;* and *Report on Potential Sites Identification*, September 1989, cited below as *ROPSI.*

NOTES

221 Experts on waste disposal . . . : Dan Grossman and Seth Shulman, "Doing Their Low-Level Best," *Garbage* 4 (December/January 1993), pp. 32–37.

221 But the public is skeptical . . . : See, for example, A. K. M. M. Haque, "Effects of Inhalation of Radon Daughters in the Lungs," in S. K. Majumdar, R. F. Schmaltz, and E. Willard Miller, eds., *Environmental Radon: Occurrence, Control and Health Hazards* (Easton, Pa.: Pennsylvania Academy of Science, 1990), pp. 282–303.

221 Despite its name, . . . : Saleska, "Low-level Radioactive Waste."

222 A decade earlier there had been . . . : White and Spath, "How Are the States Setting Their Sites?"

222 "The landfill acts a lot like . . .": Warner, "Low-Level Lowdown," p. 20.

223 Fearing that Barnwell, . . . : Susan Q. Stranahan, "The Deadliest Garbage of All," *Science Digest* 94 (April 1986), pp. 64–67, 80–81.

223 Congress narrowly averted a crisis . . . : Public Law 96-573; see Andy Plattner, "Congress Passes Low-Level Nuclear Waste Bill, Leaves Broader Solution for Future," *Congressional Quarterly Weekly Report* 38 (20 December 1980), p. 3623.

223 The other regional compacts . . . : "Regional Dumps for Low-Level Radwaste," *Science News* 123 (21 May 1983), p. 329.

224 Efforts to form a . . . : White and Spath, "How Are the States Setting Their Sites?" pp. 38–40.

224 In Massachusetts a 1982 state law . . . : Colin Norman, "High-Level Politics over Low-Level Waste," *Science* 223 (1984), pp. 258–260.

224 Alarmed by the approaching December 31 deadline . . . : Joseph A. Davis, "House OKs Regional Low-Level Waste Bills," *Congressional Quarterly Weekly Report* 43 (14 December 1985), p. 2648; and Joseph A. Davis, "Low-Level Nuclear-Waste Bill Cleared for President Reagan," *Congressional Quarterly Weekly Report* 43 (21 December 1985), p. 2714.

225 To demonstrate a sense of urgency . . . : Chapter 673, *Laws of the New York Legislature, 209th Session, 1986* (Albany, N.Y.: New York State Legislative Bill Drafting Commission, 1987), pp. 2801–2817; quote on p. 2806.

226 The first issue of its . . . : *LLRW Frontline* (New York State Low-Level Radioactive Waste Siting Commission), Winter 1988, 4 pp.

226 As instructed by the legislature, . . . : See New York State Department of Environmental Conservation, Division of Hazardous Substances Regulation, Division of Regulatory Affairs, Bureau of Energy and Radiation, *Final Environmental Impact Statement for Promulgation of 6 NYCRR Part 382: Regulations for Low-Level Radioactive Waste Disposal Facilities*, Albany, N.Y., December 1987; also see New York State Environmental Conservation Law, Title 6, § 382.21.

226 An astute mixture of common sense, . . . : New York State Low-Level Radioactive Waste Siting Commission, *Statewide Exclusionary Screening Report*, Albany, N.Y., September 1988, p. S-5.

226 It was more than mere coincidence . . . : John Dean, N.Y. Senate Energy Committee, telephone interview, 8 June 1993.

227 Relying on readily available maps . . . : New York State Low-Level Radioactive Waste Siting Commission, *Plan for Selecting Sites for Disposal of Low-Level Radioactive Waste*, Albany, N.Y., November 1988.

230 Only after the coarse, square-mile mesh . . . : Area information from Siting Commission, *CAIR*, p. S-6. The forty-acre grid is revealed in *LLRW Frontline* (New York State Low-Level Radioactive Waste Siting Commission), Spring 1989, p. 1.

231 Is GIS screening a reliably objective process, . . . : Use of transportation routes as an exclusionary factor in this example is puzzling: because screening treated access as a positive siting factor, cells with a railway or major highway were deemed favorable, and generally not excluded.

231 A particularly revealing example . . . : 6 NYCRR Parts 382.22(a) (3), (4), and (b) (1), as reported in Siting Commission, *CAIR*, p. 4-22.

232 Yet, 75 of the 105 weather stations . . . : See Table 4-12 in Siting Commission, *CAIR*, pp. 4-81–4-83.

232 Experience suggests rain infiltration . . . : Marvin Resnikoff, quoted in Warner, "Low-level Lowdown," p. 20.

233 Another example of spurious objectivity . . . : Siting Commission, *CAIR*, pp. 4-20–4-22, 4-79–4-80, and Figure 4-21.

234 In August 1988 the commission held . . . : Siting Commission, *CAIR*, pp. 3-3–3-4.

234 Weights for the fourteen preference criteria . . . : Siting Commission, *CAIR*, p. 3-15. These weights are for "aboveground/belowground methods"; different weights were used for "mine methods" of disposal.

236 To their credit, . . . : Siting Commission, *CAIR*, pp. 5-3–5-4.

236 The day after the announcement, . . . : "Albany Chooses 10 Possible Sites for Atom Waste," *New York Times*, 21 December 1988, p. B2.

236 In mid-January, a longer article . . . : Sue Halpern, "Residents Assail Plan for A-Waste Dump," *New York Times*, 20 January 1989, p. B2.

236 The other hearings must have been . . . : "Five Sites Chosen as a Possible Waste Dump," *New York Times*, 10 September 1989, sect. 1, p. 38.

237 Five days before Christmas 1988, . . . : "County on Nuke Waste Dump List," *Cortland Standard*, 20 December 1988, p. 1.

237 The following day the paper . . . : Chris Bednarski, "Dump Listing Raises Anxiety," *Cortland Standard*, 21 December 1988, p. 3.

237 An editorial titled . . . : "Not at All Reassuring," *Cortland Standard*, 21 December 1988, p. 7.

237 The county legislature, the Cortland City Council, . . . : See, for example, Kevin Conlon, "Cincy Meets on Dump," *Cortland Standard*, 3 January 1989, p. 3; and Kevin Conlon, "Solon, Cortland Join N-dump Foes," *Cortland Standard*, 18 January 1989, p. 3.

238 Two citizens' groups, . . . : "Groups Plan to Discuss N-dump Site," *Cortland Standard*, 20 January 1989, p. 3; Margaret R. McHugh, "County: State Criteria Exclude N-dump Here," *Cortland Standard*, 23 January 1989, p. 3; and Mike Mittelstadt, "EMC: Consultant Could Help Oppose Low-level Nuke Dump," *Cortland Standard*, 6 January 1989, p. 3.

238 School districts provided buses . . . : Kevin Conlon, "N-Dump Adds Foes," *Cortland Standard*, 11 January 1989, p. 3; and Margaret R. McHugh, "Busloads to Mark Vigil First," *Cortland Standard*, 24 January 1989, p. 3.

238 A *Cortland Standard* editorial . . . : "Not Here, Not Ever . . . ," *Cortland Standard*, 24 January 1989, p. 8.

238 An estimated 1,300 persons attended, . . . : Mike Mittelstadt, "1st Live Broadcast at N-dump Hearing," *Cortland Standard*, 26 January 1989, p. 3.

238 Local officials and other residents . . . : Margaret R. McHugh, "Politics Charged in Site Selection," *Cortland Standard*, 25 January 1989, pp. 3–4; and Mike Mittelstadt, "State N-dump Site Panel Says 'No Guarantees,'" *Cortland Standard*, 25 January 1989, p. 3.

238 Anticipating a long fight . . . : Barbara J. Woods, "County OKs Post for N-dump Fight," *Cortland Standard*, 25 January 1989, p. 4.

238 A second wave of bad news . . . : Constance M. Nogas, "Was Site News Leak Deliberate?" *Cortland Standard*, 11 September 1989, p. 3.

238 Although the official announcement . . . : Paul La Dolce and Richard Palmer, "2 N-dump Sites Here," *Cortland Standard*, 9 September 1989, p. 1; and "Five Sites Chosen as a Possible Waste Dump," *New York Times*, 10 September 1989, sect. 1, p. 38.

239 A report released in Albany . . . : Siting Commission, *ROPSI*, pp. S-2–S-3 and Figure S-2.

239 A team consisting of a geologist, . . . : Siting Commission, *ROPSI*, pp. 3-5–3-6.

239 As local officials were quick to point out, . . . : Constance M. Nogas, "County Charges: State Broke Own Rules in Site Choice," *Cortland Standard*, 13 September 1989, pp. 1, 3.

240 They had already read . . . : Richard Palmer, "Taylor Residents Fear Their Town Tops State's N-dump List," *Cortland Standard*, 1 September 1989, p. 3.

240 Even so, Allen's neighbors . . . : See, for example, Frances Dinkelspiel, "Family Farmer Wants to Sell, But for a Radioactive Dump?" *New York Times*, 9 December 1989, pp. 1, 30.

240 On Saturday, September 9, . . . : Constance M. Nogas, "Rally Against N-dump," *Cortland Standard*, 11 September 1989, p. 1.

242 On Monday, the *Cortland Standard* published . . . : "To Our Readers," *Cortland Standard*, 11 September 1989, p. 8.

242 On Tuesday evening at the Taylor Community Hall, . . . : Constance M. Nogas, "N-dump Opponents Plan Disobedience," *Cortland Standard*, 13 September 1989, pp. 1, 3.

242 On the following Thursday, 250 county residents . . . : Constance M. Nogas, "Taylor Looked Good Through State's Windshield," *Cortland Standard*, 15 September 1989, p. 3.

242 On November 15, as many as 5,000 people . . . : Frances Dinkelspiel, "4,800 Shout No to Dump," *Syracuse Herald-Journal,* 16 November 1989, pp. A1, A16.

242 On December 13, fortified . . . : "Inspectors Jeered at Possible Nuclear-Waste Site," *New York Times,* 14 December 1989, p. B3.

242 After another protest blocked . . . : "New York Loses Case on Protests over A-Wastes," *New York Times,* 23 December 1989, p. 34.

242 On January 19, 1990, protesters . . . : John S. Tonello, "Protesters Keep N-Dump Team Off Site," *Syracuse Herald-Journal,* 20 January 1990, p. A3.

242 On several occasions in early March, . . . : John S. Tonello, "N-Dump Office Blocked," *Syracuse Post Standard,* 2 March 1990, p. B3; and Matthew Cox, "N-Dump Panel Ponders Study Trip Abroad," *Syracuse Post Standard,* 9 March 1990, p. B3.

242 Town building inspectors . . . : John S. Tonello, "N-Dump Foes Renew Protest," *Syracuse Post Standard,* 3 March 1990, Cortland/Tompkins edition, p. B3.

243 On April 12, the county sheriff . . . : Lillie Wilson, "Sheriff: Close N-Dump Office," *Syracuse Post Standard,* 12 April 1990, p. B3.

243 A week earlier, near the other three . . . : "Allegany County N-Dump Protesters Face Hearing in May," *Syracuse Post Standard,* 19 April 1990, Cortland/Tompkins edition, p. B1.

243 Finally, on April 19, . . . : Lillie Wilson, "Charges Dropped Against Nuke Dump Protesters; State Shuts Down Office," *Syracuse Herald-Journal,* 20 April 1990, Oswego edition, p. B1.

243 Unable to carry out inspections, . . . : See, for example, Sam Howe Verhovek, "Where Will New York Put the State's Nuclear Waste?" *New York Times,* 19 August 1990, sect. 4, p. 20.

243 The new law passed in July 1990, . . . : See Chapter 913, *1990 Session Laws, New York Legislature,* pp. 2814–2821.

243 Fearing this worst-case scenario, . . . : Sam Howe Verhovek, "Waste Law Is Challenged by New York," *New York Times,* 10 February 1990, pp. A27, A29.

244 After the federal district court and . . . : Thomas Fine, "Supreme Court Date Nears for Dump Case," *Syracuse Herald-American,* 19 January 1992, p. B4; Jonathan Salant, "New York Challenges Radioactive Waste Law," *Syracuse Post Standard,* 30 March 1992, pp. A1, A8; and Linda Greenhouse, "Justices Hear Attack on Waste Law," *New York Times,* 31 March 1992, p. A15.

244 Three months later, in a 6-to-3 decision, . . . : Linda Greenhouse, "High Court Eases States' Obligation over Toxic Waste," *New York Times,* 20 June 1992, pp. 1, 10; Ruth Marcus and Thomas W. Lippman, "Court Rejects Key Part of A-Waste Law," *Washington Post,* 20 June 1992, pp. A1, A7; Andrew Yarrow, "Utilities Urged to Find Nuclear Waste Sites," *New York Times,* 21 June 1992, sect. 1, p. 33; and Paul M. Barrett, "New York State is Key Victor in Waste Suit," *Wall Street Journal,* 22 June 1992, p. A4.

244 "While there may be many . . .": 112 S. Ct. 2408 (1992), p. 2435.

244 In early 1991 siting commission members . . . : Sam Howe Verhovek, "Panel Proposes Paying Towns to Take Waste," *New York Times,* 18 January 1991, p. B3.

244 "It could be a million dollars, . . .": Ibid.

244 When the proposed "benefits package" became public . . . : Sam Howe Verhovek, "Anxious Town Might Say Yes to a Nuclear Waste Dump," *New York Times,* 28 June 1991, pp. B1, B5.

244 Negotiated secretly by local . . . : Jon R. Luoma, "Right in Your Own Backyard," *Audubon* 93 (November-December 1991), pp. 88–95.

245 Nonetheless, on the following evening . . . : Sam Howe Verhovek, "Despite Voters' 'No,' Board Allows Dump for Nuclear Wastes," *New York Times,* 11 July 1991, pp. A1, B2; and Sam Howe Verhovek, "Nuclear Dump Divides a Rural Town," *New York Times,* 12 July 1991, p. B4.

245 But the town's majority . . . : Sam Howe Verhovek, "Judge Delays Toxic Dump Town Wants," *New York Times,* 18 September 1991, pp. B1, B4.

245 On May 3, 1991, county low-level radioactive waste coordinator . . . : "Senator D'Amato's meeting with LLRW Opponents in Cortland," *Cortland County Low-Level Radioactive Waste Office Newsletter* 1, no. 3 (June 1991), p. 6; and Rebecca James, "D'Amato Asking for Investigation of N-Siting Panel," *Syracuse Post Standard,* 4 May 1991, Cortland/Tompkins edition, p. B1(C).

246 Fifteen months later, . . . : U.S. General Accounting Office, *Nuclear Waste: New York's Adherence to Site Selection Procedures Is Unclear* (report no. GAO/RCED-92-172), August 1992.

246 "For two reasons, this decision . . .": Ibid., p. 4.

246 The site should have flunked . . . : Ibid., p. 8.

246 Confusion about boundaries . . . : Ibid., pp. 8–10.

248 "Our review of the commission's actions . . .": Ibid., pp. 13–14.

248 Aware of my interest in deceptive maps, . . . : Denise Cote-Hopkins, interview, 9 June 1993.

248 Because state regulations . . . : Siting Commission, *ROPSI*, pp. 3-4, 4-39; also Appendix F, p. F-40; also see John D. Randall, Executive Deputy Chairman of the Siting Commission, letter to Denise Cote-Hopkins, 17 March 1992; New York State LLRW Siting Commission, *Site Selection Presentation, Part 2—September 11, 1991*, various pages; and New York State Low-Level Radioactive Waste Siting Commission, *Response to Public Comments Relating to the Report on Potential Sites Identification*, Albany, N.Y., November 1992, pp. 50–51.

249 "Clearly, the Commission did not . . .": Denise Cote-Hopkins, *Presentation to the New York State Low-Level Radioactive Waste Siting Commission on the Issue of Incompatible Structures*, November 21, 1991, Taylor Against LLRW, Cincinnatus, N.Y., 1991, pages not numbered; quote from section B.

249 "Why were structures whited-out . . .": Ibid., pages not numbered; quote from section D.

249 How valid was the drift-mine analysis, . . . : "Drift mine Repository Design Report," *Cortland County Low-Level Radioactive Waste Office Newsletter* 1, nos. 5 and 6 (September/October 1991), p. 5.

250 In early 1989 Concerned Citizens of Allegany County . . . : Various articles in the *Olean Times-Herald* and the *Wellsville Daily Reporter*, especially Kathryn Ross, "Citizens File Suit Against State Siting Commission," *Wellsville Daily Reporter*, 23 May 1989, pp. 1, 5, and 7; and Kathryn Ross, "Nuke Fight in Court Today in Belmont," *Wellsville Daily Reporter*, 15 June 1989, pp. 1, 5. An important source was a collection of newspaper clippings, filed in scrapbooks, in the office of John Margeson, administrative assistant to the Allegany County legislature, and maintained by Margeson's secretary, Lee Cobb. The facts reported in this section were confirmed by discussions with Margeson and various other people, including siting commission staff. Also see Siting Commission, *Response to Public Comments*, pp. 41–42.

250 "The Siting Commission publicly proclaims . . .": Petition filed by CCAC attorney David Seeger, quoted in Ross, "Citizens File Suit Against State"; also see Joan Dickinson, "Judge Denies Change of Venue," *Olean Times-Herald*, 15 June 1989, p. 1; and Kathryn Ross, "CCAC in Court in Belmont," *Wellsville Daily Reporter*, 16 June 1989, p. 1.

250 "immediately entitled to access . . .": Sue Goetschius, "Justice Orders Release of Documents," *Olean Times-Herald*, 19 July 1989, p. 1; also see Kathryn Ross, "County Wins Nuke Case in Court," *Wellsville Daily Reporter*, 19 July 1989, p. 1.

251 As ordered, the commission . . . : Joan Dickenson, "Citizens Say Computer Tapes 'Unreadable,'" *Olean Times-Herald*, 13 September 1989, p. 1.

251 The siting commission had complied . . . : Ibid.

251 Although the county legislature appropriated . . . : Ibid.; and Joan Dickenson, "Siting Commission Data Need 'Translating,'" *Olean Times-Herald*, 25 July 1989, p. 1.

251 Aware now of the differences . . . : Joan Dickenson, "Caneadea Asking Experts' Help in Interpreting Commission Data," *Olean Times-Herald*, 26 September 1989, p. 1; and Oak Duke, "New Questions Surface Concerning Weston's Credibility," *Wellsville Daily Reporter*, 8 February 1990, pp. 1, 5.

252 "The siting commission's staff told my client . . .": Dickenson, "Citizens Say Computer Tapes 'Unreadable.'"

252 In amending the state's . . . : See Chapter 913, *1990 Session Laws, New York Legislature*, p. 2819.

253 "on nights and weekends . . .": Siting Commission announcement, 13 May 1992.

253 Ironically, on a visit to . . . : John Thomas, Siting Commission, interview, 15 June 1993.

253 Only one sentence mentioned . . . : New York State Low-Level Radioactive Waste Siting Commission, Request for Proposals for Site Selection and Disposal Method Selection for the Disposal of Low-Level Radioactive Waste in the State of New York (RFP no. 001-87), no date, p. 4.

253 Nine qualified contractors . . . : John P. Thomas, Siting Commission, letter to author, 25 June 1993.

254 Humiliated a decade earlier . . . : English, *Siting Low-Level Radioactive Waste Disposal Facilities*, pp. 88–89.

254 Although New York . . . : Kevin Sack, "Albany Term: Much Undone," *New York Times*, 9 July 1993, pp. A1, B4.

8. RISK MAPS AND
ENVIRONMENTAL HAZARDS

GENERAL SOURCES

Sources for information on risk mapping include Mark Monmonier and George A. Schnell, "Natural Hazard Mapping: Status and Review," in S. K.

Majumdar, G. S. Forbes, E. W. Miller, and R. F. Schmaltz, eds., *Natural and Technological Disasters: Causes, Effects, and Preventive Measures* (Easton, Pa.: Pennsylvania Academy of Science, 1992), pp. 440–454; Robert A. Page et al., *Goals, Opportunities, and Priorities for the USGS Earthquake Hazards Reduction Program,* U.S. Geological Survey circular no. 1079 (Washington, D.C.: U.S. Government Printing Office, 1992); and Thomas L. Wright and Thomas C. Pierson, *Living with Volcanoes: The U.S. Geological Survey's Volcano Hazards Program,* U.S. Geological Survey circular no. 1073 (Washington, D.C.: U.S. Government Printing Office, 1992).

Sources for evacuation mapping and the siting of nuclear plants include Ute Dymon and Nancy L. Winter, "Evacuation Mapping: The Utility of Guidelines," *Disasters* 17 (1993), pp. 12–24; David P. McCaffrey, *The Politics of Nuclear Power: A History of the Shoreham Nuclear Power Plant* (Dordrecht, The Netherlands: Kluwer Academic Publishers, 1991); and Donald J. Zeigler and James H. Johnson, Jr., "Evacuation Behavior in Response to Nuclear Power Plant Accidents," *Professional Geographer* 36 (1984), pp. 207–215.

General references on the delineation and mapping of wetlands include Casey Dinges, "When Is a Wetland Wet?" *Civil Engineering* 61 (November 1991), p. 112; and Jon Kusler, "Wetlands Delineation: An Issue of Science or Politics?" *Environment* 34 (January/February 1992), pp. 7–11, 29–37.

Sources for the Sternglass controversy include Richard S. Lewis, *The Nuclear-Power Rebellion: Citizens vs. the Atomic Industrial Establishment* (New York: Viking Press, 1972); Richard L. Miller, *Under the Cloud: The Decades of Nuclear Testing* (New York: Free Press, 1986); and Ernest J. Sternglass, *Secret Fallout: Low-Level Radiation from Hiroshima to Three-Mile Island* (New York: McGraw-Hill, 1981).

Principal sources for the Woburn, Massachusetts, cancer cluster are Paula DiPerna, *Cluster Mystery: Epidemic and the Children of Woburn, Mass.* (St. Louis, Mo.: C.V. Mosby, 1985); and Eliot Marshall, "Woburn Case May Spark Explosion of Lawsuits," *Science* 234 (1986), pp. 418–420.

NOTES

258 My task was to examine . . . : "Design Guide for Environmental Maps" was distributed within the New Jersey Department of Environmental Protection in August 1990. For additional discussion of the project, see Mark Monmonier and Branden B. Johnson, "Using Qualitative Data-Gathering Techniques to Improve the Design of Environmental Maps," *Proceedings of the 15th International Cartographic Conference and 9th General*

Assembly of the International Cartographic Association, Bournemouth, England, 23 September–1 October 1991, pp. 364–373; and Monmonier and Schnell, "Natural Hazard Mapping: Status and Review."

261 Their first comparatively precise . . . : "California Hamlet Celebrates as Earthquake Prediction Officially Expires," *Washington Post,* 2 January 1993, p. A6.

261 Figure 8.1: Map was originally prepared for an open-file (unpublished) report by the Working Group on California Earthquake Probabilities.

262 The geographic information system (GIS) that . . . : Page et al., *Goals, Opportunities, and Priorities for the USGS Earthquake Hazards Reduction Program,* pp. 36–39.

265 Although the wind is generally . . . : Dwight R. Crandell and Donal R. Mullineaux, *Potential Hazards from Future Eruptions of Mount St. Helens Volcano, Washington,* U.S. Geological Survey circular no. 1383-C (Washington, D.C.: U.S. Government Printing Office, 1978), pp. C20–C22.

264 "within the next hundred years . . .": Ibid., p. C25.

264 When a spectacularly destructive eruption . . . : Wright and Pierson, *Living with Volcanoes,* pp. 3, 22.

266 On November 13, 1985, a month after release . . . : Eduardo Parra and Hector Cepeda, "Volcanic Hazard Maps of the Nevado del Ruiz Volcano, Colombia," *Journal of Volcanology and Geothermal Research* 42 (1990), pp. 117–127; and Barry Voight, "The 1985 Nevado del Ruiz Volcano Catastrophe."

267 Figure 8.4: For an enhanced, color version of the map, see Wright and Pierson, *Living with Volcanoes,* p. 21.

268 Hazard perception is highly subjective . . . : New York State Emergency Management Office, *New York State 406 Hazard Mitigation Plan,* Albany, N.Y., September 1986, pp. 3-3–3-4.

271 Although the minuscule risk . . . : Matthew A. Wald, "A-Plants Warned to Be Wary of Truck Bombs," *New York Times,* 1 July 1993, p. A15.

271 Around each nuclear plant . . . : U.S. Nuclear Regulatory Commission and Federal Emergency Management Agency, *Criteria for Preparation and Evaluation of Radiological Emergency Response Plans and Preparedness in Support of Nuclear Power Plants,* document no. NUREG-0654 and FEMA-REP-1 (Washington, D.C.: U.S. Government Printing Office, November 1980), pp. 10–17.

272 Before the momentous Three Mile Island incident . . . : Zeigler and Johnson, "Evacuation Behavior in Response to Nuclear Power Plant Accidents."

272 A recent study by cartographers . . . : Dymon and Winter, "Evacuation Mapping: The Utility of Guidelines."

273 A survey of U.S. nuclear-plant managers . . . : Susan L. Cutter, "Emergency Preparedness and Planning for Nuclear Power Plant Accidents," *Applied Geography* 4 (1984), pp. 235–245.

275 Geographers Don Zeigler and Jim Johnson . . . : Zeigler and Johnson, "Evacuation Behavior in Response to Nuclear Power Plant Accidents."

276 Although planning for Shoreham . . . : McCaffrey, *Politics of Nuclear Power*, pp. 33–94.

276 A U.S. General Accounting Office study . . . : U.S. General Accounting Office, *Emergency Planning: Federal Involvement in Preparedness Exercise at Shoreham Nuclear Plant*, pp. 24–37.

277 "Some looked forward to . . .": *Hearings on Oversight of the Federal Emergency Management Agency's Treatment of Emergency Planning Issues for the Shoreham Nuclear Powerplant, November 14, 1985, and April 22, 1986,* serial no. 99-136; document no. Y4.En2/3:99-136 (Washington, D.C.: U.S. Government Printing Office, 1986), p. 247.

277 FEMA's regional director Frank Petrone resigned . . . : McCaffrey, *The Politics of Nuclear Power*, pp. 128–129, 154–159, 189–190; and Matthew L. Wald, "U.S. Approves Decommissioning of Unwanted Shoreham A-Plant," *New York Times*, 13 June 1992, p. 28.

278 An editorial . . . : "Wetlands" (editorial), *Wall Street Journal*, 26 May 1992, p. A16.

279 Although consolidating wetlands regulation . . . : Frederick W. Pontius, "Federal Laws Protecting Wetlands," *Journal of the American Water Works Association* 82 (November 1990), pp. 12–16, 102.

279 One collaboration culminated in 1989 . . . : Federal Interagency Committee for Wetland Delineation, *Federal Manual for Identifying and Delineating Jurisdictional Wetlands*, U.S. Army Corps of Engineers, U.S. Environmental Protection Agency, U.S. Fish and Wildlife Service, and U.S.D.A. Soil Conservation Service, 1989.

279 The scientific community generally applauded . . . : Kusler, "Wetlands Delineation," pp. 34–36.

280 Although this charge is an obvious exaggeration, . . . : Dinges, "When Is a Wetland Wet?"

280 Other critics warned . . . : Curtis C. Bohlen, "Controversy Over Federal Definition of Wetlands," *BioScience* 41 (1991), p. 139.

280 Responding to pressure . . . : Kusler, "Wetlands Delineation," p. 29.

280 By some estimates, . . . : Dinges, "When Is a Wetland Wet?"

281 Field tests in selected states . . . : Kusler, "Wetlands Delineation," p. 34; and D'Vera Cohn, "Wetland Redefinition Shelved," *Washington Post*, 12 January 1993, p. A5.

281 Especially troubling was a shift . . . : Marguerite Holloway, "High and Dry: New Wetlands Policy Is a Political Quagmire," *Scientific American* 265 (December 1991), pp. 16–20.

281 A sixty-day public comment period, . . . : Dinges, "When Is a Wetland Wet?"; and Kusler, "Wetlands Delineation," p. 34.

281 In the end, Congress and the administration . . . : Cohn, "Wetland Redefinition Shelved."

281 Fueling this reaction . . . : U.S. Fish and Wildlife Service, *Wetlands Losses in the United States, 1780s to 1980s*, 1990.

281 *Scientific American* and various environmental . . . : See, for example, Holloway, "High and Dry," p. 20; and World Resources Institute, comp., *The 1992 Information Please Environmental Almanac* (Boston: Houghton Mifflin, 1992), p. 136.

283 Perhaps the most widely applauded environmental-health map . . . : Gerald Astor, *The Disease Detectives: Deadly Medical Mysteries and the People Who Solved Them* (New York: New American Library, 1984), pp. 41–42; Margaret Pelling, *Cholera, Fever, and English Medicine, 1825–1865* (Oxford: Oxford University Press, 1979), pp. 214–227; and L. Dudley Stamp, *Some Aspects of Medical Geography* (London: Oxford University Press, 1964), pp. 15–16.

284 Few scientific controversies . . . : For discussion of Sternglass's interests and the impact of his work, see Lewis, *The Nuclear-Power Rebellion*, pp. 59–84; and Miller, *Under the Cloud*, pp. 9, 364–365, 369–375. For an example of "Sternglass correlation," see Lewis, pp. 67, 69, and 70.

284 He was deeply interested . . . : Miller, *Under the Cloud*, pp. 364–365.

284 an article in *Science* . . . : Ralph E. Lapp, "Nevada Test Fallout and Radioiodine in Milk," *Science* 137 (1962), pp. 756–758.

285 Two years later he reported . . . : See, for examples, Ernest J. Sternglass, "Infant Mortality and Nuclear Tests," *Bulletin of the Atomic Scientists* 25 (April 1969), pp. 18–20; and Ernest J. Sternglass, "Can the Infants Survive?" *Bulletin of the Atomic Scientists* 25 (June 1969), p. 26. The September 1969 issue of *Esquire* included Sternglass's article "The Death of All Children" as a four-page insert (pp. 1a–1d) just inside the front cover. Esquire's decision to publish his work is described in Miller, *Under the Cloud*, pp. 369–375.

285 Attempts by the AEC to discredit him . . . : See, for example, Anna Gyorgy and friends, *No Nukes: Everyone's Guide to Nuclear Power* (Boston: South End Press, 1979), pp. 51, 81, and 85.

285 Sternglass also collided with . . . : For examples, see Ernest J. Sternglass, *Low-Level Radiation* (New York: Ballantine Books, 1972). In 1981 McGraw-Hill published a revised updated version titled *Secret Fallout: Low-Level Radiation from Hiroshima to Three-Mile Island*. Also see Lewis, *The Nuclear-Power Rebellion*, pp. 119–121, 213, and 268.

286 "by 1949, a pattern of excess infant mortality . . .": Sternglass, "A Reply," p. 29.

287 "Institutions have a tenacity . . .": David Rittenhouse Inglis and Allan R. Hoffman, "Radiation and Infants," *Bulletin of the Atomic Scientists* 28 (December 1972), pp. 45–52; quote on p. 52.

288 Woburn's story reached a climax, . . . : Marshall, "Woburn Case May Spark Explosion of Lawsuits."

289 Since 1853, when a local entrepreneur . . . : DiPerna, *Cluster Mystery*.

290 Leukemia patients require . . . : Paula DiPerna, "Leukemia Strikes a Small Town," *New York Times*, 2 December 1984, magazine section, pp. 100–108.

291 Released in 1981 by Massachusetts . . . : The Massachusetts Department of Public Health released several "draft" versions of the study. For an abridged version published later in a refereed journal, see Cutler et al., "Childhood Leukemia in Woburn, Massachusetts."

292 Estimated exposure (access) to water . . . : The Harvard researchers released several "draft" versions of their study. For an abridged version published later in a refereed journal, see S. W. Lagakos, B. J. Wessen, and M. Zelen, "An Analysis of Contaminated Well Water and Health Effects in Woburn, Massachusetts," *Journal of the American Statistical Association* 81 (1986), pp. 583–586.

294 Publicity surrounding Anne Anderson's discovery . . . : DiPerna, *Cluster Mystery*, p. 295.

Index

American Association for the Advancement of Science, 11
American Association of Petroleum Geologists, 164
American Broadcasting Company (ABC), 22
American Cartographic Association, 41
American Civil Liberties Union, 204
American Congress of Surveying and Mapping (ACSM), 30–32, 35–36
American Geographical Society, 183
American Heritage, 81, 86, 97–98
American Indians
  ignored in history texts, 68–70
  influence on colonial place-names, 65–66
  map use by, 107
  place-names offensive to, 48–50
  reservations exempted from radioactive waste siting, 226, 234
  territorial boundaries, 146–47
American Reference Books Annual, 38
Anatase, in ink of Vinland map, 96, 102
Ancient document, exception to the hearsay rule, 120
Anderson, Anne, 290–94
Anderson, Jimmy, 290, 292
Annals of the Association of American Geographers, 15
Antarctica, territorial claims in, 143–46
Antarctic Treaty (1959), 145–46
Arabs, renaming by, 58
Arc/Info, 252
Archipelagic waters, 138
Area, land (on world maps)
  distortion of, 16–19
  fidelity of, 23
  relevance questioned, 39
Area cartogram, 39–40, 176, 177
Argentina, claim to Antarctica, 144
Armero, Colombia, volcanic disaster, 266–67
Arno, Peter, 34
Arrigo, Victor, 84, 86
Ashford, N.Y., proposed radioactive waste disposal site, 244–45
Associated Press, 236, 237

Association of American Geographers, 181
Atlantic Monthly, 83
Atlas of Canada, 63
Atomic Energy Commission (AEC), 284, 285
Atomic testing, health effects of, 284–88
Atomic weapons, 221
Australia, claim to Antarctica, 144
Authored landscape, 114
Avataq Cultural Institute, 68
Avulsion, 127–28
Azimuthal equidistant projection, 185

Bacon, Francis, 155
Bain, Iain, 21
Baker, Howard Bigelow, 158–59, 160–63, 164, 328–29
Balkan states
  Mackinder's plan for, 178–79
  political renaming in, 62–63
Baltimore, Lord, 108, 109
Barnwell Islands, 130, 133–34
Barnwell, S.C., radioactive waste disposal facility at, 223, 243, 254
Base line, 114
Basel (Switzerland), 80
Beatrice Foods, 288–89, 293
Beatty, Nevada, radioactive waste disposal site, 223
Beaufort, Treaty of, 130, 133
Beinecke Rare Book and Manuscript Library. *See* Yale University
Benedict the Pole (friar), 76
Benvenisti, Meron, 57–59
Berry, Edward, 166
Bhopal, India, 284
Bianco, Andrea, 80, 91
Black Americans. *See* African Americans
Blackmun, Harry, 133, 135
Blouet, Brian, 182–83
Boggs, S. Whittemore, 21
Book-of-the-Month Club, 83
Bosnia-Herzegovina, 62
Boston *Gazette*, 197, 199
Boston *Weekly Messenger*, 197, 198

Boundaries
  of American Indian nations, 146–47
  on Antarctica, 143–46
  coastal front model of, 136–37
  of Exclusive Economic Zones (EEZs),
    137–41
  of national territory, 108
  of real estate, 112–26
  seaward, 136–43
  water bodies as, 123–34
  of wetlands, 278–82
Brandt, Willy, 29–30
Breckinridge, Thomas, 121
Brezhnev (Soviet city), renamed, 62
Britain and the British Seas, 172–73
British Museum, 76, 78, 81
Broad Street Pump, 283
Bulletin of the Atomic Scientists, 285
Bush, George, 280

Cabot, John, 108
Calvert, Cecil (second Lord Baltimore),
  109
Calvert, Charles (third Lord Baltimore),
  111
Cambridge University, 164
Canada
  bilingual labels on maps, 63, 68
  indigenous place-names, 67–68
  maritime boundary with United States,
    142–43
Canada, Surveys and Mapping Branch,
  Department of Energy, Mines and
  Resources, 63
Cancer registries, need for, 294
Candidate Area Identification Report, 231
Cantino world map (1502), 108
Cape Henlopen, 111
Carpini, John de Plano, 76
Carto-anthropology, 3
Cartogram. *See* Area cartogram
Case, Clifford, 84
Celestial Scenery, 158
Celli, Mario Gattoni, 86
Center for World Development Educa-
  tion, 29

Centers for Disease Control (CDC),
  291–92
Chain of title, 112
Chamberlain, Neville, 171
Charles I, King of England, 109
Charles II, King of England, 109
Chernobyl, 221, 277
Chesapeake Bay, 109
Chile, claim to Antarctica, 144
Chinese Americans, place-names offen-
  sive to, 45, 47–50, 53
Cholera, 283
Christian Aid, 28, 29
Christian Century, 35
Christianity Today, 10, 16–19
Cincinnatus, N.Y., protests at radioactive
  waste information office, 242–43
Citizens Against Radioactive Dumping
  (CARD), 238
City Beautiful movement, 118
Clark, Frank Emerson, 122, 125–26
Clay cap, 222
Clinton, Bill, 281
Closing line, 134–36, 140
Coastal front, model for seaward bound-
  aries, 136–37
Cohen, Daniel, 82
Cohen, Saul, 183, 184, 186–88
Cold war, 184–85
Colombian Institute of Geology and
  Mines, 266
Color, source of uncertainty in map design
  and map use, 260
Columbus, Christopher, 72–73, 82–85,
  87, 103, 107
Columbus Day
  national holiday, 72–73, 86–87
  Vinland map introduced day before, 5,
    73, 81–82, 85, 86
Columbus Was First, 88–90
Commission de Toponymie du Québec,
  64
Commonwealth of Independent States,
  political renaming in, 61–62
  See also Russia; U.S.S.R.
Communist bloc, 184

Division points, for riparian rights, 124
Dixon, Jeremiah, 112
Drift Mine Repository Design Report, 249
Dryer, Charles, 181
d'Urville, Dumont, 144
Dury, George, 149–51, 168–69
Du Toit, Alexander, 160, 162
Dymon, Ute, 272–73

Earthquake hazards, maps of, 260–63
Easement, 118, 122
Economist, 39, 62
Eldridge, Douglas, 238
Election reform, 219
Electromagnetic fields, 288
Elements of Cartography, 20–21
Emergency Planning Zone (EPZ), 271–75
Emergency-response maps, 267–68, 271–73
Emergency Response Planning Area (ERPA), 273
Encirclement, 184
Encroachment, 118, 122
Endecott, John, 108
England, claims to the New World, 108
English Channel, international boundary in, 143
EPA. *See* U.S. Environmental Protection Agency
Equidistance principle, in boundary cases, 137, 140, 143–44
Equivalence, in map projection, 13
Ericson, Leif, 74–75, 82–85
Esquire, 285, 351
Ethnic cleansing, 62
Eurocentric bias, in map projection, 3, 18, 29, 30–31, 41
Evacuation maps, for nuclear power plants, 271–73
Everett, Robinson, 204
Exclusive Economic Zones (EEZs), boundaries of, 137–41

Face of the Earth, The, 149–50, 168
Facts in Review, 181–82

Fairgrieve, James, 181
Fallout. *See* Nuclear fallout
Fanfani, Amintore, 85
Fax machines, 260
Federal Emergency Management Agency (FEMA), 271, 277–78
Federal Manual for Identifying and Delineating Jurisdictional Wetlands, 279–81
Ferrajoli, Enzo, 78, 99–101
Fisher, Osmund, 161
Fishing rights, 137–39, 142
Flood, The (in the Bible), 157
Flood-zone boundaries, 122–23
Folk-etymology, 65
Fortuný, Don Luís, 100–101
France, claim to Antarctica, 144
Francke, Birgit (second wife of Arno Peters), 25
Freedom of Information Law. *See* New York, Freedom of Information Act
Friedrich-Wilhelm University, 25
Friends for a Clean Environment (FACE), 292
Friendship Press, 18, 34, 35, 36
Frobisher Bay (Baffin Island), renamed, 68
Fuller, Dave, 238

Gall, James, 11–14
Gall-Peters projection, 13
    See also Peters projection
Gall's orthographic equal-area projection, 11–13
General Cartography, 20
Generals and Geographers, 183
Geographers, as expert witnesses in boundary disputes, 5, 131–33
Geographical Journal, 91
Geographical Magazine, 21
Geographical Review, 180–81
Geographic correlation, 285–86, 288–90
Geographic information system (GIS), 6, 7, 227–36
    earthquake-hazard maps with, 262–63
    emergency-response maps with, 267–68
    groundwater modeling with, 293, 295

Geographic information system (GIS) (cont'∂)
    multifactor site analysis, 231–36, 341–42
    overlay analysis, 227–29, 239
    public access to, 222, 251–54
    sensitivity analysis, 236
    software compatibility, 251–53
    spatial resolution, 230
    as a weapon, 254
Geographic names. *See* Place-names
Geographic Names Information System (GNIS), 48, 52
Geographic scope, 258–59
Geography and World Power, 181
Geological Society of America, 156
    Bulletin, 159
Geopolitics, 6, 173, 186–88
Geopolitik, 170, 181
George III, King of England, 108
Georges Bank, 142, 143
Georgia, boundary dispute with South Carolina, 130–37
German Cartographical Society, 24–25
Gerry, Elbridge, 196–97
Gerrymander, 6, 196–201
    packing, 201
    partisan, 214–15
    racial, 203–6
    shape as evidence for, 203–6, 210, 212–13
Gerstman, Marc, 55
Gnomonic projection, 20
Gnupsson, Eirik, 75, 97
Golden section, 43
Gondwanaland, 151
Goode, J. Paul, 15–16
Goode's homolosine equal-area projection, 15–16, 22–24, 43
Gorbachev, Mikhail, 62
Grace, W. R. *See* W. R. Grace Company
Green, W. L., 156
Greenhood, David, 21
Grofman, Bernard, 214–15
Groundwater contamination, 288–95
Gruber, Pamela, 29

Guinta, Francesco, 97
Gulf of Maine, disputed boundary in, 142–43

Hammond Atlas of the World, 146
Hammond Map Company, 21, 47
Harley, John Brian, 42, 107
Harper and Row, 37
Harpers, 10, 17
Harris, Chauncey, 39
Harrison, Benjamin, 51
Harvard University, School of Public Health, 292–93
Haushofer, Karl, 170, 181, 183
Heartland
    cartographic propaganda for, 173–74, 176–77, 188
    geopolitical theory, 149, 170, 175–76
    on Mercator projection, 173, 184
    Nazis believed in, 170–71
    Rimland as alternative for, 184–85
Herjolfsson, Bjarni, 74–75
Hilton Head Island, 135–37
Hispanic Americans
    place-names offensive to, 47–48
    and reapportionment, 192, 202–3, 208, 210–13
Historians, as expert witnesses in boundary disputes, 5, 131
Hitler, Adolf, 171, 181, 182
Hoffman, Allan, 286–88
Hoffman, Walter, 131–33, 136
How to Lie with Maps, 2
How to Lie with Statistics, 2
Hudson Falls, N.Y., radioactive waste hearing, 236
Huff, Darrell, 2
Hungarian Academy of Science, 25
Hypertext, 298

Ideology
    map projection and, 41–44
    place-names and, 69, 71
Igau, Dorothy, 53–54
Illinois, radioactive waste in, 222
Incumbents, and reapportionment, 206–7, 209–12

Independent Commission on International Development Issues, 29

Indians. *See* American Indians

Indonesia, 39

Industriplex site (Woburn, Mass.), 289, 292, 293

Infant mortality
atomic fallout and, 284–88
groundwater contamination and, 288–94

Influence of Sea Power Upon History, 1660–1783, The, 175

Inglis, David, 286–88

Ingstad, Helge, 85

Initiative, for political decision-making, 219

Innocent passage, 138

Institute for World History, 25

Interactive maps, 298

Intergraph, computer workstation, 253

Internal waters, 138

International Court of Justice, at The Hague, 5, 142

Interstate 85, 204, 210, 215

Inuit, place-names of, 64, 67–68

Iowa Legislative Services Agency, 215

Iowa, reapportionment in, 215–16

Isabella, Queen of Spain, 107

Islands
as evidence for riverine boundaries, 128–34
for seaward boundaries, 140

Isostasy, 167

Israel, political renaming in, 57–59

Italian Americans
American Heritage ridiculed, 86
Columbus Day significant to, 86–87
politicians supported, 83
Vinland map threatened, 83–87, 97–98
See also Musmanno, Michael

Jacob, John, commemorated by place-name, 60

Japanese Americans, place-names offensive to, 48–50, 52

Jeffreys, Harold, 164

Jelic, Luka, 97

Jensen, Oliver, 97–98

Jewish settlements, in the West Bank, 57–59

Johns Hopkins University, 149, 166

Johnson, Jim, 275

Johnson, Lyndon, 86

Jones, Gwyn, 83

Kaiser, Ward, 18, 34–37

Kartographische Nachrichten, 28, 33–34

Kentucky, radioactive waste in, 222

Kidron, Michael, 39

Kilpatrick, James J., 203

Knolls Atomic Power Laboratory, 270–71

Koch, J. P., 155

Kremer, Gerhard (Mercator), 31

Lacey, Frederick, 211–12

La Corte, John, 84, 86

Lagakos, Steven, 292–93, 352

Lake, Philip, 164

Lake Providence, La., 127–28

Lake shorelines, as boundaries, 123

Lambert, Johann Heinrich, 13

Lambert's cylindrical equal-area projection, 13–14, 43

Land survey, 113–17

Land use, control of, 119–20

Lapp, Ralph, 284

Latino Americans. *See* Hispanic Americans

Laurasia, 151

Law of the Sea, conferences on, 138

Lawrence, Peter, 21

Lebensraum, 170

Legal evidence, maps as, 5, 105–6, 120–34, 142–43

Leggett, Robert, 84

Legislatures, uneven distribution of power in, 219

Le Grand, Homer, 164

Leif Ericson Day, 73, 86–87

Leningrad, renamed, 62

Leukemia
atomic testing as cause, 284–88
groundwater contamination as cause, 288–94

Mason-Dixon line, 109, 112, 113, 116
Massachusetts, radioactive waste disposal, 224
Massachusetts Cancer Registry, 294
Massachusetts Department of Environmental Quality Engineering (DEQE), 290, 293
Massachusetts Department of Public Health, 291, 352
McCarthy, Richard, 84
McCrone, Walter, 95–96, 102
McGill University, Indigenous Names Survey, 68
McKinley, William, commemorated by place-name, 67
Median boundaries, 141
Medieval and Renaissance Manuscripts at Yale: A Selection, 98
Mercator projection, 16–22
  area distorted by, 16–19
  condemned by Arno Peters, 18
  condemned by Ward Kaiser, 36–37
  Heartland portrayed on, 173, 184
  for large-scale maps, 20
  misuse of, 20–22
  in navigation, 19
  Peters map compared with, 16–18
  popularity of, 20, 22
  Third World distorted on, 16–18, 41
Metes and bounds, 115–18, 121
Michigan Academy of Science, 161–62
Midoceanic ridges, 168
Mining rights, on ocean floor, 137–38
Minorities, voting strength of, 201–3, 207–8, 212–13, 219
Mississippi, origin of name, 66
Mississippi (state), boundary with Louisiana, 126–30
Mississippi River, as boundary, 126–30
Mollweide projection, 16
Monaco, Cindy, 245, 249
Monitoring networks, need for, 294–95
Moon, Nancy, 248
Moon, possible origin of, 160
Morison, Samuel Eliot, 82
Morrill, Richard, 215

Morrison, G. J., 21
Mother Jones, 17, 35
Mount Kenya, 171
Mount St. Helens, 264–65
Moynihan, Daniel Patrick, 276–77
Mrazek, Robert, 209
Mt. McKinley, proposal to rename, 67
Mudflows, 264, 266
Multimedia, 298
Musmanno, Michael
  Laurence Witten ridiculed by, 89–90
  Vinland map attacked by, 85, 88–90

Nakkerud, Ted, 85
Names. *See* Place-names
Names on the Land, 65
National Association for the Advancement of Colored People (NAACP), 53
National Broadcasting Company (NBC), 22
National Council of Churches, 10, 11, 17, 25, 32, 34
National Flood Insurance Program, 123
National Geographic, 146, 180
National Geographic Society, 21
National Governors' Conference, 223
Natural hazards, emergency-response maps for, 267–68
Nature, 158
Navigation rights, 137–38
Nazi propagandists, 170–71, 181–83
Nevada, radioactive waste in, 223, 224
Nevado del Ruiz, volcanic disaster, 266–67
New Cartography, The, 22, 32–33
New Hampshire, radioactive waste disposal, 224
New Jersey, radioactive waste disposal, 224
New Jersey Department of Environmental Protection (DEP), 257–59
Newspeak: A Dictionary of Jargon, 56
New Statesman, 83
Newsweek, 204
New World Order, 186

O'Connor, Sandra Day, 205, 244
Oglethorpe, James, 132
Oil companies, and wetlands, 280
Olean (N.Y.) *Times-Herald*, 252
Omni Gazetteer of the United States of America, 48–50
Oneida County, N.Y., 193–95
Oneida Nation Territory, 146–47
Onondaga Nation Territory, 146–47
Onsrud, Harlan, 122
Oregon Historical Society, 53
Origin of the Continents and Oceans, The, 152, 163
Ortelius, Abraham, 158
Oswego, N.Y., 273, 274
Our Wandering Continents, 160
Overlay analysis, 227, 229
Oxbow lake, 127
Oxfam, 28
Oxford University, 171
Oyster Bed Island, 134–36

Packing, as a strategy in gerrymander, 201
Painter, George, 76, 78, 81–83, 92, 97, 103
Pakistan, political renaming in, 60
Palestine, place-name changes in, 57–59
Pangaea, 153, 164, 167, 168, 169
Paris Peace Conference, 177, 181
Parker, John, 94–95
Parker, William, 182
Parkfield, Calif., seismic risk, 262
Parti Québécois, 63
Partisan gerrymandering, 214–15
Pax Americana, 186
Payne, Roger, 52
Pedulla, Augusto, 85
Penn, William, 108, 109, 110
Pennsylvania, disputed boundary, 109–12
Penobscot River, 66
Pentagon, 22
Peru, marine boundary dispute with the United States, 139
Petermanns Geographische Mitteilungen, 152
Peters, Anneliese (first wife of Arno Peters), 25, 26

Peters, Aribert, 25, 27
Peters, Arno, 10, 25–28
  condemned Mercator projection, 18
  criticized by cartographers, 24–25
  criticized professional cartographers, 26–27, 32–33
  dismissed Goode's projection, 22–23
  early career, 25–26
  history textbook by, 26
  map projection principles, 22–24
  New Cartography, The, 32–34
  proposed new zero meridian, 43–44
  supported by Ward Kaiser, 34–37
  unaware of Gall's projection, 27
  world atlas by, 37–39
Peters Atlas of the World, 27, 37–39
Peters projection, 3–4, 9–11, 42–44
  Africa distorted by, 34
  benefits of, 4, 44
  Brandt Commission promoted, 30
  cartographers condemned, 4, 12–13, 24–25, 31, 33–34
  compared to globe, 9–11
  golden section and overall shape, 43
  religious organizations and, 10, 25
  Third World groups and, 28–30
  shape distortion on, 10, 13, 39
Petrone, Frank, 277
Pickles, John, 41–42
Pierpont Morgan Library (New York City), 86
Pivot Area, in Central Asia, 174, 176
PIXE (particle-induced X-ray emission), used to test Vinland map, 102
Place-names, 45, 56–57, 69–71
  embarrassment over, 54–57, 71
  geopolitical significance of, 57–64
  language and, 57, 63–64, 65, 67–68
  local adoption of, 45, 54, 65–66
  political control and, 4, 57, 64, 71
  racism in, 4, 45–50, 54–56, 71
  renaming, 50–54, 57–64, 67–68
  risqué, 53
Placet, François, 156
Plane chart (map projection), 13, 14
Plat, 115

Riparian rights, 123–26, 128
Risk maps, design of, 258–60
Rivers, as boundaries, 123–34
Roberts, Sam, 210
Robinson, Arthur, 9–10, 20–21, 22, 31, 131–32
Robinson projection, 187
Rockefeller, Nelson, 84
Rogers, Rutherford, 96, 99
Romans, renaming by, 58
Rome, N.Y., 193, 195, 208
Romm, James, 158
Roosevelt, Franklin D., 171
Royal Geographical Society, 171, 172, 180
Russia, strategic location of, 149, 175, 184

Saarinen, Thomas, 41
Salt Lake City, Utah, seismic risk, 262–63
San Andreas fault, 168, 260–62
Sanchez de Huevla, Alfonso, 85
Savannah River, as boundary, 130–37
Scalia, Antonin, 137
Scandinavian Airlines System (SAS), 81
Scholastic Aptitude Test (SAT), 285
Schuchert, Charles, 165–66
Science, 284
Science 84, 10, 17, 18, 35
Science Digest, 82
Science News Letter, 82
Scott, Franklin, 93
Scottish Geographical Magazine, 11, 12
Sea-floor spreading, 168
Sections, 113–14
Sedentary species, 138
Seeger, David, 252
Segal, Ronald, 39
Seniority, of legislators, 219
Sensitivity analysis, 236
Setback lines, 122
Severe winter weather, 268, 270
Shape, as evidence for gerrymandering, 203–6, 210, 212–13
Shatterbelts, 185, 186
Shaw v. *Reno*, 205
Shoestringing, gerrymandering by, 201

Shoreham nuclear power station, 275–78
Sial, 167
Sierra Club, 222
Sima, 167
Simons, Howard, 82
Sinusoidal projection, 14–15, 16
Skelos, Dean, 210
Skelton, Peter (R. A. Skelton), 74–76, 78, 81–83, 90–92, 97, 98, 103
Skepticism, need for, 2, 298, 300
Smith, John (colonial explorer), 109
Smith, John (resident of Taylor, N.Y.), 244
Smithsonian Institution, 95
    sponsored Vinland Map Conference, 92
Snider-Pellegrini, Antonio, 156–57
Snow, John, 283
Snyder, John, 11, 31–32, 35–36
Software-data incompatibility, as threat to public access, 251–54
Sonar maps, 168
Souter, David, 205
South Carolina
    boundary with Georgia, 130–37
    radioactive waste disposal, 223, 224
    waste from New York, 243
Southeast Compact Commission, 254
Southeast in Early Maps, The, 131
South Russia, 171, 172, 180
Soviet Union. *See* Commonwealth of Independent States
Spain, claims to the New World, 107–8
Special master, 131, 210, 215
Speculum Historiale, 77–80, 93–94, 95
Spykman, Nicholas John, 184, 188
Stack Island, 128–29
Stalingrad, renamed, 62
Stanford University, 166
State of the World Atlas, 39–40
Status and Trends of Wetlands in the Conterminous United States, Mid-1970s to Mid-1980s, 282
St. Brendan the Navigator, 85
Sternglass, Ernest, 284–88
Stevens, John Paul, 137, 205

United States Bureau of the Census, 6, 216, 227, 289
United States Central Intelligence Agency (CIA), 58–59
United States Court of Appeals, 130
United States Department of Justice, 192, 196, 201–3, 205, 208, 212–13, 216
United States Department of State, 22
United States Environmental Protection Agency (EPA)
  Superfund program, 289, 292
  wetlands protection, 279–81
United States Fish and Wildlife Service, 279, 281–82
United States General Accounting Office (GAO)
  criticized New York radioactive waste siting process, 246–48
  Shoreham nuclear plant study, 276–78
United States Geological Survey (USGS), 20, 168, 169, 230, 249, 306–7
  earthquake-hazard mapping by, 260–63
  indefinite boundaries on maps of, 128
  offensive place-names on maps of, 47–50, 52, 54
  volcano-hazard mapping by, 263–65
United States House of Representatives
  New York reapportioned, 209–12
  North Carolina, 203–6
United States Library of Congress, 197
United States National Land System, 114
United States Office of Management and Budget, 280
United States Postal Service, 22
United States Soil Conservation Service, 279
United States Supreme Court, 5
  on boundaries, 130, 133–34, 135–37
  on reapportionment, 196, 202, 205–6
  struck down take-title provision, 244
University of Berlin, 151
University of Bremen, 30
University of California, 180
University of Graz, 152
University of Marburg, 152

University of Pittsburgh, 285
University of Reading, 171
University of Washington, 215
USA Today, 34, 237
U.S.S.R., presence in Antarctica , 145–46
  See also Commonwealth of Independent States; Russia
Utica, N.Y., 193, 195, 208, 218
Utility lines, 118

van der Gracht, W. A. J. M. van Waterschoot, 164–65, 167
Van Valkenburg, Samuel, 332
Vector model, of a GIS, 230
Vellucci, Alfred, 84
Vermont, radioactive waste disposal, 224
Versailles, Treaty of, 61
Vietor, Alexander, 78–79, 82, 86, 92, 95, 101, 103
Vikings, explored North America, 75, 80, 81–83
Vincent de Beauvais, 78–79, 93
Vinland map, 5, 72–105
  age of, 72–73, 91–92, 96
  chemical and spectral analysis of ink, 95–97, 102–3
  exposed as forgery, 5, 96–97
  historical significance, 81–83, 103–4
  Italian-American attacks on, 83–87
  media coverage of, 81–83, 96–97
  as medieval cartography, 80–81
  ridiculed by Michael Musmanno, 88–90
  scholars' appraisal of, 75, 82, 90–95
  Speculum Historiale and, 77–80, 93–94
  Tartar Relation and, 76–80, 93
  Yale University Press promoted, 81–83
Vinland Map and the Tartar Relation, The, 79–83
Vinland Map Conference, 92–95, 99
Volcano hazards, maps of, 263–67
Volunteered sites, 227, 239–40, 246–48
von Humboldt, Alexander, 156
Voting Rights Act (passed 1965, amended 1982), 191, 201–2, 203–4, 207, 208, 210, 212–13, 216

Voxland, Phil, 34
Vujakovic, Peter, 27, 28–29
Wall Street Journal, 278
Washburn, Wilcomb, 90, 92
Washington, low-level radioactive waste
    disposal in, 223, 224
Washington County, N.Y., candidate area
    for radioactive waste disposal, 236
Washington Post, 82
Water bodies, as boundaries, 123–34
Water column, 138
Watling Island, 75
Wegener, Alfred Lothar, 5–6, 151–58,
    162–70, 180
  as cartographic propagandist, 152–53,
    164, 169, 173, 188
  developed drift hypothesis, 153–54
  manipulated shape of continents, 164,
    165–66, 188
Weigert, Hans, 183
West Bank, political renaming in, 57–59
Western New York Nuclear Service
    Center, 226, 244
West German Press and Information
    Office, 30
Weston, Roy F., Inc., 251–53
West Valley, N.Y., radioactive waste dis-
    posal at, 222, 226, 244–45, 254
*Wetlands Losses in the United States, 1780s
    to 1980s*, 281
Wetlands mapping, 278–82
White, Richard, 122
Who's Who in the World, 25
Wilde, Kathleen, 204
Willis, Bailey, 166
Wilson Library Bulletin, 39
Winter, Nancy, 272–73
Wisconsin, reapportionment in, 217
Wisconsin Legislative Reference Bureau,
    217

Wisconsin Public Radio, 3
Witten, Cora (Mrs. Laurence), 78–79, 99
Witten, Laurence, III, 80, 92, 94, 96, 103
  attacked by Michael Musmanno, 89–90
  purchased Vinland map, 78–79
  revealed seller of map, 98–101
  at Vinland Map Conference, 92–93
Woburn, Mass., leukemia cluster, 288–94
Wolman, M. Gordon, 150–51
World Council of Churches, 28
World Development Movement, 29
World Island, geopolitical concept of,
    149, 170, 173–74, 176–78, 188
World Ring, 181
World Trade Center, bombing, 270
World War I, 61
World War II, 22, 61, 62
Worm holes, on Vinland map, 73, 77, 79,
    88, 93
W. R. Grace Company, 288–89, 293–94

Xerox effect (on map symbols), 260, 273

Yale University, 4–5, 165, 166
  Beinecke Rare Book and Manuscript
    Library, 73, 96, 97
  bought Vinland map, 78–79
  face-saving by, 98
  See also Yale University Press
Yale University Library Gazette, 98
Yale University Press, 79–83
Yellowstone National Park, 53
Young, Bruce, 291–92

Zeigler, Don, 275
Zeitschrift für Geopolitik, 181
Zelen, Marvin, 292–93, 352
Zionists, political renaming by, 58
Zoning maps, 118–20

For my parents, my first teachers.

CONTENTS

PREFACE

"Not since Adam has any human known such solitude as Mike Collins is experiencing," a NASA official commented when the *Columbia* spacecraft swung behind the moon on July 20, 1969.[1]

Collins would spend the next forty-eight minutes orbiting the far side of the moon, blocked from all radio communication with his crewmates, Neil Armstrong and Buzz Aldrin, who were on the other side of the lunar surface, as well as the rest of humanity back down on Earth. He felt this isolation "powerfully."[2] Outside his window, the vastness of space—teeming with stars—contrasted sharply with the darkness of the lunar surface below. On board *Columbia*, a spacecraft he affectionately dubbed his "mini-cathedral," Collins occupied his time busily preparing the ship for his crewmates' return. While Armstrong and Aldrin became the first humans to walk on the surface of the moon, Collins experienced the solitude of space, floating nearly a quarter million miles from Earth—alone on a spacecraft with no ability to talk to any other person.[3] He jested, "If a count were taken, the score would be three billion plus two over on the other side of the moon, and one plus God only knows what on this side."[4]

This sense of solitude would not last long. What impressed Collins most on his return to Earth was not the isolation of space travel but its unifying effects. When I spoke with Collins almost fifty years after the flight, he told me, "I thought that when we went to different countries that people would say you Americans achieved XYZ."[5] But what he discovered on returning to planet Earth was the opposite of what he experienced orbiting the far side of the moon: a profound sense of community.

Collins explained to me that everywhere the astronauts went, "It was 'we.' We human beings—'We did it, we did it.' That was the punch line, everywhere we went." As the Apollo 11 crew circled the world on a postflight diplomatic tour, touching down in twenty-seven cities in twenty-four countries, they observed the same refrain: "We did it."[6]

That use of *we* instead of *you Americans* or *you astronauts* attests to the profound sense of collective participation felt by billions of people when humans first set foot on the moon. It also hints at the growing awareness of global interconnection—or the sense that we are part of one global village—that arose alongside, and in part, because of the Space Age.

As Neil Armstrong climbed down the *Eagle*'s ladder and took "one small step" into the dusty lunar regolith, a record-breaking global audience waited with rapt attention. Never before had so many people come together to witness an event. But it wasn't just the numbers that made this audience exceptional. The sense of participation and global unity shared by billions of people around the world became one of the most significant consequences of the first lunar landing, with reverberations that affect us to this day.

During our conversation Collins added an essential point: this use of *we* around the world must have been "worth its weight in gold" for the US State Department and US Information Agency.[7] As he knew well from his years as an astronaut followed by his tenure as the assistant secretary of state for public affairs, the unifying effects of Project Apollo were not just spiritual; they were also political.

This book arose from the right combination of intention and accident. Sitting in the US National Archives on an August afternoon in 2007, I was researching how scientific programs affect culture and politics. Taking the Smithsonian Astrophysical Observatory's global network of satellite tracking stations as my jumping-off point, I spent the summer reading memos, reports, letters, and any other archival material I could get my hands on. In the archive's airy reading room,

I focused on the Smithsonian's close relationship with the Tokyo Astronomical Observatory, a Meiji-era institution established in 1888.

The material I had been reviewing told me about the day-to-day workings of the cooperative international project at the observatory, but it seemed like something was missing. I had just read John Dower's *Embracing Defeat*, a masterful study of the postwar reconstruction of Japan from within. "What matters," Dower stresses, "is what the Japanese themselves made of their experience of defeat."[8] What role did science play in the larger story of US-Japanese relations in this period? I knew that the effects of the US use of two atomic bombs during World War II echoed far beyond 1945. As an article published in the Japanese newspaper *Asahi Shimbun* less than a week after Emperor Hirohito announced Japan's surrender famously explained, "We lost to the enemy's science."[9] How were the astronomers at the Tokyo Astronomical Observatory, as well as people throughout Japan, viewing US science and technology a decade after the war? And how was the US marshaling its scientific and technological programs in support of the nation's political interests in Japan? With Dower fresh on my mind and some hours left in my day, I requested a few boxes from a collection that seemed promising.

What I found, tucked neatly in cream-colored folders, was a story that would come to dominate my life for over the next decade and transform how I understood the relationship between science, power, and globalization.

Holding up a US Information Agency field report dated September 4, 1962, I read, "The Friendship 7 Exhibit in Tokyo was held at the Takashimaya Department Store July 26th through 29th from 10:00 a.m.–6:00 p.m. daily, and was viewed by over half a million people, a crowd exceptional in size even by Tokyo standards."[10] Five hundred thousand people. Five hundred thousand people in just four days. Could a small space capsule really have attracted such an enormous crowd? I read on. Several hundred police and guides, it continued, "channeled the crowd up nine flights of stairs, zig-zagged them across the roof and brought them down nine flights of stairs to the exhibit."[11] The scale and level of interest in the exhibit are hard to comprehend. This crowd was exceptional in size not just by Tokyo standards but by any other standards. Clearly, the department-store space capsule exhibit suggests a passionate enthusiasm for the US space program in Japan. But what is the larger significance of this popularity? I knew that John Glenn became a national hero within the United States after his flight. The country conferred the status of celebrity on him. But what did his space capsule mean to people in Japan? Why did they wait in a five-hour line to walk by his small, charred vehicle? Does this story hint at something larger, something more fundamental about the ties between early spaceflight and foreign relations?

Luckily, I did not have to wait long to start finding the answers to these questions. That summer I had a fellowship at the Smithsonian National Air and Space Museum (NASM). The day after first reading about Glenn's space capsule exhibit in Tokyo, I returned to my office in the Space History Department to ask curators what they knew about the exhibit. Although a few had heard of a capsule tour, it was not an episode anyone knew about in detail. I returned to the archive to find out more.

As I delved further into those cream-colored folders, I soon discovered that this record-breaking exhibit was no chance event. Instead, this modest capsule display exposed a nuanced political tactic at the core of US grand strategy in the early 1960s. At the height of the cold war, both the US and the Soviet Union were mobilizing their vast technical and scientific resources to wield global influence. They built transnational ties through scientific exchanges and education. They attempted to influence and at times divert the trajectory of other national scientific and engineering research programs. And as in the case of the *Friendship 7* exhibit in Tokyo, they attempted to foster political alignment through demonstrations of scientific and technological preeminence. The wildly popular space capsule exhibit in Tokyo was just one part of a much larger, more extensive US initiative to spread liberal democratic values. Spaceflight spectaculars, and their promotion abroad, were by design aimed at winning over international public opinion, countering anti-American sentiment, and, most importantly, shaping the emerging global order. The Toyko exhibit was not the first time, nor would it be the last, that the Kennedy administration looked to spaceflight as an essential arm of US diplomacy.

What I learned is that the moon landing and diplomacy were profoundly and intricately allied. From the very start, US politicians weighed the soft-power potential of space exploration as they evaluated which programs to fund.[12] They argued over the psychological benefits of being "first" in space. They created a global communications infrastructure explicitly so people around the world could follow US space successes. They spent millions of taxpayer dollars on space-themed films, exhibits, press releases, buttons, lectures, and radio broadcasts to promote and leverage international interest in spaceflight. They hired polling firms on every continent that repeatedly assessed the effectiveness of space propaganda. And through years of feedback and fine-tuning, they honed a powerful message that bound global progress with US space accomplishments.

As the story of Project Apollo makes clear, people—not just advances in transportation, trade, and communication—shape and propel the process of globalization. Individuals have advanced our awareness of interconnection and have created experiences that unite us. Mike Collins cogently captured this when he explained to me, "The response we got—we human beings have landed on the moon—it was the 'giant leap,' as Armstrong put it."[13] Each person I spoke with over the past decade amplified and sharpened the significance of "we" in the story of lunar exploration. Buzz Aldrin and I laughed about his exploits with Italian paparazzi while he traveled the world after his flight. As I walked along a riverbed in western Japan with artist Michio Horikawa, he reflected on how the collection of moon rocks prompted his deeper appreciation for the rocks here on Earth. From the shiny offices of a design firm in Manhattan, World's Fair exhibit designer Jack Masey explained why space exploration was such a potent form of propaganda. In Oslo, Erik Tandberg, a television personality, told me how he became the "Norwegian Walter Cronkite." Over a long lunch at Apollo astronaut Jim Lovell's restaurant north of Chicago, I heard the story of the global Christmas Eve broadcast from the moon. And while celebrating the fortieth anniversary of the first lunar landing with Neil Armstrong at MIT, I spoke with him about the diplomatic responsibilities of astronauts and the sense of international participation shared around the globe. He told me in his quiet and unassuming way that this was an essential piece of the Apollo story that should be told. And so I took his advice.

Introduction

MOONRISE

S hortly after Neil Armstrong and Buzz Aldrin took their first
steps on the moon and before they planted an American flag
into the lunar soil, they unveiled a plaque mounted to the lunar
module and read its inscription: "Here Men from the Planet Earth
First Set Foot Upon the Moon, July 1969, A.D. We Came in Peace
for All Mankind." Above the inscription an image of the Earth's two
hemispheres—simple and solid like block prints—depicted an un-
divided planet with no political boundaries, as it is seen from outer
space. The Apollo 11 crew and President Nixon's names and signatures flanked the bottom of the plaque.[1] Crafted by the National Aeronautics and Space Administration (NASA), the White House, and
other US government staff, this inscription, the outline of the Earth,
and the signatures, although quite concise, signaled a set of values—
openness, progress, religion, inclusivity, service, and universality—
melding them into a symbolic representation of US leadership.

Well in advance of the Apollo 11 mission, NASA's head of public relations and the assistant administrator for international affairs
widely solicited advice for the plaque. Consulting with the Smithsonian Institution, the Library of Congress, the archivist of the United
States, the NASA Historical Advisory Committee, the Space Council, and congressional committees, NASA finalized a first draft of the
plaque.[2] President Nixon's advisors and speechwriters tailored the

Apollo 11 plaque mounted on the lunar module. (NATIONAL AERO-
NAUTICS AND SPACE ADMINISTRATION)

text and image to fit the interests of the administration. The Central
Intelligence Agency (CIA) notified White House staff that the Soviet
Union's robotic spacecraft might land on the moon first, beating the
Apollo 11 crew to the surface. In response, speechwriter and advisor
Pat Buchanan proposed that the plaque read "set foot" as opposed to
"landed." William Safire, another speechwriter, edited the phrase "we
come in peace" to "we came in peace," in order to disassociate it from
"something you'd say to Hollywood Indians." The addition of "A.D."
to the date, remarked Safire, was a "shrewd way of sneaking God in."[3]
The last phrase on the plaque, "for all mankind," served as the motto
of America's Space Age, populating dozens of presidential speeches,
international exhibit panels, film scripts, newspaper articles, and

radio broadcasts, and appearing in the 1958 Space Act, which framed America's space efforts.[4]

Not long after Armstrong and Aldrin unveiled the plaque on July 20, 1969, President Nixon picked up his minty-green touch-dial phone and called them from the Oval Office.[5] First, he congratulated them and then conveyed a message to the broader global audience tuning in to the broadcast that evening. It was a brief statement, but similar to the one on the plaque: the president's message signaled the major themes that US government officials had been crafting through years of public diplomacy programming. The Apollo 11 crew positioned themselves in front of the TV camera, and then Nixon articulated the significance of the first lunar landing: "Because of what you have done the heavens have become a part of man's world. . . . For one priceless moment in the whole history of man, all the people on this earth are truly one."[6]

As President Nixon's telephone conversation with the Apollo 11 crew underscored, the moon landing extended the scope of human experience in two ways: for the first time in history humans landed on another celestial body; second, they came together to witness an event in larger numbers than ever before. One fifth of the world's population watched the live television feed from the moon, while hundreds of millions more listened to the radio broadcast. In total, half of humanity followed the flight, a higher portion than for any previous event in history.

After the Apollo 11 crew returned to Earth, they watched recordings of the television coverage of their mission. Just as hundreds of millions of people had done on July 20, 1969, they saw their moon-walk alongside images of captivated audiences from around the world following the flight. "It seemed like the entire world was having a party," thought Buzz Aldrin. The event did not just take place on the moon. Recognizing this, Aldrin turned to Neil Armstrong and said: "Hey, look. We missed the whole thing."[7]

Like the message inscribed on the Apollo 11 plaque, this global audience was not spontaneous. In the 1960s the United States initiated the largest government-sponsored public relations campaign in history to encourage foreign audiences' sense of participation in US spaceflight. From organizing hundreds of thousands of space exhibits and film screenings to creating a global communications infrastructure so people could follow the first lunar landing to small gestures like US diplomats coauthoring space-themed songs with local musicians, the United States proactively cultivated a global community connected through shared experiences with spaceflight.[8]

These global connections did not start with the moon landing. Long before Apollo 11, the world had been growing increasingly interconnected. The pace of this interconnection intensified in the late nineteenth century with the development of new transportation and communication technologies. This period was also marked by the importance of Europe in world events. Industrialization paired with imperialism ensured Europe's dominance in international affairs through the early twentieth century. Although people, goods, and information traveled around the world at a quicker rate than ever before, globalization also exacerbated divides within societies, between the powerful and the weak, between the wealthy and the poor, and between the West and the non-West.[9]

As the world became more and more interconnected throughout the twentieth century, the character of globalization began to change. Two factors distinguish this era: First, the United States took a more assertive role in world affairs. Embracing its newfound position of global superpower and driven by an interest in containing the spread of Communism, the US government actively pursued political and economic influence while also promoting American culture abroad. Second, the cold war era saw a rise in the consciousness of the interdependence of humankind. Whereas Western-based technology and ideology drove nineteenth-century globalization, in the mid-twentieth century millions of people around the world participated in and shaped this process. This consciousness of interdependence did not erase diversity and division but instead enhanced our awareness that we all live in one interconnected world. These two factors—what distinguishes this moment in history—are intertwined. This is where the moon landing comes in.[10]

On May 25, 1961, just four months after assuming office, US president John F. Kennedy proposed Project Apollo at the tail end of an address to a joint session of Congress on "urgent national needs." "These were extraordinary times," Kennedy began. "Our strength as well as our convictions have imposed upon this nation the role of leader in freedom's cause." The battle was taking place in what Kennedy called the "lands of the rising people": Asia, Latin America, Africa, and the Middle East. A global revolution was under way, and if the United States did not act decisively, "the adversaries of freedom" would "ride the crest of the wave—to capture it for themselves."[11]

Kennedy's ominous warning reveals a great deal about why a nation would commit itself to sending humans to the moon, the most expensive civilian technological endeavor ever undertaken by the United States. At the beginning of the twentieth century, colonial powers governed most of Africa, Asia, the Caribbean, and the Middle East. Nearly a billion people were under colonial rule, more than a third of the world population. In the 1940s, World War II upended the global order. It left the European powers socially and economically devastated while propelling the United States to superpower status. A wave of independence movements led to the disintegration of European empires. Between 1945 and 1970 the number of nations increased nearly fourfold. In 1960 alone—the year Kennedy was elected president—seventeen African colonies became independent nations.[12] With European countries retreating from empire after World War II, and the emergence of new states, the geopolitical order of the world was undergoing a profound transformation. The United States and Soviet Union stepped in, competing to create a global coalition aligned with their respective ideologies. The United States sought to foster and spread liberal democracy, but the Soviet Union advocated Communism.[13]

Simultaneously, the introduction of nuclear weapons at the end of World War II upended how wars were waged. The US and Soviet Union staged proxy wars and launched covertly backed coups, but much of their geopolitical influence came from soft power rather than coercive force. Nuclear stalemate and US-Soviet rivalry elevated

the significance of propaganda and psychological strategy in this period.[14] As Kennedy put it in his May 25 address, the Soviet Union's
"aggression is more often concealed than open." The Soviets did not
fire missiles but sent arms, agitators, aid, and propaganda to fight this
war. The "adversaries of freedom" were consolidating their territory
and threatening the "hopes of the world's newest nations."[15]

The cold war was "a battle for minds and souls as well as lives
and territory," Kennedy told Congress and the American people on
May 25. It was not simply a matter of superpower conflict but instead a deeply ideological confrontation that pitted capitalism against
socialism. It took place on every continent, and it frequently played
out in newly independent nations.[16] The United States must act as a
model, Kennedy argued. In addition to strengthening the American
economy, the United States must invest in the economic and social
progress of other nations as a bulwark against socialist influence.

Next, Kennedy called for additional funds for "the world-wide
struggle to preserve and promote [American] ideals." During the
early cold war, Soviet and US leaders invested heavily in what today
we call public diplomacy, a combination of national advocacy and
propaganda. Mass communication in particular played a leading role
in this process. Drawing on the tools of radio, television, film, photographs, exhibits, and other forms of media, the USSR and the US
produced wide-ranging programming with the hope of shaping or
reshaping the politics, economics, culture, values, and social relationships of other nations.[17]

Finally, after more than thirty minutes of outlining an extensive plan, Kennedy added one more "urgent national need": Project
Apollo. For Kennedy, sending humans to the moon and returning
them safely back to Earth was part of the "battle going on around
the world between freedom and tyranny." It was not simply about
securing American prestige and projecting technological capability.
Spaceflight was affecting "the minds of men everywhere, who were attempting to make a determination of which road they should take." It
was time for the United States to take a leading role in space exploration, "which in many ways may hold the key to our future on earth."[18]

Importantly, Project Apollo was not the only "urgent national need" proposed to Congress and the American people on May 25, 1961. Although we usually remember this speech for Kennedy's call to send humans to the moon, the first thirty minutes reveal a context that is critical for understanding how and why the United States pursued lunar exploration. This geopolitical context does not simply add details to the story of the Apollo program; it also "changes its terms," as Mary Dudziak once wrote.[19] Project Apollo played a strategic role in US attempts to align the values and interests of the emerging, post-colonial world order with those of the United States rather than those of the Soviet Union.

Perhaps ironically, President John F. Kennedy was not a space enthusiast. Shortly after the Soviet Union launched Sputnik in 1957, over drinks at Boston's storied Locke-Ober Café, he supposedly told rocket guidance pioneer Charles Stark Draper that all rockets were a waste of money. As a US senator and then later during his 1960 presidential campaign, Kennedy paid little attention to the future of space exploration beyond its impact on his electability. But in 1961 the young president proposed the most ambitious space program in national history when he announced Project Apollo before a joint session of Congress. This quick about-face was not a reversal in Kennedy's attitude toward spaceflight. Instead, the president recognized that lunar exploration had the potential to restore America's geopolitical standing in the wake of Soviet cosmonaut Yuri Gagarin's April 1961 mission—the first human spaceflight in history—and the failure of the CIA-backed Bay of Pigs invasion during the same month. No other project, he argued, would be "more expensive" or "more impressive to mankind."[20]

Project Apollo was an enormous national investment. Sending men to the moon cost the United States the equivalent of hundreds of billions today, more than eighteen times what the country spent on the Panama Canal, more than five times the expense of the Manhattan Project, and even more than the Eisenhower administration's interstate highway system. Before the decade was out, Project Apollo had employed a workforce of hundreds of thousands, initiating a warlike mobilization of federal resources. A moon shot's broad appeal, Kennedy explained, could advance US diplomatic goals without resorting to "hard" and expensive military power.[21]

But when the United States invested in Project Apollo, it invested in more than lunar exploration. Along with developing space suits and rockets, the United States developed an accompanying public relations campaign that spanned the globe. Over the 1960s, through hundreds of thousands of film screenings, exhibits, pamphlets, books, radio broadcasts, spacecraft tours, astronaut appearances, and the distribution of space buttons and press packets, the US not only told the story of America's space program; it actively cultivated a global audience that was interested in and engaged with spaceflight. All the while, the US built up a global communications infrastructure that enabled this audience to watch the first lunar landing live—all together. Hundreds of millions of people around the world shared this first-ever moon landing experience in unison.

In the early 1960s the United States showcased its space program to demonstrate the nation's technological, economic, and political superiority on the world stage. By the end of the decade, public diplomats' approach to space programming had shifted notably. No longer were space feats employed primarily to highlight American prestige and power. Instead, the emphasis of this programming portrayed spaceflight as a global accomplishment undertaken "for all humankind." This was a direct response to the voices of people around the world whom the US tried to influence. As public diplomats listened to audiences on each continent, from cities to rural areas, from children to presidents, they shifted their messaging from a demonstration of power to one of inclusiveness, from technological and scientific might to one of humanity and unity.

Project Apollo did not win the cold war contest between capitalism and socialism. But it had a very real, concrete impact on US foreign relations. Countless reports from US ambassadors and diplomats describe how the space program—and associated public diplomacy events—created goodwill, offered opportunities to meet with local politicians, and drew record-breaking audiences to exhibits and embassy events. Apollo did not erase the effects of US militarism, civil rights injustices, poverty, and other negative marks on the nation's image abroad. But it did—if only temporarily—displace newspaper headlines that threatened to tarnish the US image overseas. As Apollo 8 commander Frank Borman later reflected, "It cast the country in a favorable light at a time when there were many things that cast it in an unfavorable light . . . the Apollo program did much more than just advance the country scientifically and technically, it advanced it in my opinion diplomatically just as much."[22] Kennedy and Nixon saw the advantages of scheduling international trips on the heels of popular spaceflights. Nixon's 1969 tour of Southeast Asia and Romania was code-named Operation Moonglow because of the expectation that Apollo 11's popularity and prestige would lend their sheen to the president's new foreign policy initiatives. Even to this day, more than fifty years after Apollo 11, a "moon shot" remains a powerful and motivating symbol, an endeavor that is simultaneously aspirational and transformational, that is bigger than one country and encompassing all humankind.

Beyond its immediate US foreign relations dividends, Apollo and its associated media campaign propelled the evolution of globalization in the twentieth century, a transformation that shaped the world we live in today. It cultivated the consciousness of global interdependence through its far-reaching circulation of icons and images—such as views of Earth from space or of the cratered moon—that formed a shared visual culture; the dissemination of space discourse to every corner of the Earth, such as the phrase "for all mankind"; and, most importantly, the creation of a global community linked by their shared experience of following the moon landing together. Looking at how the content and tone of this programming changed over time reveals that it was not static or one-sided. Instead, it resulted from a dialogue between US government officials and the populations around the world they sought to influence. Apollo contributed to a heightened awareness of global interdependence, but not social and political homogenization. Millions of people on each continent participated in the moon landing, and although they drew on shared

Apollo imagery and phrases, they did so to express their own interests and visions of the future. After the moon landing, the world was still deeply divided on racial, economic, religious, national, and ideological lines. But Apollo helped us see how we are connected, how in spite of our differences our futures are tied together, that we share one beautiful and verdant planet, suspended in outer space.

1

THE LAUNCH OF THE SPACE RACE, 1946–1957

The most important component of our foreign policy is the psychological one.

—HENRY KISSINGER, 1955

A t the launch range near Tiura-Tam, Kazakhstan, Soviet rocket designer Sergei Korolev and his team used caution. Korolev had barely slept in weeks. Anxious that the United States would beat the Soviets into space, he had moved the launch of the artificial satellite Sputnik up a few days to October 4, 1957. Just before dawn on October 3, he personally convoyed the train carrying the rocket and satellite from the assembly building.

"Let's accompany our first-born," Korolev told his staff.[1]

They joined him, walking alongside the train track as the vehicle made its way to the launchpad. Tiura-Tam was a sparsely populated stop on a rail line among Kazakhstan's desert steppes. When planners selected the site for the launch facility, the entire vegetation in the area amounted to three trees standing beside the train station. Plagued by frequent dust storms, soaring temperatures in the summer, and bone-chilling cold in the winter, it was a place to launch rockets but not much else.[2]

By the evening of October 4, the rocket was ready, but Korolev was still nervous. He had the staff check and then double-check the nearly hundred-foot-tall R-7 rocket, now standing erect on the launchpad. The stakes were too high, and Korolev was too stubborn, to let any slipups stand in his way. Earlier in the day when the unseasonably warm weather threatened the satellite's delicate instruments, staff on the launchpad covered the payload with a large white cloth, hoping that this would bring down the temperature. It did not work. Next, they tried shooting cool air from a hose into the payload fairing, which seemed to do the trick. At last, under the illumination of large floodlights, the final checks were complete.[3]

Sergei Pavlovich Korolev, or "SP" to his team, was about to realize his lifelong dream of spaceflight after laboring for years under unimaginably harsh conditions. As with many other early rocket pioneers, H. G. Wells's and Jules Verne's fantastical space exploration capers first sparked his imagination. But while his German counterpart Wernher von Braun used the workforce of Nazi concentration camps to build his beloved V-2 rocket, Korolev labored in a prison camp. In 1938, at the age of thirty-one, he had been snatched from his home in the middle of the night and sent to a gulag as part of Stalin's purges. After nearly seven years in prison, Korolev's health had diminished, but his devotion to spaceflight remained intact. As he stood on the launchpad at Tiura-Tam, about a dozen years had passed since he had been released from the prison camp. Here he was, a chief designer in the Soviet space program, about to launch the first satellite into space.[4]

A scant ten minutes before the launch, Korolev finally felt confident that everything was in order. He left the pad—already evacuated and silent by this time—and joined staff at a nearby bunker. Inside, nearly a dozen people sat at six command-and-control panels monitoring the rocket and satellite. Tensions were high. Only the operators sat while the rest of the personnel stood stiffly, waiting in anticipation. Korolev's eyes moved between the various instruments and the body language of the operators, scanning them for any sign of trouble. As a deputy in the bunker recalled, "If anybody raised their voice or showed signs of nervousness, Korolev was instantly on the alert to see what was going on."[5]

Unlike American rocket launches, the Soviet space program did not use a countdown, no three, two, one . . . blastoff. A voice over a loudspeaker announced the minutes until readiness and commands like "key to drainage." When the voice said "Pusk!" (Launch!), a young lieutenant pushed a button, initiating ignition. The rocket did not spring from the launchpad at this point. Over a minute passed as steam vented around the base of the rocket, and the engines ignited and then started their preliminary thrust process, before the operator finally announced "Pod'em!" (Liftoff!).

Vibrations from the engines pounded the walls of the small bunker. At first the rocket seemed stuck, hovering over the ground. But soon the four boosters with their hundreds of tons of thrust propelled the R-7 into the inky blackness of outer space. A little over five minutes after the launch, Sputnik separated from the core rocket booster and began its first orbit.

"It's too early to celebrate," Korolev warned his apprehensive staff.

Not until Sputnik completed its first ninety-six-minute trip around the Earth and the radio engineer picked up Sputnik's now iconic "beep-beep-beep" and shouted "It's there! It's there!" did Korolev announce that it was time to send word to Soviet premier Nikita Khrushchev.[6]

On the evening of October 4, Khrushchev had joined Ukrainian leaders and guests from Moscow for a leisurely dinner in the large hall at the Mariyinsky Palace, a grand, bright-blue baroque-style building that hinted at the lavish tastes and lifestyles of Russia's eighteenth-century czars. Once a favorite retreat of Catherine the Great, the palace now stood as a weathered remnant of another era. Khrushchev had decided to stop off in Kiev on his way back from vacation in Crimea, a popular resort destination on the temperate Black Sea. A heavy-set, complex character, with protruding ears and a volatility that left an impression, Khrushchev was a man of contrasts. He was complicit in Stalin's crimes such as the ruthless mass executions of Polish nationals but also de-Stalinized the Soviet Union. He was flamboyant, brutal, and also decent; his wife once described him as "all the way up or all the way down."[7] That evening there was an urgent matter occupying his mind, and it was not the harried preparations under way at the Baikonur space launch facility in Kazakhstan. Rather, Khrushchev was fixated on how to oust his longtime friend— or former friend—Marshal Georgy Zhukov from power.

Zhukov was a World War II hero who personally commanded the final attack on Berlin and represented the Soviet Union at the German surrender.[8] In June 1957 Zhukov, who by that time had become the most influential military man in the Soviet Union, thwarted an antiparty group's attempt to overthrow Khrushchev. But now, a mere four months later, Khrushchev was suspicious that Zhukov was planning his own coup.[9] After nearly losing his premiership in June, Khrushchev was on a mission to secure his status as the USSR's undisputed leader. To aid this mission he had gathered the secretary of the Party Central Committee, Leonid Brezhnev, who controlled the defense industry; the first deputy minister of defense and commander of ground forces, Rodion Malinovsky; the first secretary of the Ukrainian Central Committee, Alexei Kirichenko; the chairman of the Presidium of the Supreme Soviet, Demyan Korotchenko; the chairman of the Council of Ministers, Nykyfor Kalchenko; Central Committee secretaries; and his son Sergei Khrushchev.[10]

For much of the evening, the gathering of top officials discussed the harvest, new factories that needed new equipment, and inadequate capital investments. These were typical topics, meant to loosen the premier's purse strings. Near midnight, after hours of conversation, an aide interrupted the meal and whispered something in Khrushchev's ear. He nodded and then excused himself to take a phone call in a nearby room. A few minutes later the premier returned to the dining hall with a smile. He quietly sat back in his chair, paused, calmly looked around the table, and then spoke: "I can tell you some very pleasant and important news. Korolev just called."

He paused again, this time with a secretive look. "He's one of our missile designers. Remember not to mention his name—it's classified." Then Khrushchev announced the news: "So, Korolev has just reported that today, a little while ago, an artificial satellite of the Earth was launched."

The dinner guests responded with polite if indifferent smiles. At that moment no one in the room foresaw the far-reaching and long-lasting significance of Khrushchev's announcement.[11] Even the first article on Sputnik in the leading Soviet newspaper, *Pravda*, was buried below the fold with no mention of the imminent political and social fallout of the pathbreaking launch. It was brief, written in a clinical style with facts and figures like "the carrier rocket has imparted to the satellite the required orbital velocity of about 8,000 meters per second."[12]

A few minutes after Khrushchev's announcement, the aide returned to the dining hall, set up a radio in the corner of the room, and told the dinner guests that the satellite's broadcast could be picked up on the device. He turned the knob, and Sputnik's now famous "beep, beep, beep" filled the hall.[13]

Following the launch of Sputnik, US president Eisenhower tried to counteract the notion of a "space race" between the US and the USSR at every turn. But space exploration became the prime psychological battleground in the cold war. For more than a decade, and some would say right up until the fall of the Soviet Union, space competition served as the measuring stick for national strength, technological know-how, and the efficacy of political systems. How did spaceflight—an idea that once only existed in the imagination of a few scientists, analysts, and dreamers—receive such a lofty status within international relations in the mid-twentieth century? The answer does not simply lie in the popular appeal of spaceflight or the idea that it is in human nature to explore. The United States and the Soviet Union did not invest fortunes in space development solely to push forward the edges of human experience. So what accounts for spaceflight becoming the new political currency within the cold war world order?

The answer is found in the emergence of what a young Henry Kissinger—among other influential political theorists of the day—identified as the "new diplomacy." In 1955 Nelson Rockefeller, Eisenhower's special assistant for psychological warfare, asked Kissinger, at the time a recent PhD and instructor at Harvard University, to join a study panel on the "Psychological Aspects of a Future U.S. Strategy." Kissinger articulated the tenets of this new diplomacy in a secret report to Eisenhower.

Today's international relations, Kissinger wrote, required psychological strategy. Symbols, rhetoric, ideas, and images assumed new political potency in a changed geopolitical landscape. The existence of thermonuclear weapons rewrote the terms of how politics—especially diplomacy—was done. The risk of global annihilation, combined with technological innovations in mass media, revolutionized the influence of the public—especially public opinion—in both domestic and international politics. These factors contributed to a bifurcation of US diplomacy. Political negotiations were taking place on two interconnected planes. High-level political talks between governments were coupled with public diplomacy.[14] After the first gathering of Rockefeller's group, Kissinger would tell fellow panel member Walt Rostow, professor of economic history at MIT, that "the most important component of our foreign policy is the psychological one."[15]

Many historians make the claim that it was not until Sputnik orbited overhead and the world reacted that Eisenhower saw satellite development in terms of prestige. His focus before October 4, 1957, was on reconnaissance and ballistic missile strategy, they say.[16] But records from before the fall of 1957—especially from National Security Council (NSC) meetings—reveal an entirely different chronological arc. They show policy analysts attuned to the psychological implications of spaceflight years before Sputnik made its debut, and attuned to the bearing of psychology on international influence and power. Eisenhower and those who advised him, such as Rockefeller, recognized that satellites would have both military and propaganda advantages. From the very start, policy makers saw satellites as highly visible demonstrations of scientific and technological capability on the international stage.

In 1957 the small Soviet satellite—and more importantly the tenor of domestic and international response—cemented the association of space exploration and national strength for decades to come. The military undertones of space shots, the rapid spread of news coverage, the reverence for technological and scientific achievement at that time, and the political prominence of global public opinion within cold war geopolitics—the core elements of the "new diplomacy"—all played their part.

When Eisenhower took the oath of office in January 1953, the United States faced steep military and political challenges. The cold war raged. The Korean War stalled. During his 1952 presidential bid, Eisenhower charmed voters with his quiet confidence, "plain talk," and smile. And with a meteoric military record—rising from a lowly lieutenant colonel to the commander of the Allied invasion of Europe in just five years—Eisenhower seemed well-positioned to lead the country safely through swelling nuclear-war fears. More so than any president before him, Eisenhower entered office with a fully formed national security philosophy. "To amass military power without regard to our economic capacity," he warned the country at his first State of the Union address, "would be to defend ourselves against one kind of disaster by inviting another." Eisenhower's fiscally responsible national security strategy, articulated powerfully and concisely just two weeks after he assumed office, remained the defining framework of his presidential security policy.[17]

A free-market conservative, Eisenhower believed that government expenditures must be contained. He articulated this message frequently, cautioning his country against becoming a "garrison state" because of cold war fears.[18] Instead, he advocated a "New Look," a phrase taken from the fashion industry. In designer parlance the New Look referred to the lengthening of women's skirts, but in Eisenhower's defense strategy it meant "more bang for the buck."[19] Eisenhower slashed the Army and Navy budgets, arguing that the country should invest in avoiding war, not fighting one. The cold war would likely last for many years, he recognized. The US, then, must take an approach that sustained both the military and economic health of the country in the long term. And because Eisenhower was confident that the Soviet Union would never directly mount an attack on the United States, he favored investment in the psychological and political—as opposed to the military—battlefields of the cold war. He turned to the "new diplomacy."[20]

The New Look is often remembered as Eisenhower's call for the massive buildup of a nuclear arsenal to defend the country while avoiding straining the budget. These weapons would have a psychological impact, deter the enemy, and in turn avoid the need for costly direct conflict with the Soviet Union. But Eisenhower's defense policy steered a more comprehensive global governance agenda. Under the New Look, the United States nurtured international trade, funded development programs, encouraged formal and informal alliances, moderated conflicts between other countries, invested in cultural and educational exchanges, and increased overseas propaganda. The United States would pursue international influence by winning the hearts and minds of the world's public and political leaders.[21]

Eisenhower's experience on the European front during World World II had convinced him that psychological warfare figured prominently in the Allied victory. Propaganda and persuasion, he saw, were integral to power and influence. When he entered the White House, he pushed for an elaborate propaganda program to contain the spread of Communism. At his very first Cabinet meeting, on January 23, 1953, psychological warfare became a focus of discussion. In short order, Eisenhower appointed a special assistant for psychological warfare, put him in charge of the Psychological Strategy Board, and created the President's Committee on International Information Activities to assess US psychological warfare programs and to make recommendations for improving and centralizing these efforts.[22] The committee submitted a report on June 30, 1953, with forty recommendations, including the creation of the Operations Coordinating Board (OCB) within the NSC to manage the psychological aspects of foreign relations and national security. Eisenhower agreed with almost all of the committee's recommendations. However, he decided to consolidate information activities within a brand-new agency instead of under the jurisdiction of the State Department. The CIA continued its covert psychological warfare program, and the newly formed United States Information Agency (USIA) took on the organization of the nation's public information activities.[23]

Under Eisenhower, the USIA grew quickly. Employing five media divisions—press, radio, television, motion pictures, and the Information Center Service—the USIA disseminated information about the United States to cities and towns in each region of the world. By 1960, the agency ran more than two hundred United States Information Service (USIS) posts in ninety-eight countries. More than a thousand Americans managed these posts with the help of more than seven thousand locally employed foreign citizens. Larger posts were staffed by motion picture officers, exhibit officers, press officers, book translation officers, radio officers, and librarians. In some of the most populous countries, such as India, more than fifty Americans worked with nearly five hundred local employees, whereas in countries such as Sierra Leone one US public diplomat would work with three local employees. Each USIS post tailored its information programs to support the particular US political and psychological objectives in that country.[24]

The USIA also operated 164 libraries in 68 countries; distributed news releases, features, and photographs to local newspapers; organized English-language courses in 52 countries; and produced and acquired documentaries and newsreels, which ran in 42 languages for audiences of some 150 million people per week. To reach rural populations, the USIA sent mobile film vehicles equipped with kerosene generators to project these films. The agency showed programs on local television stations and circulated hundreds of exhibits around the world. From Washington, the Voice of America (VOA) broadcast radio programming in 35 languages, 600 hours per week to an audience of 20 million listeners.[25]

During the early years of the Eisenhower administration, the USIA focused on producing information programming that was factual and straightforward. "The agency struck a balance between the posture of objectivity necessary to enhance the agency's credibility and the selectivity and manipulation of information needed to further US objectives," according to historian Kenneth Osgood.[26] The

USIA director attended a monthly meeting at the White House with Eisenhower to update the president on agency activities and foreign public opinion.[27]

It was within this larger context of Eisenhower's parsimonious security strategy, and increased investment in psychological warfare, that the satellite program took shape. The president supported US satellite development but remained wary throughout his eight years in office about investing significant national resources in a space program. The real cold war threat, Eisenhower held, was economic security. Prosperity was central to the nation's soft-power appeal internationally: the economic vitality of the United States was the surest proof of the rightness of the capitalist system. By keeping budgets tight while also promoting the peaceful and progress-oriented intentions of the United States within the international arena, the Eisenhower administration looked to its satellite program as a demonstration of national superiority in the global competition for geopolitical alignment.[28]

Dreams of artificial satellites far predated Sputnik and arose far before Eisenhower articulated his New Look policy. Within the United States, a decade ahead of the space race, the newly formed R&D industry-government think tank RAND evaluated the feasibility of launching a satellite into space. In addition to an engineering analysis, a 1946 RAND report perspicaciously claimed that launching a satellite would not only have military and scientific implications but that it also "would inflame the imagination of mankind, and would probably produce repercussions in the world comparable to the explosion of the atomic bomb."[29] RAND followed this report with another analysis of satellite development four years later. The 1950 report focused on the military and psychological impact of Earth satellites even further. Pulitzer Prize–winning historian Walter McDougall called this second report "the birth certificate of American space policy."[30] Requested by the Air Force, it explained that although the primary use of satellites would be for reconnaissance—both strategic and meteorological—the US government should emphasize their scientific uses as opposed to their military applications. Allies and enemies alike would quickly realize the potential surveillance capabilities of satellites. For this reason, it was vital to US foreign relations to conduct an "open" program and emphasize—at least in public—the peaceful nature of America's space efforts.[31]

Also in 1950, on an April evening at the tail end of a dinner party in the suburbs of Washington, a group of scientists came up with a bold idea over dessert. Physicist James Van Allen had thrown the dinner in honor of the eminent British geophysicist Sydney Chapman, who was visiting the United States. Described as unassuming but also a "doer," Van Allen had been placing Geiger counters aboard V-2 rockets to record cosmic rays in the atmosphere since 1945. Although he did not know it at the time, the dinner party discussion would ultimately lead to his most noteworthy contribution to science and the first major discovery of the Space Age: Earth-circling radiation belts, later named in his honor. Rocket-based research techniques, like Van Allen's Geiger counters, were opening exciting new avenues in Earth science, capturing the imagination of his friends and colleagues as they feasted on chocolate cake.[32]

Conversations that spring evening ranged from discussions of high-altitude science to the plausibility of coordinating atmospheric research on a worldwide scale. Lloyd Berkner, a physicist, Antarctic explorer, and government advisor, suggested that a third International Polar Year (IPY) should be held. During the two previous IPYs, in 1882 and 1932, scientists from around the world banded together to conduct standardized research. In 1957, as they all knew, the sun would be at its most active. Why not organize a large-scale Earth science research program timed for the solar cycle? Described as "a big man concerned with big ideas and big things," Berkner spoke forcefully and seemed to talk "in capital letters."[33] His "big idea" on that April evening planted the seed for what would become the largest international scientific program up to that point, the International Geophysical Year (IGY), and ignite the Space Age. As the small group of scientists sat together in James Van Allen's home, they agreed that the timing was right to start planning for a worldwide effort to study the Earth and its environment.[34]

The scientific rationale for holding another international science program easily convinced the group, but Berkner had other motivations for proposing it that night. He had just completed a report commissioned by the under secretary of state, James Webb, who would later become the NASA administrator during the Apollo program. Titled "Science and Foreign Relations," the report outlined the benefits that science brings to national security, welfare, and general progress. It recommended placing scientific attachés at US embassies around the world, gathering scientific intelligence, and providing technical assistance to newly independent nations, among other initiatives. Berkner's work on the report, combined with a recent appreciation of the military significance of the polar regions, inspired him. For Berkner, scientific internationalism served US national security and global power interests.[35] By 1954, Berkner and others foreseeing the benefits of satellites to science as well as national security advocated for their inclusion in the IGY. The following year the United States and Soviet IGY committees announced satellite programs, but not without political interest outside the world of science.[36]

In the years following Van Allen's dinner party in 1950, numerous scientists, engineers, and military leaders looked into the advantages of—and issues surrounding—establishing an American satellite program. Manhattan Project scientist Aristid V. Grosse prepared a report that he shared with the top levels of the Truman administration, warning of the psychological repercussions if the Soviet Union achieved the first satellite in orbit. The report predicted that a Soviet first in space "would be a serious blow to the technical and engineering prestige of America the world over. It would be used by Soviet propaganda for all it's worth."[37] In a similar vein, German rocket engineer Wernher von Braun's 1954 report "A Minimum Satellite Vehicle" also noted the scientific benefits of satellites and recognized that "it would be a blow to U.S. prestige if we did not do it first."[38] The American Rocket Society requested funding from the National Science Foundation (NSF) for a satellite program. And throughout this period scientists involved in the IGY continued moving their satellite program proposal through the necessary government channels.[39]

As conversations about satellites permeated scientific and military institutions' hallways, populated memos, and filled reports, it was the looming possibility of a surprise attack from the USSR, and its broader implications, that troubled President Eisenhower. The Soviet Union had detonated a hydrogen bomb in August 1953. A few months later, US reconnaissance detected signs of Soviet intercontinental ballistic missile (ICBM) development in full swing. By the fall of 1954, Eisenhower wanted to know how well the US could respond to the latest Soviet weapons systems. He assigned James Killian, president of the Massachusetts Institute of Technology, to lead the Technological Capabilities Panel (TCP) in a study of "Meeting the Threat of Surprise Attack." Completed six months later, in February 1955, the highly secret "Killian Report" included an assessment of satellites for reconnaissance.

Although spy satellites seemed a technology of the distant future, Killian and others recommended the development of a civilian satellite program. They reasoned that a scientific satellite could help establish the legality of overflight of other nations' territory, essentially clearing the way for reconnaissance satellites when the time came.[40] In the mid-1950s, international law prohibited overflight of nonassenting nations. US planes flying too close to Soviet airspace were regularly fired upon. But no one had drawn a sharp line of where airspace ended and outer space began. Whereas an American aircraft flying over Soviet territory constituted a clear violation of international law, the case of orbiting satellites—beyond the reach of airplanes—was unclear. A civilian satellite, Killian and others noted, might draw that line, establishing the legality of overflight and setting an indispensable precedent for US reconnaissance satellites.[41]

The CIA's comments on the Killian Report are worth highlighting. They zeroed in on the psychological import of looming satellite launches, identifying more than the significance of spy applications: "The nation that first accomplishes this feat will gain incalculable prestige and recognition throughout the world." Five full paragraphs wrestle with the "psychological warfare value" of satellite technology. In contrast, the intelligence dimensions receive just one paragraph in this commentary.[42]

The task of evaluating the benefits of a US satellite program first fell to Donald Quarles, the deputy secretary of defense for R&D. Known as an earnest man with a "quiet efficiency and pleasant manner," Quarles worked at Western Electric and Bell Laboratories before joining the Eisenhower administration in 1953.[43] He brought his skills as a research engineer, scientist, and manager to overseeing the buildup of the US missile program. "His outstanding characteristic," *New York Times* reporter Jack Raymond observed, "was his great appetite for work."[44] He abstained from cigarettes and alcohol, and only on occasion drank watered-down coffee. His one display of anger and frustration amounted to throwing a pencil at his desk.[45] Critics found him too single-minded in his approach to defense strategy, overemphasizing bombs and missiles, but Defense Secretary Charles A. Wilson found him indispensable. Wilson relied on Quarles for his technical assessment of new weaponry development, an expertise that positioned him well when the question of establishing a US satellite program arose.[46]

Quarles held the unique position of knowing about all major satellite proposals and reports as well as the classified Air Force and TCP investigations. In addition, he was aware of the extremely secret U-2 spy plane, a program so highly classified that in 1955 only four people at the White House knew of its existence. As the TCP report stressed, the future of American intelligence efforts such as the U-2 required the "freedom of space." After being briefed on the TCP report, Quarles encouraged scientists working on the IGY to formally submit their satellite proposal to the National Security Council. NSF Director Alan T. Waterman and the others involved in the IGY remained unaware of the classified reports, and the larger concerns over American reconnaissance, as Quarles shepherded the IGY scientific satellite proposal through a complicated process of bureaucratic support to advance it.[47]

Even though three satellite program proposals waited for presidential approval—the IGY proposal and two military programs—the Eisenhower administration was slow to move forward with supporting a satellite program until US government officials felt the pressure of being second in space. It was not until the Soviet Union announced the creation of an interplanetary commission on April 16, 1955, that the Eisenhower administration began to take swifter action. Even though the announcement, made in the pages of the Moscow newspaper *Vecherniaia Moskva* (Evening Moscow), was bland and understated, the American press immediately picked up the news. Reactions to the Soviet announcement put pressure on the Eisenhower administration to finally settle on the issue of satellite development.[48]

Before Quarles submitted the proposal outlining a national policy for launching an artificial satellite as part of the IGY to the NSC in May 1955, he sent it to Nelson Rockefeller. A billionaire businessman with a longtime dedication to public service, Rockefeller had accepted the somewhat loosely defined position of special assistant for cold war strategy to President Eisenhower in January of that year. The position suited him. As Henry Kissinger once observed, "Of all the public figures I have known he retained the most absolute, almost touching, faith in the power of ideas."[49] He brought this appreciation for the "power of ideas" to international affairs, injecting questions of prestige and psychological strategy into policy discussions. Almost immediately, the ambitious Rockefeller expanded his White House portfolio, encroaching on the State Department's domain, including oversight of covert operations for the NSC.[50]

A steadfast cold warrior, Rockefeller supported a scientific satellite program, reasoning that "the stake of prestige that is involved makes this a race we cannot afford to lose."[51] In the memo he attached to Quarles's proposal, Rockefeller emphasized the "costly consequences of allowing the Russian initiative to outrun ours through an achievement that will symbolize scientific and technological advancement to peoples everywhere." He cautioned against launching a simple, unsophisticated satellite before the Soviet Union, explaining that a complex scientific satellite would be much better from a prestige standpoint. Rockefeller theorized that the Soviet Union would initiate "vigorous propaganda . . . to exploit all possible derogatory implications" of an American satellite. For this reason, the United States must pursue a program "least vulnerable to effective criticism." If the satellite was under the auspices of the IGY, he emphasized, its association with military technology would be far less likely. Furthermore, the US, Rockefeller wrote, should share with the world all the data gained through a scientific satellite. By making these data open, the American satellite program would seem peaceful and scientific, divorcing it from military associations.[52]

On May 26 the National Security Council endorsed Quarles's recommendation for an IGY scientific satellite project, incorporating all of Rockefeller's provisions as well.[53] After stating the feasibility of launching a satellite during the IGY, the draft of the policy (NSC 5520) highlighted some of the scientific data that could be collected from a satellite and then noted that "considerable prestige and psychological benefits will accrue to the nation which first is successful in launching a satellite." The significant repercussions could include the political alliance of other nations, a central concern during the cold war. The Soviet Union was already well under way on its own spaceflight program. If the United States did not act quickly, the balance of power could be at risk, the report warned. NSC 5520 summed up the threat: "The inference of such a demonstration of advanced technology and its unmistakable relationship to intercontinental ballistic missile technology might have important repercussions on the political determination of free world countries to resist Communist threats, especially if the USSR were to be first to establish a satellite." A small scientific satellite, the committee acknowledged, could also test the "freedom of space," which was seen as essential to the prospect of any future legality questions regarding military reconnaissance programs. The IGY, NSC 5520 suggested, presented "an excellent opportunity" for the United States to enter the Space Age under the aegis of a peaceful, open, scientific endeavor, clearing the way for other uses of satellite technology.[54] President Eisenhower asked the council if the United States should pursue a scientific satellite program, and everyone agreed in the affirmative.[55]

On a time line overlapping and intersecting with these satellite discussions, President Eisenhower prepared for the Geneva Summit, set to begin on July 18, 1955. The leaders of Britain, France, the Soviet Union, and the United States would gather for the first time since the cold war began. With Soviet leader Joseph Stalin's death in 1953 and Nikita Khrushchev's rise to power in the spring of 1955, improved East-West relations seemed like a possibility. Although skeptical of the summit at first, Eisenhower recognized that if he did not attend, it might appear that the United States was not truly invested in the cause of peace. For the six weeks ahead of the summit, Eisenhower's security staff prepared the president for the USSR's potential proposals, such as the reunification of Germany or demilitarizing Europe. Keeping in mind that Geneva would be on the international stage for a stretch of days in July, Eisenhower's staff struggled with the issue of nuclear arms control, the major issue at the forefront of the world public's minds.[56]

Before Eisenhower flew to Geneva in mid-July 1955, Rockefeller gathered eminent academics for a "pre-summit summit" to assess "the current international relations situation," with an eye to prestige and psychological warfare. Eisenhower's confidence in Rockefeller threatened both Secretary of State John Foster Dulles as well as his under secretary, Herbert Hoover Jr. But Rockefeller was undeterred by their prickliness, advancing his own approach to cold war grand strategy at every opportunity. Bypassing Dulles, Rockefeller's carefully chosen roomful of experts drafted "Quantico I," a document outlining an ambitious disarmament package and plans for economic development. The proposal called for "mutual aerial observation" or "Open Skies": for mutual assurance that neither country prepared for a surprise attack. Rockefeller urged Eisenhower to introduce Open Skies at the Geneva Summit.

At Dulles's advice, Eisenhower resisted Rockefeller's proposal at first: "I doubt [the Soviet Union] will agree to it . . . because secrecy means so much to them . . . they see secrecy as a great strength."

Rockefeller assured Eisenhower that no matter what happened, proposing Open Skies would benefit the United States. If the Soviets agreed, it would give "the U.S. a decided intelligence advantage." If not, then the United States would gain "a decided public opinion advantage" by putting American openness and Soviet secrecy in sharp relief. After meeting with the NSC and reviewing European public polls favoring a reduction in US-USSR tensions, Eisenhower eventually agreed with Rockefeller. Open Skies had the potential to transform the image of America from "a warmonger with an atomic bomb," as Rockefeller put it, to a peace-seeking global leader. Eisenhower would propose Open Skies in Geneva.[57]

As Rockefeller, Eisenhower, and many others in Washington had predicted, the Soviet delegation rejected the "Open Skies" proposal. But news stories printed around the world cheered the "spirit of Geneva," heralding the United States' hope for peaceful coexistence. The USIA took up this "spirit of Geneva" theme in films, glossy magazines, exhibits, pamphlets, and other propaganda programming: the United States was committed to peace.[58]

Following Geneva, Eisenhower called a meeting with Quarles and Waterman. Now he was fully committed to an American satellite program aligned with the IGY, he told them. Not only would it address the issue of "freedom of space"; this international, peaceful, and open scientific approach would also secure US prestige.[59] At the end of July, White House Press Secretary James Hagerty issued a brief statement announcing the White House's approval of a scientific satellite program as part of the US participation in the IGY. Stressing the international character of the endeavor, the statement explained that "the President expressed personal gratification that the American program will provide scientists of all nations this important and unique opportunity for the advancement of science."[60]

The news spread rapidly. A USIA opinion poll in the United Kingdom, France, West Germany, Italy, Austria, and Belgium found that awareness of the satellite announcement was widespread, although many people expressed their skepticism.[61]

Reminiscent of the response to the Soviet Union's announcement of an interplanetary commission the previous April, the US announcement of a satellite program prompted what might be considered a chain reaction within the Soviet Union. While attending a conference in Copenhagen a few days after the White House press release, Leonid Sedov, chairman of the Soviet interplanetary commission, held a press conference at the Soviet Embassy. He announced that "the realization of the Soviet project can be expected in the comparatively near future." The international press sensationalized Sedov's claims the following day. On August 3 the *New York Times* headline read "Soviet Planning Early Satellite: Russian Expert in Denmark Says Success in 2 Years Is 'Quite Possible,'" while the *Los Angeles Times* warned that "Russians Claim They'll Launch First Satellite."[62]

Since 1954, Korolev and his fellow missile designers and spaceflight enthusiasts had been campaigning for the Soviet government to support the development of a satellite program, emphasizing both the security value and scientific value of spaceflight. In historian Asif Siddiqi's assessment, news coverage of Eisenhower's announcement "provided the final weapon, international prestige, that Korolev, [Mikhail] Tikhonravov, and others needed to convince the top leadership of the importance of the satellite project."[63] They assembled the international news coverage, prepared a summary report, and claimed to Soviet leadership that their satellite could beat the United States into space. Shrewdly, Korolev added a line to emphasize the soft power potential of a satellite, explaining that it "would have enormous political significance as evidence of the high development level of our country's technology."[64] On August 8, 1955, the Politburo met and gave approval for a scientific satellite program that would use an R-7 ICBM.[65]

Meanwhile, back in the United States, the Stewart Committee, an advisory group formed by Quarles, selected the proposal for an American scientific satellite. The committee had been meeting since early July to assess potential programs. On August 3, 1955, it voted for the Naval Research Laboratory's Vanguard program over the

Army's Orbiter program. It was a close decision, highly contingent on a mix of personal agendas and happenstance. The Vanguard program appeared to use a superior satellite, was on a quicker schedule, and had the image of a civilian scientific program instead of a military enterprise.[66]

By May 1957, as the beginning of the IGY inched closer, the NSC questioned the rising costs of the satellite program. When the NSC had advanced the satellite proposal two years earlier, in May 1955, the estimated cost was $15–$20 million. The committee now faced a $110 million price tag. Eisenhower criticized the US satellite program scientists for focusing too much on instrumentation for multiple satellites while their real goal should be beating the Soviets to space, explaining that "the element of national prestige, so strongly emphasized in NSC 5520, depended on getting a satellite into orbit, and not on the instrumentation of the scientific satellite."[67] CIA Director Allen Dulles, briefing the group on the latest intelligence, observed that "if the Soviets succeeded in orbiting a scientific satellite and the United States did not even try to . . . the USSR would have achieved a propaganda weapon which they could use to boast the superiority of Soviet scientists." He also stressed that if the US canceled its scientific satellite program, it might become fodder for USSR propaganda suggesting that the United States invested in military programs, not civilian spaceflight programs. It came down to prestige, added the assistant secretary of state.[68]

The pressure of being first in space—and the stakes of prestige— also weighed on the Soviet program. Earlier in the year, after delays developing the proposed satellite—Object D—a colleague asked Korolev, "What if we make the satellite a little lighter?" In a letter to the government, Korolev explained the necessity of the revision: "The United States is conducting very intensive plans for launching an artificial Earth satellite." It was sparing no cost, he stressed. The United States "is willing to pay any price to achieve this priority."[69] The new, smaller satellites received approval. Designated Simple Satellite No. 1 (PS-1), Sputnik, as PS-1 would become known across the world in October 1957, had a better chance of beating the American

Vanguard program into space.[70] Ever attuned to politics and history, and tense from working around the clock the whole summer of 1957, Korolev yelled at the chief engineer on the assembly shop floor. But his anger was not about the quality of Sputnik's production. Instead, he was concerned about the aesthetics of the mock-up satellite's shiny surface. "This ball will be exhibited in museums!" Korolev presciently observed.[71]

2

SPUTNIK AND THE POLITICS OF SPACEFLIGHT, 1957

The satellite symbolizes an intellectual attainment that may dominate the period immediately ahead as the most powerful single instrument of national policy.

—LLOYD V. BERKNER, 1958

Before the sun had risen in Arkansas on September 4, 1957, a handful of men and women had already assembled in front of Central High School in Little Rock. A massive neo-Gothic brick building, Central High was called "America's Most Beautiful High School" when it first opened in 1927. Since then, only white students had been allowed to pass through its halls. That morning a hundred state militia troops were also at the school, some sitting on the edge of the sidewalk with their rifles lying nearby. By 8:00 a.m., a crowd of four hundred had gathered. When a fifteen-year-old African American student approached the school, the crowd closed in, jeering and yelling at her as the troops pushed them back. Eight other African American students followed. Although a federal district court had ordered their admission as part of a larger effort to desegregate schools after the US Supreme Court's *Brown v. Board of Education* ruling in 1954, the state militia denied each student entry. The images of troops and the angry mob not only appeared in national papers; they made the front pages of newspapers around the world, from the *Times of India* to the *South China Morning Post* to the *Egyptian Gazette*. The foreign press's focus on the crisis in Little Rock was so acute that US newspapers began covering the foreign press's coverage.[1]

Struggles over desegregation throughout America in the 1950s were not only a domestic issue. As the decolonization movement gained ground throughout Africa, with Ghana achieving independence earlier that spring, the issue of US racial oppression took on even more potency in international affairs. Dispatches from US embassies reported on the broader adverse impact of Little Rock on the status of American democracy abroad. A US consul in Mozambique observed that "our moral standing has been very considerably damaged and . . . any pretension of an American to advise any European Government on African affairs . . . would be hypocrisy." The Soviet Union leveraged Little Rock in a vast campaign of anti-American propaganda as part of a larger effort to sway global political alliance toward the Eastern Bloc.[2]

President Eisenhower sent federal troops to Arkansas, and the USIA developed programs to offset the negative image of the United States. Although the USIA dealt with problems of school desegregation head-on in its programming, Little Rock became a damaging reference point for international perceptions of American race relations. Secretary of State John Foster Dulles recognized that this battle was hurting "the influence of the United States abroad."[3] The unfavorable image of the United States following Little Rock and other battles of school desegregation was compounded by widespread admonishing of the savage behavior of US soldiers in Asia that summer and continued criticism of US nuclear testing. In the fall of 1957 American prestige was in need of a boost, but what it got instead on October 4 was a deep blow to national self-confidence and a sharp challenge to Americans' shared belief in the technological and scientific superiority of their nation.[4]

Exactly one month after nine students were blocked entry into Central High School in Little Rock, the Soviet Union launched Sputnik, from Kazakhstan. It was a Friday. By the time the news broke in the United States, President Eisenhower had already left Washington for a weekend of golfing in Gettysburg. Americans around the country were at home watching the premiere of *Leave It to Beaver*, unaware that the space race would soon begin. In the second-floor ballroom of the Soviet Embassy in Washington, fifty scientists gathered at a reception in honor of the International Geophysical Year (IGY), the largest scientific research program attempted in history. Shortly before 6:00 p.m., Walter Sullivan, a *New York Times* science reporter attending the event, received a telephone call from his Washington bureau chief. After taking the call, he whispered to an American IGY committee member simply "It's up."[5]

They immediately shared the news with Lloyd Berkner, the ionospheric scientist who was one of the originators of the idea for the IGY in April 1950 and a "supreme optimist."[6] Berkner clinked a glass to silence the room for attention. "I wish to make an announcement," he said. "I've just been informed by the *New York Times* that a Russian satellite is in orbit at an elevation of 900 kilometers. I wish to congratulate our Soviet colleagues on their achievement." Cheers rang out among the scientists, and vodka started flowing. Just like Khrushchev's Kiev dinner party, a shortwave radio was set up to play Sputnik's "beep-beep-beep" for guests. The Soviet scientist Anatoli Blagonrov, after listening with some surprise, confirmed "that is the voice. . . . I recognize it."[7]

Senate Majority Leader Lyndon B. Johnson had just finished a barbeque dinner with friends at his Texas ranch when he heard the news. With his wife, Lady Bird, and their close friends, he took a walk along the Pedernales River, an evening stroll that had become customary after Johnson's heart attack two years earlier. As they walked, the party looked up, hoping to catch a glimpse of the small satellite streaking across the Hill Country sky. The symbolic weight of the small Soviet satellite hit Johnson immediately, making him feel "uneasy and apprehensive." As he later recalled, "In the open West, you learn to live with the sky. It is part of your life. But now, somehow, in some new way, the sky seemed almost alien."[8] When Johnson returned to his ranch, his living-room telephone rang. It was Richard Russell, a Democratic senator from Georgia whose long history of legislative victories had earned him the reputation of the "South's greatest general since Robert E. Lee." Always the master strategist, Russell was calling Johnson about leading a Senate investigation into the launch. Not only was Johnson well versed in the issue of national security preparedness; the Texan recognized a political opportunity when he saw one.[9]

Late that evening, White House Press Secretary James Hagerty assured the press that this news did not catch the administration off guard. There was no "space race," he explained. Sputnik "was of great scientific interest," not a national security threat. Unfortunately for the Eisenhower administration, the press did not toe this line.[10] Over the coming days the historic significance of Sputnik came into focus. Within the United States, news of Sputnik overshadowed all the other breaking stories, even the civil rights battles in Little Rock. The next morning, as Eisenhower played golf for the fifth time that week, Americans were growing increasingly concerned about the broader implications of the small satellite.[11]

Both the domestic and the foreign press covered Sputnik in depth, making it the major story of the year. Most Western European newspapers devoted their front pages to coverage of Sputnik—and its broader geopolitical implications—for almost a full week. As the USIA reported, "Virtually all of this content centered around three main themes—the scientific achievement that 'sputnik' represented, the implications of the Soviet 'first' in the battle for the minds of men, and the political and military significance for the West of the recent Soviet advances in missile rocketry."[12] The international press commented that Sputnik won a major propaganda victory for the Soviet Union, especially within the developing world, the emerging focal point of the cold war psychological battlefield.[13]

The international public's awareness of Sputnik "was almost universal," according to the USIA. In remote towns and sparsely populated areas, people had heard that the satellite was traveling overhead.

Even schoolchildren in the United Kingdom parodied Perry Como's "Catch a Falling Star," singing, "Catch a falling Sputnik,/Put it in a basket,/Send it to the U.S.A./They'll be glad to have it,/Very glad to have it,/And never let it get away."[14] An American living in Brazil heard what became a common joke during the fall of 1957: "When Sputnik goes around the globe it goes, 'beep-beep, beep-beep' everywhere, except when it goes over the United States. Then it goes, 'ha-ha, ha-ha.'"[15]

For Khrushchev, Sputnik fit squarely within his larger geopolitical strategy of signaling strategic parity with the United States, his cold war rival. Two years earlier he had ordered ten bombers to fly repeatedly over Red Square, with the sole purpose of giving the impression of a "bomber gap" between the US and the USSR.[16] In August 1957 Khrushchev announced that the Soviet Union had ICBMs that could reach "any part of the globe," even though the USSR would not achieve this capability until the 1960s. After Sputnik, he exclaimed that the Soviet Union "outstripped the leading capitalist country—the United States—in the field of scientific and technical progress."[17] And when the United States eventually launched a satellite, he crowed, "It's the United States which is now intent on catching up with the Soviet Union."[18]

The political, social, and cultural implications of the Soviet satellite did not just fall into place on their own. And it was not just Soviet officials who saw the political opportunities afforded by Sputnik. Within Washington, some seized on the "Sputnik moment," spotting an opening for challenging the current Republican administration. "No sooner had Sputnik's first beep-beep been heard—via the press—than the nation's legislators leaped forward like heavy drinkers hearing a cork pop," as a Department of Defense official put it.[19] Politicians within the United States and around the world took an active hand in molding the meaning and impact of Sputnik, with consequences that would stretch into the 1960s and into today.

Johnson, the most vocal and strident of the group, almost immediately publicly called for a congressional investigation of American space and missile capability, as Russell had advised him to do on the evening of October 4. Senate aide George Reedy urged the ever-aspirant Texas senator to "plan to plunge heavily into this one," and he put the matter frankly: "If properly handled . . . [Sputnik] would blast the Republicans out of the water, unify the Democratic party, and elect you President." Over the next few months, Johnson would use the investigation to establish himself as the congressional space expert and, more importantly, as a stepping-stone to the presidential nomination.[20]

While Eisenhower would attempt to soothe the fears of a worried nation, Johnson raised them. Likening Sputnik to Pearl Harbor and the Alamo in public speeches again and again, Johnson declared that "history does not reward the people who win the battles, but the people who win the war." Sputnik, LBJ warned, was "perhaps the greatest [threat] that our country has ever known." He called on his fellow Americans to meet that challenge. The Space Age had begun. It was time for the United States to reclaim the lead.[21]

LBJ's connections between Sputnik and Pearl Harbor or the Alamo were not simply attempts to interpret the significance of the satellite; they were also about *making* it significant. He carefully selected these analogies, rejecting his aide's suggestion that he frame it as "a call to action instead of a summons to a siesta," a regurgitation of his earlier Korean War critique. No, instead, he would link Sputnik to something grander, "a disaster . . . comparable to Pearl Harbor." Sputnik was "an even greater challenge than Pearl Harbor," he stressed on another occasion.[22]

How we think of the "Sputnik moment" today, how we draw on the idea that Sputnik shocked the nation out of complacency and into action in contemporary political calls to innovate, has direct lineage to LBJ's and others' politicking. This politicking is a critical piece of the story that not only illuminates the contingency of the "Sputnik moment" but also tells us how and why spaceflight assumed the national priority status it did by 1961.[23]

It would be a busy fall in Washington: "one prolonged nightmare," as one journalist put it. "Any number of people—from the

Pentagon, from State, and from the Hill—were dashing in and out of the president's office. Each new visitor had a longer face than the one before."[24] Each day of the president's schedule was filled with meetings related to the Soviet satellite the week after the launch.[25] Early on Tuesday, on a mild but overcast fall morning in Washington, Eisenhower assembled Quarles, Waterman, and other top officials. The day before, Under Secretary of Defense Quarles had submitted a report to the president outlining the Soviet satellite launch as well as the status of the US space program. Eisenhower expressed concern. The report suggested that the Army's Redstone rocket could have launched a satellite months earlier, beating the Soviets into space. Quarles explained the decision and efforts to disassociate the American space program from military development. Not only would this approach underscore the peaceful intentions of the United States; it would also limit foreign scientists' access to information about US military rocket technology. Changing the course of action now—turning the American satellite efforts into a "crash program"—could cause tension within the Pentagon. Quarles pointed out the good news: "The Russians have in fact done us a good turn, unintentionally, in establishing the concept of freedom of international space."[26]

Later that afternoon, Eisenhower asked Detlev Bronk, president of the National Academy of Sciences, for help. He was preparing his statement for the press conference the following day, and he wanted to make sure he was setting the right tone. Eisenhower did not want to "belittle the Russian accomplishment," but he also wanted "to allay hysteria and alarm." After one or two revisions, Bronk found the statement accurate and ready for distribution.[27]

On October 9, while Sputnik passed over the Indian Ocean, 245 news correspondents packed the pressroom to hear the president's remarks. In response to question after question during the thirty-two-minute-long news conference, Eisenhower contrasted US and Soviet approaches to spaceflight. The US satellite would contribute to scientific knowledge and collect more scientific data than the Soviet Sputnik. For the United States it was not a matter of national security but science, he stressed.[28] But journalists were not convinced. The president "seeks to calm fears," but to no avail, a reporter commented. Instead of featuring Eisenhower's carefully crafted lines, journalists included statements such as "And I don't know what we could have done more" and "I didn't say I was satisfied," portraying a bewildered and unprepared leader. When the news conference ended, Sputnik had already made its way over Kodiak, Alaska.[29]

The next day, Thursday, October 10, 1957, just shy of a week after Sputnik's launch, the NSC had much to discuss. For Eisenhower, the NSC offered vital governing guidance: his Cabinet was a useful sounding board, but it was the NSC that he looked to for policy making. Unlike his predecessor, Harry Truman, Eisenhower scheduled weekly meetings of the NSC, presided over almost every meeting, and expanded its membership. Each Thursday morning CIA Director Allen Dulles would begin the meeting with a twenty-minute intelligence briefing. By noon, after moving through the agenda efficiently, the council adjourned.[30]

Dulles led off the 339th meeting of the NSC by briefing the room on the Soviet satellite launch. He came from a long line of statesmen, with three secretaries of state in his family: his grandfather, his uncle, and his brother, John Foster Dulles, who served alongside Allen in the Eisenhower administration. The CIA's covert offensive operations had increased during the Truman administration, but it was under Dulles's guardianship that it swelled, becoming a major global player in foreign sabotage, subversion, coups d'état, covert propaganda, economic warfare, and assassinations, among other interventionist activities. Extroverted and charming, sporting a mustache and with a tobacco pipe often in hand, Dulles heightened and refined covert psychological warfare in the 1950s, making it central to US statecraft.[31]

Dulles described the launch and path of the small satellite and explained that it had not surprised the intelligence community. They were expecting that the Soviet Union might launch a satellite by November 1957, just a month past the real schedule. At that point the CIA still did not know if Sputnik was sending encoded messages.

When he turned to an assessment of the world reaction to Sputnik, Dulles explained that "Khrushchev had moved all his propaganda guns into place." Sputnik, alongside the announcement of ICBM testing and a large-scale hydrogen bomb test, made up a trilogy of Soviet propaganda moves, in the CIA's estimation.[32]

In Dulles's view, the Soviets aimed this propaganda demonstration at audiences in underdeveloped countries, especially in the Middle East, as part of an effort to relate "scientific accomplishments to the effectiveness of the Communist social system." The international impact of the small satellite was "very wide and deep," Dulles concluded after reviewing reactions in each region of the world to the assembled group.[33]

Quarles spoke next. He acknowledged that much of what he could say about the satellite would already be well-known to everyone gathered in the room. Eisenhower replied that this was true and that he had started feeling "numb on the subject of the earth satellite." Although it had been just under one week since the news of Sputnik broke, Eisenhower thought public reaction had grown out of hand. Quarles reviewed the history of US satellite development, then updated the group on the progress of Project Vanguard, the Navy's satellite program. Vanguard was a science program, Quarles stressed. The US, he guaranteed the council, had long recognized the propaganda implications of being first in space, but it prioritized science and establishing the principle of freedom of outer space. "In this respect," he pointed out, "the Soviets have now proved very helpful." By October 10, Sputnik had orbited over almost every nation on Earth.

First and foremost, what Eisenhower wanted to know was if the planned lower orbit of Vanguard would negatively impact US prestige. This lower orbit would mean that Vanguard would deorbit more quickly than Sputnik, potentially prompting another blow to the image of American technological capability. Quarles once again stressed the sophisticated scientific nature of the American satellite, explaining that there was more to be learned from Vanguard than Sputnik. Later, Vice President Richard Nixon reemphasized the political payoffs of a scientific satellite. Sharing data gathered by Vanguard would be of "great propaganda advantage for the United States," he observed. NSF Director Alan Waterman affirmed Nixon's and Quarles's points: the combination of a sophisticated scientific instrumentation package and the open approach to its space program put the United States in a more competitive position in terms of prestige.

Near the end of the NSC meeting, USIA Director Arthur Larson expressed concern over the effects of Sputnik on foreign public opinion. Although Sputnik should not have caused a shock, he said, the cumulative impact of the Soviet accomplishment could be detrimental to the international standing of the United States. Before taking the helm of the USIA, Larson had been Eisenhower's trusted speechwriter, known as the "chief theoretician of moderate Republicanism." Earlier that year, a *Washington Post* journalist portrayed him as "a youngish fellow, who manages to retain an oddly hopeful look."[34] Sitting at the table with other national security experts, he likely appeared less hopeful.

"If we lose repeatedly to the Russians as we have lost with the earth satellite," he cautioned, "the accumulated damage would be tremendous." The United States should set its sights on an impressive feat—perhaps orbiting a man in space or sending a crew to the moon—to secure American prestige, he continued. NAS President Detlev Bronk responded that the United States should focus on other types of scientific breakthroughs and not simply follow the Soviet lead. But he could not deny that space exploration held unique appeal.[35]

More challenges would come in the coming months for Eisenhower. Between October 4 and the following February, countless meetings, speeches, reports, hearings, and editorials not just about Sputnik but also the role of science within American society captured national and international attention. Although dismayed and nonplussed by the public response to Sputnik, Eisenhower relented. By the end of October, he supported raising the federal budget for defense in 1958 and 1959. As sure as he was of US military superiority, Eisenhower viewed the importance of "convincing the world—presently scared by Russia—that the United States is doing what it should," that the process was worth the high cost. This was not a matter of hardware; it was a matter of signaling that he was taking action.[36]

On the same day that Eisenhower met with the NSC in Washington, Khrushchev discussed Sputnik at a Presidium meeting in Moscow. Almost in synchronization, the two world leaders were briefed on the launch and the global response. Described as "unusually animated," Khrushchev insisted that Korolev's team build a "space gift" to mark the fortieth anniversary of the Bolshevik Revolution, the second phase of the 1917 Russian Revolution in which the Bolshevik Party seized power, led by Vladimir Lenin. Khrushchev well knew that the Soviet Union could not achieve ICBM parity with the US. Instead, it would surpass America in symbolic victories, the new currency of cold war competition.[37] The time line for the next satellite would be quick. The anniversary was less than a month away, in early November.[38]

Korolev upped the stakes. He accepted Khrushchev's challenge and suggested that they fly a dog on the next mission. Even under the strain, "it was . . . the happiest month of his [Korolev's] life," recalled cosmonaut Georgy Grechko, who worked in Korolev's design bureau at the time. "He told his staff, and his workers, that there would be no special drawings, no quality check, everyone would have to be guided by his own conscience."[39]

A few days after the NSC and Presidium meetings, the USIA's Office of Research and Intelligence submitted a confidential report on world opinion of the Soviet satellite. The results were worrisome to US officials.[40] During its first few years in operation the USIA had focused primarily on information dissemination. But in 1955, as part of a larger reevaluation of basic national security policy, the NSC pushed for extending the scope of the USIA's mission to include a more robust polling and impact-assessment program to support changing national security interests.[41] These opinion polls relied on an elaborate system designed to conceal that the US government was behind each study. The USIA contracted local pollsters around the world, who in turn hired local interviewers.[42] According to the most recent poll at the time, international public opinion viewed the

Soviet Union as technologically superior to the United States, and the small satellite raised the nation's overall prestige. Western Europeans believed that the balance of military power had shifted from the US to the USSR, and they found Soviet propaganda more credible. The "World Opinion and the Soviet Satellite" report warned that the overall impact of Sputnik would be strongest among the "backward, ignorant and apolitical" populations of the developing world. Newly independent nations might be convinced of the validity of the Soviet system because of their thirst for rapid technological and economic advancement. Passing over almost every nation in the world, it was "likely to give a peculiar impact among those least able to understand it," generating "myth, legend and enduring superstition of a kind peculiarly difficult to eradicate or modify, which the USSR can exploit," the report warned. And it appeared the Soviet Union was mounting a psychological warfare campaign for marketing the supposed military implications of the satellite.[43]

The authors of the report blamed the United States for intensifying the impact of Sputnik on foreign public opinion. Not only did US officials heavily promote the nation's plans to launch a satellite during the IGY, but officials and the public alike also assumed that the country led the world in the field of space science and technology. To make matters worse, the Soviet Union capitalized on American anxiety over the small satellite in its propaganda campaign: Soviet media reported on the American public reaction, which fed into Soviet claims of technological superiority. "One moral that might be drawn," the report reflected, "is that a propagandist cannot have his crow and eat it too." The report included one positive note: fortunately, among the world leaders whose opinions were of most bearing to US global interests, final judgments about the military and technological implications of the Soviet achievement were yet to be made. In the end, the report concluded, the "most durable and useful gain" of Sputnik would very likely be the satellite's contribution to Soviet credibility. Scientific discoveries and engineering information were secondary rewards.[44]

By 1957, the USIA had become further integrated into Washington's foreign policy process. In February Eisenhower had made the USIA director a member of his Cabinet and the NSC, a distinction shared only by the director of the CIA. Newly appointed USIA Director Arthur Larson made the most of this opportunity, attending every Cabinet meeting whenever he was in Washington. In January 1957 Larson pushed for incorporating public relations into every department of the government, a plan that received support from the Cabinet.[45] In the fall of 1957, shortly before Sputnik, Larson prepared a new set of guidelines for the agency. He hoped to move the USIA away from propaganda to factual reporting. In a bold move the agency banned the use of boastful or self-righteous tones, encouraged emphasizing shared interests, and recommended that programs target "opinion formers."[46]

Although the mission of the USIA had become more refined by 1957 and the agency's influence within Washington had increased, field officers struggled to present a positive image of US policies and American culture abroad even before Sputnik's launch. After Little Rock, civil rights tensions were pulling the country apart and already undercutting US influence abroad, especially in African nations. Now the country was being mocked around the world, adding salt to the wounds of American pride. It quickly became clear that Sputnik had initiated a new era and that the Soviet Union, much to US political leaders' chagrin, could take the credit for opening the Space Age. USIA scrambled to do what it could to soften the blow.[47]

As William Traum, deputy director of the US Department of Commerce, wrote to the USIA, "[Sputnik's] signal, and then the echoes of it, will continue to be irritating and ominous facts for a long time." The best approach to dealing with the small satellite, in Traum's view, was to recognize it but then "dilute the impact" abroad. The USIA, he stressed, should emphasize the great scientific achievements of the twentieth century because this approach would put Sputnik in its place, as a single event within a larger context of predominantly American achievements.[48]

Traum's strategy mirrored what would become the US government's official response. The Operations Coordinating Board (OCB) instructed that treatment of Soviet achievements should "be held to a minimum, to avoid giving gratuitous publicity to such activities."[49] OCB guidelines for public information about the Soviet and US satellite programs directed government departments and agencies to pivot attention away from the USSR and toward the US. And when discussing the US satellite program, "Statements and releases should reflect a sense of composure, assurance, and confidence." The scientific, peaceful, and open commitments of the American program should be stressed. It was best to situate satellite development within the larger context of IGY activities, the OCB advised.[50]

In Sputnik's wake the USIA pulled exhibits, produced new pamphlets, and searched for the best way to pivot attention away from space exploration. State Department and USIA staff working on the American Pavilion for the upcoming 1958 Brussels World's Fair re-evaluated their plans. The first fair since World War II, Expo '58 was meant to usher in a new age when nuclear energy and advancements in science contributed to international cooperation and a renewed humanism.[51] After Sputnik, the US exhibit team concentrated its efforts on displaying American culture and shied away from competing directly with Soviet shows of space capability and scientific advancements. As a USIA staff member observed in November 1957, "The advent of Sputnik puts a new face on a lot of things that deal with the image which America needs to present abroad."[52] The United States might not have been the global leader of the space race in the 1950s, but pavilion staff knew they could outdo the Soviet Union in fashion, consumer products, and displays of "the good life." The fair was set to open in April 1958, leaving little time for exhibit staff to adapt their displays.[53]

Larson looked outside the space arena for other types of scientific and technological feats that could capture world public attention. He sent the Atomic Energy Commission his idea of using hydrogen bombs for slicing through mountains to create roads, exploding earth apart to dam rivers, and swiftly scooping out harbors. These types of large-scale spectacles would surely take attention away from space exploration. And, like Sputnik, they would demonstrate military and technological capability all in one blow. The OCB also searched for scientific and technological projects that would ensure that the global balance of power did not swing toward the Soviet Union: from de-salination of seawater to drilling the deepest hole that had ever been drilled in the Earth's crust, or aid-related programs such as airlifting rice to Indonesia.[54]

The CIA also proposed methods for curbing Sputnik's impact. Why not organize an "International Medical Year"? Perhaps establishing a "University of the World" or financing hospitals would shift the world's attention away from space exploration. Like Larson, the CIA's proposals included the use of hydrogen bombs for "practical applications" like obliterating icebergs that blocked polar passages or reversing the direction of typhoons and thereby stopping them. If the US needed to compete in space, then broadcasting a song from a satellite with words like "Freedom shall be yours" would "have a tremendous propaganda effect," reasoned CIA officials.[55] Although this idea was not taken up by the US space program, it would become China's approach more than a decade later. On April 24, 1970, China launched its first satellite, East Is Red 1, which played the first bars of a song about the Chinese revolution.[56]

Meanwhile, Korolev and his team of rocket engineers worked tire-lessly on their second, larger satellite, designed for a canine passenger. At the top of another R-7 rocket and below a shiny metal sphere identical to Sputnik 1 sat Laika, a dog selected for her placid demeanor and adaptability to extreme conditions. Laika and her fellow canine cosmonauts came from the streets of Moscow. Strays, often mixed-breed and female like Laika, were cheap, trainable, and suitable for studying the effects of spaceflight on living bodies. A few days earlier, on October 31, Laika was bathed, groomed, fitted with sensors, and fastened inside her spacecraft with the aid of a harness and restraining chains. She then waited inside the capsule for three days, tucked between two large cushions while instruments monitored her vitals. At 5:30 a.m. Moscow time on November 3, the R-7 launched from Tiura-Tam, carrying Laika into space. She reached orbit alive. Her physiological data, transmitted through a telemetry system, suggested that weightlessness had little or no effect on vital functions, a key discovery for any future human mission. But the cabin temperature rose far too high, past 100 degrees Fahrenheit. Within five to seven hours the telemetry data no longer indicated signs of life.[57]

The next day, Laika's flight made front-page headlines, including the *New York Times*' Panglossian "Dog in Second Satellite Alive: May Be Recovered, Soviet Hints." Walter Briggs, a reporter based in Tokyo, warned his American readers that even though "the ruthless manner in which Marshal Georgi Zhukov was ousted . . . gave many Japanese cause for criticism," the success of Sputnik 2 "outweighed this loss of prestige." Reading military significance into the launch, the Japanese press indicated that the flight confirmed Khrushchev's boasts of an ICBM arsenal.[58] Another journalist commented that "it is good that Sputnik 2 went up so soon after Zhukov went down," drawing attention to Khrushchev's political strategy.[59] In Moscow, people supposedly were greeting each other with impressions of Sputnik's "beep beep."[60]

In Mexico, many saw the satellite launch as confirmation of a Soviet lead in science in general, not just space exploration. Sputnik outweighs "the greater breadth of US scientific superiority," a confidential USIA report stated. Eight of ten people in Japan had heard the news and were aware of Sputnik's Soviet origins. And in Europe, Sputnik 1 and 2 "had a substantial adverse effect on the relative military and scientific prestige of the United States," according to USIA-sponsored opinion polling. The USIA again responded by situating the flight within the context of twentieth-century technological advances. When positioned alongside breakthroughs in American medicine, physics, and chemistry, the satellite would lose much of its symbolic weight, USIA officials hoped.[61]

A few days later, during a celebration of the fortieth anniversary of the Revolution on November 7, Khrushchev chided the United States. Dressed in a dark-blue suit, he spoke from a podium before an audience of four thousand for over three hours, on topics ranging from

Communist doctrine to satellites to proposing a conference between the "East and West": "Our satellites are circling the earth waiting for American satellites to join them and form a commonwealth of Sputniks. . . . The Soviet Sputniks proclaim the heights of the development of science and technology and of the entire economy of the Soviet Union."[62] His son Sergei recalled that his father was ecstatic about the satellite.[63] But even as he boasted over space successes, the Hungarian crisis in 1956 had seeded doubt in the Soviet leader. Khrushchev spent much of 1957 consolidating his power, fearing both internal and external threats to his leadership. The Soviet satellites bolstered his position but did not resolve his political vulnerability.[64]

On the same day, on the other side of Earth, Eisenhower spoke to his country's citizens from the Oval Office. As a reporter for *Life* magazine put it, "The two most powerful men in the world boasted at each other last week."[65] Instead of standing up at a podium at a mass event like Khrushchev, Eisenhower broadcast his remarks to sixty million Americans sitting in front of their living-room television sets. Eisenhower had turned to Arthur Larson to write the text. In addition to addressing the nation's scientific and technological position in the wake of Sputnik, Eisenhower wanted a trusted public diplomacy expert in the White House. He asked Larson to leave the directorship of the USIA and become the special assistant to the president. Larson's staunch internationalist perspective on world affairs, and his appreciation of the potential consequences of soft power, would imbue many of Eisenhower's speeches in the coming months.[66] With the hope of tempering public anxiety, Eisenhower and Larson planned a series of speeches on Sputnik's significance—or lack thereof.[67]

Speechwriting absorbed them for three days. "We are defending priceless spiritual values," Eisenhower instructed Larson. He explained that his message need not be "materialistic" but instead inspirational and optimistic. The speech was one in a series that linked the advancement of science with national security, an association reiterated again and again by the Eisenhower administration.[68]

"As of today, the overall military strength of the free world is distinctly greater than that of Communist countries," Eisenhower

President Dwight D. Eisenhower shown with scale model of Jupiter C Missile nose cone during a speech given from the White House, November 7, 1957. (SMITHSONIAN NATIONAL AIR AND SPACE MUSEUM)

assured his audience on November 7. He showed off a flown nose cone of a Jupiter missile placed on a platform to his left, a demonstration of advanced reentry technology being developed by von Braun's team in Huntsville, Alabama. But while Eisenhower emphasized American strength, he also acknowledged the need for more basic research, more science training, and more international scientific exchange to maintain global leadership.[69]

During this evening address to the nation, Eisenhower announced the creation of the new post of presidential science advisor and said that he had recruited MIT president James Killian to take the position. In his new role, Killian would establish the President's Science Advisory Committee (PSAC) to act as an intermediary between the scientific community and the executive branch. He also took on the challenge of assessing alternatives for the organization of space research within the United States.[70]

Eisenhower announced other initiatives that his administration was taking on, in hope of tempering public anxiety. He had given the head of the US missile program jurisdiction to oversee and coordinate missile programs within various branches of the military. This was an attempt to curb interservice rivalry. Eisenhower championed international scientific data exchange and advocated for a new scientific committee at the North Atlantic Treaty Organization (NATO). The State Department would appoint a new science advisor as well as science attachés at appropriate US embassies around the world.[71]

Ultimately, the military significance of the Sputniks was minimal. They proved that the Soviet Union had rockets with more thrust than the United States, but this did not demonstrate that the Soviets had the immediate capability to accurately drop weapons on the United States. It did not mean that nuclear warheads atop ICBMs threatened Americans' safety. In other areas of military preparedness—such as air power—the United States was still far ahead of the Soviet Union. And the scale of the United States' arsenal was large enough that if the Soviet Union attacked, there would be mutually assured destruction. The two superpowers were still very much locked in the cold war combat freeze. But try as he might, Eisenhower could not calm the country's nerves, especially when alarmist voices like Lyndon Johnson's were proactively stoking fears.[72]

Although it did not seem like US international prestige could sink lower after Sputnik II, it could. Before Sputnik, the Vanguard program was on schedule for a test launch later that year or in the beginning of 1958. In the initial stages of development, rocket launches often fail. Engineers in the program anticipated that they would successfully launch a satellite into orbit before the end of the IGY, but they did not assume a flawless first test launch. However, Sputnik pressure thrust what would have otherwise been a standard hardware test into a test of national technological capability. The Eisenhower administration sold the Vanguard Test Vehicle 3 (TV3) launch as America's effort to put a satellite in orbit. International media attention turned to Cape

Canaveral on December 4, 1957, but technical complications forced a delay.[73]

Allen Dulles, frustrated by the level of publicity leading up to the delayed launch, told the NSC that the United States had become the "laughing-stock of the whole free world."[74] Dulles encouraged Eisenhower to consider a new policy that would make all future launches secret until after the spacecraft was successfully in orbit, to avoid further embarrassment.[75] The blow to US prestige that followed the delay was nothing compared to what would occur on the day of the launch.

On December 6, after reaching a height of only a few feet after takeoff, the main engine failed, and the vehicle crashed back on the pad and exploded. The small payload was tossed to the ground, still beeping. The front-page coverage in the *London Daily Herald* read, "Oh, What a Flopnik!"[76] Johnson called the failure "one of the best publicized and most humiliating failures in our history."[77] Adding to this humiliation, members of the Soviet delegation to the United Nations offered the United States technical assistance as part of its program of aid to underdeveloped nations.[78] The year 1958 would be a better one for US self-confidence than 1957, if only just slightly.

3

A SPACE PROGRAM FOR ALL HUMANKIND, 1958–1960

Control of space means control of the world.
—SENATOR LYNDON BAINES JOHNSON,
JANUARY 7, 1958

"We live in a political world," observed Oliver Gale, "and no greater opportunity [than Sputnik] will ever be presented for a Democratic Congress to harass a Republican Administration, and everyone involved on either side knows."[1] As the special assistant to the secretary of defense, Gale coordinated interviews of Pentagon officials for the congressional hearing on national preparedness following Sputnik. Like the secretary of defense he served, Gale had taken on the position just a few days earlier. Both came to Washington from Procter and Gamble (P&G). Gale had been the head of public relations, and the new secretary of defense, Neil H. McElroy, had been the president. This expertise in PR likely served Gale more than he would have anticipated, as Washington—and the congressional hearings in particular—became the stage for political theater in the fall of 1957.[2]

As Gale witnessed, Democratic leaders took full advantage of the moment. With their sights set on the next two elections, leaders of the Democratic Party welcomed the opportunity afforded by Sputnik to sway the electorate. In the 1950s, powerful southern Democrats had earned a reputation for blocking civil rights legislation. Just a month earlier, their efforts to limit the passage of the Civil Rights Act of 1957 had threatened to rip apart the party. And as civil rights clashes erupted throughout the South, such as in Little Rock, the Democratic Party's chances in the upcoming elections seemed in doubt.[3] Racial segregation was an "issue that is not going to go away," a former aide outlined in a memo passed on to Lyndon Johnson. To secure the 1958 and 1960 elections, Democrats had "to find another issue which is even more potent. Otherwise the Democratic future is bleak." Sputnik might be their answer, he suggested.[4]

Democratic leadership recognized that if space exploration was going to factor into the upcoming presidential election, there were political advantages to keeping the congressional hearing on national preparedness as bipartisan as possible. They believed that Johnson had the necessary "reservoir of goodwill and an aura of statesmanlike handling of defense problems from the early Preparedness Committee days," which positioned him to take on the Sputnik challenge.[5] Johnson took a nuanced, restrained approach, with an eye to his long-term political objectives. If treated evenhandedly, or at least with the appearance of fairness, the hearings could fashion him as not only the nation's leading space policy expert but also the next Democratic presidential candidate. The stakes were high, and Johnson threw himself into the job knowing full well that his performance would affect his future in national politics. Before undertaking the congressional investigation, he assured Eisenhower in the Oval Office that there would be "no 'guilty party' in this inquiry except Joe Stalin and Nikita Khrushchev." Eisenhower found Johnson's guarantees convincing, later telling his personal secretary, "I think today he is being honest."[6]

Democratic leaders credited Eisenhower's 1953 presidential election in part to the Preparedness Investigating Subcommittee's Korean War reports. Formed in 1950, the subcommittee became the primary oversight and investigative arm of the Senate Armed Services Committee. It produced a series of more than forty dramatic and sensational reports on the Korean War, which secured Johnson extensive press coverage but undercut the Truman administration. "The Democrats really weren't very happy with it," a Democratic senate staff member recalled.[7] With Sputnik, Johnson used similar tactics as he had in the early 1950s, but this time he made sure to advance the Democratic Party's interests, not the Republicans'. To bolster the staff already in place, he assembled a strong team of lawyers from New York and scientists from Harvard, Cal Tech, and Rice Institute (now Rice University). He choreographed and controlled elaborate publicity of the hearings, from television coverage to photograph sessions to interviews. Johnson courted the press, inviting some to dinner or to his Texas ranch. For the most influential journalists, he leaked news. And it was all done with speed.[8] "Everything had to be done in a hurry," recalled Eileen Galloway, a national defense analyst aiding Johnson and the committee. "If you were working for Lyndon Johnson, everything had to be done in a hurry . . . and we were working on it from morning to night."[9]

After a nearly six-hour Pentagon briefing on US and Soviet space capability in early November 1957, Johnson publicly announced the upcoming congressional subcommittee hearings and told the press that the country needed "bold new thinking in defense and foreign policy."[10] The Soviet Union was ahead "period," Johnson told members of the press in his Capitol Building office. If the United States wanted to take the lead in space, the country needed decisive, proactive action. But in his estimation, defense officials were not demonstrating the proper sense of urgency. Critiquing the administration, he stated that "timid minds will not produce bold programs."[11]

The congressional subcommittee hearings started on November 25. Much like a conductor, Johnson orchestrated everything. He introduced the expert witnesses, led the cross-examinations, and handled the press. Johnson's aptitude for public relations shone through in his command over the meetings. Instead of starting with the national security threat of the Soviet satellite launch, he asked well-known scientists Edward Teller and Vannevar Bush, and rocket engineer and Disney star Wernher von Braun, for their testimony on the broader implications of Sputnik. Introducing the men in the public hearing, Johnson reminded those gathered that their remarks would give clarity to the significance and meaning of Sputnik. Teller talked about the future of outer space exploration having "amusing and amazing . . . consequences." Like Johnson, he compared Sputnik to Pearl Harbor. Von Braun pushed for a human spaceflight program and the creation of a "national space agency." As Johnson well knew, Teller, Bush, and von Braun not only shared their expertise; more importantly, they secured press coverage.[12]

Next came the military experts. Even though McElroy was sworn into office as secretary of defense a mere five days after Sputnik orbited Earth, he came prepared for four hours of questioning by the committee. Driven and entrepreneurial from a young age, McElroy started out in the mailroom at Procter and Gamble and quickly rose through the ranks, becoming president of the company by age forty-four. He took more than a 90 percent pay cut when he accepted Eisenhower's offer to become defense secretary earlier that week. McElroy impressed Johnson's staff, ably handling the senators' intensive probing of all areas of the nation's defenses. And he would soon impress people throughout the country as he navigated reorganizing the Pentagon with the aim of ending interservice rivalries.[13]

A major point of the questioning of McElroy and his staff focused on the Defense Department's duplication of missile development. The Army was developing the Jupiter; the Air Force had the Thor, Atlas, and Titan; and the Navy was building the Polaris. Gale explained the tension between services and the apparent overlap in roles: "There is a feeling that the military of the future will be based heavily on missiles, and the Service which has the dominant role in missile assignments will be well on top."[14]

As biographer Robert Caro points out, although Johnson masterfully heightened the sense of urgency in the hearings, he did not demonstrate the feeling himself. After the initial news of Sputnik broke, he did not hurry back to Washington but instead stayed in Texas through October 16. After a two-day visit to Washington, he returned to Texas until the subcommittee briefing at the Pentagon in

November. In total, he spent a mere six days in Washington in the six weeks following Sputnik's launch. But those days he was in Washington, LBJ was busy crafting a specific scene of "national crisis."[15] "He was really like a dynamo at that time," Galloway recounted.[16] On Capitol Hill they joked, "Light a match behind Lyndon and he'd orbit."[17]

A new year did not bring new optimism for the American public. Early polls indicated that the country felt anxious at the beginning of 1958. Not only had the Soviet satellites shaken national confidence; Americans also worried about a worsening economy. In the fall, unemployment had risen, civil rights tensions intensified, and a sharp recession jolted the nation. Leaks from the NSC report "Deterrence and Survival in the Nuclear Age," often referred to as the Gaither Report, indicated that the United States was not prepared to defend against a Soviet nuclear attack, intensifying anxieties. Eisenhower rebuffed the report's conclusions and rejected the price tag of the military buildup. And as if all of this was not enough, the cover of the January 6 issue of *Time* magazine featured a beaming Nikita Khrushchev holding up a model of Sputnik. The Soviet leader was dubbed 1957's "Man of the Year."[18]

Johnson called a meeting of the Democratic Senate Caucus on January 7, 1958. It was no coincidence that Eisenhower's State of the Union address was scheduled for two days later. As Johnson confided to Richard Russell, "I cannot overemphasize what I believe to be the importance of this meeting."[19] And he had George Reedy inform the press that this meeting would be Johnson's "State of the Union address."[20] In highly partisan comments, Johnson criticized the Eisenhower administration's New Look fiscal policy for putting the country in danger. He used the opportunity to draw the connection between the Democratic Party and a more robust space program. "Control of space means control of the world," Johnson warned. "If, out in space, there is the ultimate position—from which total control of the earth may be exercised." He continued, "From space, the masters of infinity would have the power to control the earth's weather, to cause drought and flood, to change the tides and raise the levels of the sea, to divert the Gulf Stream and change temperate climates to frigid . . . our national goal and the goal of all free men must be to win and hold that position."[21] LBJ overstated the military urgency of the Sputnik moment. Although successful for Johnson politically and for garnering support for the US space program, his alarmist rhetoric had long-term consequences.

Reedy identified one of the drawbacks of Johnson's headline-generating performance during the hearings: "In retrospect some of the material should have been examined more carefully before being spread on the record in ex parte proceedings. One of the results was the public creation of a 'missile gap'—a concept that we were hopelessly behind the Soviets in the possession of ICBMs."[22]

By the end of January 1958, the records of the congressional hearings filled more than two thousand pages. More than seventy witnesses had been called.[23] The subcommittee's unanimous report recommended a seventeen-point program, from technical developments to the creation of a new government agency. "We discovered some disturbing truths from those three months of hearings," Johnson said.[24] A memo pointedly summed up the threat facing America: "The reason the United States fell behind Russia in satellite development in the first place is because we neglected the relation between scientific achievement and international relations."[25]

On the last day of the month—January 31—the tide turned, and the United States successfully launched its first satellite, Explorer 1. The pencil-like payload contained a package of scientific instruments that collected data that led to the discovery of the Van Allen radiation belts. The launch of Explorer 1 did not receive the widespread sensational coverage that Sputnik 1 had a few months earlier. Newspapers in Europe and Asia primarily focused on the impact of the cylindrical satellite on the US position in the upcoming summit talks on nuclear disarmament. In France and the United Kingdom, journalists predicted that the United States might change its policy toward the summit talks after the success of the launch, and in India commentary suggested that the United States could now act from a position of power and strength. The balance of power, according to most coverage, had returned to equilibrium. The press in Communist countries acknowledged the launch but downplayed its broader significance.[26]

The day after the Explorer 1 launch, Saturday, February 1, was an extremely busy one at USIS posts around the world. USIS Germany's dogged efforts reveal the scope and intensity of on-the-ground activities of US public diplomats. In Germany the USIA oversaw sixty-eight libraries and reading rooms, and nearly twenty USIS posts in the late 1950s.[27] Each post created its own promotional campaign ahead of the launch. Staff reviewed their collection of scientific photographs and updated the captions for exhibits at *Amerika Haeuser*—American cultural meeting places—around Germany. They also updated two films, *Space Unlimited* and *Geophysical Year*, so that more than sixty prints of the films could be distributed shortly after the launch. USIS Germany sent film kits to posts around Germany, each filled with a prepared lecture, articles, posters, leaflets, a chronological chart, and a picture layout of the major figures in space research.

USIS staff in Bonn arrived at the embassy early on Saturday morning, prepared for a long day ahead. When President Eisenhower's official proclamation arrived, the post immediately translated it into German and then sent it via teletype to each consulate, along with instructions for the Marine guards on duty to notify the public affairs officers (PAOs) as soon as they received the message. The PAOs at each post then distributed the translated announcement to newspapers throughout Germany by late afternoon on Saturday. The USIS post in Bonn also sent the president's statement to the German wire service, all press correspondents stationed in Bonn, all foreign embassies, the Federal Press Office, the press offices of the four leading political parties in Germany, and the Bundestag and Bundesrat. By the end of the day, USIS Germany staff had produced 920 prints of 30 photographs, written 10 articles on American scientific progress, and distributed these to all the posts in Germany.[28]

The German press incorporated much of the USIS Germany material into its coverage of Explorer 1. In Bremen, for example, forty-one papers carried nine hundred inches of USIS material as well as twelve pictures, reaching an audience of roughly three million people. In this way, US government staff not only amplified the news that America launched a satellite; they also framed the story. Major newspapers and magazines throughout Germany published articles that drew heavily from USIS material, and in some instances they directly reprinted USIS articles verbatim.[29] The Voice of America distributed a radio report on Explorer 1 to all West German radio stations. Within a few days, *Amerika Haeuser* and the German-American Institute produced exhibits on the US space program that included posters, scientific books, photographs, and, at the Stuttgart *Amerika Haus*, a movable globe used to show the orbit of Explorer 1. A series of lectures by German scientists, including Dr. Julius Bartels and Richard E. Kutterer, complemented these exhibits.[30] Similar programs were undertaken by hundreds of colleagues working at USIS posts and US embassies around the world, attempting to make the most of the United States' first space shot.

With the aid of the USIA information system, the news also spread rapidly to all corners of the world. USIS Tehran created elaborate window displays with moving three-dimensional models of the satellite's path for Iranian audiences. The display created traffic jams on the sidewalk as crowds of twenty to thirty people gathered in front of the windows throughout the day.[31] In Japan, popular knowledge of Explorer 1 compared to that of the Sputniks. Although awareness was comparable, it was clear that the satellites were not equally impressive, according to public opinion polling. The Soviet Union led the space race in the eyes of the world, and many people thought the Soviets would maintain the lead for years to come. For the majority of those whom the USIA questioned, it was too late: the US was too far behind.[32]

On February 6, 1958, a few weeks after the subcommittee released its report, Johnson shepherded a resolution through the Senate for the creation of the Special Committee on Space and Aeronautics. This committee would be charged with drafting legislation for a new space agency. As expected, Johnson became chairman after a unanimous vote, selected staff, and oversaw the committee's agenda. Majority

Leader John McCormack chaired the corresponding House committee. The Senate hearings would begin in early May.[33]

By early February, it was clear that the United States would create a space agency. But the structure, location, and whether this new
agency would be civilian or military were still uncertain. During a
meeting at the White House on February 4, Vice President Nixon
pointed out that the United States' "posture before the world would
be better" if the agency was separate from the military. Later in the
meeting he reemphasized this point.[34] Members of PSAC expressed
serious concern over situating the American space program within
the Pentagon, fearing that it would limit and militarize space research. The USIA's new director, George V. Allen, recommended that
the United States create a stand-alone civilian space agency. By separating military and civilian space activities, Allen argued, the United
States would be in a better position to regain its prestige losses from
Sputnik. At roughly the same time, the President's Advisory Committee on Government Organization made a similar recommendation, which Eisenhower authorized and sent to the Bureau of the
Budget to draft legislation. Not only did it propose the establishment
of a new civilian space agency; it also gave the agency responsibility
for civilian space science and aeronautical research, and called for a
space advisory board. On April 2, Eisenhower presented this legislation to a joint session of Congress, calling for the establishment of a
new civilian aerospace agency.[35]

That same month, RAND released a report assessing the political
implications of the Space Age. It concluded that Sputnik established
that as with military power, prestige and national image mattered in
the geopolitical arena. International alliances and alignment, the report stated, relied on positive perceptions of the United States: "From
now on, the U.S. should recognize the need for restoring credibility
in U.S. superiority, stress our peaceful intentions and their aggressive
ones, and disclose and publicize U.S. outer space activities according,
first and foremost, to the effect on the U.S. international position."[36]

Many USIA staff were of like mind with their contemporaries
at RAND. They prepared extensive information programming to

highlight the second successful US satellite placed in orbit: Vanguard 1. Although Soviet premier Khrushchev derided it, calling the small metal orb a "grapefruit," American public diplomats heralded the satellite's scientific impact.[37] In its promotion of the mission, USIS Chile highlighted the work of a satellite tracking station outside Santiago. The agency sent a radio crew to the station to record the sound of Vanguard's signal as it orbited over the country. Staff airmailed this recording to radio stations throughout Chile, from Arica in the north to Punta Arenas at the southern tip of the continent. Broadcasters notified listeners that they were not just listening to the signal of the satellite, but "the actual sound of Vanguard as it passed over Chile," making a global event locally relevant.[38]

A few months later, the Soviet Union launched its third satellite— this time the massive Object D scientific satellite—on May 15. Sedov, who had surprised the world in the summer of 1955 when he announced the Soviet Union would be launching satellites during the IGY, claimed that this satellite "could easily carry a man with a stock of food and supplementary equipment."[39] The implications of a 3,000-pound satellite seemed clear to many around the world: the Soviet Union had immense rocket thrust capabilities. But as Khrushchev later admitted, the early Soviet rockets "represented only a symbolic counterthreat to the United States." In fact, the sizable Soviet rockets were the result of the inefficiencies of their design. They were unwieldy and lacked an inertial guidance system, which made them sufficient for launching satellites into orbit but not usable as strategic weapons.[40] However, Khrushchev claimed publicly, to much effect, that the USSR had "outstripped the leading capitalist country— the United States—in the field of scientific and technical progress."[41]

Congress passed the National Aeronautics and Space Act on July 16, and President Eisenhower signed it into law on July 29. The law led to the formation of the National Aeronautics and Space Administration (NASA) the following October. The 1958 Space Act— negotiating the interests of the administration, the Department of Defense (DoD), Congress, and the scientific community—became the core statement guiding US civil space policy. In addition to giving NASA direction over civilian space efforts, including the research that was undertaken at three former National Advisory Committee for Aeronautics (NACA) facilities, the Space Act called for a council to advise the president on space exploration, and it created a civilian-military liaison committee to coordinate NASA and DoD projects. Although NASA was not officially in operation, Eisenhower assigned the new agency the task of developing a human spaceflight program in August 1958. In December the newly formed Space Council granted what became known as Project Mercury "highest priority" status so that the United States could put humans into Earth orbit as soon as possible.[42]

The Space Act was internationalist in tone and vision, coupling the preeminence of US space activities with international cooperation and engagement. NASA, in concert with the State Department and Department of Defense, would proactively nurture and mold the development of space capabilities in foreign countries. As the first line of the Declaration of Policy and Purpose reads, "The Congress hereby declares that it is the policy of the United States that activities in space should be devoted to peaceful purposes for the benefit of all mankind."[43] The notion that the US space program should serve "all mankind" and not simply American citizens illustrates this internationalist vision, which had increasingly framed US grand strategy in the early cold war era.[44] The Space Act also included a provision for the widespread dissemination of information. As Johnson and other leaders envisioned, NASA would not only send satellites into space; it would also be an arm of US foreign policy. The US space program would be part of a larger effort to unite the world—politically and culturally—through American technological and scientific leadership.[45]

Nearly eight thousand people, coming from NACA research laboratories and the Navy Vanguard Project, became NASA employees on October 1, 1958.[46] Within two years this number nearly doubled. In 1958 the United States sent four civilian satellites into orbit and two probes into outer space. A year later NASA successfully launched four more satellites and sent one space probe past the moon and on toward the sun. In 1959 NASA also began training seven astronauts for Project Mercury and testing Atlas boosters and capsule instruments to prepare for human spaceflight. Two monkeys, Able and Baker, rode into space on the Army's Jupiter rocket and returned to Earth alive and well. Meanwhile, the Soviet Union launched a series of Luna probes, accomplishing the first impact on the moon and even capturing a photograph of its far side, among other firsts.[47] Near the end of 1959, NASA's formal plans adopted lunar exploration as a "long-term goal." The goal of the near future, the 1960s, would be making "manned exploration of the Moon" feasible, launching a crewed circumlunar flight, and building an Earth-orbiting space station. Landing humans on the moon would happen in the future, sometime beyond the 1970s.[48]

NASA worked closely with the USIA, the State Department, and the White House to ensure that the new civilian space program would serve foreign relations interests by promoting US space successes heavily abroad. The USIA became officially responsible for distributing information internationally about both civilian and military space activities, while NASA took charge of domestic public relations. NASA, the DoD, and the Atomic Energy Commission, along with other agencies involved in space exploration and research, supplied the USIA with material for crafting press releases, magazine reprints, pamphlets, photographs, radio and television broadcasts, films, and exhibits. The USIA also based a full-time liaison officer at NASA. Eventually, representatives from the USIA participated in some of the Space Council's activities, especially meetings that focused on overall policy.[49]

In late November 1958, USIA Science Advisor Hal Goodwin and NASA Office of Public Information Director Walter Bonney met to lay out an overall framework for space information programming. They agreed that the tone and methods that the agencies used would be critical. The agencies must present the US program as proactive and not reactionary. It was also important to frame each US launch as part of a larger program with larger aims, not as individual accomplishments. This approach was in direct response to the impact of Soviet space feats abroad and the damage that launch failures like Vanguard had inflicted on American prestige. In large part, what this plan did was offer an explanation or justification for the slower pace of US space exploration. NASA efforts, the USIA and NASA public affairs clarified, were rational, based on scientific motivations, and shared with all humankind. The USIA would not present space exploration as a race but instead as a rational pursuit of knowledge. Like research undertaken in a laboratory, space exploration was experimental, and not every "experiment" would work. Goodwin and Bonney's approach stressed that the schedule of space shots, at least in information programming, had nothing to do with Soviet progress in space. They laid out this framework "in an attempt to define a way in which the public presentation of the civilian space program would be of maximum value to the USIA."[50]

The USIA's general science program sought to strengthen ties between the United States and other countries by explaining how science and technology factored into American life and progress, not simply disseminating scientific and engineering information. Describing how science flourished under the US system and how Americans' general welfare flourished with the aid of science, USIA programming attempted to demonstrate that American democracy was in line with foreign people's "aspirations for freedom, progress and peace."[51] Shying away from a focus on space science and technology, the USIA stressed American scientific accomplishments in general. In his State of the Union address, Eisenhower called for extending the Atoms for Peace program to include other areas of science and medicine. Science for Peace, as the new initiative was named, became a priority theme for the USIA in 1958.[52]

The Soviet Union, according to the USIA, was using its satellite launches to substantiate claims about the superiority of the Communist system. The tendency to equate accomplishments in space exploration with world leadership would only increase, thanks in large part to Soviet efforts to foster this correlation. USIA officials emphasized what they perceived as the major difference between American and Soviet science: the integration of science and technology in all areas of life. To make the point, Goodwin explained that in the Soviet Union an oxcart "may be seen moving on primitive roads in the vicinity of Soviet satellite launching complexes."[53] There were fundamental differences between the use of technology in various economic and political systems, according to the USIA. Economic competition within the United States, as stated by the USIA guidelines, stimulated a diversity of technological developments, which contributed to "the good life."[54]

Even though competition lay at the core of the USIA's objectives, agency guidelines made sure that promotional material eschewed any suggestion that the United States was competing with the Soviet Union. Individual events should not be presented as markers of national superiority but instead as part of broader long-term objectives. The universal and international nature of science should be emphasized, international cooperative programs should be highlighted, and the areas of science and technology that could be applied to the modernization of underdeveloped nations were particularly meaningful and should be "presented with understanding, restraint, and a becoming humility."[55] Expressing a view consonant with modernization theory, a 1958 report on the USIA science program explained that "science and technology seem to hold special promise to the people of the less developed nations, since they see in modern science the means of bypassing the historically long, slow route to full realization of their national potential."[56] The USIA predicted that just as science and technology were developing at an exponential rate, global interest in these fields was also on the rise. For this reason, science and technology had to play a central role in USIA programming. After the Soviet Union launched Sputnik and the elaborate propaganda programs that ensued, the world became aware of Soviet capability. It was the USIA's job to make sure that the international public was not only aware of American technological and scientific know-how but also about how within the US political system these advances contributed directly to the welfare of people around the globe.[57]

The November 1958 election saw a crushing defeat for the Republican Party. Democrats won nearly fifty additional seats in the House of Representatives. The total number of Democratic victories in the Senate surpassed all previous records for any party in US history.

The lagging economy combined with anxiety over the Eisenhower administration's national security policy in the wake of Sputnik assured Democratic victory.[58] Journalists Robert Novak and Rowland Evans called Johnson's deft treatment of Sputnik that year a "minor masterpiece." Not only did he assert his role as the national space expert, inserting himself even more prominently on the international stage; he also leveraged the hearings to sow doubt about the Eisenhower administration.[59]

While Washingtonians debated spaceflight, psychological strategy, and US global power over lunch and in conference rooms, NASA and USIA staff also worked on the ground around the world, building a network of tracking stations, disseminating the latest information on the US space program, and creating exhibits that would appeal to broad audiences. Space exploration had become a major feature of USIA material. The USIA's Press and Publications Service (IPS) produced news stories, speech texts, chronologies, bylined columns, pamphlets, photographs, picture stories, cartoon features, magazine articles, and a comic book titled "Man and Outer Space." USIS posts translated and adapted this material for local audiences' tastes and interests. Foreign press drew from this material and in some instances printed USIA news releases verbatim. The USIS posts kept careful track of when and how the foreign press used the media that they distributed and reported these results back to USIA headquarters. Noting that science and technology information was often more readily used by the foreign press than other material, USIA officials observed that in places such as Cuba and Venezuela, "we never have difficulty placing scientific photos."[60]

In the late 1950s the USIA did not have the budget to create an elaborate series of space-focused exhibits, so the agency offered NASA its network of USIS posts to distribute NASA material worldwide. Requesting models, drawings, and photographs for exhibits, the Exhibits Division of the USIA hoped to distribute this material to posts ahead of NASA launches.[61] USIS posts often worked with foreign institutions, arranging for exhibits to be shown in local venues. The kit exhibits on Explorer I, a twelve-foot model of the Echo balloon satellite, and other satellite models were among the USIA's most popular displays in 1959. At the Karachi National Science Fair in Pakistan, the USIA showed the "Space Unlimited" and "International Geophysical Year" exhibits to the audience of 75,000. A "ladies only" showing attracted 2,000 women. These exhibits also drew thousands of people in Helsinki, Madras, Southern Rhodesia, and numerous other locations. The US Department of Defense exhibited models of the Explorer satellites to groups of South Vietnamese military. According to the DoD's Psychological Warfare Service, "The soldiers got a great belief in the strength and science of the free world."[62] Similarly, USIS Tel Aviv circulated the exhibit among Army camps in Israel, while USIS Rangoon created a space exhibit for the Army, Navy, and Air Force Cadet Corps of Burma's Defense Services Academy.[63]

The USIA promoted spaceflight through cultural programming, but the agency also worked with NASA and the State Department on developing a global infrastructure of tracking stations. In 1959 the US reached new agreements with foreign nations to create an expansive system to support the upcoming Mercury human spaceflight program. This network, spread across the globe in Bermuda, the Bahamas, Grand Turk Island, Mexico, the Canary Islands, Nigeria, Tanzania, the Solomon Islands, Canton Island, Hawaii, and two stations in Australia, would provide necessary tracking, telemetry, and voice communications as the spacecraft traveled overhead, while simultaneously establishing a physical US footprint in other nations.[64]

Setting up Mercury tracking stations was a complicated undertaking that required more than agreements and hardware; it also required a multipronged public relations approach that targeted both local populations and government officials. NASA ran into significant challenges in Africa, in particular. It was here that America's troubled race relations intersected with a powerful decolonization movement. Strategically, after 1957 the USIA increased programming and tailored material for the region, including films and radio features on the accomplishments of African Americans to counteract news stories about racial oppression in the US. The overall messaging stressed

America's support of African countries' independence. Secretary of State Dulles called Africa "the coming continent" and explained that "we were anxious to do all we could."[65]

US government officials made sure to avoid using the term "tracking station" in Africa because studies had shown that the phrase had a militaristic connotation there.[66] The use of military uniforms or titles by US officials, even during the Mercury flights, was prohibited. There was also general concern within the State Department over Western African criticism of American "colonial" presence in the region. Tensions arising from the tracking stations could strain the overall US position in Africa. On the other side of the continent, in Zanzibar, Communist demonstrations threatened the establishment of a station there.[67] Making matters more difficult, Moscow Radio broadcast anti-tracking features in Africa that suggested the United States was using the facilities as military bases. Peking Radio also contributed to the controversy over the NASA stations by emphasizing the "military dangers" of such "bases" in broadcasts to East Africa. Many newspapers in East Africa carried news stories about a "secret" US military base.[68]

On October 19, 1960, just a few weeks after Nigeria gained independence from Great Britain, the US reached an agreement to build a tracking station outside Kano. Nigeria was pro-West, and by that time the United States already had a history of funding development programs in the country, primarily focused on agriculture. But any sign of American empire, however informal, raised apprehension about the country's larger intentions in Nigeria. To address these concerns, USIA, NASA, and State Department officials made sure to highlight the collaborative operation of the tracking stations between NASA and the host government. NASA hired and trained local personnel, while the USIA undertook a targeted information campaign, including film screenings, the distribution of issues of *National Geographic*, and a moderate publicity buildup.[69] As one official described the situation, the United States government should "ascertain which marginal activities can be Nigerianized in interest vital [to] political factors involved."[70] The tracking stations adopted an "open-door" policy and became tourist attractions, information centers, and a counternarrative to America's troubled race relations.[71]

By 1960, the USIA determined that the United States had steadily regained prestige, although the Soviet Union still outstripped the nation in terms of space exploration. In other words, the Soviet Union was winning the space race. The effects of Soviet space accomplishments spread to other fields, including Soviet science and technology, military power, and general status. USIA officials believed that the world public saw the Soviet Union in a very different light after Sputnik. Before 1957, America had the lead in science, technology, and production; Soviet satellites and robotic probes evidenced the USSR's ability to challenge that lead. The United States, according to USIA Director George V. Allen, must "push forward vigorously with space exploration" to maintain international confidence in American leadership. The space race had broad implications; it factored into "almost every aspect" of the relationship between the United States and the world.[72]

Eisenhower formed a presidential committee to assess the USIA and America's global image and appointed as its leader Mansfield Sprague, the former counsel to the secretary of defense. A substantial section of the lengthy Sprague Report was titled "The Impact of Achievements in Science and Technology upon the Image of the United States Abroad." Scientific and technological prowess, according to USIA public opinion data, denoted both power and progress to overseas audiences. American science had the added benefit of symbolizing US values, including freedom, democracy, and pluralism. The report recommended that the president encourage federal agencies to evaluate the "international political-psychological factors" of scientific and engineering programs when allocating funding. Scientific and engineering programs promised a dual benefit: first as "status symbols in the East-West conflict, direct indices of power," and second as "promises of directly meaningful, applicable, useable instruments of progress."[73]

These conclusions mirrored what Henry Kissinger identified in 1955 as the "new diplomacy" but were updated for the current postcolonial moment. As the recommendations of the report read, "In both the new countries and the older ones going through the crisis of modernization, formal and traditional diplomacy of the predominantly government-to-government type often plays a limited role." The implications of this evolving geopolitical landscape required psychological strategy, an understanding of "public opinion in all countries, open and closed, old and new." Scientific and technological programs, especially stunning feats of prowess, were increasingly viewed as essential for securing global leadership in the rapidly evolving cold war world order.[74]

The Sprague Committee predicted that the 1960s "may prove to be one of the most convulsive and revolutionary decades in several centuries." Scientific and technological progress lay at the core of this revolution, according to the report. With half of the world's population living "under conditions of hunger, disease and ignorance," and the Soviet Union pressing for global domination, the report stressed that the United States must use economic, diplomatic, and informational instruments to contain Soviet expansionism and steer this new scientific revolution.[75]

To many people within the US government, not just the members of the Sprague Committee, the world seemed on the threshold of a scientific revolution. National Academy of Sciences Executive Director Hugh Odishaw compared the Space Age to the Copernican revolution and suggested that it could change "man's concepts of man."[76] Philosopher and political theorist Hannah Arendt considered the launch of Sputnik "second in importance to no other, not even to the splitting of the atom," and expressed concern about the larger implications for the future of the human condition.[77] USIA Science Advisor Goodwin had come to see science, and US science in particular, as the defining feature of twentieth-century society. Equating the early cold war era with the Copernican revolution and the scientific revolution, Goodwin argued that the best way to win the confidence of newly independent nations would be to demonstrate how the United States was the driving force behind the contemporary scientific revolution.[78]

The potential of spaceflight to fundamentally reshape society excited theorists and politicians alike. With widespread social and political upheavals already under way, especially the emergence of new states in the postcolonial world, spaceflight could be an instrument for shaping the ideology, aspirations, and allegiances of other nations, many theorized. The vision of a looming scientific revolution would undergird the framing of the space program's significance and broader meaning for the coming decade. In the 1960s US government officials would continue to question and reevaluate the impact of space exploration on international public opinion, the role of science and technology in foreign relations, and the best way to achieve geopolitical influence.[79]

4

IF WE ARE TO WIN THE BATTLE,
1960–1961

> What is prestige? Is it the shadow of power or the
> substance of power?
>
> —PRESIDENT JOHN F. KENNEDY, 1961

John F. Kennedy identified and leveraged the political power of image. It propelled his skyrocketing political trajectory and informed his approach to foreign relations. During the 1960 presidential campaign he presented himself as vigorous and charming, polished and self-assured, and most importantly as the forward-looking antidote to the sluggish economy and the staid policies of the Eisenhower administration. From staging public appearances to courting journalists, he carefully mediated his public persona even before he officially announced his presidential bid.[1] One journalist called Kennedy "the only politician a woman would read about while sitting under the hair dryer," suggesting that he was seeking office more as a celebrity than a politician.[2] But Kennedy did "not just look good in the media," historian Alan Brinkley has argued. "He used the media, carefully, consciously, calculatedly, not only in his campaigns but throughout his presidency."[3]

Kennedy's astute sense of the role of image in politics had been reinforced in late September 1960. Just a few weeks ahead of the presidential election, Kennedy flew to Chicago to debate his opponent, Vice President Richard Nixon. This would not only be the first televised presidential debate in history; it was also the first time that two candidates from major opposing parties came together to face off in person. The afternoon ahead of the broadcast, Kennedy prepared at the Ambassador East Hotel. He flipped through flashcards and then tossed them to the floor once he had mastered a subject. The potential rewards were too high not to prepare, not to rewrite the opening remarks his aides composed, not to think through answers to potential questions.[4] Nixon, on the other hand, did not break from his strenuous campaign schedule ahead of the debate. As Nixon's media advisor said derisively, "I was told to brief him on the most important telecast of his life while riding over to the studio in an automobile."[5]

Eisenhower had counseled his vice president not to debate the charismatic Democratic challenger. Nixon was already well-known throughout the country, respected and leading in the polls. A televised debate risked raising Kennedy's prominence. But Nixon was confident. A practiced debater, he had famously held his ground with Soviet premier Khrushchev at the "kitchen debate" in Moscow in 1959, championing the fruits of a capitalist consumer society such as dishwashers and color television sets. To Kennedy's "surprise and joy," Nixon agreed to a series of televised debates.[6]

Beginning at 9:30 p.m. eastern time, CBS Studios in Chicago broadcast the debate. Between seventy and eighty million people tuned in, roughly two-thirds of the adult population of the country. As CBS president Frank Stanton bluntly remarked, "Kennedy was bronzed beautifully . . . Nixon looked like death." Even more harshly, Chicago mayor Richard J. Daley exclaimed about Nixon, "My god! They've embalmed him before he even died."[7] Not only had Nixon declined the producer's offer for a makeup artist touch-up; he was still recovering from an infected knee that had required hospitalization. Nixon stood gangly in his oversized gray suit, the edges fuzzed into the studio's gray background. In stark contrast, an afternoon prep session on the roof of his hotel left Kennedy even more tan than usual. He stood in a dark tailored suit, confident and polished.[8]

"That son of a bitch just lost the election," Nixon's running mate, Henry Cabot Lodge, remarked after the debate.[9] Although historians disagree about who "won" that night, public opinion polls showed that to television viewers at least, Kennedy was the victor. Radio listeners had a different opinion. The spectacle of the televised debate altered the nature of politics long after the Kennedy administration. Not only a definitive event in the presidential election, the debate buttressed Kennedy's already well-developed sense that image matters in politics. The image of a poised candidate on the television screen gave the Kennedy campaign a much-needed boost that carried it through to November. Theodore Sorensen, Kennedy's counselor and speechwriter, commented that along with the subsequent debates in the series, Kennedy's performance that night was "a primary factor in Kennedy's ultimate electoral victory."[10]

Kennedy was "a man who perhaps better than any other president in our history, understood how foreign opinion worked, what molded it, what shaped it and how to shape it," observed USIA Acting Director Donald Wilson.[11] What he understood less, at least during his tenure in Congress, was the proper direction of America's space program. As journalist Hugh Sidey put it, "Of all the major problems facing Kennedy when he came into office, he probably knew and understood least about space."[12] Before the spring of 1961 he treated spaceflight as a talking point, drawing on spaceflight to depict the current Republican administration as incompetent and ill-prepared to protect the interests of the United States.[13]

Kennedy had joined the U.S. House of Representatives in 1947 and then the Senate six years later. In the eight years he served as the junior senator from Massachusetts, he paid little attention to the US space program. Instead, his focus remained firmly fixed on defense and foreign policy. As a member of the Senate Committee on Foreign Relations, he concentrated on national security and stressed the importance of international public opinion, especially in the midst of the decolonization movement. In 1960, the year Kennedy won the presidential election, eighteen new nations were established. The political map of the globe was being redrawn. Within this new world order, Kennedy ascribed to the tenets of NSC-68: the Soviet Union was an expansionist force that the US had to contain wherever its influence spread. As he described it, "This is not a struggle for supremacy of arms alone—it is also a struggle for supremacy between two conflicting ideologies: Freedom under God versus ruthless, godless tyranny."[14] Within this geopolitical landscape, Kennedy maintained, the United States must be proactively engaged internationally, confronting Communist influence where it emerged. And in Kennedy's view there was a direct correlation between supremacy and national prestige. Winning the hearts and minds of people in Asia, Africa, and Latin America soon guided his foreign policy agenda.[15]

It was not until space, and Soviet space feats in particular, became an undeniable currency for international prestige that Kennedy included the US space program within his larger grand strategy framework. On the Senate floor in mid-August 1958, Kennedy warned of national complacency, comparing the US situation to Britain's loss of Calais four hundred years earlier. Worried that "the periphery of the free world [would] be nibbled away" and that the "balance of power" in turn would "gradually shift against us," Kennedy drew on history to argue for a call to action. It was in this speech that he first used the phrase "missile gap." The United States, he said, "could have afforded, and can afford now, the steps necessary to close the missile gap." In Kennedy's able hands the phrase became a useful political tool, even if it was not based on anything but Khrushchev's exaggerated claims.[16]

Over the next two years, Kennedy would focus on American prestige in many of his speeches: the loss of it, how to regain it, how to win it. In impassioned addresses in small and large cities across the country, Kennedy cast space exploration and the "missile gap" as embodiments of what was problematic about the Eisenhower administration: passivity, lack of vitality and vision, staid and ineffectual leadership. Although he kept his speeches short—most were five minutes in length, with the longer rally speeches stretching only to twenty minutes—they were frank, they were factual, and they conveyed confidence. Underlying all his speeches, whether they were about unemployment or national security or urban housing, were the themes of "prestige," "progress," and national security. And Kennedy included "It is time to get this country moving again," which he repeated so often that it became his slogan.[17]

"We failed to recognize the impact that being first in outer space would have," Kennedy explained at a campaign stop in Canton, Ohio, in late September 1960. He warned the crowd gathered in the Municipal Auditorium that in the eyes of the world, the Soviet Union "was on the march . . . it had definite goals . . . it knew how to accomplish them . . . it was moving." What made matters worse is that it looked like the US was "standing still." Now the country needed to overcome "that psychological feeling in the world that the United States has reached maturity, that maybe our high noon has passed . . . and that now we are going into the long, slow afternoon."[18] He had a similar message for voters at a high school auditorium in Pocatello, Idaho: "They [the international public] have seen the Soviet Union first in space . . . they [have] come to the conclusion that the Soviet tide is rising and ours is ebbing." For those gathered at New York University in October 1960, Kennedy argued that "the key decision which [Eisenhower's] administration had to make in the field of international policy and prestige and power and influence was their recognition of the significance of outer space."[19]

Although he did not articulate a plan for the US space program, "space race" rhetoric and calls for renewing global stature helped him communicate a major campaign message: "a new leader for the 60s." As he accepted the Democratic Party's nomination for the presidency at the Memorial Coliseum in Los Angeles on July 15, 1960, Kennedy galvanized his party with rhetoric of a "New Frontier" and described the cold war as "a race for mastery of the sky and the rain, the oceans and the tides, the far side of space and the inside of men's minds."[20] Early the next morning, Kennedy asked Lyndon Johnson, the prime competitor in the Democratic primaries, to be his running mate, recognizing that the Senate majority leader would secure southern votes and balance the ticket. Johnson eventually agreed.[21]

Shortly after Kennedy accepted his party's nomination, Central Intelligence Agency Director Allen Dulles briefed him on the current status of national security. There was little evidence—even highly classified evidence—that there was in fact a "missile gap." After meeting with Dulles and other defense officials, Kennedy reported to Sorensen that "the briefings . . . were largely superficial anyway and contained little . . . not read in the *New York Times*."[22] A thorough assessment of the status of Soviet strategic attack capabilities would not be available until much later in the year. Roughly a month after Kennedy's briefing, a US photoreconnaissance satellite made its first successful flight. Additional missions followed, by 1961 confirming that a missile gap did not exist. But Kennedy was a political opportunist and observed the advantages that the "missile gap" idea bestowed on his campaign.[23] Nixon received the same briefing. Furious and frustrated that Kennedy chose to ignore the evidence, Nixon seethed. "I could expose that phony in ten minutes. . . . I can't do that without destroying our source, and Kennedy, the bastard, knows I can't."[24]

Kennedy challenged Nixon in the second televised debate about America's loss of prestige. The country was in decline and in need of new leadership, he argued. Nixon countered Kennedy's claim, asserting that "United States prestige is at 'an all-time high' and . . . the Soviet Union at 'an all-time low.'"[25] According to Sorensen, Kennedy's team obtained classified polls from the USIA that suggested American prestige was in fact in decline and that Nixon was misleading the country. They sent the document to the *New York Times*.[26] "U.S. Survey Finds Others Consider Soviets Mightiest" ran in a large bold font on the front page of the newspaper on October 25, 1960. Fortunately for the Kennedy campaign, there was no mention of the source of the leaked document. The poll found that although three years had passed since Sputnik's launch, the satellite still undercut confidence in US global leadership. The international public associated space-launch capability with the ability to launch nuclear weapons on intercontinental trajectories. Aligned and nonaligned nations unanimously agreed that "the Soviet Union would maintain and possibly widen its lead over the United States through the next decade."[27] In light of the release, Nixon's assurances that "United States prestige is at 'an all-time high' and that the Soviet Union at 'an all-time low'" undermined his credibility at a decisive moment in the campaign time line.[28]

Vice-presidential candidate Lyndon Johnson added fuel to the flames. He undercut Nixon's claims further by releasing a "white paper" on the status of the US space program. As lead legislator in the Senate on space issues, Johnson and the staff of the Senate Space Committee were intimately aware of the actual state of US space capability, a point that suggests the "white paper's" clear trappings of political maneuvering, evident since Johnson anointed Sputnik a "Pearl Harbor" in 1957. "The sad truth," the paper stated, "is that U.S. progress in space has been continually hampered by the Republican administration's blind refusal to recognize that we have been engaged in a space and missile race with the Soviet Union and to act accordingly. It is a fact," Johnson explained in a statement accompanying the white paper, "that if any nation succeeds in securing control of outer space, it will have the capability of controlling the earth itself."[29]

Less than two weeks later, on November 8, 1960, Kennedy narrowly won the election. He immediately established task forces to advise him on policy making and his future administration's organization. He requested studies focused on foreign policy needs to evaluate how the USIA could more effectively increase US prestige abroad. The future direction of the space program, on the other hand, received little notice from Kennedy. It may have been useful in his campaign, but when it came to the major priorities of his administration, spaceflight would not be one of them, at least at the start.[30]

But while Johnson and his colleagues in the Senate were considering the future of US space policy for the coming decade, NASA's Space Task Group in Langley, Virginia, as well as von Braun's team in Huntsville, Alabama, spent the fall of 1960 developing plans for an eventual lunar landing program. The President's Science Advisory Committee also undertook a study of NASA's post–Project Mercury future. Led by Donald F. Hornig, a chemist and part of the team that built the first atomic weapons in Los Alamos, PSAC kept an eye on spaceflight's place within society and politics.[31] The "international political situation" created the "most impelling reason" for America's current space effort, concluded the PSAC group. It was not science, the innate human thirst for exploration, or economic incentive that drove the human spaceflight program; instead, it was politics, or more precisely the particular geopolitical moment, where global superpowers competed for global leadership through demonstrations of technological superiority. Political incentive provided the "chief justification" for Project Mercury's tight time line. And the reasons to invest in human spaceflight versus robotic missions were "emotional compulsions and national aspirations." The group determined that "at the present time . . . man-in-space cannot be justified on purely scientific grounds." They presented the report at one of Eisenhower's last NSC meetings, on December 20, 1960.[32]

It turned out to be "quite a day," as the NASA administrator put it.[33] While Eisenhower expressed skepticism about future spaceflight to the NSC, Kennedy foisted off the chairmanship of the National Aeronautics and Space Council onto his new vice president–elect, Lyndon Johnson. The post promised to be impotent, restricted to an advisory role, not a legislative one. It was Kennedy's response to Johnson's efforts to expand the vice president's role within the administration. By assigning LBJ to what appeared to be an ineffectual role, Kennedy meant to temper Johnson's power from the start. During the Eisenhower administration, the Space Council had rarely met. Earlier in the year, Eisenhower had even proposed amending the Space Act to abolish the Space Council, given its nominal effect on space policy. But Johnson blocked the amendment once it arrived on the Senate floor, though likely not fathoming that he could be saddled with the chairmanship months later.[34]

Meanwhile, in the West Wing Eisenhower reviewed the President's Science Advisory Committee's "Man-in-Space" report at the NSC meeting. Accounts of the December 20, 1960, meeting describe Eisenhower's shock at the predicted expenditure. Looking a decade ahead, the report concluded that Project Apollo, a three-person spacecraft program in its nascent stages, could achieve a circumlunar flight around 1970. A lunar landing mission—potentially achievable by the mid-1970s—would require advances in rocketry and an additional $26-$38 billion in funding.[35] Eisenhower responded that he was "not about to hock his jewels" to land humans on the moon, a reference to the Spanish monarchs Ferdinand and Isabella, who funded Christopher Columbus's Atlantic voyages.[36] Even the NASA administrator found the estimate "quite staggering."[37] Eisenhower asserted tersely that he "couldn't care less whether a man ever reached the moon."[38]

During the transition from Eisenhower to Kennedy, human spaceflight seemed to be heading into an uncertain future. At the end of 1960, neither the outgoing president nor the one assuming the post saw human spaceflight as a major national priority. In his last budget message before he left office, Eisenhower stated that "further test and experimentation will be necessary to establish if there are any valid scientific reasons for extending manned space flight beyond the Mercury program." But Eisenhower originally intended a harsher, less open-ended message. In his earlier remarks he asserted that there was no military or scientific justification for human spaceflight after Project Mercury. Period. Although the NASA administrator convinced the administration to soften the statement, Eisenhower left the future of America's human spaceflight program in the hands of the next president, a bleak outcome for lunar space travel.[39]

Kennedy met with Jerome Wiesner shortly before his inauguration, on January 10, 1961. He had charged the MIT professor of engineering and member of PSAC with heading his task force on "outer space." Already well-known as an outspoken advocate of nuclear arms control, Wiesner would soon become special assistant for science and technology to the president and chairman of PSAC, arguably the most influential scientific posts at the time. Wiesner viewed science advising as an active pursuit. He took a hands-on role in shaping policy and lobbying for science, as opposed to simply offering dispassionate expertise.[40] And when given the responsibility of evaluating America's civilian space program, Wiesner and his transition team had harsh words for what they considered NASA's overemphasis on human spaceflight.[41]

"Space exploration and exploits have captured the imagination of the peoples of the world," Wiesner's report acknowledged. In the coming decade, "The prestige of the United States will in part be determined by the leadership we demonstrate in space activities." After a "hasty review" of the US space program, surveying organizational and technical issues, the report pinpointed a lagging ballistic missile program and ineffectual planning and direction. On the topic of human spaceflight, the report stated that "a crash program aimed at placing a man into an orbit at the earliest possible" moment could not be justified on scientific and technical grounds and might even "hinder" these areas of development. Recommending a reevaluation of Project Mercury and a curtailment of the popular perception that human spaceflight was the major objective of the US space program, the Wiesner Report urged the incoming president to take swift and blunt action, perhaps even amounting to the cancellation of America's first human spaceflight program.[42]

In January 1961, Kennedy ensured that the tone of his inaugural address would set the tone for his presidency. He asked Sorensen to analyze all previous inaugural addresses as well to determine what made the Gettysburg Address so powerful, so momentous. With the objective of giving the shortest inaugural address in the twentieth century, Kennedy delivered a meticulously crafted message in fewer than nineteen hundred words. He deleted all the "I"s and replaced them with "we," save for the oath and his responsibility as president. Lines such as "And so, my fellow Americans, ask not what your country can do for you; ask what you can do for your country" made their own indelible mark, striking both an air of collective duty and setting a vision of unified striving.[43] This message was not written for an American audience alone. The USIA broadcast the speech live over its global radio networks and translated Kennedy's words into French, Arabic, and Swahili for African audiences. For these listeners, Kennedy swore that the United States would "pay any price, bear any burden, meet any hardship, support any friend, oppose any foe, in order to assure the survival and success of liberty." After winning the presidential election, Kennedy's next campaign would be for hearts and minds, and for his promise to restore US prestige.[44]

Although the space program's future seemed bleak, Kennedy amplified the status and political influence of the USIA. He asked Edward R. Murrow, the famed CBS journalist with a reputation for evenhandedness and truthful reporting, to head the agency. Kennedy insisted that a representative from the USIA participate in all major foreign policy meetings during his administration. Although Eisenhower both established and championed the USIA, it was Kennedy who would increase the USIA's world public-opinion polling and advising role. For Eisenhower, polls gauged the success of the USIA's programming. But for the intensely image-conscious Kennedy, polling also guided policy making. He explicitly articulated this second role and expanded the USIA director's active participation in the formation of foreign relations policies.[45]

On January 20, 1961, plows removed eight inches of newly fallen snow just in time for the inauguration ceremony. Shortly after noon, on the crisp 22-degree day, Kennedy stood up from his seat next to the oldest president ever to serve. With the glare of the sun off the snow making it hard to see, the forty-three-year-old Kennedy, the second-youngest president the country had ever known, announced that "the torch has been passed to a new generation of Americans . . . together let us explore the stars, conquer deserts, eradicate disease, tap the ocean depths and encourage the arts and commerce." At that moment, as Sorensen later reflected, "It was time to begin."[46]

Begin it did, and likely more rapidly and with more force and trial than anyone had predicted. On February 13 the Soviet Union threatened to intervene in Congo. Less than a month later, Kennedy's aides warned him that Communist forces endangered the pro-American government in Laos. The threat prompted two National Security Council meetings and Kennedy's announcement on March 23 that the United States would intervene unless there was an immediate cease-fire. March 1961 also saw a nationalist uprising in Angola and major setbacks at the Geneva test-ban talks.[47] And then, on a cold

Monday in April, an aide whispered concerning news in the president's ear.

Kennedy on that April 11 afternoon had just thrown the opening pitch at the Senators' first baseball game of the season at Griffith Stadium in Washington. After he took his seat in the presidential box to watch the game, an aide explained that rumors that the Soviet Union had orbited a human into space were spreading. The news did not come as a surprise to Kennedy. Intelligence experts had briefed him weeks earlier on the possibility of such a mission. The Soviet Union had already sent dogs on two orbital missions in March. Days earlier, USIA Director Murrow had even crafted the president's response if the Soviets' first attempt at human spaceflight ended in failure. "Covertly," Murrow explained, "the U.S. might encourage commentators in other countries to deplore the low regard for human life which prompted the Soviets to attempt a manned shot 'prematurely' despite their earlier assertion . . . that their 'lofty humanism' demanded 'absolute certainty' of success."[48]

While Kennedy watched the baseball game and ate a hot dog, his aide, Deputy Press Secretary Andrew Hatcher, gathered more information. After a few innings he told the president that reports could not be substantiated and that the Russians were not planning on making any announcement that day. Then the Washington Senators lost their first game of the season to Chicago's White Sox.[49]

On the other side of the world, Khrushchev vacationed at his dacha on the Black Sea. On April 11 his phone rang. The head of the Military-Industrial Commission reported that the date was set. The first crewed spaceflight—with Soviet cosmonaut Yuri Gagarin onboard—would take place the next day. Straightaway, Khrushchev envisioned a grand public ceremony. The celebration would start at Vnukovo Airport with "as much magnificence as possible." Next would be an event for hundreds of thousands in Red Square followed by a reception at the Kremlin. Among the reception's fifteen hundred guests, Khrushchev would invite the diplomatic corps and foreign press.[50]

In the United States, rumors about a Soviet space shot only increased. White House Press Secretary Pierre Salinger composed a statement for the president. Science Advisor Jerome Wiesner told Kennedy he thought the launch would happen in the middle of that night.

"Do you want to be [woken] up?" Kennedy's military aide, asked.

"No. Give me the news in the morning," the president replied before retiring to the White House living quarters.[51]

Like Kennedy, Gagarin lay down for a full night of sleep on April 11. Physicians had covertly placed strain gauges under his bed, which confirmed that the first space traveler was well-rested for his historic flight. Chief rocket designer Korolev, on the other hand, tossed and turned. Anticipation gripped the father of Soviet rocketry, much like it had before he launched Sputnik into space three and a half years earlier.

Gagarin rose at 5:30 a.m. Moscow time; ate a light breakfast of meat paste, marmalade, and coffee; received a physical examination; suited up; and then rode to the launchpad. Four hours later, aboard a Vostok spacecraft, he became the first human space traveler.[52] On April 14, when he arrived at Vnukovo Airport outside of Moscow, he gave a sixty-six-word memorized account of the mission. His intonation was perfect, the result of thirty minutes of practice with an Air Force official who had scripted the report, specific details and all.[53]

Wiesner's prediction had been right: the Soviet Union captured the title of "first human in space" on April 12 at 1:07 a.m. Washington time. Seconds later, US intelligence picked up Gagarin's in-flight communications. Salinger's bedroom phone rang at 1:35 a.m. It was Wiesner on the line, informing Kennedy's press secretary that the Soviet Union had launched a rocket. Within half an hour, Radio Moscow broadcast the news. Salinger's phone rang again, this time at 2 a.m. It was a reporter from the *New York Times*, the first in a series of calls from newspapers and television networks he would receive that morning.

Wiesner called again at 5:30 a.m., this time to tell Salinger that the orbital spaceflight was a success and that cosmonaut Gagarin had returned safely to Earth. Roughly seven hours after the launch, at 8:00 a.m., Kennedy spoke with Salinger from his bedroom on the second floor of the White House.[54]

As Sorensen recounted, "The President felt, justifiably so, that the Soviets had scored a tremendous propaganda victory, that it affected not only our prestige around the world, but affected our security as well in the sense that it demonstrated a Soviet rocket thrust which convinced many people that the Soviet Union was ahead of the United States militarily." Kennedy, Sorensen observed, "thought of space primarily in symbolic terms."[55]

Unlike 1957—when Sputnik surprised the world—in April 1961, Moscow media stood prompted and primed. For four full days after Gagarin's flight, 95 percent of Soviet domestic and international radio broadcasts focused on the mission. The only comparable coverage in Moscow Radio history was Stalin's death in 1953. The celebration that Khrushchev planned for the new Soviet hero in Red Square was aired live on Moscow radio and television for viewers in the Soviet Union and fourteen European countries. Coverage emphasized that Gagarin's mission signaled the superiority of the Communist system. The USSR—not the United States—claimed yet another space victory.

Reporters gathered at the State Department at 4:00 p.m. on April 12 for Kennedy's previously scheduled press briefing. Although Kennedy did not include Gagarin's flight in his remarks—instead choosing to focus on the anniversary of the polio vaccine and his establishment of an advisory group on foreign aid—questions quickly turned to space and whether the US had plans for an armed intervention in Cuba.

"Could you give us your views, sir, about the Soviet achievement of putting a man in orbit and what it would mean to our space program, as such?" one reporter asked.

After a few customary laudatory comments, Kennedy remarked that "because of the Soviet progress in the field of boosters . . . we expected that they would be first in space."

A few minutes later, a more provocative question: "The Communists seem to be putting us on the defensive on a number of fronts—

now, again, in space. Wars aside, do you think that there is a danger that their system is going to prove more durable than ours?"

It was a question that got to the heart of the problem the administration faced: if spaceflight had become a test for political systems, was Communism the better political system? Kennedy's response likely did not inspire confidence: "Our job is to maintain our strength until our great qualities can be brought effectively to bear."

"As I said in my State of the Union Message," Kennedy portentously explained, "the news will be worse before it is better, and it will be some time before we catch up."[56]

The next morning, at the Capitol, members of the House Space Committee questioned NASA Administrator James Webb and Deputy Administrator Hugh Dryden. Representative James Fulton (R., Pennsylvania) stated succinctly: "Tell me how much money you need and this committee will authorize all you need." Representative Victor Anfuso (D., New York) expressed a similar urgency: "I am ready to call for a full-scale congressional investigation. I want to see our country mobilized to a wartime basis because we are at war. . . . I want to see some first coming out of NASA, such as landing on the moon." When the committee met the next day, Representative Carleton King (R., New York) surmised that there were three "dramatic successes" in spaceflight. The first, launching a satellite. The second, launching a person. The third, sending "the first man to the Moon and back." The Russians were leading two to zero. To King, the next step was obvious.[57]

At Kennedy's request, Sorensen arranged a meeting at the White House two days after Gagarin's flight, on April 14. As Bobby Kennedy once remarked, "If [an issue] was difficult, Ted Sorensen was brought in."[58] And this issue was undeniably difficult: how should the US respond to the Soviet space shot? NASA's top brass—Webb and Dryden—along with Bureau of the Budget Director David Bell and Science Advisor Jerome Wiesner joined Sorensen late that afternoon for a preliminary discussion. After the initial meeting, journalist Hugh Sidey arrived and was invited into the room to observe the discussion, a decision that highlights the Kennedy administration's media savvy. Sidey's behind-the-scenes account would appear in *Life* magazine the following week. The sympathetic article reassured the country that the president was working around the clock on the issue. In later years, Sorensen recalled that it was this afternoon, at this meeting, that the idea of sending humans to the moon started coalescing.[59]

Around 7:00 p.m., as the sun sank low in the sky, they took seats at the dark table in the Georgian-style Cabinet room, their footsteps softened by the thick green carpet. The room was compact, only slightly larger than the table positioned at its center. A portrait of George Washington looked down from above the mantel. A series of bookshelves on one side and a row of windows overlooking the Rose Garden on the other flanked the room.[60]

Kennedy joined the group at this point. His black leather chair, like that of all previous presidents, sat inches taller than all the others in the room. But today he took a different seat. Pulling out a chair marked with a brass nameplate that read "Secretary of the Interior, Jan. 21, 1961," Kennedy sat, leaned back, and then rested his foot on the table.

"What can we do now?" he asked the assembled group.

Each in turn responded as Kennedy balanced on the back two legs of the secretary of the interior's chair and "ran his fingers agonizingly through his hair."

"Now let's look at this," Kennedy told the men gathered around him. "Is there any place we can catch them? What can we do? Can we go around the moon before them? Can we put a man on the moon before them?"[61]

Dryden, a "mild-mannered scientist who lurked behind gold-rimmed glasses," deduced that a crash program in the style of the Manhattan Project might cost $40 billion.[62]

"How long will it take?" Kennedy asked. After hearing that decisions were one to three months away, he responded softly, "There's nothing more important." He left the room, then gestured to Sorensen to join him in the Oval Office to continue the conversation in private.[63] "The decision wasn't made then so much as the stage was set for the full-scale inquiry which would be necessary before a final and precise decision," Sorensen later explained.[64]

The next Friday, April 21, a confidential report on the world public reaction to the flight landed on McGeorge Bundy's desk at the White House. Kennedy had recruited Bundy as his special assistant for national security affairs. Gifted and driven, and in his early forties like the president he served, Bundy was one of the new administration's "best and brightest." He centralized the decision-making structure of the White House and then controlled much of the information that flowed in and out of the Oval Office.[65]

The report included translated news clippings from each continent. And as Bundy later reflected, he had quickly learned "the importance of the newspapers in the process of government."[66] The Kennedy White House released twice the number of news releases than previous administrations. Press Secretary Pierre Salinger held two daily press conferences.[67] One of the first steps of his day was keeping Kennedy "informed" on international press coverage.[68]

Gagarin's flight stole front-page headlines in newspapers and dominated radio and television broadcasts on each continent. International press coverage rivaled—if not exceeded—that of Sputnik almost four years earlier. Kuala Lumpur's China Press put it simply: "It is evident that the U.S. is losing the space race with Russia." Radio Bern suggested that "it is in the uncommitted or neutralist countries where the effects of the Soviet experiment will be felt the most." A commentator on Bogotá's Radio Continental humorously jested, "As a lady told me, 'the Russian man-in-space feat is the greatest thing in the world, but I cannot pay much attention to it, because I have something on the stove.'" Throughout Africa, coverage was "prominent, spectacular, and often sensational," according to the confidential report.[69]

The report also delineated the difference between international reaction to Sputnik and Gagarin's flight: "Intrinsic human drama and interest are great if surprise is less." Sputnik shocked the world public, but Gagarin impressed them. The tenor of commentary around the world championed the flight as an "epochal landmark in mankind's progress." Even so, the military implications were clear. As the report cautioned, "The possibility that the world may in fact face a period of decisive Soviet dominance is the uneasy thought that seems currently close to emergence."[70]

Life magazine's coverage of the mission surely did not ease American anxiety over the impact of Gagarin's flight on US prestige. The magazine ran interviews with people from around the world such as Hazumi Maeda, a Japanese student, who reflected that "I knew Russia would do it first. Socialistic science is superior to that of the western nations." Elisabeth Gulewycz, a secretary in Germany, responded with "This makes one realize Soviet boasts of ultimate superiority may not be groundless after all." And Nabil Rashad, a young man from Egypt, commented simply, "The Americans are licked."[71]

The American press responded in suit. An article in the *Washington Post* called the flight a "psychological victory of the first magnitude" and pointed out that "what people believe is as important as the actual facts, and many persons will of course take this event as new evidence of Soviet superiority." *New York Times* columnist Harry Schwartz stated that "the Soviet Union won another round last week in the psychological and propaganda war for men's minds." He continued: "The West has still not fully grasped how complete is the Soviet dedication to what Madison Avenue might call Communist institutional advertising aimed at gaining acceptance of the product—Communist ideology—by men and women everywhere." As Hanson Baldwin, the *New York Times* military correspondent, warned, "The neutral nations may come to believe the wave of the future is Russian; even our friends and allies could slough away."[72]

By the time Bundy could inform the president of the USIA report results, Kennedy and his advisors were already grappling with another new challenge to the nation's international standing: a news story even more damaging to US prestige displaced the heavy coverage of Gagarin's flight from international headlines. On April 19 Cuban leader Fidel Castro announced that Cuba had defeated a US-supported invasion at the island's Bay of Pigs that had begun two days earlier. "The worst disaster of that disaster-filled period,"

Sorensen surmised. Roughly fourteen hundred Cuban exiles, trained by the CIA, were no match for Castro's military forces.[73]

That same day, Kennedy told Johnson that he wanted an accelerated review of the status of the US space program. Sorensen drafted a memo for Johnson outlining the request. The most revealing line of the memo asks simply: "[find a] space program which promises dramatic results in which we could win." The key for the Kennedy administration was winning. How much would it cost, he asked. Were NASA employees and contractors working around the clock? Was the country making the "maximum effort"?[74] Just one page long, the memo initiated an extensive and immediate review of America's standing in spaceflight. By 10:30 p.m. that night, Johnson was on the phone with Edward Welsh, his assistant, requesting that he set up a meeting with NASA Administrator James Webb at 9:30 a.m. on Saturday morning. And he wanted a meeting with Secretary of Defense Robert McNamara later that afternoon. Weekends were no reason to rest.[75]

On Saturday morning, Webb and Dryden came equipped with answers to the April 20 memo's questions. An orbiting space laboratory would not be sufficient, they explained. Lunar flights—either orbital or a landing mission—had the potential to best the Russians. An accelerated space effort, they reported to Johnson, would cost the nation roughly $33.7 billion by 1970.[76]

Unlike NASA officials who detailed the benefits of a crewed lunar landing alongside an analysis of other potential programs, McNamara took the opportunity to analyze the political stakes in his Saturday afternoon meeting with LBJ: "All large scale space programs require the mobilization of resources on a national scale. Dramatic achievements in space, therefore, symbolize the technological power and organizing capacity of a nation." Because of this, "Major achievements in space contribute to national prestige" and "constitute a major element in the international competition between the Soviet system and our own."[77]

Both meetings that Saturday reaffirmed Johnson's view that "communist domination of space could lead to control over men's minds."

He shared this message at the beginning of a meeting that he held on April 24, laying out the stakes for the military representatives and leaders of industry and media gathered to share their reactions to Kennedy's April 20 memo. Johnson highlighted the propaganda implications of spaceflight for the future of US leadership and asked for advice from the audience. Lieutenant General Bernard Schriever, head of the Air Force Systems Command and "Father of the Air Force missiles and space programs," advocated for a moon landing for prestige purposes. Vice Admiral John T. Hayward agreed. As the Navy representative at the meeting and deputy chief of naval operations for R&D, he also encouraged practical applications of space technology, such as navigational and weather satellites. Wernher von Braun, rocket pioneer and director of NASA's Marshall Space Flight Center, followed up on April 29 with a detailed letter that addressed Kennedy's questions from a technical perspective. The United States could not compete with a space station, but there was a "sporting chance" of achieving the first soft landing of a radio transmitter station on the moon or a crewed mission around the moon. Better yet, "We have an excellent chance of beating the Soviets to the first landing of a crew on the moon (including return capability, of course)." Achieving a moon landing would require both the Soviet Union and United States to develop massive new rockets, leveling the playing field. It would also require "some measures which thus far have been considered acceptable only in times of a national emergency," an ominous closing, foreshadowing the wartime-like national mobilization that would soon follow his recommendation.[78]

A few days later, political advisor Walter Rostow warned Kennedy that "we must now, I believe, face the fact that we are in the midst of one of the great crises of the postwar years. It is a worldwide crisis . . . at a period of up-swing in Soviet military and space capabilities; and it is colored by an image of American strength and determination fading relative to the Communist thrust."[79] At the same time, Sorensen observed an evolution in Kennedy's thinking after the invasion. What the Bay of Pigs taught Kennedy was that "military ventures were not necessarily going to succeed." Instead, world problems required another approach.[80]

It was the one-two punch of Gagarin's flight boosting Soviet prestige followed in quick succession by the loss of US prestige because of the Bay of Pigs invasion that laid the groundwork for Project Apollo. As Sorensen noted, "It pointed up the fact that prestige was a real, and not simply a public relations, factor in world affairs."[81]

That same day, April 28, Johnson provided an interim response to Kennedy's April 20 request. His concise two-page review states frankly: "Other nations, regardless of their appreciation of our idealistic values, will tend to align themselves with the country which they believe will be the world-leader." And "dramatic accomplishments in space are being increasingly identified as a major indicator of world leadership." This simple formula made the stakes evident. Space spectaculars were not simply a sparring match taking place above the Earth's atmosphere. They signaled leadership at a moment when the political landscape of the Earth was shifting, new countries were being born, and power was won through alignment. Warning the president that "if we do not make a strong effort now, the time will soon be reached when the margin of control over space and over men's minds through space accomplishments will have swung so far on the Russian side that we will not be able to catch up, let alone assume leadership." Sending Americans to the moon would not only be "an achievement with great propaganda value, but it is essential as an objective." The report suggested that a lunar landing was feasible by 1966 or 1967, given "a strong effort." Although two more weeks of discussions, recommendations, and reviews followed, the benefits of setting a moon landing as the national goal were clear.[82]

A week later, Johnson laid the groundwork for Senate support. If Kennedy's announcement of a bold new space initiative was going to succeed, Congress would have to fund it. He assembled the chairman of the Senate Committee on Aeronautical and Space Sciences Robert Kerr (D., Oklahoma) and the ranking minority committee member Styles Bridges (R., New Hampshire) and their staff to meet with Webb, Dryden, and representatives from the Atomic Energy Commission, the State Department, and the Bureau of the Budget on May 3. Johnson had controlled the Senate with Kerr and Bridges as Senate majority leader during the Eisenhower administration. If he could win Kerr's and Bridges's support, the rest of the Senate would follow suit, he reasoned. Johnson set the scene: "We haven't gone far enough or fast enough. We need a new look, and to know how much it will cost." Johnson pushed on Webb: "You must not wait a month or Congress will have gone home." Unsatisfied with Webb's hesitant response, Johnson added, "We'll wait a month if necessary for people to get guts enough to make solid recommendations . . . our purpose today is to have these important Senators get the benefit of consultation and for us to have the benefit of consulting them."

Johnson also called members of the House of Representatives to ensure their backing of an accelerated space program. James Fulton (R., Pennsylvania), the ranking minority member of the Science and Astronautics Committee, discussed the matter with House Republicans and assured the vice president near-unanimous support. Owen Brooks (D., Louisiana), chairman of the House Science and Astronautics Committee, gave a similar response. He wrote Johnson that "the United States must do whatever is necessary to gain unequivocal leadership in Space Exploration." In his mind, "We cannot concede the Moon to the Soviets, since it is conceivable that the nation that controls the Moon may well control the Earth." Meanwhile, Kennedy was gauging the political feasibility of a major spaceflight program by consulting members of Congress. Once political support was in place, the next steps could be taken.[83]

The final decision hung in the balance, waiting for the results of America's first attempt at a human space shot. Project Mercury flew a chimpanzee named Ham on a test flight in January 31, followed by additional crewless test missions that spring. When the time came to launch an astronaut on a suborbital trajectory, some members of Congress as well as the president's advisors expressed concern about publicizing such a risky flight. But Kennedy recognized the larger political significance of carrying out the flight in full public view. NASA orchestrated the access of hundreds of domestic and foreign correspondents to the launch, ensuring that the mission would receive global news coverage.[84]

On May 5, 1961, after enduring several hours of weather and mechanical delays, Alan Shepard Jr. became the first American space traveler. At 9:34 a.m. EST a Redstone rocket lifted his *Freedom 7* spacecraft off the launchpad and into a suborbital trajectory. Kennedy's secretary retrieved him from an NSC meeting to watch the coverage. He crowded around a small television in his secretary's office along with Johnson, Sorensen, Jackie Kennedy, and several others. If the flight failed, the next steps of America's space program would have been put into question. But the brief flight—roughly fifteen minutes long—proved that the Mercury spacecraft was pilotable, that short exposure to weightlessness had negligible effect on the human body, and that the tracking system NASA had put in place around the world was viable. It signaled that the country might be ready for a bigger, bolder goal. Kennedy and the others around the small television "heaved a collective sigh of relief, and cheered."[85]

President John F. Kennedy, Vice President Lyndon B. Johnson, and others watching Alan Shepard's flight, May 5, 1961. (NATIONAL ARCHIVES)

"Shepard's flight dramatically illustrated the difference between the American and Soviet systems," according to Thomas Sorensen, deputy director of the USIA and Ted Sorensen's brother. "The lesson was not lost on the world."[86] To amplify this message, the USIA distributed a thick packet of material filled with scientific background pieces and photographs to the international press in advance of the flight. The USIA tracked the use of the material in eighty-three countries. BBC London picked up NBC coverage of the flight, opting to broadcast, even in British classrooms, the American network's coverage instead of its own.[87] Nearly fifty countries broadcast the USIA film *Shadows of Infinity* on television during the day of the launch. Models of *Freedom 7* went on display in Europe and Asia. Foreign newspapers published articles under Shepard's byline that were provided by the USIA. Vice President Lyndon Johnson shared a film about the flight on his diplomatic tour of South Asia. At the same time, Secretary of State Dean Rusk screened the film in Oslo at a NATO conference.[88]

On May 8, the Monday following the flight, Shepard traveled to Washington for a ceremony at the White House. In the Rose Garden, Kennedy stressed "the fact that this flight was made out in the open with all the possibilities of failure," arguing that "this open society of ours which risked much, gained much." As Webb handed Kennedy the case holding a Distinguished Service medal for presentation to Shepard, it fell to the terrace. In quick response, Kennedy picked up the medal, adding "and this decoration—which has gone from the ground up—here!"[89]

After a week of exhaustive work at NASA and DoD, a set of recommendations was ready for Johnson's review the day after Shepard's ceremony. Without crossing out one section or editing one point, Johnson sent the document to Kennedy. In a sharp about-face of Eisenhower-era space policy and even previous Kennedy-era assessments such as the Wiesner Report, Webb and McNamara recommended that "this nation needs to make a positive decision to pursue space projects aimed at enhancing national prestige." They fully recognized that major space feats have marginal "scientific, commercial

President John F. Kennedy congratulates astronaut Alan B. Shepard Jr., the first American in space, and presents him with the NASA Distinguished Service Award, May 1961. (NATIONAL AERONAUTICS AND SPACE ADMINISTRATION)

or military value" and are "economically unjustified," and they noted that the prestige resulting from space successes was "part of the battle along the fluid front of the Cold War." The central point came through clearly: space successes "lend national prestige." Ultimately, Webb and McNamara recommended that the United States send humans to the moon for this reason. In a May 10 meeting at the White House, Kennedy agreed.[90]

Six weeks after Gagarin became the first human space traveler, Kennedy stood at the Speaker's rostrum in the House of Representatives chamber in the Capitol Building's south wing before a joint session of Congress. Members of Congress sat before him in a semicircle on tiered platforms while television audiences watched from their homes. Uncharacteristically, Kennedy deviated from his prepared text. The address itself was also uncharacteristic. Billed as the second "State of the Union," Kennedy spoke on the nation's urgent needs, including economic and social progress at home and abroad, increasing funding for overseas information programs, and the importance of military buildup and disarmament negotiations. Project Apollo would be the last "urgent need" on this list.

"If we are to win the battle that is now going on around the world between freedom and tyranny, the dramatic achievements in space which occurred in recent weeks should have made clear to us all, as did the Sputnik in 1957, the impact of this adventure on the minds of men everywhere, who are attempting to make a determination of which road they should take," the president explained. Kennedy then went on to propose that the United States should commit itself to sending a man to the moon and returning him safely back to Earth "before this decade is out."[91] This goal, he emphasized, could persuade people in other countries to choose American "freedom" over Soviet "tyranny." He continued: "Now it is time to take longer strides—time for a great new American enterprise—time for this nation to take a clearly leading role in space achievement, which in many ways may hold the key to our future on earth." He added the last phrase of this sentence to the printed text with his pen.[92] Before he closed, he urged everyone to carefully consider the commitment to lunar exploration.[93]

On the drive back to the White House, Kennedy expressed doubts to Sorensen. He interpreted Congress's applause as "less than enthusiastic."[94] Kennedy may have been right to suggest that this commitment should not be entered lightly. Project Apollo would become the largest open-ended peacetime commitment in the country's history. It outstripped the Panama Canal, Eisenhower's interstate highway system, and even the Manhattan Project.[95] But there was no need for apprehension. Lyndon Johnson had made sure that the president's proposal, which he had orchestrated over the past six weeks, would be accepted by Congress. Johnson, ever the "master of the Senate," ensured bipartisan support of Project Apollo by meeting with members of Congress. Congress endorsed Kennedy's call for Project Apollo and the more than half-billion-dollar supplemental budget for NASA in the near term.[96] "The 'moon shot' was the making of America's superiority in space, and all the scientific, diplomatic, and national security benefits that followed," Sorensen later concluded.[97]

5

JOHN GLENN AND
FRIENDSHIP 7'S "FOURTH ORBIT,"
1961–1963

Friendship Seven, Munchea Com Tech. We read you.
— JERRY O'CONNOR, AUSTRALIAN
COMMUNICATIONS TECHNICIAN

The lights [in western Australia] show up very well, and
thank everybody for turning them on, will you?
— JOHN GLENN JR., MERCURY ASTRONAUT,
FRIENDSHIP 7 MISSION, 1962

T wo days after Kennedy proposed sending Americans to the
moon, the USIA sent Alan Shepard's *Freedom 7* spacecraft
to the International Air Show in Paris. USIA Director Murrow explained that "the prestige race against the Soviets is a contest
with no small importance. Over a million people at the world's largest international exposition on aviation and space will see this symbol of the latest American space success."[1] He was quick to point out
that *Freedom 7* would be on display at the same site where Charles
Lindbergh completed the first solo airplane flight across the Atlantic Ocean in 1927. An impressive crowd of 600,000 visitors saw the
small spacecraft in Paris. From France, *Freedom 7* was sent to Rome,

where it became the major draw for more than a million people at the International Science Fair. The USIA put a positive spin on the story of human spaceflight. "Two young men soared into space early this year," a USIA report to Congress read. "One was a Russian, one an American. The Russian was the first one up, but the American's achievement was more widely heard and even more widely believed." This distinction between the US and Soviet programs is idiomatic of the USIA under Murrow's leadership. What set the United States apart from the Soviet Union, the agency stressed again and again, was "openness."[2]

Not long after *Freedom 7* debuted at the air show, Kennedy flew to Paris himself to meet with French president Charles de Gaulle on May 31. "The public ceremonies were much more helpful to Kennedy than the private discussions," and the meetings were "a case study in symbol over substance," according to biographer Robert Dallek.[3] Kennedy's objective for the visit was to restore his credibility at home and abroad after the Bay of Pigs and to put Franco-American unity on display ahead of his upcoming discussions with Khrushchev in Vienna. Appearing in photos next to de Gaulle, Kennedy advanced his image as a world statesman. Both Kennedy and Khrushchev arrived in Vienna aware that the future of Berlin—a city that had been divided between East and West since the end of World War II— would be the primary focus of the summit, but they also discussed spaceflight. Informally during a meal, Kennedy suggested that the US and the USSR join together in sending humans to the moon. At first Khrushchev responded with "No" and then half-jokingly yielded: "All right, why not?"[4] Khrushchev then offered First Lady Jackie Kennedy a puppy from one of the dogs that flew in space. Although nothing came of Kennedy's invitation to cooperate on Apollo, and little came from the summit in general, a puppy named Pushinka with outer-space pedigree arrived at the White House that summer, a gift from the Soviet premier.[5]

In his yearly report to Congress, Kennedy highlighted that touching Shepard's spacecraft became a fetish for visitors at the exhibits and that the United States was lauded for the "openness" of its human space program. Overseas audiences, he continued, achieved "a high degree of self-identification with one of the greatest adventures of our time."[6] The spacecraft, small, black, and bell-shaped, evidenced the scars of atmospheric reentry on its black corrugated skin. When seen up close and in person, *Freedom 7* rendered spaceflight more tangible for these audiences.[7] Additional requests for the *Freedom 7* capsule exhibit came from cities throughout the world. USIS posts in Bonn, Turin, Karachi, Athens, London, and Kabul asked the agency to display the capsule in their cities. Foreign audiences were enthusiastic about the spacecraft display, but criticism within the United States swelled. The domestic public and members of Congress questioned NASA for exhibiting the capsule abroad before it was exhibited within the United States. This tension between prioritizing the domestic and foreign audiences was something that government officials would face for years to come.[8]

Even after the *Freedom 7* flight and exhibits, a USIA poll conducted in August 1961 noted that Western European nations were confident in the United States' world leadership but believed in the superiority of the Soviet Union's military and space program. The success of Shepard's mission, and the approach to openness in NASA and USIA programming, boosted the status of American space efforts, but in 1961 the United States was still seen as trailing behind a string of Soviet space firsts.[9]

The image of American space capability would change considerably on February 20, 1962. That morning, long before dawn, astronaut John Glenn prepared for the first American orbital spaceflight. As he sat inside his compact *Friendship 7* spacecraft, atop an Atlas rocket, the launch time was pushed back again and again. Glenn's mission had already been postponed multiple times. "The anticipation became a story in itself," he said. With the added delays, news coverage dramatized the mission, contributing "to a kind of soap opera . . . will he be launched, or won't he?" Likely for propaganda purposes, Soviet cosmonaut Yuri Gagarin publicly expressed worry for the "serious psychological and moral pressure" the delays were putting on Glenn. But at 9:47 a.m. on February 20, the countdown ended and the rocket

President John F. Kennedy and astronaut Lieutenant Colonel John Glenn Jr. look inside space capsule *Friendship 7*, February 1962. (National Archives)

launched, at first arcing over Bermuda and then placing *Friendship 7* on an orbital trajectory. Glenn circled the Earth three times. As he passed over the global network of tracking stations, telemetry sent signals to the ground, including biometric data and the condition of the spacecraft. He conducted experiments in concert with the stations, testing physical strength and vision in space. Just under five hours after liftoff, *Friendship 7* splashed down in the Atlantic Ocean. The flight had faded the stenciled American flag and the words "United States" and "Friendship 7" that were painted on the spacecraft before launch, signs of the hazards of spaceflight.[10]

Time magazine heralded the *Friendship 7* mission for putting the United States "back in the space race with a vengeance" and giving "the morale of the US and the entire free world a huge and badly needed boost." Even the Communists' "peace dove" artist, Pablo

Picasso, the article proclaimed, was moved to say, "I am as proud of [Glenn] as if he were my own brother."[11] The article stressed the importance of the openness of the mission and how everyone from "Queen Elizabeth to Bedouins in the Middle East" could follow the progress of the flight. Voice of America broadcast the flight in Hindi, Russian, Chinese, and a score of other languages. Public diplomats believed that the radio programming of the flight captured the largest audience in the history of radio broadcasting.[12]

Domestically and abroad, John Glenn's orbital flight created a public relations boon for government elites hoping to promote the strength of the US space program. Sorensen called it "a turning point in many ways."[13] The USIA sent coverage of the *Friendship 7* flight to TV stations in seventy-four countries; in Italy alone, thirteen million people viewed the telecast.[14] The *Los Angeles Times* encapsulated the far-reaching enthusiasm for the flight when it noted that convicts at San Quentin's death row and Pope John XXIII followed the flight while the chambers of the House and the Senate emptied so that members could watch the television coverage. "In Grand Rapids, Mich.," it noted, "a judge and jury interrupted a trial to watch the latest developments from Cape Canaveral. They watched on a stolen television set that had been brought to court as evidence in a case."[15]

In his weekly report to President Kennedy, Murrow noted that the press, from Africa to Asia, heralded the openness of the space program. Emphasizing the central role of the USIA in this prestige boost, Murrow explained that the heavy coverage was in large part a result of the extensive distribution of USIA material—including a fifteen-minute documentary, color photo exhibits, press packets, and motion picture footage for newsreels—before the flight.[16] Murrow, seeking to capitalize on the overwhelming response, devised his plan for an elaborate global tour of Glenn's spacecraft. He wrote to President Kennedy suggesting that "we can make a terrific impact abroad by exhibiting Colonel Glenn's 'Friendship 7' space capsule in key countries."[17] Murrow saw the exhibition of the capsule as a valuable strategic "weapon" in the psychological battlefield of the cold war.[18]

In the year following his flight, the Soviet Union sent Yuri Gagarin on a diplomatic tour of more than twenty countries. The political importance of each stop ensured back-to-back appearances, receptions, and ceremonies. "Too much politics, and nothing for ourselves; we did not even see an elephant," Gagarin remarked in India. But a million people came out to see him in Calcutta alone, a windfall for the Soviet government. After years of propaganda trips, Gagarin said he would sometimes "close my eyes and see endless queues of people with blazing eyes, shouting greetings in foreign languages." What the Soviet Union did not send around the world, however, was spacecraft hardware. Even information about the rocket designers was kept secret.[19]

At first, Murrow hoped to display the *Friendship 7* in a number of international cities, including Moscow, suggesting that "if the Soviets agree, the world will note their failure to show their capsules even to their own people, whereas we are willing to show ours even to the Russians." He believed that international audiences would see the capsule as an American effort to share scientific and engineering information, not as political propaganda. By displaying technological hardware, as opposed to sending Glenn on a tour "like a trained seal," the US space program would distinguish itself from Soviet space publicity, according to Murrow.[20]

Throughout the 1950s and 1960s the USIA displayed technological hardware so that the "facts spoke for themselves." USIA officials hoped that technology, unlike cosmonauts, would not be interpreted as a form of propaganda and could therefore more effectively carry the agency's political message.[21] This tension—namely, serving political objectives with "apolitical" objects—speaks to the relationship among science, technology, and foreign relations in the early cold war. With roots in nineteenth-century Enlightenment thought, this complicated relationship stems from a dual identity of science and technology as both tied to objective truth and as instruments of progress.[22] In the 1950s and 1960s this view of science and technology was advanced in academic circles and played out in development projects around the world, shaped US foreign policy, and influenced

A USIA space exhibit in West Pakistan, 1962. (NATIONAL ARCHIVES)

the relationship between the United States and the world. During the Kennedy administration in particular, in large part through the encouragement of Walter Rostow, a key modernization theorist as well as Kennedy's deputy special assistant for international affairs, the government invested in an elaborate array of modernization programs. The planning, execution, and later evaluation of the *Friendship 7* international exhibition are a part of this larger story of the role and impact of modernization theory on the use of science and technology in US foreign relations.[23]

Colloquially dubbed "The Fourth Orbit," a humorous acknowledgment that the international exhibition would be the fourth time the capsule circled the Earth, the tour of Glenn's spacecraft was the first large-scale space diplomacy program hosted by the USIA. During its three-month world tour *Friendship 7* visited more than twenty cities and was seen by roughly four million people, while another twenty million people watched television programs broadcast from the exhibition sites. A US Air Force cargo plane emblazoned with the words "Around the World with Friendship 7" and a map of the continents that the capsule visited that summer carried the small craft from country to country. John Williams, a member of NASA's Cape Canaveral staff and the head of the *Friendship 7* exhibit, joined the tour to answer questions from curious audiences around the world.[24] Polls, exhibit exit interviews, newspaper clippings, photographs, films, and reports reveal the myriad ways in which politics and culture, through spaceflight and diplomacy, intersected on *Friendship 7*'s "fourth orbit."[25]

The USIA and the State Department selected the exhibit locations based on a variety of criteria. The capsule visited countries that were of foreign relations significance to the United States, cities that were near American tracking station facilities, and countries where the little capsule could make the greatest impact. But the selection process was not simply one-sided. Internal correspondence at the Science Museum London reveals some of the foreign interests and motivations for exhibiting *Friendship 7*. At the Science Museum, a staff member observed that "the exhibition should arouse great interest and may be a means of acquiring further material for permanent exhibit."[26] Building a relationship with NASA through the *Friendship 7* exhibit, museum staff hoped, would give the museum greater leverage to acquire future space artifacts, which were sure to draw large crowds in the coming years. H. R. Calbert, keeper in the Department of Astronomy at the Science Museum, wrote to NASA and the USIA requesting the exhibition of the capsule at the museum. The Science Museum staff worked with the USIA to plan an elaborate exhibit that included film screenings, posters, a soundtrack from Cape Canaveral of the liftoff, and a projector with color transparencies. Museum staff from around the United Kingdom wrote to Science Museum London, requesting that the capsule visit their museums as well.[27] The relationship between US information programs and the communities they aimed to influence was dynamic and often symbiotic. The correspondence between the USIA and Science Museum London highlights the intricacies of the interplay of US and foreign interests in the support and execution of spaceflight exhibits.[28]

The Science Museum ended up being one of the first stops on the *Friendship 7*'s "fourth orbit." In early May 1962, on the first day that the capsule was on display in London, the museum turned away thousands of people because the huge crowds overtaxed the facilities. The British Broadcasting Corporation's current affairs program *Panorama* filmed a live telecast from the capsule exhibit that was watched by ten million viewers. Over the next few days two to three thousand people waited in line at all hours of the day to get a glimpse of the famous capsule before the Air Force plane carried it off to its next stop in Europe.[29]

When the USIA exhibited the capsule at the Palais de la Découverte, a science and technology museum in Paris, they had to extend exhibit hours until midnight to accommodate the enthusiastic crowds standing in five-hour-long lines to view the capsule. Nearly thirty thousand people attended the Paris exhibit in its two-and-a-half-day stopover, a record-breaking number for the well-known museum. All the major Parisian newspapers featured *Friendship 7* in their front-page headlines, and radio and television programs covered the space exhibit as well. The USIA report from Paris noted that the "few spectators recalling Soviet heavier launches [were] quickly silenced by others in [the] crowd who pointed out that at the USSR exhibit in Paris the previous year, the space craft on display was 'only [a] model, not [an] actual capsule.'"[30] For these Parisian onlookers, the US's and USSR's opposing approaches to displaying their space accomplishments had real consequences.[31]

Although Yuri Gagarin toured the world, his spacecraft, the Vostok 1, was not put on public display; before 1965, only photographs of the Vostok veiled underneath a cover were shared with foreign audiences. Historian Cathleen Lewis suggests that the lack of engineering information on the model of the first Vostok put on public display "represented a deliberate effort to conceal the actual details of the human space-flight program in the Soviet Union by carefully camouflaging details about the design legacy of Vostok and its technical properties."[32] When Glenn's capsule was put on display during its "fourth orbit," however, the USIA included engineering diagrams of its interior workings along with other exhibit components. This exhibit, as well as most US space exhibits in this period, highlighted scientific and technological information as a demonstration of openness and a symbol of liberal democratic values.[33] Later, USIA staff further articulated how they leveraged the secrecy of the Soviet program: "We capitalized even on Soviet space successes, because most of the material on the scientific and technological background of the Soviet projects was only available from our sources."[34]

Because *Friendship 7*'s "fourth orbit" was the first major US space diplomacy program, USIA continuously adjusted and revised the capsule exhibit over the course of its tour. After *Friendship 7*'s overwhelming reception in Paris, representatives from the USIA recommended that at future exhibitions of the capsule, security guards be posted to control the crowds, exhibit sites extend their hours, and a second viewing platform be set up parallel with the first to allow as many people to see the spacecraft as possible. Although curators from the Smithsonian Institution equated the capsule with the Wright brothers' plane and strongly urged the USIA to cover it in a clear plastic casing to preserve the "original, irreplaceable, priceless relic of history," public diplomats decided to keep the capsule uncovered so that eager onlookers could touch its coarse surface. The USIA also tailored information panels, lecture themes, and events at each stop.[35]

For the exhibits in Africa, like many USIA programs in the early 1960s, the agency took the opportunity to curb civil rights criticisms by presenting an alternate view of the United States.[36] *Friendship 7* made its "fourth orbit" at a time when many African diplomats were forced to live in substandard housing in Washington because white landlords would not rent to them. Just a year before the tour, the world was shocked when Alabama Ku Klux Klan members burned a Freedom Rider bus and beat the civil rights activists who rode the bus to challenge segregation. President John F. Kennedy believed that the Klan violence threatened the image of the United States in the eyes of the world. The USIA reported that the event "had dealt a severe blow to US prestige which might adversely affect its position of leadership in the free world as well as weaken the overall effectiveness of the Western alliance." In response to the incident, the *Ghanaian Times* noted that surely racism in the United States "as well as the plight of oppressed peoples in Africa and elsewhere demand much more serious attention and consideration than the sending of a man to the moon."[37] To curb some civil rights tensions and present a more positive image of the United States and its space program abroad, the USIA hired African Americans to work in Africa. Two of these employees were assigned the job of giving lectures on the space program.[38]

In late May the USIA displayed *Friendship 7* in Accra, the capital city of the newly independent nation of Ghana. The USIA estimated that twelve thousand Ghanaians viewed the capsule on the first day of the exhibit alone. The exhibit included a large photographic panel displaying general information about the US space program as well as the song "Everything Is a Go," adapted from the West African "highlife" musical genre and written in collaboration by a Ghanaian and an American for the event. According to the USIA press release, "The crowd reaction in Ghana was strongest each time the speed of the spacecraft was announced in local terms. John Glenn's 'friendship-7' an announcer pointed out 'can fly from Accra to Kumasi in 30 seconds. More than 100 miles separate the two cities.'" More than fifty thousand people viewed *Friendship 7* in Accra, a number that surpassed attendance numbers in each of the previous cities the spacecraft visited.[39]

After receiving a very favorable reception in Ghana, the capsule traveled to Nigeria, where the USIA displayed it in both the capital, Lagos, as well as Kano, the northern city near the NASA tracking facility. In preparation for its stop in Lagos, the USIA took US-Nigerian relations into account when organizing the exhibit. USIA Director Murrow stressed that it was imperative to show in the exhibit all African American "personnel working with white staff on project." He noted that any other "visually evident African contribution to operation" was important.[40] The opening ceremony in Lagos was televised, and during its thirty-hour display the capsule drew a crowd of nineteen thousand in addition to the thousands who saw it as it was towed through the city.[41]

In Egypt the *Friendship 7*'s visit to Cairo convinced skeptics that the flight had really happened and that people were really traveling to space, according to the *Washington Post*. One onlooker remarked, "I thought this space flight business was a rumor but now [that] I can see the ship I believe it." It was important for this observer, as it was for many others around the world, to see the spacecraft in person. In the mid-twentieth century, space exploration had just left the realm of science fiction. The extraordinary idea that a man had orbited the Earth was made more tangible when the spacecraft that had carried Glenn could be seen and felt.[42] A few months before the exhibit, *Al-Akhbar*, a daily newspaper, reported that "in his endeavors to invade Africa, Kennedy relies on four weapons, namely foreign aid, labor unions, peace corps, and the U.N.," a view of the US that many Egyptians were expressing at the time. The USIA reports for the exhibition of Glenn's capsule paint a very different picture that suggests the spacecraft at least—as the USIA had hoped—appeared more apolitical than Kennedy's other "weapons."[43]

Although the *Friendship 7* capsule drew record crowds in Paris and Accra, the most overwhelming response came in Asia.[44] In late June, when the capsule arrived in India, more than a million Bombay residents welcomed *Friendship 7* to their city over the course of the capsule's four-day exhibition. The USIA field office reported that "hundreds of thousands lined streets as the capsule trailered into the city from the airport for public display at the city's Brabourne stadium." Lines of fifty thousand people waited for up to four hours to see the spacecraft. A NASA scientist lectured on Project Mercury as well as US plans for lunar exploration to scientists at the Tata Institute on Fundamental Research and the Atomic Energy Establishment in Bombay.[45] The major Bombay newspapers dedicated over twenty-five hundred inches of column space to *Friendship 7*. A report summed up many of the USIA's hopes of the tour:

Friendship 7 world tour, July 1962. (NATIONAL ARCHIVES)

The fact that America's manned space program is conducted in the full light of international publicity is highly appreciated among Indians in general and particularly by those who believe in democracy and an open society. While the Russian sputniks initially made a tremendous impact upon Indian public opinion, it is believed that the showing of the Friendship 7 film and John Glenn's space capsule to the public of Bombay has demonstrated, at least in this area, which American efforts in space exploration are equal, if not superior, to the efforts made by the Soviets. Gagarin's visit was well publicized, but few Indians had an opportunity to see him. The arrival of and display of the Friendship 7 capsule captured the hearts and imagination of the people of this area and the public turnout exceeded the post's most optimistic expectations.

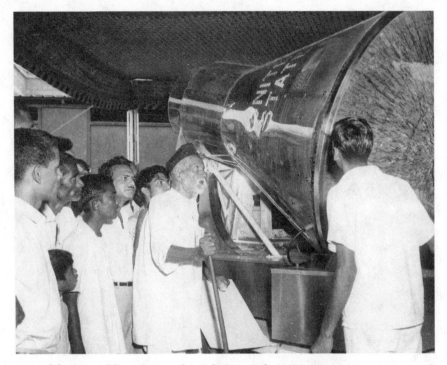

Friendship 7 world tour, Karachi, Pakistan, July 1962. (NATIONAL ARCHIVES)

In addition, the report highlighted that this reception was especially significant in light of current press attitudes and low public opinion of the United States in India.[46]

When *Friendship 7* reached Bangkok, roughly eighty thousand spectators, including the crown prince, saw the spacecraft during its three-day visit. According to a USIA report, "A Thai official described the turnout as probably the largest in Thai history to view one object exhibited by a foreign country." At the exhibit's opening ceremonies in Djakarta, the chairman of the Indonesian Council for Sciences commented, "We heartily welcome the holding of this exhibit because this event will give the Indonesian people an opportunity to witness an American apparatus which has such a great and important role in increasing our knowledge in the field of space science and technology." And in Manila the Philippine vice president,

Emmanuel Pelaez, noted that the capsule "symbolizes the aspirations of free nations in a break-away from a humdrum existence to a helpful life of progress," a commentary on the spacecraft's significance that aligned with the rhetoric of USIA information material. Priests, students, grandmothers, and Boy Scouts waded through six inches of rainwater to see the *Friendship 7* spacecraft during its first day on display in Manila. The *Philippines Herald* noted that the capsule "represents tangible proof of man's conquest of space. To the beholder, it is a connecting link that brings him within arm's reach of the reality of this conquest."[47]

Friendship 7 was exhibited in two locations in Australia, an important partner of the US space program. Australia's Weapons Research Establishment managed two Mercury ground tracking stations, one near Perth and the other near Woomera. The city of Perth had left its lights on the night Glenn orbited overhead so that he could spot the city far below on Earth. It quickly earned Perth the nickname the "City of Light." In recognition of Perth's gesture, the USIA added a brief overnight exhibit in the western Australian city to the official schedule. The *Friendship 7* tour staff kept the exhibit open all night as a thank you to the citizens of Perth. At another stop in Sydney, roughly 250,000 people saw *Friendship 7* over the course of four days. In addition to NASA officials, Sydney University students stationed at the exhibit answered questions from the eager onlookers.[48]

When *Friendship 7* arrived in Japan in late July, the press covered its visit from the moment it was unloaded at the airport. Takashimaya, the leading department store in downtown Tokyo, where exhibits were usually mounted in Japan, set up an elaborate display. In the first hour of the exhibit more than 12,000 people saw the spacecraft, and by the end of its four-day display more than 500,000 people came to the store to see *Friendship 7* in person, "a crowd," according to the USIA report, that was "exceptional in size even by Tokyo standards." In comparison, a USIA exhibit on Frank Lloyd Wright that toured four locations in Japan that year was deemed a significant success even though its attendance records were 5 percent of the attendance of the *Friendship 7* exhibit.[49]

Several hundred police and guides were called on to direct the crowd into a line that climbed nine flights of stairs, zigzagged across the roof of the building, and then descended nine flights of stairs to the first floor, where the capsule was on display. The store's eighth-floor exhibit hall was transformed into a movie theater that screened the ten-minute film *John Glenn Orbits the Earth*. The exhibit fueled favorable publicity for the US space program throughout Japan. The USIA officials were glad to report that the open information policy and the desire for international cooperation in space were widely publicized by "every communication media in Japan." According to the USIA report, "The openness of our manned space program which it symbolized was not lost on the Japanese who in several instances contrasted it favorably to the earlier presence of Soviet astronaut Gagarin."[50]

Although the response in Korea did not rival the response in Japan, within the first nine hours of its display thirty thousand people had seen the spacecraft, and the half-mile-long line showed no sign of thinning. The Korean newspaper *Hankuk Ilbo* discussed the exhibit and contrasted the Soviet and American space programs, commenting that "while the US government and its agencies have tried to make their space projects widely known throughout the world by means of television communications, tracking stations and other ways," the Soviet achievements resulted from "extreme secrecy behind the Iron Curtain." As the newspaper went on to suggest, "The mere fact that the vehicle is displayed should be enough to give stimulus to local scientists and experts."[51] Chairman Park Chung-hee, acting president of the Republic, shared a similar sentiment when he commented that the "scientific attitude of the Americans," which allowed the world to follow the flight, "demonstrates the deep-rooted spirit of true scientific democracy."[52] As in many cities the capsule visited, local organizers and events in Korea contributed to the size and scope of the event. The mayor of Seoul, Major General Yoon Tae-il, oversaw the exhibit, while the National Academy of Sciences of Korea sponsored a scientific conference in conjunction with it.[53]

In October 1962, Donald Wilson, the acting director of the USIA, told James Webb, the NASA administrator, that everywhere the display of *Friendship 7* "dramatized our willingness to make known the facts about our program to explore space, in contrast to the secrecy which surrounds the Soviet efforts in this field." Within just three months, he pointed out, the capsule was seen and frequently touched by roughly four million people. Wilson noted that as a result of *Friendship 7*'s exhibit, "Colonel Glenn's epic orbit and US progress in space have been etched more deeply upon the world's consciousness."[54]

It is worth noting that the extent to which the idea of US progress was etched upon the world's consciousness is impossible to measure. USIA officials were acutely aware of the difficulties and complexities of determining the effectiveness of their work in influencing foreign public opinion of the United States; Murrow was known to say that no cash register rang when people's minds were changed, highlighting the difficulty of adding up the impact of USIA programs.[55] The USIA evaluation studies and reports determined whether or not programming reached their intended audiences, the size of the audience, if the programming was of interest to the audience, and recommendations on how to improve distribution, but they could not ascertain the impact of the programming on these audiences.[56]

President Kennedy's report to Congress on US space activities for 1962 highlighted the significance of Glenn's flight and the subsequent capsule tour. The USIA also distributed a television program about the flight to TV stations in seventy-four countries in English, Spanish, and a number of other languages; according to the report, Glenn's flight was likely the most covered story in the history of broadcasting. A short documentary on the flight was shown in 106 countries in 32 languages followed by another hour-long documentary distributed to 71 countries in all the major languages. Describing the "fourth orbit," the report noted that "at every stop the public impact was impressive, particularly upon young people. Audiences included many chiefs of state who were given detailed explanations by accompanying NASA space scientists. Millions of people filed by to touch and examine the spacecraft."[57]

Compared to the USIA's previously most popular exhibit, a series of displays designed to complement President Dwight Eisenhower's

Atoms for Peace Program in the 1950s, the *Friendship 7*'s "fourth orbit" was an astonishingly major event. The 217 Atoms for Peace exhibits that the USIA had mounted around the world to change the way people viewed the word "atom" explained the potential peaceful uses of atomic energy with model reactors, illustrated panels, guides, and films, all while emphasizing the United States' significant role in this development. These exhibits were attended by unprecedented numbers of people, with some visitors waiting in lines of up to two hours. A public diplomat in Bonn commented that "by now it has become virtually trite to report that the Atoms for Peace exhibit achieved a greater impact . . . than any other project undertaken . . . by the US Information Service."[58] Although impressive and record breaking in the 1950s, the attendance numbers for the Atoms for Peace exhibits paled in comparison to the numbers for the *Friendship 7* "fourth orbit" tour. The audiences that attended both exhibits commented on their openness and factual approach, and found them to be educational, not political. The USIA press releases for both exhibits were written in a "journalistic style" that "masked the agency's messages, which were hidden in quotes by ordinary people, newspapers, and government officials."[59]

After his flight, Glenn wrote to McGeorge Bundy, President Kennedy's national security advisor, explaining that the *Friendship 7*'s "fourth orbit" tour "stressed the fact that [the American space program] was not just a propaganda effort before the world, but a well-thought-out scientific program that could eventually benefit all peoples of the world as the scientific exploration it is." He went on to note that Russian exhibits highlighted personal appearances of cosmonauts, while the United States emphasized scientific information via the capsule's display. According to Glenn, America's greatest advantage over the Soviet Union's space program was "the almost complete freedom to share experiences and new information." He suggested that the openness of the American program—as represented by the display of the *Friendship 7* spacecraft—stood in for the nation and its political ideology: when the *Friendship 7* capsule was laid bare before the eyes of people from around the world, it signaled

Children in Cambodia reading about Project Mercury, 1962. (NATIONAL ARCHIVES)

that the United States space program was real, benign, apolitical, and designed for the collective good of all humankind.[60]

By displaying technological hardware, US officials had hoped to fuse positive perceptions of the values and strengths of science and technology with American identity and the American political system. As one public diplomat explained, "We are striving to present the view that in an overall sense US scientific accomplishments are second to none and that US science has contributed more to the development of the world and the scientific development of newly emerging nations."[61] What distinguishes this effort is the political use of cultural attitudes about the apolitical nature of science and technology. The ramifications of this tactic were far-reaching: we can observe its impact on the types of engineering programs that the US government funded and in the transformation of artifacts from these

OPERATION MOONGLOW

programs into symbolic icons employed to sway the world public during the cold war. The spacecraft's display value was not simply tied to the function for which it was designed: it mattered that the capsule carried Glenn around the Earth, but it also mattered that the capsule was then put before the eyes of the people who came to see it in person.[62]

USIA space programming in the early 1960s centered on capsule tours and educational lectures, an approach meant to differentiate the US space program from the Soviet space accomplishments. Many early documents point to a hesitancy on the part of US officials to participate in astronaut tours because they feared it would appear that the United States was simply trying to copy Soviet cosmonaut tours and that it would be another sign that the country's space program was trailing behind the Soviet Union's impressive set of firsts.[63] But by the mid-1960s, the US revised its previous approach to astronaut tours, reasoning that if the exhibits were primarily scientific in nature, if the astronauts visited nations around the world to share scientific and engineering knowledge, then these tours would come across differently than the Soviet propaganda tours did. Unlike Soviet cosmonauts, the USIA and State Department stressed, the American astronauts would not make mass public appearances unless these were specifically designed to spread scientific information about space, which would ensure that the astronauts did not turn into "the kind of ballyhoo salesmen for space that the Soviet cosmonauts have become."[64] Although officials emphasized that these tours were organized to share scientific and engineering discoveries with the world public, in internal correspondence they were explicit that "the purpose [was] to support American foreign policy in specific situations."[65] Along with scientific information, the astronauts would become the next champions of spaceflight and American leadership abroad.

6

"THE NEW EXPLORERS," 1963–1967

From Cape Kennedy not even a white mouse can be launched without the nation and the world learning about it a few minutes later.

—NEUES OESTERREICH, 1965

In 1963, when Elton Stepherson Jr. arrived in Tananarive, the capital of Madagascar, he was already well-known, "much liked and respected" there. Born in Houston but raised in Los Angeles, Stepherson received a degree in political science from UCLA before pursuing graduate work in French languages and African studies at the University of Paris. In 1963, at age twenty-nine, he became a career foreign service officer with the USIA based in France. It was not long before the USIA sent him to NASA to train as a space lecturer for French-speaking African nations.[1] Much like other African Americans who traveled abroad as part of USIA and State Department programming in the 1960s, giving jazz performances or lectures, Stepherson was sent to African nations to demonstrate racial equality and the universal opportunities of democracy in the United States, in addition to his official duties.[2] After he lectured to an audience of Guinean intellectuals and political leaders on American space feats, the governor of Pita commented that Stepherson was living proof that "science is not the reserved domain of members of one race only."[3] This comment was exactly what US foreign affairs officials had hoped to hear when they sent Stepherson and his colleagues to Africa. Later, Stepherson would travel the world with the Apollo astronauts. But in the summer of 1963, he was sent to Malagasy Republic (as Madagascar was then known) because NASA was having trouble setting up a tracking and space communications station there.[4]

By the 1960s, the psychological battlefield of the cold war had moved from the fronts of Western Europe to the newly independent nations in the developing world. The USIA increasingly targeted African, Asian, and Latin American audiences in its programming as part of America's broader cold war strategy. The USIA budget for programming in Africa doubled between 1960 and 1964.[5] But public diplomats often encountered pushback. Newly independent nations that had just thrown off the yokes of colonialism had little interest in trading one foreign power for another. Questions about a NASA tracking station's military associations raised skepticism and even hostility. The current US reputation of racial injustice only made matters worse.[6] Nonetheless, the United States needed countries like the Malagasy Republic to fulfill its global ambitions. NASA's human spaceflight program—a primary basis of American prestige abroad—required a global tracking system. Without the cooperation of countries like the Malagasy Republic, the United States would not be able to accomplish Kennedy's moon landing challenge, with severe consequences for America's standing abroad. Given the dynamics set in place by the cold war, with two superpowers competing for the "hearts and minds" of the world, foreign governments and publics both constrained and extended America's power. So in places like Tananarive, the United States government relied on the work and expertise of African American diplomats such as Stepherson to sell the benefits of a NASA tracking station to local leaders and the public.

In 1960 the US government made an agreement with the United Kingdom to locate a NASA tracking station in Zanzibar, a British protectorate of islands off the coast of Tanganyika (as mainland Tanzania was then known). NASA also signed a lease with the government of the sultan of Zanzibar to establish two stations in Tunguu and Chwaka. These agreements would be short-lived. Like many nations throughout Africa in this period, the islands gained independence from Great Britain in 1963. Just one year later a revolution overturned the government of Zanzibar, and Zanzibar merged with Tanganyika to eventually become the United Republic of Tanzania. NASA was forced to quickly evacuate the tracking station personnel before any injury or loss of life, and it made arrangements with the government of the nearby Malagasy Republic to relocate the tracking facility to a site near Tananarive.[7]

Although the Malagasy government had already given its support to the station, the general public still needed to be won over, much like in the early stages of tracking-station establishment in the 1950s. During their first year of independence from European colonial rule, many Malagasy were rightfully suspicious of the buildup of US government presence in their country. When a US Air Force plane carrying electronic gear for the tracking station landed in Tananarive, USIA officials faced their first challenge. To disassociate the station with American military interests and thwart any suspicion of secretive activity, USIA officials decided to play up the arrival with a two-day "press party." The staff invited local press to visit the airport, explore the inside of the plane, and visit the tracking site. As one USIA staff member observed, "It was all completely open, informative, impressive. (And we got a lot of other good talk in during the waiting periods between unloading moves, about everything from the meaning of freedom to why American automobiles are so big!)." Stepherson arrived in the Malagasy Republic shortly after the electronic equipment.[8]

In preparation for Stepherson's visit, USIS staff had gathered space models, displays, and pamphlets. With Stepherson's help they arranged a space-themed exhibit in the Tananarive Chamber of Commerce called "Les Engins Spatiaux" ("The Space Machines"). They selected this location because they hoped to attract the attention of government officials in addition to teachers and students. To promote the exhibit, and in turn support the presence of the tracking station, USIS staff blanketed the city with six hundred promotional posters in French and Malagasy, Stepherson gave interviews on the radio and to newspapers, the American ambassador invited two hundred local leaders to an opening reception, and the USIS screened a French version of the film *America in Space*. During its seventeen-day run the exhibit attracted more than thirty thousand visitors, 75 percent of whom were considered "target individuals." As a report enthusiastically observed, the location of the exhibit minimized the attendance of people considered less politically influential yet it was still within easy access of government functionaries, businessmen, teachers, and students.[9]

Following the opening of the "space machines" exhibit, the American ambassador led President Philibert Tsiranana and his entourage around the tracking facility, along with a full complement of press, radio, and newsreel staff. At the station, NASA had President Tsiranana use the radio-telephone facility to speak with the Malagasy ambassador in Washington. "The result," USIS staff reported, "was not only abundant news media coverage but strong reassurance that here was something which the Malagasy could derive great pride and satisfaction to be participating in."[10] US government officials expected that this public demonstration would ensure that the Malagasy people would not only accept the tracking facility but also interpret it as an important factor in the progress and development of their country. This tracking station, US officials tried to make clear, not only served American interests; it also served Malagasy interests. Cooperation with the United States in general was good for their nation.[11]

For the next few weeks Stepherson gave lectures twice a day to groups of students, officers in the Malagasy Army, and government officials.[12] "He also did much toward enhancing the general U.S. image in such non-space-science aspects as education, race relations, cultural life and the like," commented one USIS staff member, "which in turn made his space science 'message' more meaningful."[13] Stepherson began the lectures at the exhibit, where he would explain how the NASA station functioned, using models and photos, and then he escorted the group onto a bus to take them to the tracking station. After touring the group around the facility, he would bring them to the USIS cultural center for a screening of a film on space exploration, such as *America in Space, Project Telstar,* and *Ranger VII.* When the group left the screening, USIS staff handed them pamphlets on space exploration and the upcoming US election. This pairing of pamphlets may seem incongruous, but it was standard USIA practice to co-opt the popularity of the space program to disseminate information about other areas of US politics, economics, and culture.[14]

After this series of information programs ended, the USIS staff concluded that "it is safe to say that . . . no one in Madagascar who reads the papers, listens to the radio or goes to the movies was unaware of the fact that the United States has an excellent space-exploration program going, that the nature of this program is scientific, peaceful and beneficial to all, that Madagascar in making its geography available is contributing importantly to the success of the program—and in the process stands to gain substantially both economically and in terms of technical and scientific know-how." Coverage of the NASA facility in the Communist press cut back on "sniping at NASA" and instead gave factual reports. The USIS considered this change in tone to be "the best kind of evidence of the effectiveness of all concerned."[15]

From Tananarive, Stepherson traveled to Burundi, a small, landlocked country in east-central Africa that had gained independence from Belgium in 1962. The US initiated diplomatic relations with the country in June of that year and established an embassy in Bujumbura, a city on the shore of Lake Tanganyika.[16] The embassy had arranged a big "kickoff" event on November 22, 1963, Stepherson's first day in Bujumbura. Just as Stepherson finished his lecture on the space program and was about to take questions, one of the staff members nicknamed Doe rushed into the room and spoke to the ambassador. He announced that there was an emergency, that the program was ending early. Once the audience left, the ambassador gathered the American staff together and shared the news: President Kennedy had been assassinated.[17]

That afternoon Kennedy had been shot while riding in a motorcade in Dallas, Texas. Almost immediately, the Voice of America spread the news around the world. In Burundi, Doe had been listening to the radio when the story came over the airwaves. US Foreign Service staff—including Stepherson—stayed up all night listening to the radio, waiting for more news. Later that day Lyndon Johnson assumed the presidency.[18]

In the coming months Johnson would shoulder and modify many of the programs and policies that Kennedy had set in motion during his abbreviated term in office, including Project Apollo and US information efforts. Like Kennedy, Johnson believed that the image of America abroad was critical to the nation's geopolitical standing and political alliances. Also like Kennedy, Johnson appreciated the space program's role in enhancing the nation's image. But in Lyndon Johnson's hands, space exploration became an even more multifaceted political instrument. In addition to enhancing prestige and supporting national security, Johnson's advocacy for space exploration was tied to his faith in liberal internationalism; he expected that a robust space program would contribute to global economic and social progress.[19] In his memoirs, Johnson reflected that "space was the platform from which the social revolution of the 1960s was launched. We broke out of far more than the atmosphere with our space program. . . . If we could send a man to the moon, we knew we should be able to send a poor boy to school and to provide decent medical care for the aged. In hundreds of other forms, the space program had an impact on our lives."[20]

Johnson's commitment to the broader applications of the space program, as well as science and technology more generally, had roots in his experience with New Deal development projects. When he was elected to the House of Representatives in 1937, Johnson pushed for federal funds to dam the Colorado River to supply power to central Texas. His dedication to the political value of science and technology carried through his tenure as the Senate majority leader, as the vice president, and then as the president. Programs that could both support US national security and improve the standard of living were of particular interest to him.[21] During his presidency, Johnson enthusiastically supported exporting the New Deal model for the Tennessee

Valley Authority to the Mekong Valley in Vietnam. In Johnson's view the development project provided justification for the US presence in Vietnam. Much to Johnson's surprise, North Vietnamese leader Ho Chi Minh rejected the offer.[22]

Johnson and other government officials folded space exploration into the administration's larger project of building the Great Society. As Vice President Hubert Humphrey articulated, "An adequately funded, well-directed space program is an integral part of our nation's commitment to its future, to its greatness." He added that "we can put a man on the moon at the same time as we help to put a man on his feet," stressing that the space program was consistent with civil rights and social concerns, not in opposition to them.[23] Spaceflight, officials hoped, could symbolize civil rights progress, counter the nation's growing warmongering image after the further Americanization of the Vietnam War, and serve as a much-needed positive arm of US foreign relations.

But for some US public diplomats working in the field, this expectation seemed too optimistic. Spaceflight, and its popularity among foreign audiences, could only do so much. Polls showed that negative impressions of American race relations and militarism dominated America's overseas image in 1964. Although the Johnson administration made progress with the passage of the Civil Rights Act in July, calls for US withdrawal from Vietnam grew louder around the world.[24] When NASA attempted to build up additional tracking facilities to support the upcoming Project Gemini spaceflights, American personnel experienced opposition. As the USIS Tananarive explained in a secret field message, "Just as an attempt to chrome-plate corroded metal would come to nothing, so would a good job of publicizing the specifics of the U.S. space effort be of little use if the general image of the U.S. itself were a tarnished one." The willingness of the Malagasy government and people to cooperate with the United States, to support an expansion of US tracking activity in the country, required confidence in America as "a worthy leader of the free world." It was not just a matter of promoting and heralding the American space program. Space achievements could not "chrome-plate" over more-general impressions of American politics and race relations.[25]

In 1964, with no crewed missions on NASA's agenda for at least a year, the USIA focused on promoting the significance of the photographs taken by lunar probe Ranger 7 as well as emphasizing the applications of communication satellites. In July 1964, Ranger 7 sent more than 4,300 pictures of the moon to Earth.[26] USIS posts placed these photographs on their walls, circulated NASA-produced pamphlets, and broadcast television programming featuring the mission.[27] President Johnson sent special volumes of Ranger 7 photographs to foreign heads of state and leading scientists around the world in ceremonies that were heavily covered by local journalists and television crews. The gift included a letter noting that "the people of the United States hope that this knowledge will work to the ultimate benefit of mankind in all lands." Nevertheless, a worldwide opinion survey taken a few months before Ranger 7 indicated that the USSR still led the space race.[28]

Unbeknownst to the world, the Soviet Union did not officially join the Americans in a race to the moon until August 1964. As part of a larger five-year plan, the USSR pursued two lunar programs: a crewed circumlunar program and a separate lunar landing program. Driven primarily by rocket designer Sergei Korolev's enthusiasm, not Premier Nikita Khrushchev, the Soviet lunar program aimed for a landing in 1967 or 1968 to mark the fiftieth anniversary of the Bolshevik Revolution.[29]

On October 12, 1964, the Soviet Union launched the three-person Voskhod spacecraft, once again demonstrating the program's superior heavy-lift capacity. Normally, the test range would call Khrushchev after a launch, updating him on the results. That day, no one called. Little did Khrushchev know at the time that this was a sign that secret plans were already under way to overthrow him. "One of the first rules for an autocrat on his last legs," Khrushchev biographer William Taubman explained, "is not to leave rivals minding the capital." This is exactly what Khrushchev had done, spending 170 days away in 1963 and 150 days away in the first ten months of

1964.[30] Khrushchev spoke to the cosmonauts in orbit, bobbing "up and down in unashamed excitement and pride." A few hours after the call, Khrushchev was put under house arrest. The blurry television broadcast of his conversation with the cosmonauts would be Khrushchev's last public appearance. He had told the cosmonauts about a homecoming celebration he was planning, but in the end he would be barred from attending it. Khrushchev's fears from 1957 had come true. But it was not Zhukov who overthrew the Soviet premier.[31]

Two days later, the Soviet Union announced a new period of "collective leadership." Leonid Brezhnev and Alexei Kosygin assumed power. Under this new regime, the organization and authority of the Soviet space program cleaved as well. Khrushchev-era initiatives were overturned, but the public relations repercussions of this upheaval would take more time to manifest.[32] Although Khrushchev's swift ouster displaced most other news, Voskhod was still a public relations boon for the Soviet space program. Voskhod was the Soviet Union's follow-on series of spacecraft to its first human spaceflight program, carrying three cosmonauts into Earth orbit. However, what the broader world public did not know at the time was that the spacecraft itself was the same size as the earlier Vostok series. Cosmonauts forewent space suits so that all three could squeeze into the compact vehicle. Little was known about the new program. The dimensions of the spacecraft, its capabilities, and the technical details remained a mystery for the general public. As one of the engineers working on the program later commented, "The program made no contribution whatsoever to the further development of space research. Sending three people into space together was done purely for prestige."[33] The first launch of Voskhod achieved the prestige that the Soviet Union was after, and the United States fell behind again in foreign public opinion polls. According to the USIA, "The capacity of the Soviets to launch larger spacecraft in manned flight remained the negative factor affecting foreign opinion of US space activities."[34] Later that month, Brezhnev hosted the celebration in Red Square that Khrushchev had promised the cosmonauts. The next mission the following spring, Voskhod 2, would add momentum to the Soviet lead in space when cosmonaut Alexei Leonov performed the first spacewalk in history.[35]

In 1965 the US could compete in human spaceflight again when NASA started launching crewed missions in quick succession to prepare for lunar exploration. NASA leadership selected the "lunar orbit rendezvous" approach to the moon in 1962. This would involve launching a multistage rocket and a modular spacecraft. After a brief stay in Earth orbit, the spacecraft would speed toward the moon in what was called translunar injection. Once the spacecraft reached lunar orbit, a lander with two astronauts inside would descend to the surface of the moon while one astronaut in a separate command module (and service module attached) remained in orbit. Once the astronauts completed their moonwalk, they would launch and rejoin their crewmate in the orbiting spacecraft and then return home to Earth. This multistep process saved on fuel, weight, and the length of time needed to develop a rocket, but it required the development of new techniques and capabilities such as spacewalking, rendezvousing, and docking. That is where Project Gemini came in. In a series of ten two-person missions over two years, NASA mastered these tasks while also testing the effects of long-duration spaceflight on the astronauts' physiology and psychology. The Gemini spacecraft resembled the Mercury vehicle: blunt-ended and bell-shaped. With a white adapter module attached, it could carry enough oxygen, water, power, and propellants for two weeks in space. NASA recruited two new classes of astronauts to fill its ranks for the Gemini flights as well as the upcoming Apollo missions. This decision also meant that there was a greater number of astronauts who could promote the space program and its successes.[36]

The second crewed Gemini mission launched in June 1965. Two astronauts—Edward H. White and James McDivitt—spent four days in space, orbiting the Earth sixty-two times, conducting the first American spacewalk, and performing a series of experiments to prepare for future missions. The Voice of America (VOA) featured the mission in more than ten hours of radio programming. In Latin America alone, four hundred radio stations in eighteen countries

updated audiences on the flight. The USIA provided posts with stories about the flight, features on the astronauts and their families,
pictures, posters, and the exhibit "Man Maneuvers Moonward." The
Times of India said that the Gemini expedition "in many ways, particularly in daring and scientific precision, equals the flight of Voskhod 2 last March."[37] Forty-seven countries carried television coverage
of the flight, including the program *Conversation with Astronauts
White and McDivitt,* prepared in English but translated into Spanish, French, and Arabic. More than a hundred countries screened the
USIA film *The Flight of Gemini 4,* which was made available in more
than twenty languages.[38]

A USIA assessment of international press coverage observed that
"the very manner in which the American astronauts chatted and
joked while performing feats of courage drew considerable comment, and appears to have subtly encouraged a measure of self-
identification with the U.S. space effort," a goal of US foreign policy.
The Swedish newspaper *Svenska Dagbladet* called the mission "re-
freshingly human and winning," while the Hong Kong *South China
Morning Post* commented on "a certain carefree and cheering lack
of regimentation . . . a pleasing combination of lightheartedness and
high courage." International journalists appreciated that the astronauts snuck a sandwich on board and that White was reluctant to
finish his spacewalk.[39]

According to a USIA report, openness "provided the main peg
for contrasting the US and Soviet political and social systems" after
the Gemini 4 flight. Throughout Latin America in particular, praise
for the openness of the American space program spilled over and
became praise of the United States' open democratic system. Media
coverage of the space race in the mid-1960s continued to stress the
"openness" of the United States space program in contrast to the Soviet Union's more closed program. As Vienna's *Neues Oesterreich*
playfully commented, "From Cape Kennedy not even a white mouse
can be launched without the nation and the world learning about it a
few minutes later. Cape Kennedy symbolizes democracy."[40] In 1965
the State Department viewed the policy of openness and information

dissemination of civilian space activities as "desirable from the viewpoint of our foreign relations."[41]

Shortly after they returned to Earth, McDivitt and White dined at the White House. As Johnson told the story, he looked at the astronauts and their families sitting in his living room and thought, "What finer representation, what greater ambassadors, what more appealing personalities could this country send out to the world than these astronauts?" It happened to be the same week as the Paris Air Show. Inspired, he asked them to fly to Paris that very night to represent the United States at the opening day of the air show. The astronauts and their wives had planned only for a dinner at the White House and not a trip to Paris, so they had not packed the necessary clothing. Lady Bird Johnson solved this problem by taking the astronauts' wives to her and her daughters' closets, where she gathered up enough dresses and gowns for their trip. A NASA photographer was called in to the White House to take pictures for impromptu passports. The astronauts' five children stayed behind and slept at the White House, swam in the president's private pool, and watched Disney movies in the White House theater. The astronauts and their wives boarded a 3:00 a.m. flight to Paris and, according to Johnson, "performed a very valuable service to their country."[42]

The Soviet Union had already made arrangements for cosmonaut Yuri Gagarin to attend the Paris Air Show. According to satirist Art Buchwald, "The belief here in Washington is that the president made his decision [to send the astronauts to the air show] because the Russians had sent cosmonauts Gherman Titov and Gagarin to Paris and this put the Russians ahead of us in ground travel."[43] A model of the Vostok capsule accompanied the cosmonauts, making this the second time that the Soviet Union revealed substantial information about the spacecraft to the general public. Much as with earlier displays, the Vostok model on display in Paris was not meant to share technical details with the audience. The model was a pared-down, simplified version of the capsule. In interviews, Gagarin stressed that the new Voskhod spacecraft was "of [an] entirely different design," a misleading statement meant to suggest that the Voskhod was the product of major technological advances, even though it was simply a version of the earlier Vostok retrofitted to carry an additional two people into space.[44]

Large crowds greeted the Gemini 4 astronauts and their wives, along with Vice President Hubert Humphrey, when they arrived at Le Bourget Airport on the morning of June 19, 1965. The astronauts spent the day speaking to the press, attending a large reception hosted by the US ambassador, and meeting with French president Charles de Gaulle. The French press gave extensive coverage to the astronauts' spontaneous visit to the Paris Air Show and ruminated over its broader political ramifications for Franco-US relations. The success of this brief visit prompted President Johnson to support an expanded astronaut tour program.[45]

Many US embassies enthusiastically embraced President Johnson's decision to send astronauts abroad in 1965. Embassy staff throughout the world outlined a number of positive political outcomes of astronaut tours: (1) they would foster support of the local government's relations with the US, strengthening international alliances; (2) they would demonstrate American interest in particular countries; (3) they would offset negative reactions to other US foreign policies and international interventions; and (4) the astronauts would demonstrate the ideal virtues of American citizens.[46]

In countries with tracking stations, such as Nigeria, embassy staff explained how these tours could be "unparalleled" public relations events for the national governments that hosted the stations. The government of Nigeria, the US Embassy in Lagos explained, could "gain credit [for] their participation [in] space scientific program thru [the] Kano tracking station."[47] In Kenya, US Embassy staff described how an astronaut visit to Nairobi would "have a great impact on Kenya and East Africa showing that US recognizes importance of this part of the world at the highest level" and that it "would get more attention in black Africa than any other combination of programs or events imaginable." The US Embassy in Manila, vying for an astronaut visit, pointed out that it would "provide [a] considerable psychological boost to Philippine ego and do much to counter

Gemini 5 crew members Charles "Pete" Conrad Jr. and L. Gordon Cooper Jr. show President Jomo Kenyatta of Kenya photographs from their mission, 1965. (NATIONAL ARCHIVES)

Philippine attitude that [the] US always take Fils for granted." The US Embassy in Ankara presented its case to the secretary of state for scheduling an astronaut visit to Turkey in August 1965: the "psychological impact" of such a visit would be "extremely useful [for] this NATO partner which directly confronts USSR, and which since Cyprus dispute has experienced some frustration with NATO." The previous summer, the US had tried to broker an agreement between Greece and Turkey over control of Cyprus but ultimately failed to find terms that both countries could agree on, prolonging the ongoing dispute. The US Embassy added that an astronaut visit to Turkey could also have the benefit of counteracting the successful improvement of Turkish-Russian relations since the Turkish prime minister's recent visit to Moscow.[48]

In a May 1965 public opinion poll conducted in Morocco and Nigeria, the USIA found that "in science and space we are clobbered both north and south of the Sahara," among what the agency considered the elite in primarily urban areas. In Morocco respondents preferred cosmonauts to astronauts by a margin of 51 percent to 13 percent. In Nigeria the split was even larger. Cosmonauts were favored by 75 percent to the astronauts' 7 percent. On the other topics covered, the Soviet Union usually came out on top. Almost everyone polled knew about the civil rights injustices in the United States, with two-thirds commenting on how "the problem had darkened the U.S. image in their eyes." The United States was moving too slowly on addressing these issues, those polled agreed. "In Morocco we are in worse order even than the South Africans!" exclaimed the USIA report. In addition to providing a hard look at why the US was faring as badly in perceived race relations as apartheid South Africa, the poll prompted the major conclusion that "clearly, we must devise dramatic new departures in our space propaganda."[49]

In the fall of 1965, Johnson sent Gemini 5 astronauts Gordon Cooper Jr. and Charles "Pete" Conrad Jr. on an elaborate two-week-long goodwill tour of Athens, Thessaloniki, Izmir, Istanbul, Ankara, Addis Ababa, Tananarive, Nairobi, Lagos, a handful of smaller cities in Nigeria, and Las Palmas.[50] The astronauts spoke to scientists and engineers at the International Astronautical Federation Congress, met with cosmonauts in Athens, appeared on television programs, were received by foreign heads of state, appeared before large crowds of people, and visited NASA tracking stations.[51] Favorable reports on the tour filled the State Department mailbox.[52] An aide reported back to President Johnson that the response to the astronauts in African countries was especially positive.[53]

The American ambassador in Tananarive found the visit of Gemini 5 astronauts Cooper and Conrad the "most impressive demonstration [of] US-Malagasy friendship ever achieved here." During the two-day program in mid-September 1965, Cooper and Conrad took a motorcade through the city with an estimated seventy thousand people lining the route. The American ambassador reported that it rivaled the reception given French general Charles de Gaulle in 1958, when he presented the Constitution of the Fifth Republic, leading to the colony of Madagascar's independence from France. President Tsiranana and his wife greeted the astronauts "with unusual warmth" and took great interest in learning about the details of their mission. US Embassy receptions attracted most of the Malagasy Cabinet and "everyone else who could beg or borrow an invitation." Before they left, the astronauts presented awards to American and Malagasy staff at the NASA tracking station.[54]

When Cooper and Conrad visited Nigeria, public diplomats attempted to make "the Nigerian mass aware that the US considers Nigeria of sufficient importance" to schedule the longest stay of the tour in their country. During this stop, the astronauts visited cities throughout the country; received elaborate gifts, including ostrich feather fans and hand-embroidered robes; gave countless lectures on spaceflight; appeared on television shows; learned the popular West African "highlife" dance; and met with Nigerian officials.[55] At the tour stop in the northern Nigerian city of Kano, the location of one of the US tracking stations, the party "received tumultuous welcome, reminiscent of 13th century feudal celebrations, given by Emir of Kano. Waves of horsemen carrying spears swept across open courtyard in traditional salute . . . for half a mile up to gates of palace."[56] This stop played an important role in ensuring continued cooperation with Nigeria in running a tracking station from Kano. Agency officials asked the astronauts to "take every opportunity to thank Nigeria for its contribution to the success of US space exploration. To underline this 'partnership,' they were asked to share the scientific results of space exploration with the educated Nigerian elite."[57] It is worth pointing out how the public diplomat used quotation marks, suggesting that he viewed the "partnership" between the United States and Nigeria as more theoretical than literal. This attitude was not unusual among Americans involved in cooperative space programs in this period.[58]

While the Gemini 5 crew toured Europe and Africa, their spacecraft toured Latin America. The capsule spent six weeks in each of three countries: Brazil, Argentina, and Mexico. At its exhibition in Buenos Aires the USIA set up the capsule under a brightly colored tent near the Costanera. After waiting in line typically for over an

Gemini 5 crew members Charles "Pete" Conrad Jr. and L. Gordon Cooper Jr. show Emperor Haile Selassie of Ethiopia a space helmet, 1965. (NATIONAL ARCHIVES)

hour, visitors would pass through another exhibit arranged by the Argentine Air Force, followed by photographic panels taken during Gemini missions and objects such as space food, a space suit, and human-waste-disposal equipment. According to a USIA report, "This panorama of people [visiting the exhibit] represented the humble and the great," from mailmen to nuns to Arturo U. Illia, president of Argentina.[59] During the exhibit, USIA/NASA Liaison Harry Kendall gave prepared slide lectures that he described as "pure Cold War rhetoric, reflecting Washington's determination to convince the world that the United States was way ahead of the Soviet Union, not just in space but also in every aspect of science and technology." Kendall decided to temper the tone of the lecture and include more technical information about the planned lunar landing.[60]

Later, Kendall would explain that by the 1960s, the USIA and NASA had two different types of space-themed information campaigns: one for "industrialized societies" and one for the "Third World." In European countries and other countries considered "industrialized societies," the USIA made sure that local science writers and scientific organizations had access to information about the American space program. In contrast, Kendall recalled that "in most Third World countries materials had to be more refined." For these countries the USIA produced press articles, films, and radio and television programs to distribute directly to foreign audiences. USIA programming in Africa required an even "more direct approach."[61]

A month after the Gemini 5 astronaut and spacecraft tours, *Friendship 7* astronaut John Glenn visited a number of European cities in October. NASA, the Department of State, and the USIA organized Glenn's travel arrangements based on requests from US embassies. The President's Report to Congress observed that Glenn's tour "gave a real boost to public and official support in Western Europe for the US space effort."[62] USIS Naples commented that Glenn was a "continual lesson in diplomacy." The Italian newspaper *Il Giornale d'Italia* suggested that Glenn's "human appeal is irresistible, almost contagious . . . he is an unpretentious man, but a first-rate man with an exceptional, almost rare technical background."[63] American ambassador Margaret Tibbetts enthusiastically reported that the visit "contributed to a bettering of US image in Norway, particularly indirectly offsetting some of [the] reverses we have had here lately stemming from [the] Viet Nam issue." After the tour, USIA Director Leonard Marks assured President Johnson that he would continue to program astronaut tours with NASA whenever possible.[64]

A few months later, Glenn traveled to Rangoon for the "American Progress in Space Science and Technology" exhibition. Burma had come under military rule in 1962 after gaining independence from Great Britain fourteen years earlier. The new Revolutionary Government of the Union of Burma (RGUB) sought to make the country a socialist state. The RGUB's severe control of foreign propaganda necessitated that the USIA plan only low-key, education-themed events. All the material that the USIS distributed to local newspapers required the RGUB's stamp of approval before it could be printed.

The RGUB allowed the space exhibit and Glenn's visit because, according to one American public affairs officer, "the United States exhibit was apolitical. Its soft-sell contents appealed to the RGUB's reviewing committee, which noted the complete absence of anti-Soviet and anti-Chinese propaganda pictures."[65]

Glenn was an astute politician; he sold US space exploration by downplaying competition with the Soviet Union, emphasizing the theme of unity, and explaining the local relevance of space exploration to the Burmese people. This approach not only promoted an idea of American-led progress; it also navigated the strict restrictions on information dissemination in Burma. In each of his lectures, Glenn praised the Soviet Luna 9 spacecraft, a gesture that encouraged neutralist leaders to allow Burmese press to cover the event in full detail. In both Glenn's presentations and the USIA's programming, the focus on global benefits, the dissemination of scientific and engineering information, and the use of inclusive rhetoric not only projected a particular image of American space accomplishment but also enabled the United States to sponsor a propaganda program in a country to which it otherwise had extremely limited access.

Thanks to Glenn and the exhibit, "the American 'image' looked more positive and glowing in Rangoon newspapers" than it had since the RGUB assumed power in 1962, according to a USIS field message.[66] Glenn's appearances and the space exhibit drew more than 250,000 people in just a handful of days, an impressive crowd even according to Burma's anti-American press.[67] A telegram from the US Embassy in Rangoon to Secretary of State Rusk described the reason for the success of Glenn's visit: "Glenn's discussion of space exploration de-emphasizing 'space race,' complimenting USSR for its achievements, depicting exploration as [an] adventure for all mankind, relating exploration to Burma—has been most effective here where more heavy-handed trumpeting of US achievements could have alienated neutralist-minded Burmese."[68]

In 1966 the US sent Gemini 7 astronaut Frank Borman and Gemini 6 astronaut Wally Schirra to tour Asia. After his trip, Borman wrote President Johnson a firsthand account of his experience. The crowds and individuals received them warmly, he reported. In a number of private conversations, government officials expressed their support for US policy in Vietnam. Borman felt that the trip was so successful that "interest in the American space program . . . might be great enough to attract additional visits to countries with which we do not enjoy the same friendly climate that we do with the eight countries our trip included."[69] USIA Director Marks wrote to President Johnson to let him know that he was "convinced that the programming of astronauts is the best way to call attention to our superiority in science."[70] After the tour, Dean Rusk sent a letter thanking James Webb for allowing the astronauts to take time away from training to support US foreign relations. Rusk used this letter as an opportunity to stress the political significance of the upcoming astronaut tour of South America, pointing out "the importance of that area to our space program and that its development into an economically viable region, friendly to the United States, continues to be one of our major foreign policy objectives."[71]

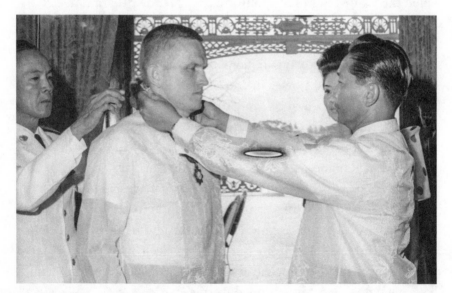

Frank Borman, on an eight-nation goodwill tour of the Far East with fellow Gemini astronaut Wally Schirra, is presented with the Philippine Air Force Aviation Badge by President Ferdinand Marcos as First Lady Imelda Marcos watches, 1966. (JUANITO PARDICO © STARS AND STRIPES)

Not long afterward, Gemini 8 astronaut Neil Armstrong and Gemini 11 astronaut Dick Gordon crisscrossed South America with George Low, NASA Manned Spaceflight Center's deputy director. Low kept a travelogue of the journey. In one entry he enthusiastically commented that "I have no doubt whatsoever that our accomplishments in space have a profound influence on our relations with South America and how we are viewed by the South American people and their governments. This impact far exceeded my expectations and is perhaps a most powerful tool that the United States had in our international relations, pursued for the purpose of peace." Before each lecture on the trip, Low gave an opening statement about how the group members were the official representatives of President Johnson sent to South America to share scientific and engineering information. Low recalled that in private conversations following these talks he learned that "during a visit to South America by the Soviet cosmonauts, the heroic efforts of the cosmonauts in space were hailed; our visit, on the other hand, was looked upon as an official visit by a team of scientists as well as space heroes." This was an important distinction to Low, as it had been to public diplomats since the launch of Sputnik in 1957. Although he observed that it might not make a difference to the general population of South America, government officials, scientists, and the press applauded this distinction.[72]

Low had been particularly impressed by Armstrong's performance in South America. Before the tour, Deputy Assistant Secretary for Inter-American Affairs Bob Sayre briefed Armstrong and Gordon on the political, economic, and social context of South America. Country directors then briefed them on the history, ideology, economics, and culture of each particular country. Armstrong put extra time into learning Spanish as well as some Guarani for their stop in Paraguay, he incorporated local history into his speeches, and he studied encyclopedias to learn about the countries he would be visiting. These gestures, Armstrong biographer James Hansen has suggested, may have been influential in 1969 when Armstrong was selected to be the first man on the moon.[73] Low commented that Armstrong "had a knack of making short little speeches in response to toasts and when getting medals, in response to questions of any kind, and in each of those speeches he never failed to choose the right words."[74]

The astronauts presented letters from the president to each of the chiefs of state of the countries they visited. The letter emphasized the United States' interest in sharing knowledge gained from space exploration with the "world scientific community."[75] Many of the astronauts' speeches highlighted national connections, promoted the sharing of space information, and frequently used the word "we." These symbols of inclusiveness, like Armstrong's efforts to link his mission with local heroes or history, became increasingly important features of astronaut tours.[76] These gestures of cultural unity were not one-sided. At many stops, people integrated the astronauts and the space program into their culture and historical narrative through speeches, the presentation of keys to cities, including the astronauts in the lyrics of traditional songs, and numerous other gestures. For instance, during their stop in Panama the astronauts were given a colorful *mola* (appliqué) made by the Kuna indigenous community. The mola depicted their Gemini spacecraft.[77] These small deeds and exchanges from both the astronauts as well as the populations they visited were part of a larger process that fostered a sense of interconnection. The astronaut tours emphasized commonality, shared history, and global unity through American-led space exploration: fundamental elements in the United States' cold war "new diplomacy."

The USIA produced a film on the Latin American tour titled *The New Explorers* in English, Spanish, and Portuguese, a title that hints of a colonial subtext in much of the space-themed public diplomacy programming. The film began with a narrator recalling sixteenth-century Spanish expeditions to "the new world," where across uncharted waters these explorers arrived in "the newly discovered lands of the Americas." These Spanish explorers "opened up the Western Hemisphere and doubled the size of the world as man knew," the narrator added. This introduction positioned space exploration, as well as the astronauts' tour of Latin America, as contemporary examples of sixteenth-century colonial expeditions. The narrator went on to emphasize the "unity" of the Western Hemisphere. The astronauts, he explained, did not see national boundaries from space, and when they visited Latin America, they came to learn as much as they came to teach.[78]

The title as well as the content of this film exposes a paradox of space program messaging. Government officials used words like "unity" and the phrase "for all mankind" in a way that was similar to the manner in which the American diplomat in Nigeria used quotation marks to refer to a "partnership" between NASA and the Kano tracking station. Historian Jenifer Van Vleck called this perspective "a global imaginary that represented the world as one but also endowed the United States with exceptional national characteristics and unique entitlements to global power."[79] Space exploration created fertile ground for public diplomats and astronauts to promote this vision of the globe, united under US leadership. Images of the Earth from space with no political boundaries, the US network of tracking stations spread across the world, and the repetition of phrases like "for all mankind" were part of the larger project of building a US-led global coalition.

In the mid-1960s, even after enhanced programming in Africa and countless new publicity strategies, officials within the State Department worried that the United States was not doing enough to take advantage of the potential impact of the space program on the country's international standing. Secretary of State Rusk urged Vice President Humphrey, chair of the National Aeronautics and Space Council (NASC), to have the council assess the US space program from the viewpoint of foreign policy objectives: "We have not yet recovered from the blow to our prestige and the burden imposed upon our diplomacy by the early Soviet sputniks and the continuing Soviet lead in manned space flight."[80]

Before joining Kennedy's Cabinet in 1961, Rusk served as the president of the Rockefeller Foundation. He kept the post as secretary of state when Johnson assumed the presidency. Described as "dogged, durable, unfailingly courteous and considerate," he won both Kennedy's and Johnson's loyalty, even as antiwar protesters called him a "warmonger."[81] According to Rusk, the international standing of US science and technology hinged on the standing of the nation's space program, and US military credibility hinged on the image of the nation's science and technology capability, making future planning of space activities critical to maintaining the nation's international posture: "The credibility abroad of our will to assure our own national security, and to assist effectively in preserving the security of the Free World, rests on the belief that we will not again fail to match major technological breakthroughs in space." The United States had to demonstrate a robust space capability, even if these programs had very little to do with the actual military balance of power. Rusk was concerned that the United States was not investing enough in the types of space research and activities for long-term foreign relations dividends.[82]

During the next Space Council meeting, Humphrey told the council that he wanted to discuss the international aspects of the space program. As Rusk commented, "The mutuality of these cooperative projects, and the opportunities they present to further identify the interests of others with our own, constitute a new asset in our foreign relations." This statement pinpoints the complexity of the space program's role in foreign relations. Space exploration activities not only demonstrated US power and prestige; they could also, if properly employed, "increase the identity of foreign interests with our own [and] strengthen the fabric of common action based on mutual interest and commitment."[83] Space exploration, Rusk argued, could cultivate an idea of global interdependence in direct support of US foreign relations interests.[84]

Humphrey inquired about public opinion polls and how to improve the United States' standing in space activities relative to the Soviet Union. The vice president was also interested in enhancing international cooperation in space and the effectiveness of the policy regarding the release of information about space activities. Humphrey wanted to know if NASA-themed exhibits had been "increasing understanding of our space program abroad" and "what the plans [were] for improving this part of the program." His final question

reflected Rusk's thinking: "What space programs should be emphasized in our current planning so that the United States does attain
and maintain a world leadership position in the future?"[85]

Rusk submitted the State Department's responses to the vice president's questions. Based on reactions abroad, Rusk's report concluded
that the United States must successfully complete Project Apollo to
counter the Soviet lead in space. The United States government must
also enlarge its space-themed public relations efforts and invest in
international cooperative programs. Increasing NASA participation
in events abroad like the Paris Air Show, broadening the number
of space exhibits and space lectures, and organizing astronaut tours
would have a significant impact on the image of US space capability,
and in turn the geopolitical standing of the country, he argued. Even
when the United States trailed the Soviet Union in space feats, public
relations could emphasize the significance of the "openness, breadth
and purpose" of the American space program. Rusk concluded by
explaining that cooperative space programs would also serve foreign
relations interests by encouraging foreign identification with American space efforts and offering another, and perhaps more credible,
avenue for foreign publicity. Rusk's main message was clear: all agencies should be doing more, not less, to exploit American space exploration in foreign relations.[86]

On January 27, 1967, in the early evening, Lyndon Johnson spoke
to officials from Great Britain, the Soviet Union, and fifty-seven other
nations. They had gathered in the East Room of the White House for
the signing of the United Nations Outer Space Treaty, which laid out
the basic framework for international space law. The treaty restricted
the use of weapons of mass destruction in outer space and any claim
to sovereignty over celestial bodies, among other legal principles.
Johnson and the State Department also saw the treaty as a step toward the "de-fusing of the space race" and, in turn, a way of saving
money that could be spent on domestic needs, "thus mitigating the
strain of the war in Vietnam."[87] Speaking to his guests assembled in
the classical-style room, Johnson called it "an inspiring moment in
the history of the human race. This treaty means that the moon and

our sister planets will serve only the purposes of peace and not war." Signing ceremonies were also held in London and Moscow.[88]

A little over an hour later at Cape Kennedy, a fire ignited underneath astronaut Virgil Grissom's couch inside the AS-204 spacecraft. Grissom and his crewmates, Edward White II and Roger Chaffee, were conducting a standard prelaunch test of an Apollo command and service module. Flames spread rapidly in the pressurized, pure-oxygen environment while thick black smoke engulfed the command module, impeding ground technicians' rescue efforts. The three astronauts were trapped inside, and they perished from asphyxiation of toxic gases. The mission was later assigned the name Apollo 1, in honor of the crew. The horrific accident cast doubt on the competence of NASA management, the reliability of the spacecraft, engineering protocol and training, and the future of the Apollo program.[89]

NASA not only faced a tragedy; it also faced the challenge of experiencing this loss in full view of the public. NASA's policy of "openness" meant that news of the accident spread around the globe in the following days. The State Department programming highlighted the astronauts' courage, their confidence in the scientific and technical personnel working at NASA, and included a quoted statement from Grissom: "It is worth it to risk one's life for the sake of space exploration."[90] The international press responded with support and goodwill. The USIA reported that "the accident at the Cape has brought us no problems—only sympathy and compassion."[91] The Greek newspaper *Kathimerini* stated that the astronauts' names would become part of "the legend of a new era." Ceylon's independent newspaper *Dinapathi* praised the policy of openness, noting that "similar accidents might have occurred in the Soviet Union. But the world has never had the opportunity of knowing the results."[92]

President Johnson later said that "the shock [of the accident] hit me like a physical blow." He found comfort in the "outpouring of sympathy from all over the world."[93] Many foreign leaders sent condolence letters. Thanking them for their support, President Johnson conveyed a universalist message in reply: "These men were truly envoys of mankind. In their memory we rededicate ourselves to the task of achieving, together with all nations, an understanding of our common space environment and its successful exploration for the mutual benefit of all peoples of earth."[94]

Johnson "grieved [not only] for the men and their families but [also] . . . for the space organization. I felt very sad and sorry for Jim Webb and all his loyal employees."[95] Over the coming months a NASA review board exposed the circumstances that led to the accident. Congressional investigations and hearings left NASA with technical recommendations as well as the go-ahead to pursue its lunar landing goal. Over the next year and a half, NASA postponed launches, instated major redesign modifications, and revised testing procedures, the manufacturing process, and quality control procedures. Meanwhile, the State Department and the USIA amplified their message of "openness" and the unifying effects of the US space program. When the time came to send astronauts into space in 1968, the images they took of the Earth, the broadcasts from missions, and an array of public programming offered American diplomats an opportunity to enhance this message once again.[96]

7

"RIDERS ON THE EARTH TOGETHER," 1968–1969

We were all with them during those five days.
—*ŻYCIE WARSZAWY*, 1968

The year 1968 began with a women's antiwar march in Washington on January 15. Two weeks later came a major turning point in the Vietnam War: the Tet Offensive. In a series of surprise attacks during the lunar new year, or Tet, North Vietnamese forces infiltrated deep within South Vietnam, reaching the US Embassy in Saigon, the Long Binh US Army Base, and other sites thought to be part of a US stronghold. *Washington Post* journalist Don Oberdorfer called Vietnam "American's first television war." The Tet Offensive, he figured, was "America's first television superbattle."[1] As millions of people around the world watched this "superbattle" unfold on television sets in their homes, confidence in US leadership eroded. The war started looking unwinnable. After assessing the situation firsthand in Vietnam, CBS news anchor Walter Cronkite warned of the potential loss of "American lives, prestige, and morale."[2] After Tet, polls showed that the majority of Americans now thought it was time to end the war. Fighting in Vietnam was not only driving a wedge between American politicians and the public; it also called the moral character of the country into question for many observers around the world, damaging the status of American global leadership.[3]

Throughout 1968 the stability of American democracy would be called into question again and again. When Reverend Martin Luther King Jr. was assassinated in Memphis in April, riots erupted throughout the United States. USIA Director Leonard Marks told President Johnson that the "confidence of America's allies and friends around the world" had been shaken. "We have suffered a blow from which it will take a long time to recover."[4] Two months later, on the other side of the country, presidential hopeful Robert F. Kennedy was shot shortly after he made his California Democratic primary victory speech. Then, in late August, violent clashes between protestors and police at the Democratic National Convention broke out in Chicago, casting more doubt on the US political system. Parallels were quickly drawn between the Chicago riots and the Soviet Union's suppression of the Prague Spring that same month.[5] At the end of the year the USIA concluded that the Vietnam War, protests, assassinations, and upheaval throughout the country led "many persons abroad to question whether the vaunted American system might be on the verge of decay and disintegration."[6]

Tear gas, body counts, protests, and riots all appeared on television sets around the globe and in international newspapers. By the end of the year, the House of Representatives Foreign Affairs Subcommittee observed that "the mental picture that many foreigners have of our nation is increasingly that of a violent, lawless, overbearing, even sick society."[7]

Then, in late December, at the tail end of the year, the Apollo 8 crew traveled farther and faster than any humans in history. They saw what no other eyes had seen: the far side of the moon, and the Earth from a great distance, blue and white and shining. They became the first humans to ride the mighty Saturn V rocket, break the bonds of Earth's physical pull, and enter the gravitational field of another celestial body. But the mission, and the program more generally, "did much more than just advance the country scientifically and technically," its commander, Frank Borman, argued. "It advanced it—in my opinion—diplomatically just as much. It cast the country in a favorable light, at a time when there were many things that cast it in an unfavorable light."[8]

Apollo 8 offered an antidote: an image of a nation striving for grand goals, inclusive and focused on peace and unity. The crew's broadcasts from the moon captured the attention of a billion people worldwide. Inclusive language during the broadcasts, as well as the soon-to-be-iconic photo *Earthrise*, amplified the USIA and State Department messaging that the American space program was "for all humankind." When the world felt divided—between democracy and Communism, among generations, races, and gender—Apollo 8 offered a moment of unity and a sense of connection.

From the start, Apollo 8 commander Frank Borman understood his flight and then later promotion of the space program abroad as part of his service to the country, not as a purely scientific pursuit: "If you think I would've devoted that much of my life simply to exploration or science, I wouldn't have, I'm not built that way, that's not my thing."[9] The cold war threatened the security of the United States, and his role as an astronaut was part of confronting that threat, lessening Soviet influence on the geopolitical landscape.

Borman joined the astronaut corps in September 1962. Along with James Lovell Jr., who would fly on Apollo 8 as well, Borman was selected for Gemini's long-duration mission: Gemini 7. The two astronauts spent nearly two weeks together in space, confined to a capsule smaller than a compact car. After 206 Earth orbits, a rendezvous with Gemini 6A, and logging medical data, the crew returned with information that NASA needed for upcoming moon missions.[10] Michael Collins, Apollo 11 astronaut and Apollo 8 CAPCOM (capsule communicator), noted that at times Borman could annoy others with "the crisp and arbitrary military precision with which he ran the operation, and the merciless ribbing he applied to any who might disagree with him."[11]

On a Sunday afternoon in August 1968, Frank Borman was hard at work at the North American plant a few miles southeast of Los Angeles. He had spent much of 1967 and 1968 at the plant, working seven days a week to ensure that the new and improved command module—the block II—was safe to fly. Few were as intimately aware of the hardware flaws and political and cultural shortcomings that contributed to the Apollo 1 disaster in January 1967 as Borman. After serving on the accident investigation committee, he headed the spacecraft redefinition team. The hatch needed a major redesign, which added weight. In turn, the command module required a new parachute system. Within the spacecraft, engineers improved wiring, removed combustible materials, and filled the atmosphere on the launchpad with a mix of oxygen and nitrogen. In early August Borman was back at the plant, running checks, when Deke Slayton, chief of the astronaut office, called him on the phone. NASA needed Borman in Houston, immediately.[12]

NASA had flown three crewless missions, Apollo 4, 5, and 6, to test hardware between the fall of 1967 and spring of 1968. Next, according to NASA's plan, Apollo 7, the first crewed flight, would assess the command and service module (CSM) in orbit. Then a flight would test the lunar module and CSM in high orbit. Only then, after both crewed missions proved successful, would humans venture beyond the Earth's gravitational pull toward the moon. But in the summer of 1968, NASA officials worried that lunar module delays could push the lunar landing into the 1970s, past the goal Kennedy had set in 1961. So as Borman learned once he arrived in Houston, George Low, manager of the Apollo Spacecraft Program Office, found an audacious solution: send an Apollo mission to the moon without a lunar module. The CSM was ready, the Saturn V was ready, and if Borman and his crewmates Lovell and William Anders agreed, they would be moved up to the Apollo 8 mission. If Apollo 7 succeeded in October, Apollo 8 would become a lunar orbital flight.[13]

NASA leadership was not only facing the pressure of making Kennedy's deadline; it also faced the threat of a Soviet victory in the space race. Throughout the 1960s the US spy satellites Corona and Gambit captured photographic evidence of Soviet rocket development, while ground stations in Turkey and Iran listened for telemetry signals.[14] In early April 1968 the CIA produced a top-secret report suggesting that the Soviet Union might send a crewed spacecraft around the moon by the end of 1968, in part "to lessen the psychological impact of the Apollo program."[15] Then in mid-September, from the steppes of south-central Kazakhstan, the Soviet Union launched Zond 5 on a loop around the moon. Built to carry a human crew, this space-craft instead ferried turtles and insects. It captured photographs of the moon and the Earth at a great distance. During Zond 5's return to Earth, a failure with altitude control set the spacecraft on a course for ballistic reentry, which humans could not have survived. Even so, the mission suggested that the moon was likely within reach of the Soviet space program.[16]

Shortly before his launch, as Borman engrossed himself in train-ing for the new mission time line, his phone rang again. It was Ju-lian Scheer, NASA's deputy administrator for public affairs. "Look, Frank," Borman recalled Scheer explaining. "We've determined that you'll be circling the Moon on Christmas Eve and we've scheduled one of the television broadcasts from Apollo 8 around that time." He pointed out that more people would hear the crew's voices than had heard any voice in history. NASA estimated that a billion people around the world would be following the flight. He then added the simple but imposing instruction: "So, we want you to say something appropriate."[17]

Borman turned to his friend Simon Bourgin, the USIA science ad-visor, for help. The two had become close during the Gemini 7 diplo-matic tour of Asia. When Borman prepared for interviews, he would ask Bourgin for advice. "He came to trust me," Bourgin recalled, "and I dined at his and Susan's home on a number of occasions."[18] A week before the flight, Bourgin wrote that "I think it would be a mistake for me to write a script . . . what you say has to be *all* Frank Borman." He encouraged him to avoid reflections on the "transcendental signifi-cance of all of it."[19] And added that it would be a "mistake to do the Christmas tree thing. It would degrade the image of the mission."[20]

Instead, Bourgin suggested a simple and short broadcast. "With six television transmissions, you are overexposed . . . and with that much time you could be tempted to pad, ham it up, or try to

153

entertain. Avoid all of these." In other words, he explained, "Keep your audience hungry." For the Christmas Eve broadcast, start with a description of what you see, he suggested: "I have a feeling that any direct message that you might compose reflecting on Christmas eve, conditions on Earth, and the way you feel about it at the moon, could get awfully sticky; it would be difficult not to sound pretentious or patronizing."[21] In its place, end with a quotation.[22]

Bourgin had called his friend Joe Laitin, assistant to the director of the Bureau of the Budget, and his wife, Christine, for advice. Christine came up with the idea of reading Genesis. "Why don't you begin at the beginning?" she asked. The first ten verses of Genesis from the Old Testament would have "universal appeal and a sense of reverence that is called for," agreed Bourgin. As he told Borman, "About the only thing I can think of to match the majesty of the occasion, and the evening, is to read the opening lines of Genesis."[23] When Borman shared the idea with Lovell and Anders, they agreed. The passage, typed on fireproof paper, was inserted into the Apollo 8 flight plan.

"Don't forget," Bourgin reminded Borman, that "to use 'one world' or 'peace' more than once is to begin to lose credibility, and beginning of being political, which you wouldn't want." He added that "the astronauts are respected for being non-political and having no axe to grind, and it's terribly important to observe that . . . so, don't be preachy, say it in your own way, say what has universal appeal, and say what you think needs to be said and no more (above all, don't pad)."[24]

Watching the early-morning launch of Apollo 8 was the first thing on the president's agenda on December 21, like much of the nation.[25] At 7:51 a.m. EST, Borman, Lovell, and Anders became the first humans to ride the huge Saturn V rocket into space, one of countless firsts that the astronauts would claim on the mission. Susan Borman, Frank's wife, found it "awesome . . . like watching the Empire State Building taking off."[26] A little over ten minutes later, the Apollo 8 crew was orbiting Earth. Then, at 10:42, the third stage burned again, sending the astronauts on their way to the moon.

As the spacecraft glided out toward the stars, the astronauts departed the Earth and stopped experiencing sunrises and sunsets.

Another first. On the second day of the flight, the crew took a live television audience around the interior of the CSM and treated them to a view of Earth from 140,000 miles away. Back on Earth, Julie Nixon married David Eisenhower. The next day, now more than 200,000 miles out, they aimed the camera toward the window again, capturing the Western Hemisphere, at once luminous and familiar. The crew pointed out Baja California, Cape Horn, the North Pole, and clouds dotted over the eastern coast of the United States.[27]

The mission proved a boon for American ambassadors and other officials, who were invited by local media for interviews on the flight. "An excellent opportunity to get positive exposure through a variety of media in many countries," the USIA assessed. USIS posts recorded the heaviest placement of agency media material in memory, providing hundreds of photos, thousands of feet of TV film, and "reams of copy" to local newspaper, radio, and television outlets around the world. The Voice of America provided live coverage of each stage of the mission, from launch to splashdown, in English, Chinese, Russian, Spanish, Portuguese, and Arabic.[28] American embassies in Eastern Europe assembled exhibits in their windows with pictorial explanations and a step-by-step schedule of the flight. As the crew completed stages of the mission, embassy staff would post announcements. The US Embassy in Sofia, Bulgaria, reported that the window display "drew exceptionally large crowds, despite cold and snow."[29] In warmer climes, inhabitants of Martinique followed radio coverage of the flight so carefully that consulate personnel reported walking down the street and hearing status updates from shopkeepers and acquaintances.[30]

After three days, Apollo 8 reached the moon. The crew fired the service module engine, slowing the spacecraft down just enough to put it into orbit around another celestial body, another first. On the fourth orbit, Borman rotated the CSM, tilting the nose of the capsule back toward Earth. The spacecraft's small windows framed the Earth seemingly rising above the lunar horizon. The view caught the crew by surprise, even though mission planners had anticipated that the moment would come.[31]

"Look at that picture over there!" Anders called out. "Here's the Earth coming up. Wow, is that pretty!"[32] With a Hasselblad camera in hand, Anders snapped a photo. Most of the photography scheduled for the flight focused on the moon. NASA needed detailed images of potential landing sites for future missions. As Anders watched the Earth rise above the lunar horizon, the black-and-white film magazine mounted to the camera's boxy body would not do. Only color film could capture the contrast of the gray moon and the bright-blue Earth that Borman called "the most beautiful, heart-catching sight of my life."[33] Anders called out, "You got a color film, Jim? Hand me that roll of color quick, will you . . . hurry up!" After a swift swap of film magazines, Anders started snapping again.[34]

He caught the Earth above the gray-chalky lunar horizon, the sun illuminating parts of Africa and South America. Eddying clouds suggested an alive, dynamic planet. *Earthrise*, as the photograph would come to be known, amplified the beauty—and rarity—of humans' home planet. Shortly after the crew splashed down a few days later, this photograph would grace the front page of newspapers around the world and become one of the most iconic images of the Space Age.[35]

The crew discovered that the food packed for them included a special feast, tied up in fireproof plastic green ribbons and labeled "Merry Christmas." Inside they found turkey with gravy and a fruitcake coated with gelatin to prevent crumbs from floating into the CMS's systems.[36]

At 9:30 p.m., during the second-to-last lunar orbit of the flight, the crew began their last broadcast from the moon. Taking a cue from Bourgin, the crew turned the camera toward the moon and took turns describing their perspectives. Borman called the moon a "vast, lonely, forbidding-type existence, or expanse of nothing, that looks rather like clouds," and Lovell agreed, commenting that "the vast loneliness up here of the Moon is awe inspiring, and it makes you realize just what you have back there on Earth." Anders added, "The sky up here is also rather forbidding, foreboding expanse of blackness, with no stars visible."[37]

Earthrise, taken by Apollo 8 astronaut William Anders. It is displayed here in its original orientation, although it is more commonly reproduced with the lunar surface at the bottom of the photo, December 1968. (NATIONAL AERONAUTICS AND SPACE ADMINISTRATION)

"We are now approaching lunar sunrise," Anders explained to the television and radio audiences around the world. "For all the people back on Earth, the crew of Apollo 8 have a message that we would like to send to you." Minutes before the spacecraft slipped behind the moon for the last time, the crew took turns reading from Genesis.

"In the beginning, God created the heaven and the Earth," Anders read.

Borman ended the passage, adding "and from the crew of Apollo 8, we close with good night, good luck, merry Christmas, and God bless all of you—all of you on the good Earth."

Television sets around the world glowed with the broadcast. One in four people on Earth—roughly a billion people spread among sixty-four countries—listened to the reading. Within twenty-four hours, recorded broadcasts of the address from the moon reached people in another thirty countries.[38] Audiences in North and South America as well as Europe tuned in live thanks to the recently launched Intelsat-III F-2 satellite. COMSAT put the satellite into operation a week ahead of schedule so that international audiences could follow the flight.[39] Even though it was costly, Portuguese government-controlled TV purchased Eurovision coverage of the mission. In villages throughout the country, people gathered around "TV sets in places open to the public," drawing connections between the audacity of Portuguese explorers and the Apollo 8 crew. But unlike ocean and polar exploration of the past, people had an immediate portal onto the action aboard Apollo 8, unfolding live.[40]

Added to the long list of "firsts" realized by the mission—such as traveling farther and faster than ever before—Apollo 8 achieved diplomatic "firsts" in the Communist Bloc. According to the USIA report to Congress, a Eurovision-InterVision hookup enabled live and delayed television transmissions to reach audiences in the Soviet Union. In Yugoslavia, for instance, millions of people watched live coverage of the flight. A local commentator provided narration over the Eurovision feed. Complementing the live coverage, the Yugoslav national television network broadcast a documentary made in collaboration with the USIA, depicting Yugoslavs visiting the Kennedy Space Center and speaking with Apollo astronauts.[41]

At first, Frank Borman felt skeptical about including heavy television equipment on missions because weight and time were at a premium. But the broadcast, and world reaction, changed his mind. He reflected that it was "probably [the] most important part of space . . . in view of [the] impact on people of the world."[42]

Reactions to the telecast were unprecedented, and the USIA won a significant public diplomacy victory with the carefully chosen, inclusive wording of the Christmas Eve address.[43] A BBC correspondent commented that the reading "struck on instantly as a stroke of genius."[44] In Latin America alone, 1,353 stations carried the VOA broadcast, breaking records. Even Radio Havana picked up VOA coverage, an anomaly for the official Cuban-government-run station known for transmitting programming created by the North Vietnamese, North Koreans, and Russians. The station cheered the mission as "a total success."[45] Borman received some 100,000 letters of appreciation for the Christmas Eve broadcast from around the world, and only 34 letters of complaint.[46]

The next day the front page of the *New York Times* carried an essay by Archibald MacLeish: "To see the earth as it truly is, small blue and beautiful in that eternal silence where it floats, is to see ourselves as riders on the earth together, brothers on that bright loveliness in the eternal cold—brothers who know now they are truly brothers."[47]

Other public intellectuals reflected on the larger significance of the flight. Science fiction author Arthur C. Clarke believed that "the Apollo 8 mission marks one of those rare turning points in the human history after which nothing will ever be the same again. . . . We no longer live in the world which existed before Christmas 1968." Dr. Isidor I. Rabi urged people to focus on how Apollo 8 represented "the cooperation of hundreds of thousands of people over a period of years" instead of being concerned about the cost of the program. It reflected the "profound desire of mankind to prove to itself that it had the knowledge and the ability to overcome its earthbound limitations."[48]

Just before dawn on December 27, Apollo 8 splashed down in the Pacific Ocean. At Mission Control, miniature American flags started waving, cigars were lit, and both relief and exuberance overtook the usually sedate room. Shortly after the crew arrived at the

USS *Yorktown*, stationed a few miles away, live television recorded them answering a telephone call from President Johnson. During the brief five-minute call from the Fish Room—which would soon be renamed the Roosevelt Room by Nixon—Johnson told the crew, "My thoughts this morning went back to more than 10 years ago . . . when we saw Sputnik racing through the skies, and we realized that America had a big job ahead of it . . . it gave me so much pleasure to know that you men have done a large part of that job."[49] An hour earlier he had spoken with the crew's wives from the Oval Office, reflecting to them, "When sputnik came over the ranch many years ago, we had dreams of something like this but we never thought it could be so perfect."[50]

Johnson fielded praise and congratulations from leaders around the world, as was customary. "A great milestone in man's continued search of the unknown," wrote Ethiopian emperor Haile Selassie. The mission demonstrated America's "courage and the high level of scientific technology," noted Japanese prime minister Eisaku Satō. And Pope Paul VI expressed "thanks to God for the successful completion of the magnificent enterprise."[51] The US Embassy in Moscow reported receiving letters and telegrams of praise from enthusiastic Soviet citizens.[52]

The splashdown played live on Bulgarian television. The weekly publication of the Bulgarian Union of Journalists, *Pogled*, ran a long article illustrated with photographs provided by the US Embassy. Although praiseworthy, the article "somehow managed to avoid any reference to the fact that Apollo Eight was American." And at the Bulgarian deputy foreign minister's yearly reception for foreign correspondents on December 27, the conversation quickly turned to the Apollo 8 mission. The chief of the Press Department found a television for party attendees to view coverage of the splashdown. Guests spent much of the following hour surrounding the TV set, with the Vietnamese and Chinese correspondents positioned closest.[53]

International newspaper coverage was "almost embarrassing in treatment," a reporter based in Uganda mentioned. The headlines across the top of the *Argus*, Kampala's "normally staid publication," were larger than any in memory.[54] The embassy in Warsaw amassed hundreds of press clippings. "We were all with them during those five days," an article in Polish *Życie Warszawy* observed.[55] The top story in Madrid's *El Alcazar* on December 30 reflected that "there has been no attitude of exclusiveness in the whole thing—but the desire to let everyone participate." The tone of the coverage, the astronauts' "understated attitude and sense of humor," and the lack of "triumphalism" impressed correspondents and commentators throughout Madrid. And the USIA tracked the placement of 12,800 column inches of agency material in South African newspapers.[56] The *Times of India* dubbed Apollo 8 "the most magnificent achievement in space to date . . . decidedly the most daring adventure man has ever undertaken."[57]

American ambassador Walter J. Stoessel observed that "it is stating the obvious to say that, on the world generally, the impact of Americans winging around the moon and preparing to land on it has been great. In Poland, however, it has been immense." Stoessel found the appetite for Apollo-related material in Poland remarkable, and worth capitalizing on. The Polish press eagerly devoured Apollo diagrams, flight plans, and press photos within hours of their arrival at the embassy. This material then appeared in magazines, scientific journals, and newspapers throughout the country. Polish-language pamphlets on Apollo 8 quickly became the most popular in the embassy's history.[58]

In Hong Kong the *Star* newspaper cheered "our boy" in a front-page feature, adopting Bill Anders, who had been born in the city while his father was stationed there in 1933.[59] The government of Chad issued a bright-blue Apollo 8 commemorative stamp with a lushly colored Earth and the CSM setting off for the moon.[60] The crew received two offers for automobiles, one from the French race-car company Matra and the other from Iranian carmaker Peykan, "in gratitude for their courage and epic achievement."[61] A mother in Sogamoso, a small city perched in the Columbian Andes nicknamed the "City of the Sun," gave birth to triplets that night. She named them Frank Borman Aguilar, James Lovell Aguilar, and William Anders Aguilar.[62]

For Johnson, the mission set a new tone for the end of his presidency, mitigating the negative effects of the war and the challenges of the past year. A week after the mission, as he was preparing to leave office, National Security Advisor Walter Rostow encouraged him to send farewell letters to heads of state and chiefs of government, along with a copy of *Earthrise*. The photograph should be sent along with a personal card to "Mao, Castro, and all." As Rostow explained, "The idea is that the astronauts saw our planet as one world." LBJ agreed and told him to "get some personal touches in there."[63]

The photographic processing team at NASA Manned Spaceflight Center in Houston had been working around the clock. "I wanted to get this information out to the world," explained Richard Underwood, the technical monitor of Apollo photography. He painstakingly developed each film roll by hand, finding automatic processing too risky for such valuable negatives. Once the film was developed, technicians spread photographs from Apollo 8 out on tables. The crew, along with NASA staff, selected a set of images for public release. They cropped and rotated the photo that would eventually become known as *Earthrise*, making the planet larger and the perspective more familiar.[64]

Johnson's letter, composed by the State Department, stressed how Apollo 8 embodied the increasing interconnection of the world and the role that American science and technology played within that process. "As the enclosed photographs of our recent lunar flight suggest, this shrinking globe is rapidly becoming a single neighborhood," referencing the *Earthrise* image. "Countries are learning that we all must work together for common ends if any are to survive and prosper in the new world of interdependence which science and technology are helping to create."[65]

Yugoslavian president Josip Tito "examined with great interest the photographs" from Apollo 8 and expressed his agreement that "the countries of our globe are rapidly becoming a single neighborhood thanks to the extraordinary progress of science and technology." He explained how grateful he was for Johnson's "friendly attitude towards Yugoslavia and its people . . . with regard to the independence and economic development of our country."[66]

Amazed by the positive response, Johnson later recalled receiving a response from North Vietnamese leader "Ho Chi Minh thanking me for sending him this picture, and expressing his appreciation for this act. I think the appreciation of our space effort is universal."[67] *Earthrise* played an important role in visually symbolizing global unity, as well as creating an opportunity to demonstrate Johnson's message of "peace" before he left office.

In the White House East Room on the morning of January 9, 1969, or "Astronaut Day," Johnson stood before Borman, Lovell, and Anders, remarking that "the flight of Apollo 8 gives all nations a new and most exciting reason to join in man's greatest adventure." After praising the crew as "history's boldest explorers," he highlighted the international-relations dimensions of the US space program, referencing cooperative projects with more than seventy countries and efforts to expand international partnerships. He awarded Borman, Lovell, and Anders NASA Distinguished Service medals, the highest awards bestowed by the agency.

In return, the crew presented the outgoing president with two gifts. The first was miniature copies of recent international space treaties that the astronauts had carried to the moon. Presenting the second, Borman joked, "Jim Lovell has a picture of [your] ranch I think you would like to have," as Lovell handed the Texan president the *Earthrise* photograph. Johnson laughed "delightedly."[68]

When the crew arrived at the Capitol, their next stop that day, they received a standing ovation. Borman, "smiling, relaxed and speaking from rough notes," addressed hundreds of legislators, Cabinet members, foreign diplomats, and all nine Supreme Court justices. "The one overwhelming emotion that we carried with us is the fact that we really do all exist on the small globe," he explained in his brisk twelve-minute address. When Borman kidded that "one of the things that was truly historic was that we got that good Roman

Catholic, Bill Anders, to read from the King James Version," laughter rang through the House chamber. He then looked toward the line of Supreme Court justices who had just ruled against reading the Bible in public schools and added: "But now that I see the gentlemen in the front row, I am not sure we should have read from the Bible at all." After the laughter and clapping subsided, he paraphrased Newton, explaining that the mission's success rested on the "shoulders of giants." Underscoring this message, he stated that "if Apollo 8 was a triumph at all, it was a triumph of all mankind."[69]

When Borman met with the United Nations Security Council, Secretary-General U Thant, a strong critic of US involvement in Vietnam, described Borman and his crew as "the first universalists." Borman responded, "Apollo 8 was a triumph for all mankind."[70]

Ensuring political advantage before excitement waned, US diplomats screened a film based on the Apollo 8 mission at embassies, in their homes, and in local theaters, universities, and other institutions. Embassy officials and USIA officers also fielded requests for the film from leaders such as the governor of the Bahamas and the president of Haiti, who held a special showing in his palace for government officials and "leading intellectuals." After the screening, *Le Nouveau Monde*, the Haitian-government-run newspaper, reported that the event removed "any doubt that might exist on the determination of Washington to stabilize and deepen its relations with the chief of the Haitian state." US Embassy staff interpreted the glowing news coverage of the event as a sign of President Duvalier's hope for increased economic assistance and a shift in US policy toward Haiti with the new US presidential administration.[71] The Soviet Embassy in the Democratic Republic of Cameroon even requested from their counterparts at the US Embassy the opportunity to view *Apollo 8*. In a confidential telegram to the secretary of state, the US Embassy in Cameroon proposed loaning the film to the Soviet Embassy.[72]

The US Embassy in Warsaw produced a Polish-language version of *Apollo 8—Journey Around the Moon*—and invited top-level officials to special screenings followed by receptions in the embassy library. As part of the event, American diplomats handed out a translation of

MacLeish's *New York Times* essay. These high-level events were soon followed by public screenings that attracted standing-room-only audiences, "the largest number of Polish guests ever to attend a single Embassy screening." Over the following weeks, hundreds of Poles crowded into the twice-daily screenings. Even a month into showing *Apollo 8*, audiences of 300–500 people arrived at the embassy to view the film each day. In some instances, Communist Party officials intervened, canceling planned screenings of the film at local organizations and clubs. But regular embassy and consulate screenings went uninterrupted.[73]

The US Embassy in Prague found a similar reception to the film, requesting three more prints from the State Department to keep up with demand. Enthusiasm for the American mission was so high that just a few months after the Soviet invasion of Czechoslovakia, a restaurant in Prague served an ice cream dessert named after Project Apollo.[74] A Czechoslovak youth newspaper, *Mlada Fronta*, extended an invitation to the Apollo 8 crew to visit the country. A USIA official stationed in Eastern Europe recommended a tour of Czechoslovakia, Romania, the Soviet Union, and Poland: "The psychological impact of such a visit to these countries would greatly advance our political interest and would reinforce our protestations of interest and support for the peoples of these countries."[75]

President Tsiranana of the Malagasy Republic hosted a screening of *Apollo 8* for ministers, Malagasy ambassadors, and other high-level officials. After the film, American ambassador David S. King presented Tsiranana with a gift from President Johnson: reproductions of *Earthrise* and a photograph of the moon. The evening, US officials reported, was an "outstanding success" in a country where the presence of a US tracking station always seemed at risk. Other showings of the film "generated a renewed and generally favorable image of our NASA tracking station." Overall, the space-themed cultural programs in Malagasy "have done a great deal to blunt leftist critics," the embassy staff concluded.[76]

A group of teachers from Poland's Poznan Polytechnic Institute who attended a screening of *Apollo 8* approached their university's

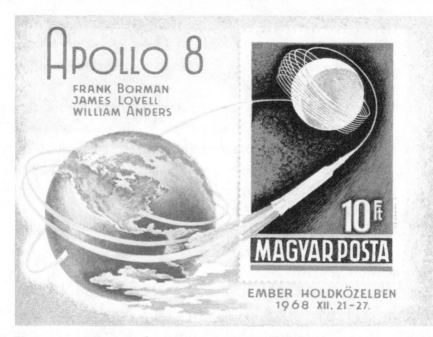

Hungarian stamp issued to commemorate Apollo 8. (AUTHOR'S COLLECTION)

rector about showing students the film. He suggested a two-day US space film festival open to all students. Although US consulate staff were happy to oblige, the idea provoked objections from Party members until organizers agreed on including Russian space films as well. An estimated two thousand students attended the festival.[77]

Less than a month after the Apollo 8 mission, Richard Milhous Nixon took the oath of presidential office on January 20, 1969. Borman had been surprised when he received an invitation to Nixon's inauguration. He had heard that the incoming president felt reticent about the space program. But when the Bormans found their seats on Inauguration Day, they were on the main platform behind the podium. It was at this moment that Borman said he discerned Nixon's genuine curiosity about spaceflight as well as the role that the astronauts would take on for the new administration.[78]

As Nixon stood at the podium, just after noon, with temperatures hovering around freezing, he immediately invoked space imagery:

"In throwing wide the horizons of space, we have discovered new horizons on earth." Project Apollo and the presence of Apollo 8 commander Frank Borman at the inauguration helped the president deliver his message to the American people and the world: the Nixon administration would renew domestic morale and national prestige.

Under "leaden skies," a seventeen-minute address followed. Nixon, "reaching widely for inspiration," according to *New York Times* reporter Robert Semple Jr., referenced poets, presidents, and the civil rights anthem "We Shall Overcome." But to Nixon, the main message of his address was clear: peace. "The greatest honor history can bestow is the title of peacemaker," he said. Drawing on the well-known story of Apollo 8, he illustrated his point: "Only a few short weeks ago we shared the glory of man's first sight of the world as God sees it, as a single sphere reflecting light in the darkness. As the Apollo astronauts flew over the moon's gray surface on Christmas Eve, they spoke to us of the beauty of earth." Quoting MacLeish, Nixon continued: "To see the earth as it truly is, small and blue and beautiful in the eternal silence where it floats, is to see ourselves as riders on the earth together. . . ."

As he ended the address, underscoring his peace theme, Nixon turned to the Apollo 8 story again: "In that moment of surpassing technological triumph, men turned their thoughts toward home and humanity—seeing in that far perspective that man's destiny on earth is not divisible."[79]

Although Nixon repeated MacLeish's line about being "riders on the earth together" again in his closing sentence, many people around the world and within the United States still felt deeply divided in January 1969. Peace seemed an unlikely prospect. Protesters lined parts of Nixon's inaugural parade route shouting "Ho, Ho, Ho Chi Minh, the NLF [Viet Cong] is going to win." The new president looked through his limousine window and spotted the Viet Cong flag lifted above the crowd of protestors. Before the procession reached the White House at 1600 Pennsylvania Avenue, members of the crowd threw rocks and beer cans. Never before had an incoming president received such treatment.[80]

When Nixon entered office, the Vietnam War dominated headlines. Three hundred American soldiers were dying every week. More than half a million US troops in Vietnam were fighting what was rapidly becoming a more and more unpopular war at home. Foreign policy, not domestic affairs, was Nixon's primary interest when he entered office, describing domestic policy as "building outhouses in Peoria."[81] He set his sights on ending the Vietnam War, improving relations with the Soviet Union, and diffusing tension in the Middle East. In the long run, he knew US-Chinese relations also had to improve. One of his first steps, and his first public action as president, was a tour of European capitals, where he hoped to informally discuss policy with foreign leaders and reinforce the stature of the United States. But before he traveled to Europe himself, he decided to send Frank Borman to pave his way. After the moon heroes' enthusiastic reception in European capitals, Nixon would follow days later while the celebratory hum still filled the air.[82]

In the warm afterglow of the Apollo 8 mission success, Nixon saw the benefit of identifying himself and his administration with America's space program. Although it had been proposed by one of his fiercest rivals—John F. Kennedy—Nixon shrewdly co-opted Project Apollo in support of his foreign relations agenda. When in the company of astronauts, "The color comes to [Nixon's] face and a bounce to his step," observed *Life* writer Hugh Sidey. It was a phenomenon that journalists nicknamed Nixon's "moonwalk." Nixon's fondness for the astronauts was akin to that of a father or the "all-star teammates he never knew." Sidey speculated that the Apollo astronauts distilled "what Nixon considers to be the best in this country . . . in his single-minded manner, he seems to be trying to assess and grasp the spirit of the astronauts, almost as if he wanted to bottle it and merchandise it."[83]

Nixon took to Borman immediately, making him the space program's "symbolic representative."[84] As Bill Safire, Nixon's speechwriter, explained, Borman was "the kind of well-organized, highly motivated, intelligent serviceman Nixon admired, the product of a mission the President identified with. . . ." He also exemplified "the

The Apollo 8 crew and President Richard M. Nixon before astronaut Frank Borman's European diplomatic tour, 1969. (NATIONAL ARCHIVES)

old-fashioned virtues, what the President thought of as the American Character."[85]

The Nixon administration planned for the Apollo 8 commander and his family to visit nine cities in eight European countries over the course of eighteen days.[86] The objective of the trip was clear: "to support US foreign policy." Secondary motivations included demonstration of the peaceful nature of the US space program and the interest in sharing knowledge far and wide. The tour should emphasize the "scientific, technological, and educational values of [the] US space program," the planning documents outlined.[87]

Ahead of Borman's tour, the secretary of state fielded requests from US embassies in Europe. "Belgrade, Vienna, and Berlin" were seen as ideal candidates "because of their proximity to Eastern Europe where a demonstration of American interest would be particularly appreciated." This type of tour could be especially fruitful in dramatizing US concern in the "post-Czechoslovakia period," reasoned USIA officials. The Department of State sent the request to

NASA Acting Administrator Thomas Paine even before the Apollo 8 crew had splashed down in Pacific waters, but the final selection of stops included Western European nations only.[88]

On February 2, a frigid Sunday, the Borman family departed Andrews Air Force Base bound for London. They "were never given a script," according to Borman. Instead, they received State Department diplomatic and protocol briefings and then "were simply told you are an ambassador . . . try to do your best." As he saw it, his main objective was to "cast the country in a favorable light," so he promoted the idea of "one world" united through the US space program during press conferences, lectures, film screenings, and dinners. Simon Bourgin and Nicholas Ruwe from the State Department's Office of Protocol escorted the Borman family, providing guidance along the way. US embassies in each city planned the party's itinerary, down to hotels, meals, and meetings.[89]

Shortly after his arrival in London, Borman, "completely self-possessed and in breezy form," lectured to seven hundred scientists at the Royal Society, spoke with the press, and toured London. When Susan Borman stopped in at Fortnum and Mason for a tea cozy, she was recognized and asked for her autograph.[90] During their audience with Queen Elizabeth at Buckingham Palace, Borman presented her with a framed reproduction of *Earthrise* as well as a scale model of the Saturn V rocket. Prince Andrew, who was soon to be nine years old, and his younger brother, Edward, immediately unwrapped and then claimed the model for themselves. After they took it apart, Borman helped them put it back together, a gesture that "seemed to disarm [his] hosts."[91]

"The problem is to get people thinking as Earthmen, rather than as Americans or Russians . . . because we really are," Borman explained on BBC's *Panorama*, a current affairs television program. The Ministry of Technology in London had a similar message for him. During his meeting with thirty top representatives from the defense sector, the Ministry of Technology, and universities, Borman was told "[we are] immensely proud of you—we regard you as a citizen of the whole world." At the US Embassy, he gave NASA achievement awards to the station manager of the NASA Communications Switching Station in London and the director of the United Kingdom's post office telecommunications system.[92] And at the press conference, Borman made a point of praising the British engineer who had developed the predecessor to the fuel cells used on Apollo 8.

"Admittedly," he later granted, "this was a little diplomatic buttering-up gesture, the kind that made the State Department happy, but I didn't mind. Those fuel cells were excellent." Borman's praise of the fuel cells also boosted his message: international cooperation in the peaceful exploration of outer space.[93]

Between giving a lecture at the Royal Society to press conferences and meetings, the Bormans attended *Canterbury Tales* at the Phoenix Theatre and dined at the Ivy Restaurant, a mainstay in London's Theater District. The manager surprised the family with a cake cut into the shape of a half moon. The entire restaurant started singing "For He's a Jolly Good Fellow." As Bourgin recalls, they arrived back at the Hilton Hotel "at 1 am full of champagne and cheer."[94] From London, the Bormans flew to France.

"Frank was the biggest thing in Paris since Lindbergh," Bourgin concluded after the two-day stop in the City of Light. French radio gave updates on his movements throughout the city, enabling thousands of people to gather at each of his stops, some getting "close enough to shake his hand, ruffle his close cropped hair and rip two buttons off his overcoat," reported the *New York Times*. Parisians gathered at the Élysée Palace shouted *"Nous vous adorons"* as he passed. As he left the palace after his meeting with French president Charles de Gaulle, a crowd blocked his exit. While the guards cleared a path for the vehicle, Borman hopped out of the car and shook hands. Later, another crowd of two hundred greeted him at the Hotel Crillon, an eighteenth-century luxury hotel at the foot of the Champs-Élysées. Closing in around him asking for autographs, the crowd cheered the space traveler when he finally let the embassy aides whisk him away. At City Hall, Jules Verne's grandson gave Borman a copy of *From the Earth to the Moon* and remarked on the similarities of the voyages. In return, Borman presented a framed engraving of the

Apollo 8 splashdown with a note that read "Jules Verne is one of the pioneers of the space age," a gesture that bound the two nations' histories together.[95]

The stakes were high for Borman's Paris visit. Nixon saw de Gaulle as key to next steps in his new foreign policy strategy: ending the war in Vietnam and establishing a relationship with Communist China. Given that France maintained diplomatic relations with North Vietnam and the People's Republic of China, Nixon reasoned that Paris was the logical location for establishing secret communication channels with Hanoi and Peking. But de Gaulle was a notoriously prickly figure. In 1966 he removed the NATO headquarters from France.[96] US ambassador Sargent Shriver concluded that "it's an undeniable fact that with de Gaulle as with French scientific elite and the public at large, Borman's visit was a very good thing for US prestige in France."[97]

In France as in England, Borman had an internationalist message: "I hope this will blend people of the world together so that we will all begin to look at ourselves as Earthlings, rather than as Germans, the Dutch, or Americans. . . . " In Bonn, after an *Apollo 8* film showing in the Beethoven Hall to a crowd of fifteen hundred students and government officials, he made a very similar point: "I believe this research will teach us that we are first and foremost not Germans or Russians or Americans but earthmen."[98]

The next stop "paid public relations dividends in this country beyond our expectations," reported US Embassy officials in Brussels. Enthusiasm for the Apollo 8 flight in Belgium had been so heightened, even among critics of the American government, that embassy staff "looked forward to Col. Borman's visit," expecting that it would "extend and enlarge upon the favorable aura created by the flight itself." Not only did the visit affect formal US-Belgian relations; it also paid "exceptional dividends in terms of . . . the Belgian attitude toward things American." The anti-American slogans that usually appeared before dignitaries' visits were nowhere to be seen.[99]

The Borman family, "so eminently marketable," impressed the Belgian press and public. News coverage of the visit was "unvaryingly

Apollo 8 astronaut Frank Borman gives a lecture in Germany, 1969.
(National Archives)

and, in some cases, extravagantly laudatory." The US Embassy observed that leftist media must have "recognized that any critical shafts
directed at the skipper of Apollo 8 would be poorly received here."
As Bourgin described it, the Brussels press conference was "fairy-tale
like," without "one sour note." Borman even received a standing ovation after one of his public lectures, a rarity in Belgian culture.[100]

Queen Fabiola and King Baudouin, whom the State Department
labeled a spaceflight enthusiast, hosted the Borman family at the palace for an intimate dinner on February 9. After the meal, the king
invited his staff and their families to view *Apollo 8* with his American
guests. As at all screenings of the film, Borman provided commentary
and answered questions. At similar events throughout Brussels, US
Embassy staff were impressed with how Borman projected "both an
extraordinary knowledge of his field, a willingness to get down to detail on all questions, and an innate modesty." When he did not know
the answer to a question, he would respond with "that is somewhat

beyond my field of competence, but I will get the information for you." According to Bourgin's report, the "visit helped put many Emb[assy] officials—and certainly USIS—in working relationship with Belg[ian] counterparts whose closeness made almost simultaneous gearing up for Nixon visit remarkably smooth."[101]

A winter storm grounded the party in Brussels. Instead of waiting for the icy runways to clear, the Bormans boarded a train for The Hague, the next stop on their itinerary. And although a member of the US Embassy staff overheard someone shout "Yankee, go home" when the family arrived in the Netherlands and the press found the shifts in schedule frustrating, on the whole the embassy deemed the visit another "solid success."[102]

At The Hague Congress Building, an imposing modernist structure, Borman screened *Apollo 8* and lectured to an audience of 550, including Queen Juliana, Prime Minister de Jong, other high-ranking officials, scientists, and the public. Standing in the center of the stage, without the aid of a lectern, Borman held a microphone in his hand, "self-assured and unassuming, relaxed and wide awake," as he described his voyage around the moon. When he presented the framed copy of *Earthrise*, he modified the joke he had told President Johnson: "And here I have for you a photograph of the Netherlands." The *Nieuwe Rotterdamsche Courant*, the Dutch equivalent of the *New York Times*, was impressed by the sizable audience and conjectured that "people did not only come to listen to Borman, but also to experience his presence; to be personally involved, more directly yet than by looking at television, in the unique experience which he personifies."[103]

Ahead of his meeting with Pope Paul VI, Bourgin showed Borman a "fancy-language talk that Washington suggested he use in [his] address." Borman replied he "better use his own improvised remarks," quipping that he read that the pope thought the Apollo 8 crew "were very simple men" and he "wouldn't want to disappoint him. . . . "[104] But Borman was wrong about the pope, who personally took the initiative to schedule a longer meeting with the astronaut. He had been a space enthusiast for years. NASA cultivated a relationship with the

Vatican throughout the 1960s, hand-delivering photographs from the Ranger probe and inviting a papal delegation to the Apollo 8 launch.[105]

A Sunday-morning audience in the pope's private library was followed by a tour of the Vatican Observatory in the Alban Hills, south of Rome, and dinner. Borman presented a papal medal that the Apollo 8 crew had carried around the moon, and Pope Paul VI gave him rare copies of the Bible. The Apollo 8 crew's Christmas Eve reading from Genesis "was a big hit in Rome," recalled Borman. The pope praised the flight, noting that "for that particular moment of time, the world was at peace."[106]

In Italy a forty-five-minute television special replaced the major weekly news program. It not only captured the attention of ten million viewers, but according to Bourgin, it was also "quite a thing, considering Italian Communist Party Congress, [the] Berlin crisis, Arab-Israel, etc."[107]

In Spain, Susan Borman met with sixty members of the press in the *salon rojo* (red salon) at the Hilton Hotel in Madrid. She was asked a range of questions, primarily focused on how she felt during the flight. When a reporter wanted to know if the Apollo 8 flight had changed her husband, she replied that "he returned more humble. Only three men have seen the world so small and you can't have that experience and not feel humble and insignificant." Mirroring the messaging of the tour, Susan Borman expressed how women throughout the world could relate to how she felt during the flight: "A woman who loves her family—her husband and children—has shared similar anxieties in one way or another."[108]

Borman placed a wreath at the statue of Christopher Columbus, symbolically linking not only space and ocean exploration but the United States and Spain as well, much like other diplomatic tours had done.[109]

One of the most important takeaways from the tour, according to Bourgin, was that the US underestimated the impact of Apollo 8. "It's still going," and if "properly exploited," according to Bourgin, it's "still [the] greatest source for policy support abroad." Borman noted of the diplomatic tour that he found "extreme identification of people in all walks of life in Europe with our flight. They were very well informed about it and looked on us as representatives of Earth."[110]

In addition to incorporating certain themes and phrases into public diplomacy material, USIA officials also monitored the adoption of these rhetorical devices by the audiences they hoped to influence. When judging the impact of Borman's tour of Europe, Bourgin noted positively that the "vocab [was] thoroughly absorbed."[111] A foreign leader's inclusion of the phrase "for all mankind," or a newspaper article that described a space feat as part of a greater good, proved to many USIA officials the effectiveness of USIA and State Department programming.

President Nixon began his eight-day European trip on February 23, immediately following Borman's tour, with an itinerary that loosely mirrored the astronaut's travels: Brussels, London, Paris, Bonn, Berlin, and Rome. He explained he "wanted to show the world that the new American President was not completely obsessed with Vietnam, and to dramatize for Americans at home that, despite opposition to the war, their President could still be received abroad with respect and even enthusiasm."[112] Reminiscent of Kennedy's trip to France on the heels of the *Freedom 7* exhibit in 1961, Nixon saw a European tour at the beginning of his presidency as a way to present himself as a master statesman on the global stage.

Ahead of the trip, Nixon expressed worry that protests might erupt in Europe, prompting negative press at home. He wanted Americans to see how respected and experienced their new president was when he traveled abroad. And although he hoped to correct the impression that he was obsessed with Vietnam, it was on the top of mind during the transatlantic flight. Dressed in a maroon smoking jacket, he sat alone for most of the flight, reviewing briefing material. He took one break, calling over National Security Advisor Henry Kissinger. The day before, on February 22, North Vietnamese forces had begun an offensive on South Vietnam from sanctuaries in Cambodia. The president instructed Kissinger that it was time to plan a secret bombing campaign, even though Cambodia was a neutral country. Nixon believed that such a campaign would not only send the message to the North Vietnamese that the new American president was a more assertive, bellicose adversary than Johnson; it would also strong-arm them into peace negotiations.[113]

When *Air Force One* touched down in Brussels, King Baudouin greeted Nixon at the airport, reflecting that "during this year, which will perhaps be that of man's first landing on the moon, we are more than ever conscious of the gulf between the wonderful possibilities open to us and the obligations which burden the world because of war, want, injustice and inequality." The next day at the North Atlantic Council, the assembly of NATO's permanent ambassadors, Nixon quoted MacLeish in his remarks. "We are all 'riders on the earth together'—fellow citizens of the world community," he recited, while also emphasizing the importance of strong alliances.[114] He posed questions about the future direction of NATO, hoping to demonstrate his appreciation of how the organization's priorities had evolved since World War II and his willingness to engage with European allies: "The ties that bind our continents are the common tradition of freedom, the common desire for progress, and the common passion for peace."[115]

While the public-facing Nixon championed "peace" and Apollo 8's lessons of unity and connectedness at NATO, his top advisors gathered at Brussels Airport. They met on *Air Force One*, sitting on the tarmac, because aboard the plane they could form a strategy for bombing Cambodia without the threat of electronic surveillance. The campaign, which would be named Operation Menu, was delayed at first out of concern that the administration would not be able to keep it secret. But by mid-March, after he returned from Europe, Nixon ordered an immediate attack. Over the ensuing fourteen months, B-52s dropped 110,000 tons of bombs, destroying property and food supplies, killing troops and civilians alike, and turning more than 100,000 Cambodians into refugees. The campaign would not have the effect Nixon and Kissinger had intended. It did not speed along peace negotiations, and it would eventually undermine Nixon's public-facing strategy of "peace with honor."[116]

The rest of Nixon's European tour was generally considered a public relations success, even if the long-term impact proved intangible. Stops in London, Bonn, Berlin, and Rome received positive coverage in the international and domestic press. Nixon believed that his meetings with de Gaulle were the "high point of the trip." The two leaders agreed on the need for détente with the USSR and diplomatic relations with China. Nixon asked for de Gaulle's opinion on the Vietnam War and indicated an interest in initiating direct conversations with Hanoi, a message that he "felt confident . . . would be passed to the North Vietnamese Embassy."[117] Although Nixon attempted to portray himself as the "peacemaker" in public and augmented this message by referencing the Apollo program, his private actions like Operation Menu would undercut this image. He would ride the political momentum of Apollo's popularity again in August, when the first lunar landing provided him an even more efficacious opportunity to advance his new foreign policy agenda.

8

MAKING APOLLO 11 FOR ALL HUMANKIND, 1969

One of the benefits of the space program is that we were getting favorable publicity rather than the negative publicity of Vietnam.

—APOLLO ASTRONAUT FRANK BORMAN

"The moon landing will be such a massive achievement, and attract such wide regard and admiration, that to blow a horn about it could hurt the US abroad," warned John E. Reinhardt, an assistant director of the USIA, to public affairs officers stationed around the world. Drawing on his extensive experience, he cautioned public diplomats about how they should frame the first moon landing. It would be necessary to strike the right balance between nationalism and globalism to advance US foreign relations interests. Described as "cautious, courteous, intelligent, and quietly modest," Reinhardt labeled himself as a person who "believes in the power of ideas." In 1956, the year of the Montgomery bus boycott, Reinhardt became one of the few—"a handful . . . like five"—African American foreign service officers in the USIA. He had spent years on the ground as a foreign affairs officer first in Japan, then the Philippines and Iran, before becoming the ambassador to Nigeria. He knew firsthand why "in today's increasingly interconnected world, where regional issues quickly transform into global challenges, the value of public diplomacy has never been greater."[1]

It was already well-known that Apollo was an American program, Reinhardt noted. Emphasizing that point would be counterproductive: any "self-congratulatory, bragging . . . , could only irritate people abroad and detract from their otherwise sure-to-be-favorable impressions of the event." He clarified that he was not saying that USIS posts should do nothing about Apollo 11, the first moon landing mission, scheduled for July 1969. "The caveat," he explained, "[was] about tone." Instead, "Set the entire feat in perspective and interpret its significance to mankind," he suggested.[2]

An episode in Kenya that spring substantiated Reinhardt's guidance. At screenings of the documentary *Apollo 8: Journey Around the Moon*, a public affairs officer watched the audience's reactions closely. "When the narrator intoned words like 'product of American skill and American sweat' or the camera panned to a close-up of the flag or the letters 'USA' on the space craft an embarrassed laugh would run through the audience," he reported. This critical commentary captures the growing antipathy toward demonstrations of American power and might among international audiences. And the report evinces how public diplomats fine-tuned their messaging and framing of the space program based on reactions on the ground. The agency's objective was to draw the audience in, to make them identify and empathize with American spaceflight, and in turn the United States. Any emphasis that Project Apollo was an American accomplishment was coming across as chauvinistic. As NASA edged closer and closer to landing humans on the moon, USIA material left behind images and messages of American dominance in science and technology, and instead focused on the deeper implications of the lunar program on humanity and human experience. If Project Apollo was going to advance US foreign relations interests, it would do so by being more inclusive, by presenting Apollo as an American-led achievement of all humankind.[3]

In January 1969, after the stunning success of Apollo 8 in December 1968, NASA announced the crew selection for Apollo 11. The mission would likely occur in late July or August, just six months away.[4] Apollo 8 commander Frank Borman profiled each of the Apollo 11 astronauts for President Nixon, consolidating their biographies and capturing their personalities in a few short phrases. Mission Commander Armstrong hailed from Ohio, flew combat missions in the Korean War, then became a test pilot for NASA before joining the astronaut corps and flying on Gemini 8. Borman described him as "quiet, perceptive, [a] thoroughly decent man, whose interests still turn to flying." Lunar module pilot Edwin "Buzz" Aldrin was "very athletic, aggressive, hard charging . . . [, with a] serious personality." Born and raised in New Jersey, Aldrin also flew in the Korean War before becoming an astronaut and flying on Gemini 12. His PhD research at MIT contributed to NASA's rendezvous techniques. Command module pilot Michael Collins was the best handball player of the whole astronaut corps, Borman claimed. Born in Rome, Italy, Collins came from a military family, flew on Gemini 10, and was "in some senses skeptical . . . more inclined toward the arts and literature rather than engineering."[5]

But before Armstrong, Aldrin, and Collins traveled to the moon in July, two more Apollo missions had to prove that NASA was ready. First, Apollo 9 in March 1969 tested the lunar module (LM) in Earth orbit. The crew qualified both the LM and CSM for lunar operations by conducting a complete rendezvous and docking sequence, an inter-vehicular crew transfer, a CSM/LM consumable assessment, and a spacewalk. The ten-day flight proved that NASA was ready for Apollo 10, the next step: a "dress rehearsal" of the moon landing that May.[6] Meanwhile, the crew of Apollo 11—Armstrong, Aldrin, and Collins—spent hundreds of hours in simulators and hundreds more being fitted for space suits and enduring high G forces in centrifuges, while also engaged in public relations, among other preparations. Collins recalled being "busier than I had ever been in my life."[7]

While the astronauts trained and NASA prepared, the USIA, the State Department, and the White House planned the largest public relations campaign in world history. USIA Science Advisor Simon Bourgin praised "Colonel Borman's good sense and aplomb in public affairs" after his tour of Europe, and he assured his colleagues that Apollo 11 commander Neil Armstrong "is of a similar stripe." But they were "uneasy" about not planning formal and extensive programming for what they predicted would be "the sensation of this or any other century."[8] The moon landing "offered the [USIA] its greatest opportunity in its 16-year history to inform the world about America's scientific genius, industrial and technological skills and the personal courage that made it all possible," as USIA staff saw it.[9] In order to make the flight an effective instrument of US foreign relations, government officials downplayed nationalistic jargon, stressed that the mission was "for all humankind," and established an infrastructure to encourage global "participation" in the flight, employing messaging that linked unity and progress with American science, technology, and global leadership. This proactive and deliberate global public relations campaign lay at the heart of how and why the lunar landing touched the lives of so many and why it became such a global phenomenon with a lasting legacy.[10]

Project Apollo had already offered many public diplomats a unique opportunity to connect with foreign leaders and publics who were otherwise critical of the United States and unreceptive to many other types of diplomatic gestures. When Nixon became president, the situation in the Middle East was volatile. For the USIS post in Iran, "Space on the basis of popularity and broad interest [was] the number one subject that USIS [had] to offer in Iran."[11] A USIS post in Turkey reported that it was having "difficulty communicating persuasively with some of [the] important elite groups in Istanbul including those who control most public media in Turkey and generate much of [the] anti-American poison being spread around this country." Project Apollo missions were "luckily . . . one US activity in which these groups have displayed intense interest and this gives us an opening to them." Posts around the world wrote to the USIA headquarters in Washington, asking for help developing a plan to take advantage of the rare political opportunity that the first lunar landing offered.[12]

The USIA, recognizing that the upcoming Apollo 11 mission could be "one of the stories of the century," responded with a thorough examination of how the agency should cover the mission. In addition to striking the right tone, as Reinhardt instructed, public diplomats created a set of guidelines. Commercial and foreign media would cover the mission in great detail, and they would look to the USIA for press briefings and other material. This is where the agency could play a major role in shaping the narrative. All agency material would (1) treat the mission as an accomplishment of all humankind, (2) emphasize that exploration is an essential component of great nations, (3) explain how Project Apollo was built on the innovations of scientists from around the world, (4) identify the astronauts as "the envoys of mankind in outer space," (5) describe Apollo's contribution to progress and solving problems on Earth, and (6) avoid highlighting the fact that the mission was being undertaken by the United States. Like the public diplomacy framing of American space exploration throughout the 1960s, Apollo 11 programming emphasized values of universalism and progress through science.[13]

In February 1969, NASA Administrator Thomas Paine established a Symbolic Activities Committee to ensure that the symbolic gestures carried out on the mission reaped the greatest political rewards for the Nixon administration. The committee coordinated and assessed recommendations from the USIA, State, members of Congress, and the White House.[14] One idea that did not make the cut was sending an international volume of commemorative poetry to the moon. The United Nations, on behalf of the United States, would invite each country to select a poet to compose a new work specifically for the landing. Each poem would be printed in its original language "to avoid the appearance of disparaging any culture." The idea behind the volume was that it would "reflect the richness and diversity of world culture as well as its common longing for peace and brotherhood, at the beginning of the extra-terrestrial era of man." Not only that, but "several obvious benefits would also accrue to United States foreign policy by this generous gesture." As Armstrong placed the bound volume on the lunar surface, the scene would be broadcast on worldwide television: "The United States will demonstrate dramatically that it wishes to involve all the world in the historical landing."[15] Another idea that was overruled involved depositing a microfiche copy of the basic documents of all major religions, also with the intention of sending an item of "significance for all mankind."[16]

Whether to raise an American flag on the moon was a less clearcut issue, so much so that a highly contentious debate broke out in Washington throughout the spring. In 1967 the United Nations Outer Space Treaty codified that no nation could claim sovereignty of the moon. The State Department warned the Nixon White House that raising an American flag on the moon might suggest "conquest and territorial acquisition" and could spark international controversy. Instead, State recommended that an American flag adorn a plaque, "providing a visible symbol of U.S. accomplishment while avoiding the possibility of being misconstrued." This would be the best course for taking "full advantage of this accomplishment both to enhance our posture abroad and to encourage other countries to further identify their interests in the exploration of space with our own."[17]

Numerous politicians echoed the State Department's sentiments, including Senator Charles Mathias Jr. (R., Maryland). He made the case that Armstrong and Aldrin should carry miniature flags from every nation: "We must signal clearly to the world that Apollo 11 will carry through space not only America's pride of accomplishment, but also America's bright offer of hope and progress for all the world."[18]

The USIA weighed in, arguing that the American flag patches sewn on the astronauts' space suits and the flag at the base of the lunar module were sufficient national identification. To "neutralize the effect of the American flag," they recommend leaving a silver globe with the continents in relief or a box filled with soil from every continent on the moon, or even casting the Earth's soil "symbolically over the lunar surface." The agency also proposed carrying to the moon and back miniature flags from other nations, which could become presentation items to chiefs of state.[19]

Members of Congress vehemently rejected the idea of a UN flag. Indeed, Congress threatened to cut NASA funding if a UN flag became part of the Apollo 11 mission. Because US taxpayers funded Project Apollo, Congress argued, the astronauts should plant an American flag, not a United Nations flag, on the moon.

On June 10, 1969, roughly a month before launch, NASA announced that Armstrong and Aldrin would raise an American flag, and only an American flag, on the moon. Later that day, the House of Representatives approved the NASA appropriations bill, but only after amending it with the following provision: "The flag of the United States, and no other flag, shall be implanted or otherwise placed on the surface of the moon, or on the surface of any planet, by members of the crew of any spacecraft . . . as part of any mission . . . the funds for which are provided entirely by the Government of the United States."[20] Astronaut Michael Collins agreed with the final decision: "Since the landing was being financed by the American taxpayers alone, I felt they were entitled to one show of nationalism amid the international totems."[21] His comment highlights just how international in tone the first lunar landing had otherwise become.

In late June 1969, NASA invited heads of state to compose messages of goodwill, which would be shrunk down and then etched onto a small silicon disk the size of a fifty-cent piece. Armstrong and Aldrin would deposit the disk on the moon along with other symbolic objects. Many of these messages expressed sentiments that reflected the inscription on the plaque designed for the lunar module along with hope for a new era of peace on Earth. "May the great achievements of space research inaugurate an era of peace and happiness for all mankind," wrote Kristjan Eldjarn, president of Iceland. "I fervently hope that this event will usher in an era of peaceful endeavor for all mankind," added the prime minister of India, Indira Gandhi. President Ibrahim Nasir of the Maldives wrote that "this message of peace and goodwill from the people of Maldives came with the first men from planet Earth to set foot on the Moon."[22] NASA Administrator Thomas Paine believed that this effort "greatly enhances the sense of international participation in the Lunar Program."[23]

The Symbolic Activities Committee sent its final recommendations to President Nixon in early July, shortly before the mission. The package included the plaque mounted to the leg of the lunar module *Eagle*, the silicon disk, fifty small US state flags, a United Nations flag, small flags of each of the countries that belonged to the United Nations, as well as a large American flag to plant in the lunar soil.[24] During the flight the inclusion of international flags on the mission made headlines. A story on the Ghanaian flag taken to the moon, for instance, made the front page of a major newspaper in Accra.[25] The Apollo 11 crew also carried mementos from the three astronauts and two cosmonauts who had perished on earlier missions. When astronaut Frank Borman visited the Soviet Union shortly before the mission, he was given medals from Vladimir Komarov and Yuri Gagarin to include on the upcoming lunar flight.[26]

Although not planned by the Symbolic Activities Committee, the Apollo 11 mission emblem adopted similarly all-encompassing imagery. The Apollo 11 crew decided unanimously to eschew tradition and opt for inclusion by leaving their names off the emblem. They wrote out "11" rather than spelling out the word "eleven" so people from non-English-speaking countries could understand it more easily. Collins traced a bald eagle from a National Geographic bird book and then added "a pockmarked lunar surface," with Earth in the distance. The eagle's talons looked too aggressive, so Collins placed an olive branch in them: a traditional symbol of peace. With the national bird at the center, the mission patch communicated subtly that it was a US program. The naming of the lunar module *Eagle* followed the emblem design.[27]

On May 18, 1969, NASA launched Apollo 10. Widely regarded as the "dress rehearsal" for the moon landing, Apollo 10 performed all the mission events that Apollo 11 would carry out in July, except for the descent, landing, moonwalk, and ascent from the lunar surface. The Apollo 10 crew brought the lunar module within 47,400 feet of the moon's surface, passing over the planned primary landing sites for Apollo 11. As they passed each site, the astronauts tested radar, made visual observation of lunar lighting, took stereo photography,

The Apollo 11 mission emblem. (Smithsonian National Air and Space Museum)

and performed the phasing maneuver using the lunar module's descent engine. The ascent engine performed an insertion maneuver, and shortly docked with the CSM in lunar orbit.[28]

The eight-day Apollo 10 mission also served as a dress rehearsal for the USIA's global public relations campaign planned for Apollo 11. The television broadcasts from the mission broke records in Chile, surpassing any other single broadcast in the country's history. Other newspaper coverage commented on the absence of American gloating. London's *Sunday Telegraph* mentioned approvingly that "we have been spared any trace of propaganda."[29] As M. V. Kamath of the *Times of India* described it, "The Apollo 10 astronauts swept into the lunar orbit yesterday and beamed back the first colour telecast of the moon in sunshine and earthshine and the pictures were stunningly beautiful . . . the details were so sharp and distinct that one almost seemed to peer over the brims of some of the craters oneself."[30]

For the Voice of America (VOA), Apollo 10 acted as a test run for determining staffing needs and language service requirements. Radio coverage of Apollo 10 also promoted the upcoming Apollo 11 mission and offered free giveaways: a NASA color fact sheet and the Raytheon Corporation's "Mission Analyzer," both detailing the stages of the Apollo 11 mission. By June, the agency had already received fourteen thousand letters requesting the items.[31] It quickly became apparent that the scale of Apollo 11 would not only outstrip Apollo 10 but that it would be larger than any other projects undertaken in VOA's twenty-seven-year history.

By June, prelaunch promotion programming was in full swing. VOA supplied foreign national networks and radio stations with more than five thousand individual programs. As part of the preparatory activities, the VOA asked posts to mount publicity campaigns for the upcoming Apollo 11 coverage by putting up posters and placing advertisements in local press and magazines. USIS Singapore designed its own posters encouraging people to "Follow Apollo over VOA" and distributed them to radio shops, schools, community centers, and other locations. In the weeks before the flight, newspapers from Lagos to Kabul contained advertisements promoting Apollo 11 and VOA coverage. The BBC made an agreement to broadcast VOA coverage on its domestic service throughout the United Kingdom. In order to reach audiences in China, Korea, and Russia, the VOA planned satellite circuits to relay stations in the Philippines, Thailand, and Okinawa.[32]

In mid-June the Apollo 11 launch date was set. On July 16, just one month later, the United States would send humans to the moon. At this point, the astronauts moved into the Merritt Island crew quarters with, as Collins described, "my bottle of gin and my bottle of vermouth, and a heavy load removed from either shoulder. One month to get ready, one month working at maximum capacity with minimum interference."[33] The accommodations were simple: a living room, windowless bedrooms, a gym, sauna, dining room, kitchen, and briefing room. Each day they got up early, quickly ate breakfast, and then headed off to the nearby simulators to train. After dinner each night they reviewed phases of the mission before bed and then awaited another day of training.[34]

Also in mid-June, the USIA in cooperation with NASA opened an Apollo News Center (ANC) in Paris. The ANC supported foreign newspaper coverage of the launch, handled telephone and in-person inquiries, distributed printed material, and hosted media correspondents from countries where live television coverage was not available.[35]

Prelaunch coverage of Apollo 11 was particularly extensive because public diplomats recognized that commercial media would not focus on the mission until just before the launch. Together, the USIA and State Department produced a barrage of information programs to heighten anticipation and excitement for the mission, sparing no expense to take advantage of this public relations opportunity. The USIA alone spent millions of dollars on Apollo 11 programming, a large percentage of the agency's annual budget.[36] Leading up to the flight, the VOA broadcast hundreds of features and documentaries on space exploration in all regions of the world; films on space ran in hundreds of theaters and on television. The USIA also supplied material to support locally produced exhibits; in Bonn, for instance, the USIA provided material for two large German department store chains, which dedicated 180 window displays throughout the country to the story of Apollo 11.[37]

By July, all individual USIS posts became small-scale space resource centers. The USIA sent posts kiosk exhibits, space-food samples, pamphlets in various languages, more than 900 models of the Saturn V rocket and the Apollo spacecraft, 10,000 lunar maps, 275 lunar landing charts, 240 16-inch moon globes, and countless publications.[38] The USIS post in Taipei became a particularly key reference service for the Taiwanese media by organizing interviews with scientists, distributing pamphlets and articles, and acting as an Apollo 11 clearinghouse. The USIA Press and Publications Service created background articles, features, photos, and picture stories to distribute to foreign newspapers. The foreign media, according to USIA reports, picked up and published much of the material that the USIA news branch supplied. In Ethiopia the USIS post distributed astronaut

USIA Apollo exhibit in Kenya, 1969.
(NATIONAL ARCHIVES)

food, including soup, chicken, and crackers, for the news media to sample. The USIS post in Tunisia covered its facade with large-scale photo murals of astronauts, the moon, and Earth, in addition to a large sign that renamed the post "Space Information Center."[39]

Space souvenirs became an important part of the USIA's promotion of Apollo and its engagement with local populations. In downtown Belgrade, for instance, ten thousand people visited the USIS post, with the majority requesting Apollo 11 buttons. Each button depicted a graphic lunar module superimposed on the facepiece of an astronaut's helmet with the word "APOLLO" appearing in bold. The buttons came in the colors of the US flag—red, white, and blue—but they did not include any direct reference to the United States.

After the post's supply of buttons ran out, staff printed thousands of photos of the astronauts to give away as a substitute. The post in Warsaw distributed 100,000 buttons by July 21. A *Washington Post* correspondent reported that thousands of people wore the buttons around the Polish capital.[40]

Space exhibits reached millions more around the globe ahead of the mission. In Japan alone, nearly a million people visited the thirty-six Apollo 11 exhibits scattered around the country. For an exhibit in Delhi, the agency constructed a full-scale model of the lunar module, which drew an average of thirteen thousand visitors a day.[41] At the First International Fair in the Democratic Republic of Congo, the United States joined twenty-seven other nations from mid-June through mid-July. Project Apollo, the major focus of the US pavilion, was featured alongside a Department of Labor–sponsored exhibit on nation building. Attendance reached nearly 300,000 people, including Congolese president Mobutu Sese Seko, who visited the US display three times over the course of the fair.[42] During the Apollo 11 flight, locally trained Congolese narrators fielded questions from the audience. The Apollo portion of the pavilion included models of spacecraft, space food samples, and films on other missions, while the nation-building exhibit included displays of American poultry, wheat, and rice, promoting US spaceflight and industry simultaneously.[43]

In Southeast Asia another large-scale exhibit, the American Pavilion at the Djakarta International Fair, also focused on the Apollo lunar landing. A twenty-five-foot-high model of the upper stage of a Saturn V, scale models of spacecraft and launch vehicles, a space suit, photo transparencies, and films captured the attention of more than a million people, including President Suharto and other high-ranking Indonesian officials. The USIA had trained Indonesian college students and graduates to be guides, and a leading Indonesian science commentator gave two presentations on space exploration. Live radio and video telecasts were broadcast from the site to reach an even wider audience.[44]

The US pavilions in Djakarta and Kinshasa were split into two parts: half of the display was devoted to the American space program, and

USIA Apollo exhibit in Iceland, 1969. (National Archives)

the other half was dedicated to commercially oriented development exhibits. The second half of these pavilions relied on the sponsorship of US private firms. Both of these pavilions, like other American exhibits abroad, recruited and trained local students to work as guides. Apollo displays became a useful technique for drawing crowds to the pavilions and exposing foreign audiences to non-space-themed programming. Many public diplomats knew that spaceflight, especially in the summer of 1969, would guarantee large crowds. Pairing spaceflight exhibits with other content became a useful way to promote general US products and policies abroad.

American space documentaries played in movie theaters around the world, while television stations broadcast space-themed features in more than a hundred countries.[45] During the two weeks before the launch, a Santiago TV studio produced six shows featuring Project Apollo. A Spanish-speaking USIA information officer and the director of a local NASA tracking station appeared on the programs. The agency organized outdoor viewing areas, from public squares to sports stadiums, allowing people who did not have their own televisions to view the preflight screenings of films and television shows as well as the moon landing coverage. For example, the USIS post in Seoul set up a projection system with a 19-foot-by-16-foot screen fastened to a building and streamed USIA space films for ten days before the Apollo 11 flight. USIS Delhi showed the Apollo 10 film on an outdoor screen to more than 68,000 people during a series of evening screenings. The USIA also organized private invitation-only film showings for local leaders, journalists, and academics. Some of these were held in the homes of American ambassadors or USIS officers. Newly appointed Venezuelan president Rafael Caldera invited a public affairs officer to his mansion to screen space-themed films and offer detailed narration for his Cabinet.[46]

By 1969, the number of television sets around the world had grown twentyfold since the mid-1950s, reaching at least 150 million. NASA, the USIA, and other US government agencies had been planning for years to ensure that Apollo would reach these viewers.[47] In 1961, when President Kennedy proposed Project Apollo to Congress, he also asked the nation to support the development of a worldwide satellite communications network. The following year Kennedy signed a bill for a public-private satellite communication partnership, which would be overseen by the president.[48] The Communications Satellite Act of 1962 created the American Communications Satellite Corporation (COMSAT), which would construct, manage, and operate a commercial communications satellite system in cooperation with businesses and foreign governments. When it became clear that COMSAT could not turn a profit without larger government investment, the NASA contract for the Apollo communication service went to COMSAT instead of the Department of Defense. This decision provided important revenue for COMSAT and supported the extension of the network to Asia, Africa, and Latin America.[49]

In the early 1960s, the national television networks broadcast television programs that were sent by landlines or microwave networks or were replays of a tape or film relay. For instance, John Glenn's 1962 flight was shown live on American television sets, but audiences in Europe and other parts of the world had access only to recorded tapes and film.[50] In August 1964 COMSAT, in cooperation with ten other nations, formed Intelsat, an international organization to create a constellation of geosynchronous satellites. Nations interested in broadcasting Intelsat feeds could set up large Earth stations with dishes aimed at one of the satellites. This system allowed continuous feeds without the earlier issue of relays ending abruptly when a satellite in a lower orbit traveled beyond the horizon. In 1965 Intelsat launched Early Bird, a commercial geostationary satellite for transatlantic television transmission. Thanks to Early Bird, re-christened Intelsat I, North Americans and Europeans could watch Pope Paul VI's visit to the United States, coverage of the Gemini program, news, and sports.[51] Senator John Pastore (D., Rhode Island), chairman of the communication subcommittee of the Senate Commerce Committee, reflected that satellites such as Early Bird "can become the show window through which America can and will be seen throughout the world. And indeed, it will be the mirror of our own image."[52]

With the growth of the Intelsat network, live coverage of space-flight missions started reaching audiences in Asia as well as Europe. National networks ran many space-themed broadcasts because, according to one USIA assessment, science was viewed as politically neutral. In Indonesia, "space films [are] especially popular, probably a continuing hangover from days when that was the only material they dared use. [It is] still too soon for use of controversial political clips," stated a USIA report in 1967.[53]

By 1969, a series of Intelsat III satellites completed a truly global network just in time for the first lunar landing. Intelsat III F-3 was moved to geostationary orbit over the Indian Ocean. When one of the three satellites in the network failed, the Intelsat I satellite (Early Bird) was put back into operation for coverage of the mission.[54]

Press conference for Apollo 11, 1969. (NATIONAL AERONAUTICS AND SPACE ADMINISTRATION)

Putting the global satellite network into operation in time for the landing also reflected well on US prestige. The official newspaper of Ivory Coast, *Fraternité-Matin*, commented that "the very fact that they installed two communications satellites instead of one to allow 600 million television viewers to see man's first steps on the moon proves that the children of Uncle Sam will do nothing in secret which could compromise the progress of men on earth." Although unctuous in tone, the article highlights the larger ramifications of creating a global communications infrastructure in time for Apollo 11.[55]

While a physical communications infrastructure was being built up over the 1960s, the USIA fostered the careers of local spaceflight experts. USIS Belgrade arranged for Milivoj Jugin, a television journalist, to spend five days at Cape Canaveral and provided him with a documentary film team. Yugoslav television broadcast the film during the Apollo 7 launch in the fall of 1968, which quickly cemented Jugin's role as the country's NASA expert. The USIA arranged and paid for his subsequent trips to Houston, Huntsville, and space laboratories in California, making sure that he would be home in Yugoslavia in time to cover the moon landing on national television.[56] Erik Tandberg, a television personality from Oslo, had a similar story. After he became a technical consultant on space matters for NRK (the Norwegian Broadcasting Corporation), with US government support Tandberg traveled to NASA centers and met with personnel such as rocket pioneer Wernher von Braun. He took his firsthand experience touring aerospace centers in the United States on the air when he cohosted Norway's moon landing coverage.[57]

Aware that duplicating this widely available coverage would be of little political value to the United States, the agency instead focused much of its attention on reaching audiences outside of the established global communications network. USIA worked with foreign television networks to ensure that live coverage of the lunar landing would reach every potential TV set. In areas where live coverage was not available, the USIA shipped foreign television networks copies of TV clips of the major phases of the mission as well as a final wrap-up after splashdown.[58] In 1969 Venezuela did not have a satellite ground station to receive the live television coverage of the lunar landing. Private stations in Venezuela as well as Colombia invested $230,000 to rent a portable ground station, while the USIA, along with help from the US Air Force, transported the station and equipment to Maracaibo. This collaboration between private industry and the USIA ensured that 1,500,000 television sets in Venezuela and Colombia could pick up Apollo 11 coverage, making the lunar flight the first international event to be seen live by these audiences.[59] Israel's television service planned the most extensive programming in the network's brief history: a three-hour special on the historical, philosophical, religious, and technical facets of the moon landing. The network did not have access to the live satellite feed, so the USIA helped arrange a recording of the landing to be flown from London, ensuring a shot of the landing for Israeli audiences just a few hours after it occurred.[60]

As the agency was the first to recognize, "Obviously, the world would have learned about the moon landing without the USIA." The international press would have covered the story, run large head-lines and photographs on their front pages, and kept their readers informed. But it was the USIA's efforts, over years and years, that enhanced the impact of Apollo 11. They broadened and deepened the reach while cultivating worldwide attention ahead of the flight.[61]

9

ONE GIANT LEAP, JULY 16–JULY 24, 1969

For one priceless moment in the whole history of man,
all the people on this Earth are truly one.

—RICHARD NIXON, JULY 20, 1969

Nearly one million people descended on Florida's Space Coast in sweltering July heat and humidity for the launch of Apollo 11. Small towns dotting the coast swelled past capacity. By July 14, gas stations' and liquor stores' stocks had been drained. "Through the whole of Florida," the *Times of India* detailed for readers, "it is now impossible to obtain a car on hire; hotel reservations are about as scarce as moon dust." Many hotels purchased cots and even deck chairs to accommodate people who could not find housing, while others camped along the beaches and beside the road in tents and trailers. Cocoa Beach bars created two special drinks in honor of the flight: "the Liftoff Martini and the Moon lander, which consisted of vodka, crème de menthe, crème de cacao, soda and lime." Hundreds of European tourists arrived as part of a "moon shot tour" organized by Trans World Airlines.[1]

NASA hosted thousands of special guests; the agency's invitation to "the entire diplomatic corps" was accepted by almost every ambassador but Soviet ambassador Dobrynin. Former president Lyndon

Johnson arrived at the Cape along with Texas congressman Olin Teague and former NASA Administrator James Webb. Representatives, senators, governors, and members of Richard Nixon's Cabinet were also in attendance. Pulitzer prize–winning novelist Norman Mailer covered the launch for *Life* magazine, while actor John Wayne, comedian Jack Benny, television host Johnny Carson, and aviator Charles Lindbergh were among those spotted at the Cape.[2]

Protesters arrived too. Organized by the Southern Christian Leadership Conference (SCLC), a group of five hundred marched with mule-drawn wagons to demonstrate the discontinuity between NASA's high technology and the basic needs of American society. The protest received both national and international coverage, including an article in the Soviet newspaper *Izvestiya*.[3]

On July 16, the day of the launch, the Apollo crew woke at 4:15 a.m. After final physical exams they ate the customary prelaunch breakfast: steak, eggs, toast, coffee, and orange juice. Artist Paul Calle, who was part of the NASA art program, sketched them while they ate. Next it was time for suiting up before boarding a white transfer van for a seven-mile ride to launchpad 39A and the 320-foot elevator to the "white room." Slightly before 7:00 a.m. they climbed into the *Columbia* command module. With 210 cubic feet of habitable space, the cabin was roughly the size of three telephone booths. Armstrong sat on the far left, Collins was on the far right, and Aldrin took the middle seat. More than 500 switches, 24 instruments, 40 event indicators, and 71 lights were mounted in the cockpit control panel.

Columbia sat on top of a Saturn V. The giant rocket measured 363 feet tall, and 92 percent of its nearly 3,000-ton weight was fuel. During the next two and a half hours closed-circuit cameras monitored the spacecraft, and more than 450 engineers sat at their consoles in the firing room performing preflight checks.[4]

While a million people converged on the Space Coast to watch the launch in person, hundreds of millions more experienced it through television coverage. Walter Cronkite and the CBS *Man on the Moon* special mediated the way that much of America and the world viewed the first lunar landing. Nearly 94 percent of Ameri-

USIS Radio Prague at Cape Kennedy, July 1969. (NATIONAL ARCHIVES)

can households with television sets tuned in. More people watched CBS coverage than the two other national networks combined. The international audience also tuned in to CBS. From the BBC in the United Kingdom to the Panamanian television network to South Vietnamese television, all supplemented their own programming with CBS's coverage. A major feature of the program featured CBS correspondents stationed around the world conveying foreign reactions to Apollo 11.[5]

Known affectionately as "the most trusted man in America," Walter Cronkite, with his avuncular voice and calm delivery, became a national fixture in the 1960s and for all the launches since the early days of Project Mercury. From the CBS News Studio at Cape Kennedy, Cronkite told viewers that "some 4,000 members of the press registered here—press, radio, television, magazines . . . more I believe than for any other event in history. . . . Some one-fourth of those at least are from the foreign press representing over 50 nations." The largest representation by far came from Japan, with 120 journalists. USIA staff provided local support and translators, and also recruited members of the local foreign-language clubs to volunteer as escorts and interpreters.[6] Cronkite continued, "David Schoumacher is over at the press site with a report on that foreign press."

Schoumacher, a correspondent who regularly covered the space program, introduced two journalists "who had been covering space shots about as long as Walter Cronkite has." Ruggero Orlando from RAI, Italy's state broadcasting system, was the best-known television personality in Italy. Louis Deroche of Agence France Presse specialized in spaceflight, knew the astronauts personally, and was the first French person to set foot on the South Pole when he traveled to Antarctica to cover the International Geophysical Year.[7]

When asked if it was the biggest story in France as it was in the United States, Deroche replied, "Certainly, I think it is a big magnificent story all over the world."

Schoumacher followed up with Orlando: "Is this considered truly a feat of mankind, or do you resent the fact that it is an American accomplishment?"

"No, no, no," assured Orlando, "in Italy it is absolutely connected with the idea of universality."

"I think it truly is an international feat," Deroche chimed in.

Did the planting of an American flag bother them, Schoumacher asked.

"No, inasmuch as there is the other rather universally inspired declaration side by side [on the plaque], which remains linked to the descent stage of the lunar module. . . . In Italy, they are really anxious following every moment."[8]

The message for both international and domestic audiences was clear: Apollo was a global event. Television, radio, and newspaper coverage did not just focus on the astronauts' voyage to the moon; it also featured stories about people on each continent following the flight with rapt attention. When a family watched the television broadcast of Apollo 11, part of what they saw on their screens was

Launch of Apollo 11 on July 16, 1969. (NATIONAL AERONAUTICS AND SPACE ADMINISTRATION)

the faces of other people watching the flight in far-off countries. In this way, media coverage further amplified the sense that the moon landing was a globally shared experience.

At 9:32 a.m. more than 7.6 million pounds of thrust lifted the massive Saturn V rocket off the launchpad. The first two stages of the Saturn V produced enough energy in a few short minutes to power the whole of New York City for over an hour. When the roar of the engines reached the viewing stands, "The earth began to shake and would not stop." Norman Mailer described it as a "nightmare of sound," so powerful that it could be heard a hundred miles away.[9] The craft reached a height of over forty miles in under a minute.

Once the propellant in the first stage of the Saturn V—the S-IC—was expended, the stage fell away and splashed in the Atlantic Ocean. Next, the S-II—the second stage—fired for six minutes, pushing Apollo 11 into nearly orbital velocity, before falling away itself. It was then that the third stage—the S-IVB—burned its engine briefly, putting the Apollo 11 spacecraft in orbit, 118 miles above the Earth.[10]

As the astronauts orbited overhead, former president Lyndon Johnson left the NASA bleachers and joined Cronkite in the CBS studio. With sweat still collected on his forehead from the Florida heat, Johnson expressed the thrill of seeing a launch live for the first time, the sense of unity and possibility brought about by Apollo, and recalled lines from the 1958 Space Act that the intention of the American space program would be for bringing peace to all humankind:

> I don't believe there is a single thing our country does, our government does, our people do that has greater potential for peace than the space effort. As I walked out from the blastoff I saw that special section of Ambassadors there from all the nations of the world, all taking such great pride in America's effort, all entertaining such great hope for the success of this mission. And I recall that after Apollo 8 I sent to the leaders of the world a picture of the earth taken from that mission and the response was universally favorable, and hopeful, and they all expressed great admiration for our people.

Spiro Agnew and Lyndon Johnson watch the Apollo 11 liftoff on July 16, 1969. (National Aeronautics and Space Administration)

As the interview ended, Johnson stressed the importance of the openness of the program, of the decision to work with the media and put everything in front of the camera. This openness, in Johnson's view, reflected the strength of the US political system.[11]

The Voice of America (VOA) radio broadcast the launch live in fourteen languages, including Burmese, Georgian, Urdu, and Estonian. VOA correspondents around the world helped provide a constant flow of news to their audiences and communicated foreign reaction reports back to Washington.[12] Whenever notable mission events occurred during scheduled broadcasts, the VOA would interrupt planned programming to update the audiences on the astronauts' latest movements. The VOA instructed broadcasters "to avoid boasting and the excessive use of adjectives." Apollo 11 "was big enough to carry itself without being self-congratulatory in a way that foreign listeners might find distasteful." This approach was warmly received, with USIA Director Frank Shakespeare noting that "foreign listeners have applauded the tone as well as the quality of VOA Apollo coverage."[13]

As the astronauts made their three-day journey to the moon, television and radio stations around the world asked USIS officials to be guest experts on Apollo-themed broadcasts. Two USIS officers appeared nightly on each of the Colombia television channels. On one unfortunate evening, a USIS officer showed up to the studio in a space suit and almost suffocated on camera when the helmet stuck to the suit. In the Philippines a USIS Apollo specialist spent forty-six hours on Manila television. An Arabic-speaking diplomat in Algiers became a space commentator for that country's television network. In Chile a USIA officer along with a NASA tracking station director, both fluent in Spanish, appeared on television for hours during the flight to comment on the mission and to answer call-in questions. Another USIA officer became an anchorman for the thirteen-hour television broadcast in Hong Kong. In Ethiopia a USIA staff member discussed the landing in Amharic, and in Taiwan the US ambassador spoke to the public on the radio.[14]

During CBS's continuous thirty-two-hour moon landing coverage, Cronkite received reports from correspondents in London, Rome, Paris, Amsterdam, Manila, Tokyo, Saigon, Belgrade, Bucharest, Mexico City, Montreal, Lima, Buenos Aires, and the International Arrivals Building at New York's John F. Kennedy Airport. Requiring the largest foreign and domestic remote coordination effort in CBS history, the network had to set up a separate control room in New York to handle the increased number of television feeds from abroad.[15] Reporting from London, correspondent Mike Wallace explained to the record-breaking audience tuning into CBS that "the traditionally stolid British seem to be as captivated by the fact and implications of the flight of Apollo 11 as we Americans." The correspondent in France noted that Parisians had a mood of "relaxed anticipation" but that, "as everywhere else, the French press is dominated by Apollo 11, and what if Armstrong should encounter living beings." Daniel Schorr, a CBS correspondent in Amsterdam, noted that the national television network audience was estimated at 80 percent of the Netherlands' total population of twelve million. Schorr reported to audiences other evidence of Dutch enthusiasm for the flight: "Gas stations in Holland are distributing Moon maps, instead of road maps."[16]

CBS correspondents stationed around the world reinforced a vision of international enthusiasm for Apollo, a sense of shared experience and unity, and foreign admiration of the US space feat.[17] Audiences in Bucharest, one CBS correspondent explained, "have become 'space bugs' overnight. . . . The average Romanian thinks of Apollo 11 a little bit as his own personal adventure. . . . Apollo 11, they keep telling the visitor, is for everyone, for all mankind." He suspected that President Nixon's upcoming trip to Romania heightened the response to the flight and affected the government's tolerance of Apollo 11 coverage in the country. Immediately following the Apollo 11 splashdown, Nixon would fly there on his Operation Moonglow diplomatic tour. It would be the first visit of an American president to an Eastern European country since World War II. This diplomatic trip might have also inspired the correspondent's decision to describe Romania as a "maverick East European country." Apollo 11 gave the average Romanian "another reason for his feeling good about his increasing independence in the East." This was a comforting picture to paint for American audiences concerned about the grip of Communism on much of the world. The cold war contest for hearts, minds, and political alignment played out on television sets as Armstrong, Aldrin, and Collins approached the moon.

On July 19, the fourth day of the mission, Apollo 11 was less than ten thousand miles from the moon. Not long after the astronauts woke, the spacecraft entered the moon's shadow. With the sun's bright rays blocked, the sky no longer appeared black and nearly barren, but instead filled with stars. For the first time on the mission, the astronauts could recognize constellations. The moon became three-dimensional. "This cool, magnificent sphere hangs there ominously," wrote Michael Collins, "a formidable presence without sound or motion, issuing us no invitation to invade its domain." Armstrong called the view "worth the price of the trip."[18]

Mission Control reviewed the daily news from Earth for the crew: "First off, looks like it's going to be impossible to get away from the fact that you guys are dominating all the news back here on Earth. Even *Pravda* in Russia is headlining the mission and calls Neil 'The Czar of the Ship.'" Mission Control added that West Germany had already announced Monday would be "Apollo Day," schools in Bavaria were canceled, and TV sets were installed in public places around Frankfurt. In London, BBC considered measures for a special radio alarm system in case the timing of the moonwalk changed. Mission Control also updated the astronauts on their families, politics, and the latest baseball scores. Later, when Mission Control asked Armstrong to make an adjustment in the spacecraft, Collins joked, "The Czar is brushing his teeth, so I'm filling in for him."[19]

The time for lightheartedness would be brief. One of the riskiest maneuvers—lunar orbit insertion—was the next phase of the mission. A mistake or computer error could potentially ruin the mission, crashing the spacecraft into the moon or sending it on a trajectory toward the sun. Apollo 11 performed two engine burns perfectly. The first burn put the astronauts in orbit, and the other refined their trajectory, making their path tighter and more circular.

On July 20, after a restless night, Armstrong and Aldrin put on their extra-vehicular-activity (EVA) space suits. Bulky and pressurized, these suits acted like spacecraft, micro-habitats that would keep the astronauts safe during their moonwalk. Once dressed, they took their places inside the lunar module and sealed the hatch. The *Eagle* lunar module was utilitarian in every way. It needed to be as light as possible, so engineers dispensed with items that added unnecessary weight like covers over circuit breakers or even seats. Armstrong and Aldrin stood shoulder-to-shoulder in front of two triangular windows that flanked the control panels. Collins pushed a button, and the spring action in the docking mechanism pushed *Eagle* away from *Columbia* and on its way to the moon.

At 50,000 feet above the lunar surface, Aldrin and Armstrong began the powered descent. At this point, they were just over twelve minutes away from landing. *Eagle* flew horizontally over the lunar surface, legs first. At about 35,000 feet an alarm flashed. With subtle urgency in his voice, Armstrong told Mission Control: "Program

People in Japan watch television coverage of the Apollo 11 mission, July 20, 1969. (NATIONAL ARCHIVES)

Alarm." Later he confirmed it was the 1202 alarm, indicating that data was overloading the guidance computer. Mission Control assured the astronauts that they should go ahead with the landing, but almost immediately the alarm appeared again. With another reassurance, Armstrong and Aldrin continued. At about 7,500 feet the guidance computer pitched *Eagle* upright, allowing the descent engine to slow both the forward and downward trajectory of the spacecraft. Now Aldrin and Armstrong were just four minutes from touchdown. Aldrin called out, "Program alarm. 1201." It was the same issue, so Mission Control gave them a go-ahead again.

The moon was in the waxing crescent phase, with the sun sitting ten degrees above the horizon. At this angle, the sun's rays put the lunar topography into relief. Armstrong spotted the planned landing site: a broad crater surrounded by a field of massive boulders, some more than nine feet wide. There seemed no smooth surface to land. Armstrong put the spacecraft into manual control, pitching *Eagle* forward. Just 350 feet from the surface, it flew over the boulder field in search of a less dangerous landing site. The engines consumed almost all of the last remaining fuel. Less than 200 feet from the surface, Armstrong finally found a level, safe place to land in the Sea of Tranquility.

With less than a minute of fuel remaining before they would be forced to abort the mission, Armstrong set *Eagle* down gently, so gently in fact that *Eagle*'s legs, whose honeycomb structure was designed to compress at impact, remained long. This would make Armstrong's eventual step from the ladder to the footpad of the lunar module a bit of a leap. "Contact light," Aldrin told Mission Control once the long metal probes mounted to the footpad struck the moon. Armstrong shut off the engine. With the LM safely resting on the moon, Armstrong added, "Houston, Tranquility Base here. The *Eagle* has landed." It was 4:17 p.m. EST on July 20, 1969.[20]

In Rio de Janeiro, church bells started ringing throughout the city. Hundreds of Brazilians gathered at the Museum of Modern Art's "space center" stood in silence at first, then cheered "*bem-sucedido*" ("successful"). In Tokyo it was at 5:17 a.m. on a hot and overcast Monday morning. Cheers of "*Apollo juichi-go banzai*" (roughly translated as "Long live Apollo") were heard. Riders on the bullet train between Tokyo and Osaka followed the flight on the public-address system. In Nice, well-heeled guests had gathered at the Le Negresco, a palatial hotel on the French Riviera, for an elaborate gala in honor of the first lunar landing. Live news coverage of the Apollo 11 mission played on a dozen televisions set up throughout the banquet hall. Patrons dined on a space-themed menu and danced a special step named after the lunar module.[21] In Kinshasa, Democratic Republic of Congo, thousands assembled in a public square to listen to the radio over loudspeakers as they watched space films projected on a large outdoor screen. Millions more met in public squares in South Korea and Ethiopia and in countless other cities and towns throughout the world to witness the lunar landing with others. Given the popularity of the event, the Korean government extended the city curfew by three hours. Thousands of people from Montreal to Brussels, who could have watched the coverage on their own televisions at home, filled parks and squares to view the lunar landing with others. In Paris, 36,000 joined together at the Palais de la Découverte to watch the landing together. A New Yorker who stood with thousands of other people in Central Park during a rainstorm to watch the landing on a gigantic Eidophor screen explained, "Well, it was something that we wanted to see together with a lot of people."[22]

Television and radio crews had a handful of hours to fill between the lunar landing and the revised estimate for the moonwalk, which initially was 9:00 p.m. EST. As the astronauts prepared for the next phase of their mission, CBS aired British prime minister Harold Wilson giving a speech from his study in London: "I suppose it's an achievement which incorporates all of the work, all of the discoveries of the mathematicians, and the scientists and the space experts, almost from the earliest days of mathematics and science. An incorporation that acknowledges the experience of many nations." Cronkite then asked correspondents in London, Amsterdam, Paris, Rome, and Belgrade to report on foreign reactions to the touchdown. According to reporters, audiences in London and Amsterdam watched quietly and then gave a "sigh of relief." A woman in Paris said, "I think this is a very important day and a wonderful day as far as the whole universe is concerned, because it's the landing on the moon." The CBS correspondent reporting from Rome added that Pope Paul VI watched television coverage from the Vatican Observatory after spending much of the evening viewing the moon through a telescope. In Belgrade, the correspondent posited, "Yugoslavia has adopted the three American astronauts as its own heroes."[23] Another report came from the USIA Apollo exhibit in Indonesia, which was packed by a massive crowd.[24]

After a light meal of bacon bars, peaches, sugar cookie cubes, coffee, and a pineapple-grapefruit drink, the astronauts began the cumbersome task of suiting up. At 10:28 p.m. they vented the cabin's oxygen and opened the hatch. Armstrong knelt and moved backward through the opening. He paused on the porch of *Eagle*. Before he would set foot on the moon, Armstrong made sure that the scene would be captured on TV. Pulling a D-ring with his left

People in Norway watch television coverage of the Apollo 11 mission, July 20, 1969. (NATIONAL ARCHIVES)

hand, he released the LM's equipment bay, where a television camera was mounted. The footage soon lit up television sets of more than half a billion people back on Earth, one-fifth of humanity. Hundreds of millions more were at their radios. On reaching the last rung of the ladder, Armstrong jumped onto *Eagle*'s golden footpad. He told those listening in that the "footpads are only depressed in the surface about 1 to 2 inches, although the surface appears to be very, very fine grained." With an arm outstretched holding the ladder to steady himself, Armstrong lifted his left leg. At first he placed his boot cautiously into the lunar soil, testing its give, before fully taking his first step. It was 10:56 p.m.[25]

"That's one small step for [a] man, one giant leap for mankind," he proclaimed.[26]

Armstrong's first words on the moon relayed the universalist ambitions of America's space effort. It echoed the "We Came in Peace for All Mankind" message inscribed on the plaque mounted on the lunar module, intended to be left forever on the lunar surface, and the countless pamphlets, films, radio broadcasts, and television interviews. Neither NASA public relations nor the USIA composed this message. As it had been for the Apollo 8 crew, Armstrong was left with the sole responsibility of scripting his own first words. The first words spoken on another celestial body resulted, in part, from the US space program's policy of openness. The larger effort to differentiate the United States from the Soviet Union, to signal the values and benefits of a democratic society through how the country conducted Apollo, permeated every gesture, every step of the mission. Years later, historian Stephen Ambrose told Armstrong that "everybody in the world knows the line, and everybody in the world is so grateful you didn't say, 'that's one giant leap forward for the United States.'" This comment identifies the power of inclusive language, capturing the geopolitical implications of Armstrong's message.[27]

Nineteen minutes later, Buzz Aldrin joined Armstrong on the moon. They unveiled the plaque mounted on *Eagle*'s leg and read the inscription for the global audience. The astronauts then moved the camera to a tripod a few dozen yards from the lunar module. In front of a live global television audience, they unpacked the American flag from its protective steel case, drove its pole into the lunar regolith, and extended the horizontal telescoping rod. The flag caught on the rod, preventing the astronauts from extending it fully. This created a rippling effect, making the flag look as if it was flapping in the wind. Next, they spoke with President Nixon, deployed scientific experiments, and collected lunar samples. Before returning to the LM, the astronauts pulled out a packet containing the Soviet medals in honor of two deceased cosmonauts, an Apollo 1 patch in honor of their deceased colleagues, and a gold olive-branch pin that signaled the peaceful nature of the mission, and left it in the lunar soil. After less than three hours on the moon's surface, Armstrong and Aldrin were back inside *Eagle*.[28]

"Absolutely bloody marvelous," exclaimed Anthony Crowhurst, a solicitor's clerk in London. He had set his alarm clock for 2:00 a.m.

Astronaut Buzz Aldrin descends the ladder on the lunar module during the Apollo 11 mission, July 20, 1969. (NATIONAL AERONAUTICS AND SPACE ADMINISTRATION)

Like many people throughout the city, he made sure that no matter what hour of the day, he would witness the moonwalk. "I leaped out of bed feeling pretty awful and I just smoked away watching TV until 4:45 am . . . it was worth waiting for." While people danced in the streets of Santiago, Chile, reports from other cities—such as Mexico City, Oslo, Belgrade, and Rome—describe downtowns as deserted because millions were watching landing coverage in their homes. In Bangkok prisoners who were to be released that day supposedly refused to leave because the jail was showing coverage of the moon landing on a TV set. A judge in Wollongong, Australia, reportedly had a television brought into the courtroom so he could watch coverage while he heard cases. In Beirut the announcement "Ladies and gentlemen. The Moon is now within man's grasp" interrupted coverage of a major air battle over the Suez Canal that was part of the War of Attrition between Egypt and Israel. Instead of covering the battle, the rendition of "Oh Moon I Am with You" by the popular singer Feirouz played on television sets.[29]

Lella Scalia, an elementary-age child in Italy, knew "it was one of those special evenings, really special . . . because it was special for everybody." She recalled watching the moonwalk on a black-and-white television set on a hot July night. Her mother wore a flowered dress, and her father play-acted walking on the moon. She nodded on and off throughout the night, with her parents waking her up, "and the scene was always the same: lights, voices, talks, the word [LM], a spider in the belly that had the world's three heroes in it." And then, at last, her parents shook her awake and said, "It's time." She remembered Ruggero Orlando reporting from the United States. Armstrong descended the ladder and spoke his famous words. "'Now I could sleep,' said my mother, a little moved. I could leave the hot velvet couch and go to bed. To dream about the Moon, not lonely and empty anymore." Along with Scalia's family, twenty million people in Italy watched the moon landing. Supposedly, no robberies or muggings occurred throughout the country that night.[30]

Millions of Romanians stayed up through the night to follow the flight as well. *Scînteia*, the leading Communist Party newspaper, called Apollo 11 "undoubtedly one of the most extraordinary events in the whole history of mankind" and noted that Armstrong's message "aroused a tremendous wave of enthusiasm in the whole world, which held its breath while watching the grand mission of the daring explorers."[31]

A woman in the capital of Somalia gave birth to a baby boy and named him Armstrong Abdurahman Osman after the first man on the moon. The US Embassy explained that it was "publicized over the radio and in the press, [and] had the coffee house devotees buzzing." Babies born in other corners of the world, including Lebanon,

Scotland, and Tanzania, were also named after the Apollo 11 crew. Seven new mothers in Lima who gave birth during the Apollo 11 mission named their sons Neil, and a baby girl in the Peruvian capital was named Selena after the Greek goddess of the moon.[32]

During Armstrong and Aldrin's moonwalk, electric utilities throughout the world recorded an unprecedented upsurge in consumption. The VOA broadcasts in thirty-six languages reached an audience of 750 million, more than twice the number that followed John Glenn's flight in 1962. Five million of Switzerland's total population of six million watched coverage on their television sets.[33] The BBC's domestic and foreign services, thirty-seven other national networks, and approximately two thousand individual radio stations in Latin America relayed and rebroadcast VOA programs.[34] All India Radio relayed VOA prerecorded programs and dedicated over twelve hours of airtime to the live VOA broadcast, reaching an estimated audience of more than 85 million. Radio Djakarta kept its studio open late to broadcast VOA coverage, a break in its decade-long political refusal to broadcast American radio.[35] With the help of 1,400 radio stations in Latin America, another estimated 500 million people heard VOA programming.[36] More than 3,600 foreign television and radio stations broadcast the relayed transmission of the lunar landing to audiences in more than sixty non-Communist countries. It was the first time in Laotian history that radio listeners heard news about a world event as it was taking place.[37]

An early-morning television press conference featured Japanese prime minister Satō praising the landing as "epochal in the history of mankind." Other foreign leaders around the world expressed their congratulations to the astronauts and the United States, using the global rhetoric cultivated by years of US information programming. Far away—both geographically and politically—Romanian Communist leader Nicolae Ceauşescu deviated from his prepared speech to praise the first lunar landing, interrupted periodically by boisterous applause from his audience. The achievement, he reflected, was "in the interest of all mankind," phrasing that mirrored US public diplomats' Apollo 11 messaging. The Algerian representative on the

People in South Korea watch Apollo 11 coverage on a large outdoor screen, July 1969. (NATIONAL ARCHIVES)

UN Security Council used similar language when he congratulated the American delegation: the lunar landing "is to the glory of man, above all."[38]

In addition to providing live coverage of the flight, the USIA's Motion Picture and Television Service worked quickly to create clips of all phases of the flight, transmitting them via satellite to Tokyo, where they were made into prints and then distributed to television stations throughout Asia.[39] Foreign television networks drew on USIA material, including older television clips, to fill out their live reporting on the lunar mission. Television news anchors often used USIA- and embassy-supplied visual aids, including charts, models, moon globes, and space suits. The walls of the Prague television studio were covered with embassy-supplied space photographs and posters, Czech television anchors used the USIA's models during their telecasts, and

NASA press kits proved to be a useful source of information for the station. Austrian television broadcast twenty-eight hours of continuous moon landing coverage, which drew on USIS films, models, photographs, a space suit, and USIA text material.[40]

On July 21, leaders of Venezuela and Chile, among other countries, announced national holidays.[41] The headline of Beirut's *an-Nahar* read "Moon Age—First Day."[42] Many people expressed their enthusiasm and congratulations to local embassy and USIA officials. As USIS Rabat reported, "We continue [to] receive countless phone calls praising [the] successful Apollo journey . . . as well as visits from individuals offering congratulations."[43] A headline in one of Romania's Communist papers called the flight "a brilliant victory of man."[44] Reports from some countries even noted that American expats were stopped on the street and congratulated.[45] In Warsaw people laid bouquets of flowers on the lunar module model on display at the US Embassy, and in Tokyo people left paper cranes at the US Embassy for good luck.[46] Many nations, including the Congo and Algeria, released Apollo 11 stamps.[47] Letters, telegrams, poems, and drawings filled USIA posts' mailboxes.[48]

Although widely lauded, Apollo did not receive universal acclaim. As a CBS journalist stationed in Tokyo reported on the lunar landing, a crowd of 3,500 Japanese students filled the city streets, protesting the US government.[49] The US Embassy in Dar es Salaam described a scene of three youth leaders at the local university using bullhorns to protest Vietnam and call for the "slaughter [of] leading imperialist gangster [President Nixon]" even as the astronauts walked on the moon. A newspaper in Niger questioned the high cost of lunar exploration, given more pressing needs on Earth. The Libyan anti-American weekly publication *al-Rakib* quoted a Cuban diplomat who commented, "Sending a man to the Moon is a useless luxury."[50]

The Soviet Union restricted live coverage of the lunar landing, but Apollo 11 did feature in subsequent press, radio, and television programming. For the most part this coverage balanced enthusiasm for the mission with descriptions of the importance of the Soviet robotic probe Luna 15, which orbited the moon as Apollo 11 landed. Planned as a simple return mission, Luna 15's transmissions cut out within minutes after it began powered descent on July 21. Later analysis found that the spacecraft had crashed into the side of a mountain in Mare Crisium.[51] Soviet news called Apollo 11 and the Soviet Luna 15 mission "interrelated wonderful achievements." An article in *Pravda* referenced Frank Borman's visit to the USSR in early July and his statement that Project Apollo was made "possible through the common efforts of all mankind," including Russians from Konstantin Tsiolkovsky to Yuri Gagarin.[52]

Among the Soviet Union's closest Eastern European allies, Hungary offered the most praise for the mission. All Eastern European countries—save East Germany, Bulgaria, and Albania—carried live coverage of the Apollo 11 mission. Almost all of the front pages of Czech newspapers featured the landing.[53] The East German *Der Morgen* newspaper argued that "the moonwalk of the US astronauts is a success which no nation can claim for itself" because it relied on accomplishments of mankind "from Galileo to Einstein." In a similar vein the Polish paper *Trybuna Ludu* called the astronauts "the ambassadors of humankind" and stressed that the success of the mission was based on the efforts of many countries over time. *Życie Warszawy*, another Polish paper, commented that "today we rejoice together with the astronauts, with their families, with all Americans and all mankind." The overarching message in the Soviet press was that Apollo 11 was a shared human achievement as opposed to a national mission. The media in China, North Korea, and North Vietnam did not acknowledge the flight, while Cuban media covered some of the mission. The VOA attempted to reach audiences by broadcasting on two channels to circumvent Chinese government jamming.[54]

In the days following the mission, reports from American embassies and USIS posts reinforced how "worldwide reaction to the Apollo 11 mission has been unprecedented." "Morocco held its collective breath," raved Ambassador Henry Tasca. "Never in all my years in the service of the United States abroad have I witnessed an event with anything like the psychological effect of [the first lunar landing]." Apollo 11 "was a unifying experience," captivating everyone in Rabat, from the "King and the street beggar" to the young and old. Perhaps overconfidently, Tasca believed that there was "no doubt that the international position of the United States in all its aspects has been deeply" and, he hoped, "irreversibly changed." He predicted a host of potential effects, including the perceived superiority of "the American system and values," the prestige of the US educational system, the erasing of "any image of weakness" resulting from Vietnam, a widespread identification with the lunar landing in the global South, and having the West's reactions to US policies and leadership become more unified in the future.[55]

The USIA reported record-breaking use of the agency's press material by foreign newspapers and publications. Of the thirty-nine articles sent to Hong Kong, newspapers printed thirty-two. The agency estimated that in India between 70 percent and 80 percent of all photographs and articles were sourced from USIA material. The Moroccan press "ran virtually all items sent," while one of the major Belgrade papers used material from the USIA for five of the six pages of its special color supplement on space exploration.[56]

The USIA also kept track of instances when agency rhetoric filtered into the foreign press or speeches. In Rome, the title of one of the USIA pamphlets, "L'Uomo sulla Luna," became the title of most Italian media special inserts as well as radio and television programs on space. Although this title, translated into English as "Man on the Moon," did not contain the universalist rhetorical underpinnings of other USIA material, the act of adoption demonstrated the potency of agency material in the framing of nonagency lunar exploration coverage.[57]

A weekly public opinion poll conducted in Rio de Janeiro by the *Jornal do Brasil* found that Apollo 11 was followed enthusiastically through all forms of media and that only 18 percent of interviewees were not "absorbed" in moon landing coverage.[58] An opinion poll conducted in Britain and France found that "the U.S. space standing, unlike U.S. military standing, had a definite relationship . . . to both the U.S. general strength image and confidence in the U.S. handling of world affairs." These results suggested again to USIA public diplomats that spaceflight was a "particularly useful subject for US communications overseas."[59]

There was also evidence of smaller-scale local impact. "In addition to effecting changes in attitudes around the world," one foreign affairs officer based in South Africa related, "the Apollo 11 mission may bring about some more practical changes." The South African government had not authorized the establishment of a television network, prompting some people to charter flights to other countries to watch the moon landing live. Widespread disappointment about not having access to the live coverage of the moon landing fed into "mounting pressure for the establishment of a television service."[60] An American journalist based in Japan observed that Apollo "added ammunition for the pro-American policies of the conservative Liberal-Democratic government of Prime Minister Eisaku Sato." Opposition parties—socialists advocating "unarmed neutrality" and Communists who argued for the withdrawal of US military presence—"lost face."[61]

A few days after the mission, a group of poets in the Republic of China (Taiwan) gathered in Taipei to write poems about the lunar landing. The State Department translated a selection of these poems and sent them to President Nixon.[62] A poet in Israel composed a piece for President Nixon that expressed hope that the "white queen" would "bestow understanding on the people."[63] Both Pablo Neruda and W. H. Auden, moved by Apollo 11, wrote poems to explore the significance of lunar exploration. Neruda, the politically outspoken writer and member of Chile's Communist Party, penned a poem dedicated to the astronauts, expressing hope that "the new visitors to the Moon will leave up there a portrait of that good poet Jules Verne who showed them the way."[64] He had examined space exploration in poems throughout his career, including "Lazybones," which he composed shortly after Sputnik 1 orbited the Earth. The two poets did not herald the astronauts' accomplishment or the impressiveness of America's scientific and technological know-how. Instead, they considered the ways that lunar exploration could expand human imagination and heighten expectations for a better life on Earth.[65]

From Hungary to Guyana, leaders congratulated Nixon and the Apollo 11 crew on the historic mission. Many of these letters adopted US public diplomats' framing of the significance of the moon landing, including phrases like "for all mankind." But what was even more notable than the inclusion of global rhetoric was the widespread prevalence of the word "hope." The majority of congratulatory letters contained the word "hope" and expressed the wish that the lunar program would benefit "all of mankind" by bringing the world peace, prosperity, and other markers of progress promised in years of US public information programming. President Duvalier of Haiti told Nixon that it was his "hope that this historic milestone will contribute to the well-being of all mankind." President Kenyatta of Kenya wrote that it was his "hope that this momentous event will serve to strengthen the cause of international peace, security and co-operation for the benefit of all mankind." President Tito of Yugoslavia noted that it was his "hope that this achievement of human brain and ability inspire all countries and peoples in the world to work together." And the king of Morocco used the phrase "all mankind" three times in his short congratulatory telegram. These messages, and the hundreds of other letters sent to Nixon, articulate very similar sentiments: they describe enthusiasm within their country for the lunar landing, and they convey unfulfilled wishes for social and political progress.[66]

Whether or not space exploration was actually viewed as "for all mankind" or "all humankind" by foreign leaders, the press, and the world public, this language was a central feature of how the intentions, accomplishments, and benefits of Apollo were articulated. Language is appropriated for many reasons, from an expression of goodwill and alliance to an assertion of power. Often, world leaders used the space rhetoric promoted by US diplomats to cloak broader discussions of national and international needs, to push back on US policies, and to articulate their own definitions of progress. The use of this global rhetoric and the reiterating of the phrase "for all mankind" do not necessarily represent a demonstration of shared beliefs and interests. But it clearly played a performative role by facilitating communication about national and international interests in an increasingly interwoven but diverse global community.[67]

The day after the moon landing, Nixon felt energized and inspired. He gathered his closest advisors and expressed the "need now to establish the mystique of the presidency." Chief of Staff H. R. Haldeman recorded in his notes that Nixon was so "impressed by the astronauts last night" that he had an idea, a new leitmotif. He wanted to start using the word "'GO' as the theme" for his presidency. To Nixon, it meant "all systems ready, never be indecisive, get going, take risks, be exciting." What he was less enthusiastic about was the *New York Times* article "Nixoning the Moon," published shortly after the launch. It criticized the president for taking any credit for the moon landing when it was Kennedy and Johnson who initiated the Apollo program. "Mr. Nixon's attempt to share the stage with the three brave men on Apollo 11 when they attain the moon appears to us rather unseemly," the *Times* editorialized. Even worse, Nixon's behavior evoked "the type Khrushchev used to indulge in. It strikes us as unworthy of the President." Livid, Nixon told Haldeman to put a White House press ban on the *New York Times*.[68]

Nixon's response on July 21, both his animosity toward negative press and the inspiration he took from the moon landing, demonstrates his sensitivity to and appreciation for public image. As Eisenhower, Kennedy, and Johnson before him, Nixon understood the role of public relations in politics. However, the difference was in how it played out. The moon landing and the conflicting press coverage that surrounded it convinced Nixon that he should leverage the popularity of spaceflight and control the media narrative, an effort that did not produce wholly positive consequences. Historian Kathryn Brownell argues that Nixon's meeting with his advisors on July 21 initiated the centralization of Nixon's public relations operation. What would become known as the Committee to Re-elect the President (CREEP) drove this effort over the next few years. Taking a cue from the moon landing, CREEP sold Nixon as a "man of action" with a bold vision.[69] The moon landing taught Nixon that he should be "looking for a sense of history and a sense of drama," believing it was a key to the success of his presidency and the assertion of a global leadership position.[70]

10

OPERATION MOONGLOW,
AUGUST 1969

In Nixon's eyes, the American "spirit"—as exemplified
by the Apollo mission to the Moon—was the most important psychological weapon that could be used.

—WILLIAM SAFIRE

T he sun was rising over the Pacific Ocean as *Columbia*'s
three main parachutes opened, slowing the spacecraft to a
relatively gentle 22 miles per hour. They held *Columbia* at
a slight angle so its leading edge would cut through the water first.
The impact attenuation system—a series of crushable ribs made of
four-inch-thick corrugated aluminum within the structure of the
command module—combined with the crew's suspended couches,
absorbed some of the force of impact. But still, Michael Collins recalled, "Splat! Like a ton of bricks, we hit."[1]

President Nixon and his advisors had taken a spot on the flag
bridge of the recovery aircraft carrier USS *Hornet* for a better view.
"We saw the fireball (like a meteor with a tail) rise from the horizon and arch through the sky, turning into a red ball, then disap-
pearing," Haldeman wrote in his diary. Landing about thirteen miles
from the *Hornet*, and over 800 miles southwest of Hawaii, on July 24,

President Richard Nixon on the USS *Hornet* watching the Apollo 11 splash-down, 1969. (NATIONAL AERONAUTICS AND SPACE ADMINISTRATION)

Apollo 11 fulfilled the last step of Kennedy's challenge: returning the astronauts safely to Earth.[2]

At first, *Columbia* bobbed upside down in the ocean swell. But before long, large beach-ball-like inflation bags righted the craft. By the time green dye released into the water signaled the location, recovery crews were already hovering overhead in helicopters. Swimmers jumped from the helicopters, stabilized the spacecraft, deployed inflatable life rafts, and prepared for Armstrong, Aldrin, and Collins's decontamination process. A recovery net lifted the moon travelers dressed in biological isolation garments up one by one to the helicopter to be transported to the aircraft carrier. All the while, the USS

Hornet powered closer toward the landing site, stopping just half a mile from the spacecraft.[3]

President Nixon watched the scene unfold. "Exuberant, really cranked up, like a little kid," Nixon "soaked it all up," wrote Haldeman. He waved and even showed off his "fancy binoculars." At Nixon's request, the *Hornet* band played "Columbia, the Gem of the Ocean," a nod to the spacecraft's name, when the astronauts arrived on deck.[4]

The Apollo 11 crew swiftly made their way to the mobile quarantine facility (MQF), an Airstream trailer modified to prevent the spread of lunar microorganisms, however unlikely their presence might be. After changing into more comfortable NASA flight suits and receiving medical exams, the astronauts assembled in front of the trailer's window to speak to the president.[5] Nixon found it "hard to contain [his] enthusiasm or [his] awe."[6] In front of a global television audience of 500 million, he told the astronauts that he was the "luckiest man in the world." Nixon immediately put the accomplishment into its larger geopolitical context. The White House had already received over a hundred congratulatory messages from foreign heads of state, he told the crew. These messages "represent over two billion people on this earth—all of them who have had the opportunity through television to see what you have done."

After a little light banter, with Nixon asking the astronauts to a celebratory state dinner in Los Angeles and inquiring whether they got seasick on the moon with all that "bouncing around," he impulsively exclaimed that "this is the greatest week in the history of the world since the Creation." "Because as a result of what happened this week," Nixon continued, "the world is bigger infinitely. And also, as I'm going to find on this trip around the world . . . as a result of what you've done the world's never been closer together before." He thanked the astronauts for their role in making the world united. And he noted that "all of us in government, all of us in America . . . as a result of what you've done, we can do our job a little better."[7]

The job that Nixon hoped to do "a little better," in particular, was establish a new Asia policy and ultimately advance the United States'

President Richard Nixon speaks to the Apollo 11 crew members while they are still in quarantine, July 24, 1969. (NATIONAL AERONAUTICS AND SPACE ADMINISTRATION)

standing in the world. He and his advisors anticipated that the moon landing would not only lend its sheen to the United States in general, but it could have a targeted strategic impact. Nixon timed a diplomatic tour explicitly to take advantage of the international popularity of the moon landing while it was still ripe. His eight-country trip, aptly named Operation Moonglow, officially sought to demonstrate the administration's concern for Asia and Eastern Europe, its commitment to finding peace in Vietnam, and "a desire to listen" and connect with the world, with the pretext of "the unifying symbol of the moon voyage." Nixon would incite the "Spirit of Apollo," promoting the message that "if mankind can send men to the Moon, then we can bring peace to the Earth." As one journalist put it, "The way has been paved for him by a deluge of publicity over the feat of Apollo 11."[8]

Moreover, by Nixon's own admission, Moonglow also provided the "perfect camouflage for National Security Advisor Henry Kissinger's first secret meeting with the North Vietnamese. It was arranged that Kissinger would go to Paris, ostensibly to brief French officials on the results of my meetings. While there he would meet secretly with [head of North Vietnam's Paris delegation Xuan] Thuy."[9]

Photographs of Nixon speaking with the Apollo 11 crew through the window of the MQF made front-page headlines around the world. Although these photos were most often paired with quotations from Nixon's welcome message, some journalists considered the impact of the scene—and the moon landing more generally—on his upcoming diplomatic tour. As London's *Daily Telegraph* commented, "Certainly his words to his hosts during his subsequent visit to Asia, to Bucharest . . . will carry that much more weight." The front page of Vienna's *Wiener Zeitung* similarly noted that "President Nixon and his advisors are convinced—and diplomatic observers agree—that the prestige America has gained by the Apollo 11 success will favorably affect the diplomatic outcome of his trip to Southeast Asia and Eastern Europe." In Japan's *Asahi* a correspondent predicted that Asian countries would receive Nixon more warmly because of "the Apollo mood." An editorial cartoon published in Bombay depicted Nixon in a space suit traveling to planets labeled with the names of the countries on his diplomatic tour.[10] A *Washington Post* column predicted that "as President Nixon sets forth on his world tour, Apollo's triumph becomes a handy tool with which to ask all that any great nation can expect: not that others love us but that they respect us and, hopefully, that they trust us just a bit more than before."[11]

Until the first moon landing, Nixon had difficulty getting any traction on two of his major foreign policy objectives: finding resolution for the war in Vietnam and normalizing relations with China. Operation Moonglow, he hoped, would advance these two objectives. Much like his European tour in March 1969, he planned Operation Moonglow to ride on the coattails of space enthusiasm.

Nixon had attempted to contact North Vietnamese leadership throughout the spring of 1969, with the hope of ending the war. In

April he composed a one-page document with three points outlining the US position that he had Kissinger share with Soviet ambassador Dobrynin. Although Dobrynin told Kissinger that he would give the document to the North Vietnamese, Washington never received a response. Nixon tried again. On May 14, in a televised speech, he presented a comprehensive peace plan. And again, no response came. In June, during remarks on the White House South Lawn, Nixon explained how the plan he presented in May and the beginning of US troop withdrawal "left the door to peace wide open." He invited "the leaders of North Vietnam to walk with us through that door," but still no response came.

In July Nixon tried yet again. But before he did, he had a series of long conversations with Kissinger. They developed an "elaborate orchestration of diplomatic, military, and publicity pressures," and they set November 1 as a deadline for an agreement to be reached. On July 15, the day before Apollo 11's launch, Nixon sent a letter to North Vietnamese leader Ho Chi Minh through Jean Sainteny, a retired French diplomat and Kissinger's acquaintance. On July 16 Sainteny passed the letter to Xuan Thuy. Within days, Nixon received word that the North Vietnamese were interested in arranging a secret meeting between Kissinger and Xuan Thuy. Luckily, Operation Moonglow afforded the perfect opportunity for such a meeting to take place surreptitiously in Paris.[12]

In parallel, Nixon pursued Sino-American communication and better relations. China had severed ties with the Soviet Union, leading Nixon and Kissinger to see the potential of playing the "China card": they would use Sino-American relations to put pressure on the Soviet Union. In June the president supported Montana senator and majority leader Mike Mansfield's request to visit Peking. Chinese premier Zhou Enlai rebuffed the overture, however. Nixon then initiated a relaxation of economic controls on trade with China that summer. But still more had to be done to foster this relationship.[13]

Inspired by Borman's warm welcome in Europe, Nixon told Kissinger in early June, "Henry, I believe we could needle our Moscow friends by arranging more visits to the Eastern Europe countries."

What Borman's success had demonstrated was that people in the Eastern Bloc, "if given a chance," would enthusiastically embrace Nixon's Cabinet. "It is time we start causing them some trouble," he added with emphasis.[14] Within a few weeks Nixon decided that he should be the one "needling" Moscow. Romania should be added to the Operation Moonglow itinerary, he told Kissinger. In 1967 Nixon received a more gracious reception in Romania than any other Eastern European country he visited. "Nixon never forgot courtesies of this kind," Kissinger commented.[15]

"By the time we get through with this trip they [the Russians] are going to be out of their minds that we are playing a Chinese game," Nixon told Kissinger.[16]

On June 21 Kissinger called Romanian ambassador Corneliu Bogdan. Nixon would be meeting the Apollo 11 astronauts at splashdown, then embarking on a diplomatic trip, he explained. "Would it be convenient for the President to stop in Bucharest on August 2 or 3?"[17] The immediate response was positive although, as one would expect, Bogdan mentioned that he needed to confer with his colleagues in the Romanian government.[18] Forty-eight hours later, the response came: the Romanian government welcomed Nixon's visit. It required postponing the Romanian Party conference, a sign that Kissinger took as proof of the "importance Romania attached to a separate opening to Washington."[19] According to Kissinger, Soviet leaders Brezhnev and Kosygin canceled their attendance at the rescheduled Romanian Party conference because "they did not want to compete with President Nixon's trip."[20]

By 8:10 a.m. on July 24, the ceremony on the USS *Hornet* had finished. Nixon boarded his helicopter and took off for Operation Moonglow. When he arrived in Guam, Nixon gave remarks at the Top O' the Mar Officer Club for members of the press. Outlining the "Nixon Doctrine," the president explained that the United States would supply arms and assistance to threatened nations but would expect that these nations supply their own personnel for military defense. Later he clarified that he was not suggesting a full US withdrawal from Asia. The United States was a Pacific power, he stated, and would remain so. But unless a major nuclear threat arose, the United States would support other nations with only material, economic, and military aid in the future.[21]

Nixon stunned his advisors. They had expected the president simply to give a few informal remarks. It was 6:30 p.m., the tail end of a full day, after all. But the Apollo splashdown affected Nixon profoundly. "I do not think that Nixon intended a major policy announcement in Guam," Kissinger reflected. "But, perhaps carried away by the occasion, Nixon, in an effective and often eloquent statement, spelled out his concerns and approach toward Asia."[22]

London's conservative *Sunday Telegraph* called the Nixon Doctrine "an almost complete reversal of what his predecessor used to say," while the *Manila Times* wrote that Nixon "declared America's readiness to scrap old special relations and to develop a new one based on mutual respect." The Nixon Doctrine received positive coverage internationally, but Moscow Radio broadcast programming throughout South and Southeast Asia to counter this message. It warned that Nixon had plans for a "strategic belt" that stretched across Asia from the Pacific to the Indian Ocean. US military-political plans for the area were a serious threat, it alerted listeners ahead of Nixon's diplomatic tour.[23]

A description of Operation Moonglow prepared for Kissinger noted that Apollo "dramatically demonstrate[d] the interdependence of men and nations travelling through space together on this planet." The "earthly interdependence" made apparent by Apollo and the astronauts' space-based perspective of Earth "unites Americans with peoples of all countries—whether they be new allies in Asia or an old ally in Europe, large neutral nations in Asia or a small Communist one in Eastern Europe." Throughout the rest of his trip—which took the president to the Philippines, Indonesia, Thailand, Vietnam, and Romania—Nixon drew on space-themed rhetoric to couch larger foreign relations issues, especially the US role in Vietnam.[24]

Operation Moonglow officially began in Manila. In anticipation of the visit, Filipino newspapers expressed mixed expectations. One welcomed the visit, seeing it "as an auspicious signal of all the forthcoming benefits to the Philippines and to the human race of U.S. aerospace adventures," while another paper suggested problems in Asia would remain "when the present euphoria has worn off." Everyone was clear-eyed about Nixon taking "advantage of the new spirit of friendship toward America [amid Apollo 11 acclaim] to lay groundwork for departures in his Asian policy," as the *Manila Chronicle* observed.[25]

Three television stations gave full coverage of Nixon's visit to the Philippines, repeating a new precedent that was set during the Apollo 11 moon landing for significant events. At every stop, Nixon encountered "placards among the crowds, moon shot exhibits in the streets and toasts of foreign presidents celebrat[ing] the U.S. triumph in space." In Manila people held signs reading "Fly us also to the moon, Mr. President" and even "Viva Apollo, viva America."[26] But alongside this warm welcome, a critical cross-current could be found in the Philippines. Although massive demonstrations never materialized, cartoons like one featured in Manila's *Daily Mirror* depicting Filipinos holding signs that instead read "Stop Abusing Bases Pact!" and "Remove Your Poison Gas Stockpiles" indicated the tenuousness of Nixon's warm reception.[27]

Not long before, a secret CIA memo had warned that a recent chemical weapons leak in Okinawa would lead to "the possible embarrassment of the US during President Nixon's Asian tour. . . . [Japanese leftists] may well be tempted to try to give this present incident a good propaganda ride."[28] Okinawa had become the primary staging post for the US military during the war in Vietnam. Although island residents were well aware of the B-52 bombers flying in and out of Kadena Air Base and US ships transporting troops and supplies through Okinawa's ports, the full extent of munitions— including hundreds of nuclear warheads and chemical weapons— was unknown. Even though the Geneva Protocol prohibited the use of chemical weapons in 1952, both the US and Soviet Union built stockpiles during the cold war. Over the course of the 1950s and 1960s, for example, the US amassed the extremely toxic chemical weapons sulfur mustard, sarin, and VX near the Kadena Air Base.

On July 8, 1969, a chemical leak sent twenty-four Americans to the hospital. Then, spread across the front page of the *Wall Street Journal* on July 18—two days into the Apollo 11 mission—the headline "Nerve Gas Accident: Okinawa Mishap Bares Overseas Deployment of Chemical Weapons" prompted sharp criticism around the world. Although US officials assured the Philippines government that there was not a similar stockpile in the Philippines, skepticism greeted Nixon on his arrival.[29]

To counteract the impact of planned anti-Nixon demonstrations, concern over rumored US nerve gas stocks in the Philippines, and critique of US foreign policy, Nixon drew on the moon landing in each of his speeches in Manila. His staff prepared arrival remarks claiming that the world has "entered a new age, the Age of the Moon." Nixon then referenced Apollo to introduce his new Asia policy: "I am convinced that if mankind can send men to the Moon, then we can also bring peace to the earth. If we can travel across the reaches of space to explore newer worlds, we can also travel across the surface of the earth to build a better world." He concluded by expressing the significance of unity inspired by Apollo: "A great French writer has said that true affection develops not from sitting and looking at one another but rather from 'looking outward together in the same direction.' It is when we look outward together—as we now look out into space—that we appreciate more fully our common condition and our mutual destiny."[30]

At a toast in Manila, Nixon's prepared remarks read "I am sure that during this historic week, your thoughts—like mine—go back frequently to the inspiring fact that men have actually landed on the surface of the moon." He then went on to tell a story of how a "fellow citizen" became involved in US spaceflight. Recalling First Lady Marcos's Washington visit in May 1968, Nixon playful remarked how a member of her party showed great interest and knowledge in Apollo, so an escort brought him to Cape Kennedy, where he "exchanged views with some of our space experts and told them that he would like to be the first Filipino to go to the moon." In a classic storytelling twist, Nixon revealed that the "fellow citizen" was President

Marcos's son Bongbong: "I might add that in answer to his request, we are reserving a seat for him on the first passenger ship which goes to the moon. . . . Provided of course that he obtains the consent of his parents."[31]

Before boarding *Air Force One* for Indonesia, Nixon explained that "the people of Asia must seek "Asian solutions to Asian problems." Then he added that "the imagination of all mankind has been fired by the success of man's first landing on the moon." Using Apollo 11 as a model, he suggested that "it is that same human spirit, applied in different circumstances, which is even now bringing economic and social progress in this part of the world."[32]

Tens of thousands of people welcomed Nixon to Djakarta for his twenty-two-hour visit in Indonesia. President Suharto and his wife met the Nixons at the airport for a ceremony where Nixon referenced Apollo 11: "It is not important what country the men came from who performed this act—what is important is that they represented all mankind." Later that day, Nixon promised Suharto "a piece of the moon rock as a souvenir."[33]

At a white-jacket dinner in the palace pavilion, Suharto, echoing Armstrong's first words on the moon, toasted "these small human steps which form a great leap to mankind."[34] Nixon responded with "you very appropriately referred to the fact that here on earth, too often, the steps for mankind are very, very small . . . we will be thinking of how those steps can be larger, how they can become, finally, a giant leap for mankind on earth at a time that we have made a giant leap for mankind in expanding his knowledge beyond the earth toward the heavens."[35] In private, Nixon assured Suharto sizable economic aid but stressed the need for "Asian solutions to Asian problems."

Thankful for the atypically positive press coverage, one US official based in Djakarta commented, "What a relief to have news like the moon shot after the usual Vietnam, Vietnam, Vietnam day after day."[36] Other press coverage of the visit noted the "hundreds of thousands" of people on the motorcade route cheering. The *Suluh Marhaen*, a paper that the USIA described as typically hostile to US interests, welcomed the Nixon Doctrine, agreeing that "the problem of the defense of the countries of Asia is the basic task of the countries themselves." This article ran with the simple headline "U.S. Understands Indonesia's Wishes."[37]

Against the Secret Service's warnings, Nixon and his entourage flew to Saigon on July 30. This would be the one and only time Nixon visited Vietnam during his presidency. Flying low over the countryside from Thailand, they spotted triangle-shaped fortifications dotting the landscape. Instead of a welcome party at the airport in Saigon, they were met with a military helicopter to transport them directly to the palace. It "seemed to go straight up out of range of possible sniper fire and then plummeted like a stone between the trees of Thieu's offices," Kissinger recalled. Nixon and South Vietnamese President Nguyen Van Thieu gave remarks on the front steps of the palace before withdrawing for private talks.[38]

The first official remarks that Nixon made in South Vietnam referenced Apollo: "I am happy that the moon landing, which in its universality signifies a symbolic drawing together of all mankind, has provided an occasion for me to meet with President Thieu in the capital of his country." Nixon then articulated the Nixon Doctrine, stressing the progress already made in the political and security participation of the South Vietnamese. He described how the United States' "purpose is peace" before introducing Thieu, who expressed his agreement with Nixon and US policy goals.[39] After speaking with Thieu for two hours, Nixon and his advisors helicoptered north some twelve miles to a small Army base. Touring the base by jeep and by foot through the muddy red clay, Nixon made a point of shaking hands, "chatting in folksy tones about their favorite baseball and football teams," pinning Distinguished Service Crosses on soldiers, and referencing the moon landing.[40]

Most South Vietnamese press reported positively, recognizing the implications of the visit for propping up Thieu's position in the region while simultaneously expressing optimism about the Nixon administration's interest in peace. But some, like *Tin Sang*, expressed skepticism: "It is difficult for Asians to believe that 'Asia belongs to Asians' when the number of American troops in Southeast Asia increases daily, and especially when U.S. military bases are still here with warehouses full of poisons."[41]

Ahead of his stop in Pakistan, Karachi's *Morning News* likened Nixon's visit to Lahore with the astronauts' visit to the moon, noting that both were twenty-two-hour stays. The journalist predicted that the talks would have something do to with US policy toward China and with helping achieve a "new equilibrium" in Pakistan-American relations.[42] In Pakistan Nixon compared the lunar landing with local modernization efforts. Both, his speech argued, were "manifestations of that same human spirit."[43] But his primary objective there was passing a message through Pakistan leader Yahya Khan that the United States was interested in normalizing relations with China. While Nixon met with Khan, Kissinger received a briefing on the current state of Chinese politics. Chinese leaders, Kissinger was informed, were considering ending the period of diplomatic isolation.[44]

Nixon left Pakistan, filled with anticipation for his next stop: Bucharest. No American president had stepped foot in the Eastern Bloc since 1945. Nixon's visit to the Romanian capital signaled the new presidential administration's interest in altering the trajectory of US foreign relations in a broad, public forum. This diplomatic tour stop, in particular, received heavy worldwide attention.

Nixon interpreted the enthusiastic crowds that welcomed him in Bucharest as "proof that deep difference in political philosophy cannot permanently divide the people of the world." He expressed this alongside statements about how Apollo 11 had brought the world together, how the "spirit of Apollo transcends geographical barriers and political difference." It was the moon landing that brought "the people of this world together in peace," Nixon emphasized. He had expected a warm welcome after the success of Apollo 11, but he did not anticipate the scale and excitement that greeted him in Bucharest. "At one point Ceausescu and I were literally swept up by the dancing in the streets," he remembered. The *Washington Post* theorized that "President Nixon came here today not only as the envoy of capitalist America but also as the man from the moon. This is the most obvious explanation for the remarkable welcome he received—warmer, many observers suspect, than the Rumanian Communist regime wanted."[45]

Romania had been one of the few Eastern Bloc countries that carried the lunar landing live on television. Armstrong's first step was broadcast around 6:00 a.m. local time. Maria Mihaly, who worked at a cooperative store, recalled that "I had such an emotion when I saw the astronauts on the moon that I stood up in my chair—I did not know what to do." The moon landing stirred her interest "in seeing Mr. Nixon and learning more about the United States," she told a reporter. Eugen Bituleani, a forty-year-old economist, agreed: "Of course [the moon landing] increased our interest in Mr. Nixon." This unprecedented decision to broadcast the coverage suggested a potential thaw in relations, especially given that the country had yet to broadcast live coverage of Soviet space missions.[46]

Thousands of people lined the motorcade route. Stopping spontaneously at the Arch of Triumph, Nixon stepped out of his host's black Mercedes limousine and into the crowd. After shaking hands and then having the Secret Service escort him back into the limousine, he told Ceaușescu, "These are very warm people—very warm." Estimates suggest that half of this crowd had been recruited by the Romanian government. The small Romanian and American flags they waved were handed out by Romanian authorities ahead of Nixon's arrival. As Septimiu Bratu, a history teacher, explained, he and his colleagues were given free time and encouraged to gather in a particular location to greet the American president.[47] Kissinger recognized the Romanian government's hand in staging the reception but still found it "an extraordinary demonstration of Romania's independence from the Soviet Union." Furthermore, "It would have been difficult if not impossible for any government to create the emotional, joyful, human quality of the public outpouring . . . it was profoundly moving."[48]

Nixon brought up the theme of Apollo and global unity in his toast in Bucharest: "People all around the world watched the television pictures of the landing on the Moon. And thoughtful men saw the earth in a new perspective—as the home of the human mind—

where our similarities and our common interests far outweigh our differences."[49] Once the evening reception wore down, Nixon returned to the guesthouse where he and his party were staying. It was a large, modern facility, with a swimming pool, banquet room, and lake.

After 11:30 p.m., Nixon called for Haldeman and Kissinger. The three men, with Kissinger already dressed in pajamas, went for a walk in the garden. For over an hour they walked and talked and smoked cigars as they evaluated the day. "Feels history was made," Haldeman wrote. Stopping at the edge of a lake in the gardens, Nixon took a seat on a concrete bench. He "expounded on all this . . . [and] sees the great historical first of United States P[resident] here, topped by the fantastic reception of the people, as highly significant."[50]

Although the public remarks and toasts advanced US policy objectives, it was Nixon's private meeting with President Ceaușescu the following day that distinguished the visit. After touring the city, Nixon and Ceaușescu sat down for a private conversation with only two interpreters and Kissinger also in the room. "Romania had good diplomatic relations with the North Vietnamese, and I knew that anything I said would be repeated to them," Nixon stated frankly in his memoir.[51] But it was not only the North Vietnamese that Nixon planned to communicate with via Ceaușescu.

Nixon acknowledged "the problem in Vietnam is very difficult. We must recognize that neither side can win or suffer defeat." Nixon went on to explain why the United States could not simply pull out of Vietnam. Too many troops were committed: more than 500,000. "To withdraw and let Hanoi take over would be a defeat." He added, "Look at our position: We stopped the bombing, have offered to negotiate seriously, will withdraw more troops, have offered elections in which the VC [Viet Cong] have an equal chance. Yet Hanoi has given absolutely no indication that they are willing to talk substance."[52]

After noting that an agreement between the United States and North Vietnam could improve relations between the US and China and maybe even the Soviet Union, Nixon firmly stated that "I want peace, but I will never accept defeat and will not have the U.S.

humiliated by Hanoi. What may be necessary here is to open another channel of communications."

Kissinger jumped in, adding, "We have no thought of humiliating Hanoi. . . . But at every meeting they treat us as if we are school boys taking examinations in their 10 points, and discuss nothing else." The National Liberation Front (NLF, also known as the Viet Cong) had issued a "ten-point overall solution" in May. It included US troop withdrawal and the creation of a coalition government. Nixon countered with his own eight-point plan on television the following week, calling for mutual troop removal and accepting the prospect of a neutral South Vietnam.[53]

Ceaușescu asked if the United States was ready to accept "a coalition government of South Vietnam to resolve the problem of South Vietnam."

As Nixon responded, "I want to emphasize why we cannot agree to a coalition (a hard word for us). . . . "

"Change the word," Ceaușescu offered.

Nixon said the US would withdraw all troops "tomorrow" if North Vietnam did the same. He was unwilling to bend when asked if he would "dump Thieu and form a coalition government." As Ceaușescu observed, "This is no solution." They ended the conversation with plans for Kissinger and Ceaușescu to establish a direct channel of communication through the Romanian Embassy.[54]

Although journalists considered various theories for why Nixon visited Romania, they agreed that it signaled Nixon's commitment to détente. The *Times of India* went as far as to comment that "Rumania, because of friendly relations with Peking and Hanoi, can prove of considerable assistance to him," a prescient observation.[55] London's *Sunday Telegraph* suggested that Operation Moonglow tour "may well go down in history as the most momentous American Presidential voyage since Roosevelt's journeys to the great wartime summits." Not only that, the paper considered it the "end of America's attempt to make her physical and political presence dominant throughout the noncommunist world." It was an end to the Domino Theory, the paper suggested.[56]

Kissinger told Nixon that "the visit was seen as evidence of a fresh and positive initiative in American policy towards the Soviet Bloc." Not surprisingly, "The Soviets and their close friends . . . could scarcely be expected to applaud," but foreign reaction in general was "very positive and frequently perceptive." There were even a number of attempts to draw connections between the visit and the United States' China policy.[57]

On August 4, the day after Nixon's private meeting with Ceaușescu, Kissinger had his first secret meeting with the North Vietnamese. In his memoir, Nixon calls the story "an extraordinary one, full of classic cloak-and-dagger episodes, with Kissinger riding slouched down in the back seats of speeding Citroens, eluding inquisitive reporters, and putting curious embassy officials off the scent."[58] The official pretext for the meeting hinged on Operation Moonglow: the claim that Kissinger must brief French president Georges Pompidou and prime minister Jacques Chaban-Delmas on Nixon's Apollo-themed diplomatic tour.[59]

From Romania, Kissinger flew directly to France. Late in the day he told staff at the American Embassy that he wanted to see the Parisian sights. Instead, he headed straight to Rue de Rivoli, passing by the Musée de l'Orangerie with its remarkable collection of impressionist and postimpressionist art. Journalists were not covering his movements at this point, so he made his way undetected. Just half a mile from the embassy, and overlooking the Tuileries Gardens, Kissinger stopped at number 204. He opened a wooden door wedged between a souvenir shop and a jewelry store. After making his way along a tiled passageway, he passed a classical-style statue before ascending a red-carpeted staircase. A small elevator took him to the second floor.[60]

Jean Sainteny, the retired French diplomat who passed Nixon's messages to the North Vietnamese in July, welcomed Kissinger to his apartment. Kissinger had arrived thirty minutes early because he was nervous. This would be his first time as the principal in a negotiation. Sainteny showed him to the living room, which was decorated in artifacts from Vietnam. A set of sofas faced each other. "I hope if you disagree you will not throw the crockery at each other," Sainteny kidded.

Xuan Thuy arrived right on time. Described as a "slight, smiling man," both articulate and the consummate diplomat, Thuy headed North Vietnam's negotiating team. He and his interpreter sat facing the window overlooking the gardens while Kissinger and his interpreter took their place directly across from them.[61]

The meeting left Kissinger optimistic. He told Nixon it was "business-like and serious, but conducted in a fairly easy manner." Over three and a half hours, Kissinger laid out a proposal for the mutual withdrawal of troops, free elections, and the opening of a secret communications channel to aid negotiations, all while emphasizing his appreciation for the honor of the North Vietnamese people. The US wanted to end the conflict by November 1, Kissinger explained. Although Kissinger's proposal mirrored what Nixon had told other leaders on the Operation Moonglow tour, it was far more detailed and comprehensive. Thuy proposed the complete withdrawal of US troops as well as the removal of the president, vice president, and prime minister of South Vietnam from office, followed by the creation of a coalition government with the Communists. These were terms the United States could not accept, but Kissinger found Thuy's connection of the withdrawal of North Vietnam's forces with the withdrawal of US troops "clear and significant."[62]

This meeting did not end the war. It did not even alter the course of the war. But it did initiate a channel of secret communication between the United States and North Vietnam. Many more secret meetings would follow, eventually settling on negotiations that led to the agreement ratified at the Paris Peace Accords in 1973 to establish peace in Vietnam.[63]

While Kissinger spoke in secret with the North Vietnamese, Nixon flew back to Washington on August 3. The pouring rain at Andrews Air Force Base did not dampen his spirits. To a crowd of three thousand he said that "America has millions of friends in this world." Thrilled by his reception in Romania the day before, Nixon gave a short speech about how "people [in Bucharest] were out by the hundreds of thousands, not ordered by their Government, but cheering and shouting, not against anybody but simply showing their affection and friendship for the people of the United States." The experience moved him, he said.[64]

Nixon felt strongly about sustaining the momentum of Apollo's popularity and political impact for as long as possible. Once the Apollo 11 crew members left quarantine, they were cheered by millions at ticker-tape parades across the country. On August 13 Nixon threw an elaborate dinner in the crew's honor at the Century Plaza Hotel in Los Angeles. Members of Congress, foreign representatives, and Nixon's Cabinet rubbed elbows with Hollywood stars, astronauts, and aerospace executives. Nixon, "in an exuberant mood," hosted what was the "most publicized state dinner in history." Military bands played moon-themed songs, such as "Fly Me to the Moon," and for dessert guests were served "Claire de Lune, a concoction of marzipan, raisins, and meringue, with an American flag on top." In front of his many guests and for the broader television audience watching the dinner live, Nixon said that on his recent Operation Moonglow tour, he learned "what the astronauts meant to the world." The crowds in Bucharest, he recounted, held up pictures of Armstrong, Aldrin, and Collins: "That is certainly the way to bring the world together."[65]

11

GIANTSTEP: THE APOLLO 11 DIPLOMATIC TOUR, 1969

The visit of our astronauts abroad constitutes one of the effective policy vehicles available to us.

—HENRY KISSINGER, 1969

"The President is most anxious that the Apollo 11 astronauts commence their world-wide trip as soon as possible," wrote Peter Flanigan to NASA Administrator Thomas Paine in mid-August, 1969.[1] Flanigan was one of Nixon's most trusted and influential aides. His dominant role in the Nixon administration led to nicknames like "mini-president" and the "most evil" man in Washington.[2] With each subsequent memo that Flanigan wrote, the high priority the White House placed on the diplomatic tour, as well as Nixon's growing frustration, became increasingly apparent. The stakes were high, and Flanigan soon let NASA and the State Department know that the White House was taking over Apollo 11 tour planning. "The President has given the White House staff the responsibility for reconstructing this schedule," he explained to Julian Scheer, NASA assistant administrator for public affairs.[3]

Over the next few weeks, the president and his most trusted aides, including Kissinger, Flanigan, and Haldeman, took a hands-on role in selecting each stop on the Giantstep worldwide diplomatic tour.

This cadre, made up of some of the most powerful men in the world in 1969, attests to how tour planning was geopolitical chess at the highest level. Despite other pressing issues, such as weekly anti–Vietnam War protests, the Vietnam War itself, the Manson family murders, and the Woodstock festival's music electrifying upstate New York, the Giantstep tour captured Nixon's attention in August 1969. The president's irritated response to Hungary's refusal to host the astronauts would indicate how seriously Nixon treated the Giantstep tour. Against advice from the NSC, Nixon demanded reassessing bilateral relations with the Soviet Bloc country. Nixon's reaction, and the negative impact of the tour on US-Hungarian relations, suggests that space diplomacy not only reached the top levels of Washington but that it also altered the course of US foreign relations.

A few months earlier, requests and suggestions for tour stops from ambassadors, public affairs officers, senators and representatives, and even foreign leaders poured in to the White House, the State Department, and NASA. For example, Illinois representative Paul Findley wrote to Secretary of State William Rogers urging the State Department to start scheduling a post-mission diplomatic tour "as soon and extensively as possible." Although Findley had originally voted against Project Apollo—as well as Social Security, rural electrification, the Peace Corps, Medicare, and any other federal programs he could vote against—not only did he come to recognize the foreign relations potential of the lunar landing, but he also became a vocal champion for space diplomacy.[4] As he stressed, "Across the world, the enormity of the accomplishment [lunar landing] has created a reservoir of goodwill toward the American people." Take advantage of this goodwill, Findley argued to Rogers, especially within the Western Hemisphere: "To the south, priority attention from our astronauts would heighten the prestige of the United States as could no amount of foreign-aid dollars and help erase the memory of the disappointments and disagreeable events of recent years."[5]

A similar call came from politicians looking to foster goodwill toward the US in Africa. John E. Reinhardt, the USIA's assistant director who had outlined the proper tone and approach to promoting the moon landing abroad for all USIA posts, sent a memo listing capitals in Africa that would have the "highest potential for favorable impact in support of U.S. policy goals" by the astronauts' visit. Reinhardt would soon become one of Kissinger's closest advisors on Africa policy. He saw the astronauts' visit not only as a way to boost US prestige but also as an effective means for boosting the status of American-friendly regimes in Ethiopia, the Democratic Republic of Congo, Tunisia, Ghana, and Morocco.[6]

At Under Secretary of State U. Alexis Johnson's request, a Committee on Astronaut Travel, composed of State Department, NASA, and USIA officials, began planning an itinerary. Johnson, who would later shepherd the first strategic arms limitation treaty with the Soviet Union, had experience in space policy matters and participated in the White House's post-Apollo planning. Over the course of August 1969 the committee debated the guidelines for the tour, selected the cities the astronauts would visit, and coordinated with foreign embassies. The tour selection involved a number of criteria, including foreign relations' relevance to the United States, the potential enthusiasm of the public in various countries, the location of cooperative scientific programs or tracking stations, and a manageable itinerary. The first region of priority was Latin America, followed by Africa, Europe, the Middle East, and Asia. The committee recommended that the astronauts visit each city for roughly two days, which would allow for a brief rest between public events.[7] NASA would supply a "chief of mission," a "mission director," and eleven additional supporting staff, accompanied by four people from the USIA and one from the State Department.[8]

Before plans were put in place, the State Department wrote to diplomats stationed in potential astronaut tour stops to see how the visit would be managed there, what the crowd reaction would be like, and when heads of state would be available to receive the crew. The American Embassy in Rawalpindi assured the State Department that the astronauts would receive a warm welcome in Pakistan and an especially "tumultuous welcome" in Dacca. The embassy in Tehran guaranteed the State Department that given the "almost emotional interest generated here by Apollo 11, we [feel] certain [that the] astronaut presence will be received with tremendous popular acclaim." The shah of Iran had a number of upcoming trips scheduled, so it would be best if the Apollo 11 crew could visit before October 2, according to the embassy. The embassy in Mexico also eagerly encouraged an astronaut visit to Mexico City, explaining that the "intensity of Mexican admiration derives from special kinship engendered by saturation [of] television/press/radio simultaneous coverage of Apollo exploits by popular Mexican commentators" as well as the tracking station located in Guaymas.[9]

Enthusiastic requests arrived from Eastern Europe, the Middle East, and Asia. "Coming at a time when the Arabs believe the US is turning its back on them," a USIA official explained, a visit of the astronauts to Lebanon would have "a powerful and favorable effect on Arab public opinion from the UAR to Iraq, particularly in those nations with whom we do not have diplomatic relations but who nevertheless gave wide-spread, enthusiastic coverage of the Apollo 11 mission." And in India, a visit "would represent convincing evidence of our recognition of that country's important place in the world community of nations."[10]

Astronaut John Glenn, who had just visited Japan, weighed in on a postflight tour as well. He told Nixon that it was crucial for the US to send astronauts to Japan, in particular, to counteract negative public opinion toward the United States. Nixon thanked Glenn for his "valuable and perceptive remarks about the usefulness of a trip of this nature" and assured him that the recommendation would "receive very careful consideration" by the State Department and NASA.[11] The USIA supported the suggestion but recognized that it should "not expect that criticism from Japanese opposition groups relating to Viet-Nam, Okinawa and the US-Japan Security Treaty will diminish significantly as a result [of the visit]." Public opinion polls taken in Japan indicated harsh criticism of certain US policies but general admiration for the United States, suggesting that the astronaut tour would have little impact on the acceptance of these policies.[12]

By mid-August, Nixon was "dismayed" by the tour planning. He felt that many of the countries on the list were "unimportant." Nixon asked Flanigan twice within a few days to make it clear to NASA and the State Department that the White House staff would be taking over the job. He had strong opinions about the cities the astronauts should visit and whom they should meet. As Nixon complained to Kissinger, "If you leave things in their hands like this, they come out with an utter disaster."[13] Kissinger solicited the help of Apollo 8 astronaut and White House liaison Frank Borman, a favorite of the president.

Borman was dismayed too. There was no "black Africa stop on the proposed tour," he pointed out. Kissinger should "decide this issue," he instructed.[14] At first, plans were made to add Cairo to the list, but these were canceled because of a "deteriorating situation in [the] area as a result [of] Israel air attacks on Egypt." The War of Attrition— with the attacks during the night of the moon landing—made Cairo too risky of a choice politically. But as the USIA stressed, "In Africa our problems are largely in the psychological area. It would be much enlarged if Africa turned out to be the only continent omitted from the greatest world-wide public relations gesture we have ever undertaken." Not only could it be viewed as a "racist slight"; political adversaries could also use it as fodder to undercut America's position in the region. Later, the Democratic Republic of Congo was put on the Giantstep itinerary.[15]

Some appeals for astronaut visits also came from foreign leaders eager to host the Apollo 11 crew in their country. Dutch prime minister Piet de Jong wrote to President Nixon on September 2, requesting that he intervene to ensure that the crew visited the Netherlands. The State Department recognized that de Jong was a "space buff."[16] NSC staff member Don Lesh wrote to Kissinger, confirming that a stop in the Netherlands was a wise decision because "the Dutch undoubtedly would have raised an even bigger stink over their omission from the astronaut tour than they did last February when the President visited Europe and did not find time for the Netherlands." Kissinger conveyed Lesh's suggestions to the president but softened the message, explaining that "the Dutch took their omission from your tour of Europe last February and March very much to heart, and I felt it would be best to try to accommodate the strong desire of the Dutch to play host to our astronauts." Amsterdam was soon placed on the tour itinerary.[17]

When US officials approached the Hungarian government, they were turned down. Hungarian deputy minister Beli Szilagyi was "obviously ill at ease" while reading from a prepared text, recounted US ambassador Alfred Puhan. Szilagyi explained that "the planned world tour has propaganda character and creates unwanted pretenses to which we do not want to contribute . . . we cannot contribute to the strengthening of such pretenses." Puhan suggested that the reason Szilagyi read the response was to "dissociate himself from the wording," and he "turned beet red" as he spoke. Szilagyi joked that the astronauts should visit East Berlin and then used the comment as a way to explain how such a visit would upset West Germany in the same way that a visit to Budapest would upset Hungary's "friends."[18] The State Department speculated that the Hungarian government likely coordinated its decision with the Soviet leadership.

Nixon understood the reason why the Hungarian government declined, but he found "the language employed by the Hungarians invidious and uncalled for." Kissinger explained to Secretary of State Rogers that after Hungary's announcement the president wanted to review "the substance of possible additional bilateral arrangements with the Hungarians" before any additional negotiations were made, which suggests that Nixon was considerably offended by the statement.[19]

"It is inevitable that the course of US-Hungarian relations will be affected," explained Under Secretary of State Elliot Richardson to the US Embassy in Budapest. "You should take steps to effect an appropriate cooling down." Not only was the White House concerned by the rejection of the proffered visit by the astronauts, but the tone and words were viewed as "a source of concern to the White House." Although the White House and the State Department had no plans to make the rejection public, the "cooling down" of bilateral relations "should be brought clearly home to Hungarians that they relate to the cancellation of [the] astronauts."[20]

Although Nixon pressed for a visit to Poland as an alternative to Hungary, the Department of State worried that this gesture would make it "look as if we were shopping for an East European stop" and that if Soviet leadership prevented Hungary from receiving the astronauts, it was unlikely that permission would be given to Poland. The popularity of Apollo 11 in Hungary and neighboring Warsaw Pact countries "quite clearly did not please Soviets," observed Puhan. The State Department eventually reached an agreement with President Marshal Tito of Yugoslavia, making it the only Eastern European stop on the tour.[21]

The White House announced the final itinerary on September 19, 1969. The astronauts would visit twenty-nine cities in twenty-two countries, beginning with Mexico City on September 29 and ending with Tokyo on November 4. They would return to the United States in time for the Apollo 12 launch. If something went wrong on the next lunar voyage, the astronauts should not be abroad, planners concluded. After it was assured that the Apollo 12 mission went smoothly, the crew would fly to Canada for two additional stops in early December.[22]

A day before the tour was scheduled to start, Nixon was still concerned that it might not serve his foreign relations interests. He called Armstrong on the phone with instructions. Drawing on a set of talking points that Borman had prepared for him, the president counseled Armstrong to make it clear to the leaders of all the countries he was about to visit that the tour symbolized "the interest of the United States in maintaining space exploration as a project of peaceful benefits for all nations of the world." Nixon added that Armstrong should consider repeating the message from the Moonglow tour that "the success of the Apollo 11 mission belongs to all the people of the earth and not just the people of the United States."[23]

On Monday, September 29, the Apollo 11 crew, their wives, and a large support staff boarded *SAM 970*, one of two modified Boeing 707s that became *Air Force One* when the president was onboard.

251

It was outfitted with rows of typical airline seats, a curved green table for meetings or card games, cream-colored curtains, and couches where travelers would often nap between stops. Rest was hard to come by, given the grueling schedule. A gray haze of cigarette smoke frequently hung throughout the cabin, and glasses were usually full of something strong: scotch, bourbon, or gin.[24]

Before the astronauts landed at each new destination, they received country briefings. NASA staff outlined US space agreements with each country, but it was USIA Science Advisor Simon Bourgin who told them about everything else: "Don't drink the water, history of the people and country, what the city's like, and how the moon landing did there."[25] Bourgin was tall and thin with a wry sense of humor, and was liked by the astronauts. In 1963 Bourgin's friend Edward R. Murrow asked him to leave his journalism job in California to join the USIA in Washington. He excitedly took on the role of working with the Apollo crews. "After the narcissism of many of the denizens of Hollywood," he reflected, "these men were true superstars." Between reading the newspaper, watching movies, and "long talks relieving the boredom of flights between stops," Bourgin told them what to expect at the next destination. He also wrote home to his wife, Mariada, applying his wit and journalism background to each letter.[26] Their first stop: Mexico City.

At 11:00 a.m. local time that day the Giantstep party touched down on the tarmac at Mexico City International Airport. Armstrong, speaking in Spanish, told the gathered crowd that "it is a great pleasure to join you on the first stop of a tour of the world to share our experiences in space."[27] Armstrong was remarkable when it came to engaging with audiences everywhere they traveled, Collins later said. By the time Armstrong "got to a particular capital, he had done his homework about the place, he knew some of the local problems, had a feeling for the local ambiance, and he would make a very short but impassioned effective speech." He connected with his audiences and made them want to "crawl right on board and go on into space."[28]

Describing the motorcade into the city center in a letter to his wife, Bourgin recounted that "half the kids in town chased us through

The Apollo 11 crew members visit Mexico, September 1969. (NATIONAL AERONAUTICS AND SPACE ADMINISTRATION)

the police lines and in and out of the cars, with everyone wanting to touch, touch and to embrace." The mayor gave the astronauts keys to the city during a ceremony at City Hall, which was decorated by large Mexican and US flags constructed from flowers. People dressed in colorful costumes cheered while a band played the Mexican National Anthem.[29] President Gustavo Díaz Ordaz then hosted a dinner with two hundred social and political leaders.[30] Billed as the "Conquistadors of the Moon," the Apollo 11 crew enjoyed an overall enthusiastic reception in Mexico, reaching an audience of more than eighteen million television viewers.[31] Robert McBride, the US ambassador to Mexico, commented that the visit was "highly successful" and that it was the "best USIS operation that I have seen in my many years in the Foreign Service."[32]

Early the next morning, the Giantstep party reboarded *SAM 970*. The plane took off at 9:00 a.m. for Bogotá, a stop that one public affairs officer called "a 23-hour wonder."[33] During the Apollo 11 mission, all of Colombia's 230 radio stations broadcast coverage. The launch on July 16 became the first live satellite telecast in the country's history. The government-owned nationwide television network picked up the feed from a ground station in Maracaibo, while newspapers and magazines throughout the country detailed the astronauts' progress. So it likely came as no surprise to the crew when crowds surpassing the warm reception in Mexico embraced them physically and metaphorically during their brief visit.[34] These events were picked up by Venezuela television via microwave link between the neighboring countries and were carried live via radio and television throughout Colombia. Newspapers adapted material from the 350 press kits the USIA had distributed ahead of the visit. "For a day—and this is a relief—everyone loved the United States once again," a USIS report stated.[35]

Next, the airplane touched down in Brasilia for a quick refueling stop before flying on to Buenos Aires and Rio de Janeiro for a day each. One out of every two Argentines watched live television coverage of the astronauts' visit to Buenos Aires. The viewership outstripped all previous records for nonsport events in the country's history, but not even the moonwalkers could outshine soccer in Argentina. Six hundred journalists overcrowded the ballroom of the Plaza Hotel, a grand historic hotel—which had previously hosted guests such as Charles de Gaulle, Teddy Roosevelt, and Luciano Pavarotti—to ask the astronauts questions.[36]

Bourgin described the motorcade back to the airport as "sheer bedlam: fightingest competition to stay close to astronauts' car . . . [with] horns tooting, confetti falling, and these handsome people waving and reaching from sidewalks."[37] The astronauts' reception in Rio de Janeiro mirrored their previous stops: swelling crowds, eager journalists, and countless toasts and speeches. In a letter home, Bourgin described "an unofficial official lunch that was epic." The "King of [a] publishing empire hosted instead of [the] Foreign Office, and host's giant Dalmatian ate Mike Collins's dessert."[38] After a day in each city they departed for the Canary Islands for a small break before starting the European leg of their tour.

"Is it possible we have been on road only a week?" Bourgin asked his wife in a letter as the party flew over the Atlantic Ocean. "Impossible. Seems like a month . . . are we really going to do this 23 more times? That's impossible too."[39]

After their short rest on the Canary Islands, the crew flew to Madrid. Stopping in Spain had initially raised concerns within the White House and State Department. "We believe it would be a serious mistake from a political and public affairs standpoint," warned Albert Hemsing, USIA area director for Europe. Francisco Franco's dictatorship over Spain was criticized throughout Europe, and any sign of American support risked souring European political and public opinion of the United States.[40] But the crew did visit Spain, an important ally for the US space program. NASA had established three tracking stations there during the mid-1960s. In Madrid the crew received what the *Chicago Tribune* claimed was the largest reception that foreigners had been given since President Eisenhower visited the country in 1959. As Bourgin described it, "crowd so swollen at [the] Airport upon arrival tonight they exploded the plate-glass windows at the door in a great surge."[41]

The astronauts laid wreaths at the Christopher Columbus monument in Columbus Square, acclaiming Columbus as their colleague in exploration. As with many stops on the tour, national and American flags decorated streets filled with cheering crowds. In advance of their visit, the weekly top-circulation picture magazine, *La Actualidad Española*, distributed two posters of the astronauts superimposed in Spanish scenes. The first was a parody of a bullfight cartel, and the other depicted the crew in the Spanish matador's *traje de luces* (suit of lights).[42] These gestures—both the adoption of the Apollo 11 crew into local costumes or traditions and the crew's efforts to link their mission with local heroes or history—were fundamental parts of the process of creating cultural connections, national alliance, and a sense of global community. It reflected US government officials' interest in making the lunar landing "for all humankind" and also reflected the interest abroad in participating in the event.

"Warm autumn sun and clear blue skies" greeted the Giantstep party in Paris along with a celebration that rivaled Charles Lindbergh's historic reception in 1927.[43] After a motorcade following Lindbergh's parade route, Armstrong spoke to thousands gathered at Paris City Hall: "As our flight extended the boundaries of man's physical world, Paris has, for a thousand years, extended the boundaries of men's minds." The USIA post reported that "with the astronauts on view in the flesh, the exploit was instantly brought back to human terms. Armstrong's boyish face, his unfailing good nature, his lack of pretense helped to reinforce the feeling that the moon landing was a human feat that, for various good reasons, was first accomplished by Americans." Furthermore, the astronauts "succeeded in dispelling the notion that the moon landing was simply the result of the US winning a space race and they emphasized that the moon landing should serve the cause of peace by giving all humanity a new perspective on our planet and its problems."[44]

After Paris it was on to Amsterdam and then Brussels, all within the same day. Instead of a traditional motorcade in Amsterdam, the crew embarked on a "watercade" down the canals of the city. Thousands of Amsterdamers saw the astronauts in person, while another estimated three million watched the event on television in the Netherlands as well as in Austria, Belgium, Ireland, Luxembourg, Spain, Switzerland, Tunisia, and Yugoslavia. According to a number of sources, the astronauts' motorcade in Belgium evoked memories of the Allied Liberation.[45]

On October 10 the Giantstep party flew north to Oslo. Before a rest day in the Norwegian mountains, the crew paraded through the city, attended a press conference for journalists from throughout Scandinavia, and then dined with the royal family. "People infinitely polite and clean," Bourgin observed. USIA staff reported that "the Norwegian Government, from his Majesty on down, took great pride in the fact that the Astronauts were visiting Norway (and only Norway of the Nordic countries)."[46]

The Apollo 11 crew's "watercade" in Amsterdam, October 1969. (NATIONAL ARCHIVES)

In Norway, as in many countries the astronauts visited, USIA staff reported that it was seen as a great honor to be part of the Giantstep tour and was taken as a mark of the country's significance in geopolitics and US foreign relations and friendship. Norwegian television journalist Erik Tandberg ran the press conference where the astronauts were asked questions about their mission, its impact on their view of the Earth, and the impression that the crew was humorless. Collins suggested that if the audience thought astronauts were humorless, they "should get to know the Apollo 12 crew." Next came a question about how they planned to spend their time in the Norwegian mountains. "We hope there is some fishing," Collins responded, "because I am a very good fisherman and [Armstrong] is terrible and I would like to have a chance to show off that fact."[47]

After stopping in Berlin, the party visited London, where Collins told Queen Elizabeth and others gathered at Buckingham Palace that he would like to "take all the world's political leaders up about 100,000 miles, tell them to look back and see how there are no borders and how small the differences between nations really are." He continued, "The earth seems like a jewel in the sky . . . it is a shame how people are fouling up the place."[48]

After twenty-four hours in London the party continued to Rome and then Belgrade, the only city visited in Eastern Europe.[49] The State Department viewed Yugoslavia as the most liberal state in Eastern Europe and believed that it might be moving toward becoming a free-market open society. Josip Tito, the Yugoslav president, was "avidly interested and well informed on space matters," according to the US Embassy. Police estimated a crowd of half a million on the motorcade route, often twenty people deep, with many waving small American flags, which was a record-breaking welcome, according to one Yugoslav official.[50]

During a dinner with President Tito, "As the small talk got smaller and smaller," Michael Collins recounted, "I could see Madame Broz [First Lady of Yugoslavia] totally frozen . . . so things were not well at this formal dinner and about that time I saw Neil get up out of his chair."[51]

The Apollo 11 crew members visit Belgrade, Yugoslavia, October 1969. (National Archives)

Armstrong deviated from the prepared text and instead reflected on his personal connection to the country, explaining that this visit was a particular pleasure for him. "My first scientific experiment . . . in school . . . I produced a replica of the invention of Nikola Tesla. We enjoyed great fun and mystery by holding up a light bulb and illuminating it without any wires."

Jovanka Broz was particularly pleased: "All of a sudden she brightened up. Big smile." Tesla was her relative. Collins explained that "Marshal Tito noticed and that changed the entire complexion. From then on, we were all big buddies."[52]

Bourgin called the stop in Belgrade the "high point of [the] trip so far," noting that dinner with Tito and Jovanka at the White Palace was "unforgettable in its way." The USIA post in Belgrade reported that "the Apollo-11 Astronaut visit served our foreign policy interests by further demonstrating to both the government and people of Yugoslavia our friendly interest in the country and providing them with an opportunity to show in an apolitical context the admiration they feel for US achievements." *Politika*, a Belgrade paper, featured a front-page story of the Giantstep tour, and the visit was covered in all local papers, many with multipage features on the astronauts. The newspaper *Borba* noted that on the astronauts' hunting trip with President Tito, Armstrong proved himself to be "a good astronaut but poor hunter (he bagged only five ducks)."[53]

After their stop in Belgrade, the Giantstep party enjoyed a brief visit to Ankara, where, according to a Turkish newspaper, "By the time the motorcade reached Ataturk Boulevard, the people on either side of the road were three deep, undisciplined and excited, straggling deeper and deeper into the road in spite of the police motorcyclists who patrolled to keep them at bay." The newspaper contrasted this reception with the previous year, when "anti-American feeling led to the burning of the US ambassador's car and American sailors being tossed into the sea," whereas the Giantstep tour "produced only smiles and welcome, from the curious and the lunch-break crowd."[54] According to the US ambassador, the tour was a "uniquely successful Presidential Mission" and "an outstanding success from the point of view of our relations with Turkey." He suggested that the visit was an important factor in improving the image of the United States in Turkey and in emphasizing friendship between the two countries.[55]

From Turkey the crew flew to Kinshasa, their only stop in Africa. In July the USIS in Congo had "pulled out all stops in exploiting the Apollo 11 moon-landing mission," the astronaut briefing report explained. The US documentary *One Giant Leap for Mankind* had been playing in Kinshasa's commercial theaters since mid-September.[56] Members of the Giantstep party described the welcome as one of the warmest that they had experienced on their trip. Almost the entire population of the city greeted the astronauts when they arrived, and the news coverage of the event was extremely heavy. President Mobutu remarked that the people "appreciate highly the honor and the privilege that was reserved to them by the US people and government by inscribing Kinshasa as the African stop of your world tour, and they see, in this attention, a new testimony of the sincere and deep friendship which unites your great country to theirs." Similar to the stop in Oslo and many other cities, the visit was treated as an honor and mark of distinction. A Congolese newspaper noted that the visit was "testimony to the confidence the Congo enjoys at the present time in the eyes of the world" and a recognition of the "privileged place occupied by the Congo in Africa."[57]

From Africa, the crew flew to Iran and then India and Pakistan. Although Iran practiced what the US State Department called "independent foreign policy," the country was viewed as an important ally in the region. In Tehran, "close to a million persons, jamming sidewalks, rooftops, balconies and windows poured out their welcome" during the motorcade. That evening, the astronauts and their wives attended the shah's fiftieth birthday party. The crew received the Order of the Taj, Persian carpets, and a model of the Saturn V constructed and presented by nine-year-old Crown Prince Reza Pahlavi. In return, the astronauts offered a reproduction of the plaque and silicon disk left on the moon. At the press conference, Armstrong again stressed that "we feel that [the] outerspace program is of and

The Apollo 11 crew members visit Kinshasa, Democratic Republic of Congo, October 1969. (NATIONAL AERONAUTICS AND SPACE ADMINISTRATION)

for everybody around the world," reaching a television audience in the tens of thousands.[58]

When the astronauts arrived in India, they were greeted by more than two million people and the kind of sweltering humidity where—according to NASA support staff member Gennie Barnes—"you could see the heat rising, literally rising, kind of like a cloud."[59] The principal objective of this two-day visit was exposing the astronauts to "the maximum number of people in the limited time available," a goal easily accomplished in a country where enthusiasm for space travel ran high. Leaving the Santa Cruz airport at 3:10 p.m. on October 26, the astronauts made their way on a fifteen-mile route into Bombay.

For years, the United States had been courting India—one of the most prominent and populous nations of the nonaligned movement—through the promotion of its space successes and with cooperative space projects.[60] During the planning stage of the tour,

William Thompson, an officer from the Near East and South Asia (IAN) Bureau of the USIA, had made the case that "the nomination of India is, we believe, self-explanatory. It is the largest country in the non-communist world and is of paramount importance in Asia. A visit by the Apollo 11 crew would dramatically support our policy objectives in India and would be particularly useful in strengthening our relations with the Indian scientific community, one of the most sophisticated in Asia."[61]

As Thompson alluded, the United States was pursuing tighter relations with the Indian scientific community as an instrument of diplomacy and influence. One project in particular, the Satellite Instructional Television Experiment (SITE), promised to strengthen these ties while simultaneously diverting Indian attention away from the development of nuclear weapons, a major US objective.[62] The newly established Indian Space Research Organisation (ISRO) teamed up with NASA to develop SITE, a direct broadcasting communications satellite that would give rural communities access to educational television. In 1969 India's population had swollen to nearly 550 million, a number higher than South America and Africa combined. Seventy-five percent of this population resided in rural areas dominated by agricultural production. Strategically, the astronauts repeatedly referenced the new educational satellite project aimed at this population, as well as instances of American-Indian space cooperation that contributed to development. Even though their visit was short, the astronauts made sure to emphasize the shared interests and friendship of the two countries whenever possible.[63]

In advance of the astronauts' visit, the USIS post distributed 100,000 motorcade route maps, 5,000 of which were used by the Bombay Police Department for traffic planning. Major Bombay newspapers reprinted the map, and All India Radio broadcast details about the route. As the party's motorcade slowly made its way down the fifteen-mile stretch, crowds of people five to ten deep lined the road while "balconies, rooftops, car-tops, trees and every conceivable vantage point was packed with cheering waving citizens," according to a report. The *Indian Express* described the scene as "people, people all the way, 15 miles of cheers. The cheers, claps, bravos, whistles, and applause . . . were tremendous."[64]

Bourgin had a slightly different take on the motorcade. As he wrote to his wife from the hotel that night, "My god what a day. Two million people in streets of city and countryside so totally different [than] any I've seen, I still can't believe. . . . Downtown Bombay [is] regal with great style. Suburbs a nightmare of people."[65]

The public event at Azad Maidan, an expansive triangular-shaped sports ground, was "standing-room only," reported the *Times of India*. It was the "biggest-ever welcome in the city," outstripping the previous popular visits by the shah of Iran and Pope Paul VI. The mayor of Bombay, J. K. Joshi, began the ceremony by welcoming the astronauts to the city and introducing them as "a team of gallant men, illustrious space pilots and distinguished citizens of America." Buzz Aldrin gave the kind of down-to-earth response that astronauts were becoming well-known for: "We thank you all for coming out on this Sunday afternoon to see us."[66]

A full-scale replica of the lunar module *Eagle*, almost twenty-three-feet high, sat waiting for the Apollo 11 crew on a circular platform designed to mimic a "Moon basin" in the center of the Azad Maidan. The tour photographer thought that the astronauts wouldn't climb the ladder of the model and pose for photos. But he was wrong. Armstrong, Aldrin, and Collins graciously ascended the ladder, waved to the hundreds of thousands of Bombay residents, and smiled for the newspaper cameras as "a thunderous ovation" spread among the crowd.[67]

Armstrong playfully commented that "only three months ago, our lunar module *Eagle* landed on the moon—but somehow it seems to have found its way to Bombay," referencing the model built by Indian engineers. Although the joke was scripted, prepared by foreign affairs officers before the visit, it was so well received that it appeared in nearly every newspaper article covering the event.

The crew's wives and their hosts sat atop another circular platform a short distance from the "Moon basin." This gray circular structure represented Earth. American ambassador Kenneth Keating reaffirmed

The Apollo 11 astronauts wave from a model of the lunar module in India, October 1969. (National Archives)

the US diplomatic framing of Project Apollo when he referenced those "eight days in July when the world grew smaller, and for one golden moment we were all brothers as we shared in the suspense of the moon voyage."[68]

Then Armstrong deftly emphasized the international participation in US space exploration, as he had done since his first diplomatic tour in 1966. Seventy-four nations contributed to Project Apollo, he explained to the scores of people gathered at Azad Maidan. Just three months earlier, he and Aldrin had left a disk on the moon that was inscribed with Indian prime minister Indira Gandhi's message "Peace for all mankind." Further underscoring this theme, he stressed that the United States was resolved to make international cooperation wider and deeper in the coming years. Armstrong declared that "space research and technology belong not to one—or two nations alone—but to all the countries of the world. For there are no boundaries in space—and this is as it should be."[69]

Aldrin augmented this message by describing the "magnificent gathering" as a "symbol of the spirit of international adventure." The year "1969 will be remembered as the year man set foot on [the] moon," he reflected, but "I think it will also be remembered as the year when the world grew smaller and international co-operation increased."[70] Aldrin went on to describe how all humans have the same needs, the same aspirations. Space research, he explained, had the potential to improve people's lives, especially when undertaken cooperatively between nations. Aldrin's prepared script ended with a quotation from Mahatma Gandhi: "We may call ourselves Christians, Hindus, or Mohammedans . . . but whatever we may be, beneath that diversity, there is a oneness which is unmistakable." Bourgin later assessed that "Aldrin [was] by far [the] best speaker in the park."[71]

Before leaving the event, Armstrong gave the mayor a photograph of the first footprint on the moon. In exchange, the crew and their wives received sandalwood garlands. From the event at the Azad Maidan the Apollo crew went to a press conference, where two hundred media representatives were gathered. As they often did, the astronauts provided the narration for a twenty-minute film that took the press through each stage of the Apollo 11 mission. Sitting in front of an expansive Apollo photomural, the astronauts answered questions before rushing off to their next appointment at the Tata Institute of Fundamental Research to meet with scientists. At Tata, Bourgin implored the crew not to overstay their scheduled twenty-minute visit.[72]

They had a night to recuperate, "get washed," and "get some food" at the palatial Taj Mahal Hotel, constructed in 1903. Simon Bourgin called it a "fascinating hotel, straight out of [the] days of British Raj."[73] At 8:00 the next morning, the Giantstep party piled back into their "drag-bag of vintages, plus only open car in city," and began another motorcade along a new route to the airport. Hundreds of thousands of people gathered in the streets to watch them slowly drive by. In the following days, as the astronauts touched down in East Pakistan (soon to be an independent Bangladesh) and then went on to Bangkok, their visit to Bombay made the front pages of all major Indian news outlets, reaching a readership of nearly nine million people.[74]

Aldrin later commented that it was a "terribly depressing country and it was also an awkward stop from a diplomatic standpoint," but "if Bombay was depressing, Dacca was even more so . . . I can see why a revolution was inevitable."[75] Although the crowds were generally enthusiastic, a group of political, student, and religious leaders called Aldrin a "member of the Zionist movement" and insisted that the government of Pakistan ban the crew from visiting the country. According to a USIA memo, "The rationale of the statement is really a compliment to the astronauts: we cannot permit America to erase anti-American feelings in the minds of Pakistanis through the American moon men." In response, the USIA released a story that stressed Aldrin's commitment to Presbyterianism and his work with the Boy Scouts. Even with criticisms from some groups, the visit to East Pakistan was considered "extremely successful and most productive in terms of serving US Foreign policy interests and USIS objectives in Pakistan."[76]

The Apollo 11 crew members visit Thailand, October 1969. (NATIONAL ARCHIVES)

The Apollo 11 crew members visit South Korea, November 1969. (TAE WON CHUNG © STARS AND STRIPES)

From East Pakistan the party flew to Bangkok, where one editorial claimed that "no American goodwill ambassadors have achieved such great success in promoting friendship, good understanding and a feeling of joint responsibility among all human beings as these astronauts."[77] From Thailand, the party flew to Darwin, where they refueled and then went to Sydney, Guam, Seoul, and Tokyo.

In Sydney, according to a telegram from the US Embassy, the "image of [the] United States was enhanced through the personalities and actions of the astronauts themselves," and in Seoul the visit "increased bonds of friendship already strong with Korea, added to strength [of the] image of US as [a] strong ally way out ahead in space effort and by analogy strong in [a] military sense."[78]

In Tokyo they were welcomed by a cheering crowd of more than a hundred thousand people throwing confetti, an audience with the royal family, and the honor of being the first foreigners to receive Japan's Cultural Medals. A USIS telegram from Tokyo reported that the "astronauts' visit to Japan was [an] unparalleled success in highlighting and humanizing [the] image of US scientific achievements in space research." The reception in Japan was so positive that Armstrong expressed his regret that they were not able to stay longer, especially in a country with such scientific sophistication and strong mass media.[79]

When Bourgin presented his last briefing on November 5, 1969, the Apollo 11 crew and their wives had already been traveling for weeks: "Here are a few helpful reminders. 1. The water is drinkable, although it is not the most popular native drink. 2. You can always expect student demonstrations. 3. Never turn your back on the president. 4. Never be seen with the vice president. 5. If you leave your shoes outside the door, they will be stolen. 6. It is unsafe to walk on the street after dark. 7. Do not discuss the following sensitive issues with the natives: Vietnam War, Budget, Foreign Aid, Import-Exports. 8. Rate of exchange is .05 cents per one dollar (American)."[80] The moonwalkers likely laughed when he finished. They were bound for Washington, DC, the last stop of the tour. Although they were not visiting a foreign country and it was doubtful their shoes would be stolen at the White House, Bourgin's playful briefing held true weight. The US was facing significant challenges at the end of 1969. The "sensitive issues" that he highlighted in the mock briefing were the very topics to avoid when meeting with the leader of the United States. Although the first lunar landing had marked the cutting edge of human technological achievements and was in many ways the pinnacle of engineering and managerial success in this period, it could not be divorced from the fact that the US was also struggling to cope with poverty, crime, racial tension, environmental problems, and the Vietnam War.

Two days earlier, on November 3, Nixon had given his famous "Silent Majority" speech, calling on his "fellow Americans" to support his Vietnam policy. In October, tens of thousands of protesters had stormed Washington while Nixon sat contemptuously inside the White House and watched a football game on television. Combined with millions of marchers attending rallies around the country, the October 15 antiwar protest was the largest in US history. Roughly two weeks later, on a Monday night at 9:32 p.m., Nixon went on national television and radio from his desk in the Oval Office to outline his plans for the Vietnam War and ask "the great silent majority of my fellow Americans" for their support. In a stern but calm voice, Nixon explained that on November 3, "North Vietnam cannot defeat or humiliate the United States. Only Americans can do that."[81] Now, on November 5, he was about to welcome the Apollo 11 crew—in his mind, the "ideal Americans"—back to the United States.

By the time the crew arrived at the White House on November 5, they were exhausted, and it showed on their faces. After traveling 44,650 miles in just 38 days, directly on the heels of their lunar mission, what the astronauts needed most was rest. Nixon, perhaps sympathizing with his weary guests, kept the celebration in Washington small. First, they would attend an intimate gathering on the South Lawn of the White House for members of the Cabinet and White House staffers and then join the president for a quiet dinner followed by an overnight stay. "We think that after all they have done publicly, it is time they had an evening by themselves—and why not in 'everybody's house,' the White House," Nixon melodramatically explained.[82]

In the middle of a cold November afternoon, the three astronauts stood hatless and coatless in the wind as the Army fife and drum band, dressed up in Revolutionary War costume, played on the South Lawn. From the red-carpeted platform, President Nixon exclaimed that their tour was "the most successful goodwill trip in the history of the United States."[83] "We are trying to indicate to you our appreciation, not only for what you did in your travel to the moon but also what you did for the cause of peace and better understanding through your travel on this earth," Nixon explained. "Certainly, the first men ever to land on the moon have demonstrated that they are the best possible ambassadors America could have on this earth."[84]

Reports and commentary on the tour emphasize that it was a resounding success. Bourgin's review indicated that American am-

Buzz Aldrin, Michael Collins, Neil Armstrong, and President Richard Nixon at the White House, November 1969. (NATIONAL AERONAUTICS AND SPACE ADMINISTRATION)

bassadors observed a new flexibility in tackling a "host of problems whose approaches were otherwise frozen." He went on to explain that "by their modesty, expertise, and warmth the astronauts projected an image of the kind of Americans other nations would like us to be."[85] A diplomat in Bogotá reported that "New contacts were made, and old contacts were reinforced. The American hegemony in the field of science, although never overtly expressed, was implicit by necessity in all of the activities." Under Secretary of State for Political Affairs U. Alexis Johnson suggested that "the visits of the Apollo 11 astronauts to many nations . . . have helped greatly in extending and deepening the sense of personal involvement of the peoples of the world in our space program."[86] Giantstep drew record crowds, prompted extensive media coverage, and eased political relationships between the United States and many countries.

After the ceremony with the president, the astronauts and their wives spent the rest of the day at the White House. Buzz Aldrin took the opportunity to have the president's dentist replace a filling that had come loose on the tour.[87] Later that evening, over cocktails, Nixon wanted to hear all about the tour. They handed him a large book of photos and used it to illustrate their stories.[88] "He was very interested in everything we had to report about the tour, about the various leaders we had met, what their reaction was and what did they say," Armstrong recalled. Nixon thanked them for acting as his ambassadors and, according to Armstrong, told them that "he had been trying for years to get a meeting with Romanian President Nicolae Ceausescu and after leaving the USS Hornet he was able to get an appointment. President Nixon said something to the effect, 'That meeting alone paid for everything we spent on the space program.'"[89]

Although Nixon's comment to Armstrong is clearly an overstatement—a remark meant to entertain as much as communicate—it reveals important elements in Nixon's evaluation of Project Apollo and spaceflight more generally. He fully recognized the public relations value of the moon landing, and he leveraged its popularity to serve a host of foreign relations needs. As historian Steve Wolfe argues, the tour was "a highly political and carefully choreographed event designed to reward friends, snub foes, and promised a flood of positive foreign headlines that would for a while help offset the dreary news from Vietnam and elsewhere." Days later, Kissinger would echo Nixon's sentiment, claiming that "the visit of our astronauts abroad constitutes one of the effective policy vehicles available to us."[90] What both savvy leaders understood was the variety of the space program's political spin-offs. The astronaut tour, both hoped, would amplify the message that the moon landing was "for all humankind," a message designed to herald the liberality of US global leadership.

That evening the president and Armstrong sat at the head of the dinner table in the upstairs family dining room. White House staff had put out Woodrow Wilson's china, rimmed with a dark cobalt border and gilded trim. It was a small, intimate meal, just the six exhausted travelers and the Nixons. The Apollo 11 crew's first dinner back in the United States was a French feast of a timbale of lobster Americaine, breast of chukar partridge Veronique alongside a bottle of Chateau Mouton Rothschild (1966), and for dessert soufflé au Grand Marnier.[91] After dinner, Pat Nixon gave them a tour of the White House's private quarters, including her porcelain collection. Then Neil, Buzz, and Mike retired to their sleeping quarters, where they found a bucket of ice and a bottle of Scotch waiting for them. Two more hours passed quickly, as they talked and drank and tried to make sense of everything they had experienced together over the past four months.[92]

AFTERGLOW

A little over five hours after launch on December 7, 1972, the Apollo 17 crew looked back and spotted the whole Earth through a window of their command module. Stunning and bright, the Earth seemed to hang with "no strings hold[ing] it up . . . out there all by itself," observed Apollo 17 commander Eugene Cernan. They captured this world on film, but the political divisions, and the turmoil taking place over 18,000 miles away, were not recorded on the emulsion.[1]

When NASA technicians processed the film from Apollo 17 in Houston, one photograph in particular caught their attention: the whole Earth, without the night side obscuring any part of its surface, against an inky backdrop of outer space. Much like the *Earthrise* photograph, the image that became known as *Blue Marble* quickly circulated around the globe, appearing on the covers of newspapers and magazines. It coalesced conversations about the need for world peace and environmental action. Reminiscent of the plaques that the Apollo crews left on the moon, *Blue Marble* revealed a planet devoid of political boundaries. For years, public diplomats had been cultivating the consciousness of global unity and, by 1972, thanks to the Apollo 17 crew, they had a photograph that embodied this message. *Blue Marble* translated a concept that had once been abstract and ethereal into one that was visually concrete.[2]

Cernan contended that *Blue Marble* should be seen "from the philosophical point of view." Holding up a copy of the photograph in his hand, he explained: "This is us. We're looking at ourselves."[3]

Blue Marble, taken by the Apollo 17 crew, December 1972. (NATIONAL AERONAUTICS AND SPACE ADMINISTRATION)

The viewing of the Earth from space, whether it was the astronauts' firsthand perspective or later by looking at their photographs, has been said to recast our understanding of humanity and the planet. In 1966 counterculture writer Stewart Brand started campaigning for NASA to take a photograph of the whole Earth, arguing that "no one would ever perceive things in the same way" if they saw the Earth in its entirety.[4] Brand's contemporaries also expressed the notion that viewing the Earth from space would prompt a revolution, causing people to understand Earth and civilization anew. Once Apollo astronauts returned to Earth, many recounted a transformative experience in space, perhaps expressed most succinctly by Apollo 8 astronaut Bill Anders: "We came all this way to explore the Moon, and the most important thing is that we discovered the Earth." Reactions to the lunar flights around the world echoed Anders's experience.[5]

Since the Apollo 17 crew left the moon in 1972 and brought *Blue Marble* back to Earth, countless efforts have attempted to understand

and articulate the meaning of Project Apollo. They often assert that the lunar missions created the experience and consciousness of global unity. Sheila Jasanoff suggests that the *Blue Marble* image "perhaps more than any other has come to symbolize the Western world's heightened perceptions of connectedness." Others have likened the space-based optic to the Copernican revolution for its transformative impact on human self-perception.[6]

Decades after Project Apollo, astrophysicist and science promoter Carl Sagan proposed that the photographs taken by the Apollo crews "helped awaken our slumbering planetary consciousness." Critical of the politics of spaceflight, Sagan reflected that "whatever the reason we first mustered the Apollo program, however mired it was in Cold War nationalism and the instruments of death, the inescapable recognition of the unity and the fragility of the Earth is its clear and luminous dividend, the unexpected final gift of Apollo."[7]

Sagan contrasts cold war nationalism and planetary consciousness, treating them as an unlikely pairing. But the planetary consciousness that Sagan observed was not simply a spontaneous, unanticipated consequence of Apollo. As the story of the Apollo 11 plaque—or the countless exhibits, press releases, film screenings, and diplomatic tours, among other programming—demonstrates, the idea of "planetary consciousness" was cultivated and marketed to advance US national interests. Although the linkages between nationalism and "planetary consciousness" may have become obscured by 1969, the story of Project Apollo shows us how they are historically intertwined.

After Apollo 11, the USIA sent a report to the White House assessing the impact of the US space program on foreign public opinion. It noted that more than ten years earlier, the launch of Sputnik cemented the Soviet Union's role as a superpower in the minds of many people around the world. In the 1950s, scientific and technological prowess became equated with all facets of national power. "The USSR was seen as not only holding a commanding lead in space; it was viewed as able to offer a credible challenge to the U.S. in any field where it chose to compete," the report stated. Two factors amplified the impact of Sputnik: "its unexpectedness" and the "drama" and "innate appeal" of spaceflight, which ensured heavy news coverage. Sputnik II, Gagarin's flight, and all the other USSR space firsts only reinforced the public's perception of the Soviet Union's superpower status: "The clamor of domestic debate in the US reverberated through the world press, and was assiduously cited by the USSR," fueling the association of spaceflight and national power even further.[8]

It was an uphill battle at first. The United States came from behind, but eventually a "see-saw pattern" emerged, with the "latest or most spectacular feat" determining the lead. The report stated that a single factor had a profound effect on the impact of the two programs: openness.

Although at first CIA Director Allen Dulles and others initially questioned the advisability of the openness of the US civil space program, it paid off exceptionally well. According to the 1969 USIA evaluation, this open approach "fostered a sense of vicarious participation" throughout the world. It was in this arena that the US outmatched the Soviet Union. The essence of the difference was that the Soviet space program did not "escape the shadows of secrecy, concealment, and nationalistic possessiveness." The US approach to space information dissemination—especially the policy of "openness"—bore fruit. It not only became a symbolic illustration of democratic values; it also enabled people to engage thoroughly with the US space program, from learning technical details of spacecraft to feeling the anticipation that comes with following a risky mission in real time.[9]

By 1969, when Neil Armstrong and Buzz Aldrin took their first steps on the moon, US public diplomats had come to recognize that beating the Soviets in space was no longer their top priority. Instead, the greatest geopolitical rewards of Project Apollo came from the international public's sense of personal participation in this American-led endeavor. This is a lesson learned from years in the field, observing what did and did not resonate with international audiences. Public diplomats adapted their presentation of Project Apollo abroad, in part to ensure its political impact. Initially, framing the US space program as "for all mankind" helped suggest that the modified ICBMs lifting satellites into orbit were for peace and global prosperity instead of national security. At the beginning of the Space Age, diplomats touted US space hardware and scientific expertise. When public opinion polls and feedback from foreign posts revealed that international audiences did not respond well to the heralding of American greatness and technological strength, this message was dampened and subsequently replaced with one emphasizing global unity and international participation. Critique of US involvement in Vietnam and civil rights tensions prompted US officials to stress even more strongly themes of peace, inclusion, and a shared mission. The framing of Apollo was not wholly born within the United States but through a dialogue between US government officials and the audiences they hoped to influence.

A report from the US Embassy in Chad distilled this lesson: "The psychological success of Apollo in Chad is largely due to Washington's decision to stress the fact that Project Apollo is a venture of all mankind and not just of Americans . . . [the] soft-pedaling of nationalistic sentiments succeeded in establishing America's technological supremacy and in adding to our overall national prestige much more effectively than a 'hard sell' ever could have." For public diplomats the triumph of Apollo lay in the widespread identification with an American accomplishment.[10]

Although the reach and intensity of foreign audiences' responses to the moon landing were record-breaking on many metrics, what struck public diplomats the most was "the general tendency of foreigners to claim the feat as an <u>achievement of all mankind</u>." Not just an achievement *for* all mankind, an achievement *of* all mankind: "The deed seemed too important to bear a national label."[11]

The openness of the program, combined with the broad access afforded by television, radio, and newspaper coverage, "permitted a world audience simultaneously to share the human drama . . . [heightening] the participatory involvement inherent in the nature of the moon landing."[12]

Public diplomats asked if the sense of unity brought about by Apollo 11 programming signaled "an emerging new dimension in the international political process, comparable to the experience of unity that might be expected to emerge from such global disasters as a world epidemic, a meteor collision, or a nuclear accident." This sense of unity or "solidarity of the human community" was "heightened by awareness that the whole of mankind was sharing the emotions and exhilaration of a single experience." US government and commercial TV coverage of the flight had promoted the sense of global participation in the moon landing by streaming images of audiences around the world viewing the flight in unison. A fundamental element of the experience of watching or listening to the lunar landing was the awareness that the rest of the world was simultaneously following each stage of the mission with you.[13]

Political theorist Benedict Anderson described a very similar phenomenon, arguing that political and geographic boundaries have little to do with nations and nationalism. In Anderson's view, the globe resembles the *Blue Marble* photograph or the Earth as it was depicted on the Apollo 11 plaque. A nation is "an imagined political community," he explained. These communities are imagined because, like strangers reading the same morning newspaper, citizens may never meet one another even though they participate in an abstract communal event. Common language, common experiences, common history, and an awareness of interdependence make communities cohere.[14]

The story of spaceflight's role in US diplomacy reflects the process that Anderson described. NASA and US public diplomats inundated the world population with global iconography like the Apollo 11 plaque and mission emblem or the *Blue Marble* and *Earthrise* images. They explicitly and intentionally spent years building up a communications infrastructure that gave global access to flight coverage, thereby ensuring that the moon landing would be a common experience, creating a shared history for people around the world. All the while, US messaging and media compounded the sense of participation and the awareness of interdependence. Together, these efforts not only promoted spaceflight; they also fostered an imagined global community.[15]

It is important to note that Apollo did not create a united world, peaceful and serene like the one caught on film by the Apollo 17 crew.

USIA staff knew that Project Apollo could not erase the growing criticism of government spending, civil rights tensions, urban unrest, the war in Vietnam, or the other challenges facing the United States in 1969. And they recognized that enthusiasm for Apollo 11 would fade with time. For although the moon landing may have heightened a consciousness of global interdependence, it also raised awareness of the disparity between America's mastery of space and the faltering attempts to address problems on Earth. Many questioned the cost of Apollo. The expense of human spaceflight highlighted "the chasm between the superpowers and all other nations," the USIA evaluation acknowledged.[16] As with earlier phases in the process of globalization, some divisions were exacerbated while new connections were formed.

Just as Kennedy's May 25, 1961, address to Congress included a series of "urgent national needs" in addition to Apollo, the United States pursued numerous major national commitments throughout the 1960s. And many foreign relations initiatives, unlike the moon landing, did not have broad appeal. Under Kennedy the United States expanded the military and the nation's nuclear weapons arsenal and increased involvement in Vietnam. The CIA backed regime change in Cuba, the Dominican Republic, Laos, and Brazil. When Johnson became president, he escalated the Vietnam War. The number of American military personnel in Vietnam jumped from fewer than 20,000 to more than 500,000 during his administration. One year of fighting in Vietnam by 1969 cost the United States roughly the equivalent of the entire Apollo program. The Nixon administration signed the Anti-Ballistic Missile Treaty and initiated a new era of relations with China, but it also intervened in Chile, Cambodia, and Bolivia and ordered major bombing campaigns in Southeast Asia. The space program had a tangible and immediate impact on US foreign relations, from creating "apolitical" opportunities for US diplomats to meet with local leaders, to displacing negative newspaper headlines about the US, to Nixon's Operation Moonglow tour. But any expectation that the moon landing could have brought unequivocal peace and unity to planet Earth, or a full embrace of the United States and its policies, misunderstands the broader geopolitical context that begot Apollo and that Apollo existed within.[17]

After Project Apollo ended, the geopolitical landscape did not reflect the image of the borderless Earth inscribed on the Apollo 11 plaque or depicted in *Blue Marble*; the world was still very much parceled into nation-states, and the US and the USSR radiated their separate spheres of influence. Nonetheless, as embossing helps us see patterns on paper, images such as the *Blue Marble* made it easier to see what unites us even in a divided world. This was Apollo's afterglow: our heightened awareness of global interdependence.

During the fiftieth anniversary of the moon landing, Michael Collins reiterated a point he made years earlier. When he traveled the world with Armstrong and Aldrin after their flight, in every country they visited "everywhere people said 'we did it,' we humans—humanity finally left this little dinky planet and set foot else elsewhere."[18]

As Collins elaborated, "That trip around the world kind of changed [me], opened my vista. I would not swap the US for any other place but I think when we are in the business of foreign policy, the technology that goes into foreign policy, the use of that technology, how it manifests itself, and how we treat other countries, I think it's important that we try not to be—not the dominant leader. I think we ought to bend over backwards to have a unified, worldwide approach."

The message that resonated with people around the world was not of US greatness and strength; it was of sharing and community and openness. The global power that Eisenhower, Kennedy, Johnson, and Nixon all pursued through the American space program required "bending over backwards to have a unified, worldwide approach," as Collins put it. It required forgoing the message of nationalism in favor of global connectedness. For Apollo to "win hearts and minds," to advance US national interests, it had to be an achievement *of* and not *for* all humankind.

ACKNOWLEDGMENTS

I have had the good fortune of accruing immense debt from generous colleagues, friends, and family during the long life of this project. First, I want to acknowledge and thank the participants in Project Apollo public diplomacy who shared their time and recollections with me: Buzz Aldrin, Neil Armstrong, Frank Borman, Simon Bourgin, Ken Bryson, Charlie Duke, Don Eyles, Conan Grames, Beverly Gray, Fred Haise, Michio Horikawa, James Lovell, Jack Masey, Beverly Payeff-Masey, Harrison Schmitt, David Scott, and Erik Tandberg.

A special thanks is owed to Michael Collins, who answered all of my space diplomacy questions over the years with his characteristic graciousness and good humor. As should be readily apparent in these pages, his insights and stories shaped my thinking about Apollo and its relationship to the world. I feel immeasurably lucky for our conversations. I thank Ann Collins Starr and Kate Collins as well, for their support of this project and all the work they do to preserve the legacy of Apollo.

This project began at MIT more years ago than I would like to count. It benefited from the History, Anthropology, Science, Technology and Society program as well as the wisdom and work of John Durant, Mike Fischer, Karen Gardner, Stefan Helmreich, Merritt Roe Smith, and Larry Young. Rosalind Williams, David Kaiser, Leo Marx, and David Mindell guided this project, lending encouragement and good advice at each stage. I am deeply appreciative of their support and mentorship.

During the life of this project I was fortunate to receive fellowships and scholarly homes at the Royal Institute of Technology (KTH); the Consortium for the History of Science, Technology and Medicine; the Daniel and Florence Guggenheim Foundation; the Adler Planetarium and Astronomy Museum; and the American Institute of Physics (AIP). For their hospitality, thoughtful questions, and conversations, I thank my friends and colleagues at KTH and the University of Gothenburg, including Anna Åberg, Dag Avango, Mats Fridlund, Johan Gärdebo, Sabine Höhler, Arne Kaijser, Sverker Sörlin, and Nina Wormbs. The consortium provided an ideal community for a budding historian of science and technology. I wish to thank the many friends and colleagues I gained in Philadelphia, in particular Simon Joseph, Julia Mansfield, Rebecca Onion, Sheila O'Shaughnessy, Emily Stanback, and especially Babak Ashrafi. From the Adler's Webster Institute I thank Lauren Boegen, Marv Bolt, Jennifer Brand, Misty DeMars, Sara Gonzales, Jodi Lacy, Jill Postma, and Bruce Stephenson for sharing their expertise in astronomy and public history. From AIP, I thank Melinda Baldwin, Charles Day, Greg Good, Paul Guinnessy, Stephanie Jankowski, and Melanie Mueller whose collegiality and feedback shaped this project.

Like a bad penny, I have kept on returning to the Smithsonian National Air and Space Museum for more than a dozen years. The five positions I have held at the Smithsonian have not only been formative to this project, but I owe much of my knowledge of and appreciation for the role of spaceflight in society to the opportunities afforded by being part of such a vibrant community. I would especially like to thank Chandra Bhimull, Chris Browne, Luca Buvoli, Paul Ceruzzi, Pete Daniel, Jim David, David DeVorkin, Jim Fleming, Alexander Geppert, Hunter Hollins, Monique Laney, Roger Launius, Jennifer Levasseur, Cathy Lewis, Neil Maher, Esperanza Mayobre, Patrick McCray, Valerie Neal, Allan Needell, Anke Ortlepp, Richard Paul, Tony Reichhardt, Matt Sanders, Matt Shindell, Ellen Stofan, Jim Thomas, Deborah Warner, Margaret Weitekamp, Collette Williams, and Lisa Young. I add an additional note of appreciation to Martin Collins for his thoughtful and charitable mentorship.

Michael Neufeld generously read and edited every chapter of this book, offering both his expert eye and encouragement. He is a model colleague, and I am deeply indebted to his encyclopedic knowledge, precision, and insights. John Logsdon helped by sending me dozens of archival documents, providing opportunities to speak with Project Apollo participants, sharing his vast expertise, and providing his incisive chapter edits.

Much of the material in this book had its first run in conference presentations and lectures, and benefited immeasurably from the thoughtful critique of colleagues. In particular, I wish to thank Bill Barry, Laura Belmonte, Nicholas Cull, Brian Etheridge, Steven Grundman, Scott Knowles, John Krige, Bhavya Lal, Hallvard Notaker, Kathy Olesko, Kendrick Oliver, Irene Porro, Marc Rodriguez, Giles Scott-Smith, Eran Shalev, Asif Siddiqi, David Snyder, John Soares, Jenifer Van Vleck, and Audra Wolfe.

The research for this book relied heavily on the expertise and generosity of archivists and colleagues at the National Archives and Records Administrations, the John F. Kennedy Presidential Library and Archives, the Lyndon B. Johnson Presidential Library and Archives, the Richard M. Nixon Presidential Library and Archives, the Boston University Archives, the Smithsonian National Air and Space Museum Archives, the Library of Congress, the MIT Archives, and the Smithsonian Institution Archives. I am especially thankful to the archivists and librarians at the NASA History Program Office, including Colin Fries, John Hargenrader, and Elizabeth Suckow; Patti Williams at the NASM archives; Nobumichi Ariga for guiding me through Japanese archives; Anja Kolzsch and Helmuth Trischler for hosting me at the Deutsches Museum; Rory Cook at Science Museum London; Karl Erik Andersen at the National Library of Norway; and Frode Weium at the Norsk Teknisk Museum. I am grateful to colleagues at the US State Department who have supported this project in various ways, including Dinah Arnett, Jane Carpenter-Rock, Heiko Herold, Molly Kress, Hillary LeBail, Wibke Reincke, and Eitan Schiffman.

I am grateful to Howard Yoon, my literary agent at Ross Yoon Agency, for his advocacy of this project. At Basic Books I thank

Rachel Field, TJ Kelleher, and Katie Lambright. An additional note of gratitude is owed to Melissa Veronesi and Donald Pharr for their patience and generosity.

To my many friends and colleagues who have offered encouragement, guest rooms, conversation, feedback, tours of their cities during research trips, company at launches, humor, and kindness, thank you. In particular, I wish to acknowledge Poppy Alexander, Etienne Benson, Martins Blums, Jonathan Coopersmith, Elise Crull, David Deen, Angela Deen, Nate Deshmukh, Aaron Divine, Chris Divine, Rebecca Dobrow, Ethan Dobrow, Xaq Frolich, Emily Gibson, Matt Hersch, Chihyung Jeon, Patricia Kanngiesser, Toby Elliman, Angelina Long Callahan, Jason Callahan, Yanni Loukissas, Larry McGlynn, Diane McWhorter, Lisa Messeri, Zara Mirmalek, Emily Miraldi, Canay Ozden-Schilling, Tom Ozden-Schilling, Matt Ogle, Rebecca Perry, Claire Scoville, David Singerman, Leo Slater, Eden Savino, Ellan Spero, Michaela Thompson, John Tylko, Janice Norcutt, Ben Wilson, Uri Mariash, and Beatka Zakrzewski.

Profound gratitude is due to my family, my unstinting champions: Ayr, Brooke, Clementine, Blue, Violet, Asa, Alex, Amos, Arla, and Arrow. And to Zeke Emanuel for his endless wells of confidence and optimism for me and my work, and his willingness to edit draft after draft of this book.

Finally, my greatest debt—as always—is owed to my parents, Michael and Rebecca. I dedicate this book to them in appreciation for all the lessons they have taught me.

NOTES

PREFACE

1. Norman Mailer, *Of a Fire on the Moon* (1969; reprint, New York: Random House, 2014).

2. Michael Collins, *Carrying the Fire: An Astronaut's Journey* (New York: Ballantine, 1975), 409.

3. "EP-72 Log of Apollo 11," NASA History Division, https://history.nasa.gov/ap11ann/apollo11_log/log.htm.

4. Collins, *Carrying the Fire*, 408.

5. Michael Collins, interview with author, March 23, 2016.

6. Collins interview.

7. Collins interview.

8. John Dower, *Embracing Defeat: Japan in the Wake of World War II* (New York: W.W. Norton, 1999), 30.

9. *Asahi Shimbun*, August 20, 1945, quoted in Dower, *Embracing Defeat*, 494.

10. USIS Tokyo to USIA Washington, September 4, 1962, Box 258, Folder "Outer Space, 14.B.5, Outer Space Exhibits, June–December 1962, Part 2 of 2," Entry 1613, RG 59, NARA.

11. USIS Tokyo to USIA Washington, September 4, 1962.

12. Joseph Nye Jr. coined the phrase "soft power," defining it as attractive or co-optive power, as opposed to "coercive power" or "hard power." Governments often use both in concert. See Joseph Nye Jr., *Bound to Lead: The Changing Nature of American Power* (New York: Basic Books, 1990), and Joseph Nye Jr., *Soft Power: The Means to Success in World Politics* (New York: PublicAffairs, 2004).

13. Collins interview.

INTRODUCTION: MOONRISE

1. Robert C. Seamans Jr., *Project Apollo: The Tough Decisions* (Washington, DC: National Aeronautics and Space Administration, 2005), 113.

2. Courtney Brooks, James Grimwood, and Loyd Swenson, *Chariots for Apollo: A History of Manned Lunar Spacecraft* (Washington, DC: NASA Special Publication-4205, 1979), chapter 13.

3. Dwight Chapin to H. R. Haldeman, "Status Report on Apollo 11 Project," July 1, 1969, Outer Space-3 File, Richard Nixon Presidential Library, Yorba Linda, CA (hereafter RNPL); William Safire, "Of Nixon, Kennedy and Shooting the Moon," *New York Times*, July 17, 1989, A17.

4. Chapin to Haldeman, "Status Report"; Safire, "Of Nixon"; Seamans, *Project Apollo*, 113; Brooks, Grimwood, and Swenson, *Chariots for Apollo*, chapter 13.

5. National Archives Identifier 6922346, Richard Nixon Presidential Library and Museum Identifier 1979-12. The phone is part of the collection at the RNPL.

6. The astronauts thanked the president, commented on the privilege of representing all mankind, and then spent the rest of their stay on the lunar surface deploying a science package, collecting moon rocks and soil, and taking what would become iconic photographs. Brooks, Grimwood, and Swenson, *Chariots for Apollo*, chapter 14.

7. Buzz Aldrin, *Magnificent Desolation: The Long Journey Home from the Moon* (New York: Crown Archetype, 2009), 54.

8. Looking at Project Apollo through the lens of diplomatic and transnational history reveals how the production and transfer of knowledge is a malleable, variable process that was connected to the local, regional, and global, and was not simply a diffusion of Western science and engineering information. For a discussion of the circulation of (techno)science, see John Krige, ed., *How Knowledge Moves: Writing the Transnational History of Science and Technology* (Chicago: University of Chicago Press, 2019).

9. This process of globalization has roots in sixteenth-century voyages of exploration and commerce. See Alfred Eckes Jr. and Thomas Zeiler, *Globalization and the American Century* (New York: Cambridge University Press, 2003), 6. On the increase of global society, see Emily Rosenberg, ed., *A World Connecting, 1870–1945* (Cambridge, MA: Harvard University Press, 2012).

10. Akira Iriye, ed., *Global Interdependence: The World After 1945* (Cambridge, MA: Harvard University Press, 2014), 5, 847.

11. Papers of John F. Kennedy, Presidential Papers, President's Office Files, Speech Files, Special message to Congress on urgent national needs,

May 25, 1961, John F. Kennedy Presidential Library and Museum, Boston (hereafter JFKL).

12. Mark Philip Bradley, "Decolonization, the Global South, and the Cold War, 1919-1962," in *The Cambridge History of the Cold War*, ed. Melvyn P. Leffler and Odd Arne Westad (Cambridge: Cambridge University Press, 2010), 464-465.

13. The broad front of the competition—especially aggressive interventionist tactics—was often discordant with the ideological claims of each superpower. Words and deeds did not always align. Odd Arne Westad, "The Cold War and the International History of the Twentieth Century," in *The Cambridge History of the Cold War*, ed. Melvyn P. Leffler and Odd Arne Westad (Cambridge: Cambridge University Press, 2010), 4.

14. See Nye, *Bound to Lead* and *Soft Power*.

15. John F. Kennedy, "Message to Congress on the Nation's Urgent Needs," May 25, 1961, Office of the Clerk, General Records, 1791–2010, 87th Congress (1961–1963), Records of the United States House of Representatives, NARA.

16. Historian Odd Arne Westad has called it a competition "for the society of the future." Westad, *The Cold War* (New York: Basic Books, 2017), 4.

17. Historian Jason Parker usefully describes public diplomacy as "a multifront media war, launched by the superpowers in pursuit of strategic and psychological gains, to win [people] over." Parker, *Hearts, Minds, Voices: US Cold War Public Diplomacy and the Formation of the Third World* (New York: Oxford University Press, 2016), 3. For a history of US public diplomacy, see Nicholas Cull, *The Cold War and the United States Information Agency: American Propaganda and Public Diplomacy, 1945–1989* (New York: Cambridge University Press, 2008).

18. Kennedy, "Message to Congress on the Nation's Urgent Needs."

19. Mary Dudziak, *Cold War Civil Rights: Race and the Image of American Democracy* (Princeton, NJ: Princeton University Press, 2000), 17. Newspapers, astronaut memoirs, historical studies, philosophical essays, and popular publications have all detailed the astronauts' experience on the lunar surface and considered its significance. And while these authors may gesture to the global "participation" in the flight, it has surprisingly been left unexplored. Recent examples include Douglas Brinkley, *American Moonshot: John F. Kennedy and the Great Space Race* (New York: Harper, 2019); Andrew Chaikin, *A Man on the Moon* (New York: Viking, 1994); Charles Fishman, *One Giant Leap: The Impossible Mission That Flew Us to the Moon* (New York: Simon & Schuster, 2019); Stephen B. Johnson, *The Secret of Apollo: Systems Management in American and European Space Programs* (Baltimore: Johns Hopkins University Press,

2002); Roger D. Launius, *Apollo's Legacy: Perspectives on the Moon Landings* (Washington, DC: Smithsonian Books, 2019); Neil Maher, *Apollo in the Age of Aquarius* (Cambridge: Harvard University Press, 2017); and Matthew D. Tribbe, *No Requiem for the Space Age: The Apollo Moon Landings and American Culture* (New York: Oxford University Press, 2014).

20. John Logsdon, *John F. Kennedy and the Race to the Moon* (New York: Palgrave Macmillan, 2010), 6; Kennedy, "Message to Congress on the Nation's Urgent Needs."

21. Logsdon, *John F. Kennedy*, 2–3; Casey Drier, "How Much Did the Apollo Program Cost?," accessed October 24, 2019, www.planetary.org /get-involved/be-a-space-advocate/become-an-expert/cost-of-apollo -program.html. Inflation in 2019 dollars adjusted using NASA's New Start Index for aerospace projects.

22. Frank Borman oral history interview with author, December 10, 2012.

CHAPTER 1: THE LAUNCH OF THE SPACE RACE, 1946–1957

1. Asif Siddiqi, *The Red Rockets' Glare: Spaceflight and the Soviet Imagination, 1857–1957* (New York, Cambridge University Press, 2010), 350.

2. Mike Gruntman, "From Tyuratam Missile Range to Baikonur Cosmodrome," *Acta Astronautica* 155 (February 2019): 350–366; Siddiqi, *Red Rockets' Glare*, 350.

3. Siddiqi, *Red Rockets' Glare*, 350–352.

4. James Harford, *Korolev: How One Man Masterminded the Soviet Drive to Beat America to the Moon* (New York: John Wiley, 1997), 1; Walter A. McDougall, . . . *The Heavens and the Earth: A Political History of the Space Age* (New York: Basic Books, 1985), 36–40.

5. Siddiqi, *Red Rockets' Glare*, 353.

6. Siddiqi, *Red Rockets' Glare*, 353–356; McDougall, *The Heavens and the Earth*, 61.

7. William Taubman, *Khrushchev: The Man and His Era* (New York: W.W. Norton, 2003), xx; Geoffrey Roberts, *Stalin's General: The Life of Georgy Zhukov* (New York: Random House, 2012), 258–259.

8. His service gained him universal acclaim among Soviet citizens, which in turn made him a threat to the paranoid Soviet premier Joseph Stalin. Instead of awarding Zhukov for saving the Soviet Union from Hitler's Nazi forces, Stalin relegated him to an obscure regional command. When Stalin died in 1953, new Soviet leadership saw the advantages of Zhukov's popularity among the armed forces, and he was able to rise quickly to the rank of minister of defense. Roberts, *Stalin's General*, 5–8.

9. Taubman, *Khrushchev*, 362.

10. Sergei Khrushchev, *Nikita Khrushchev and the Creation of a Superpower*, trans. Shirley Benson (University Park: Pennsylvania State University Press, 2000), 259.

11. Khrushchev, *Nikita Khrushchev*, 260.

12. "Announcement of the First Satellite," *Pravda*, October 5, 1957; F. J. Krieger, *Behind the Sputniks* (Washington, DC: PublicAffairs, 1958), 311–312.

13. Khrushchev, *Nikita Khrushchev*, 260.

14. Although the impact of Rockefeller's particular report on Eisenhower's policy making is debatable, the ideas articulated within its more than two hundred pages reflect a larger trend in political theory that framed policy making in the 1950s. In many ways it reflects Eisenhower's own appreciation of psychological warfare within the larger cold war battlefield. Henry Kissinger, "Psychological and Pressure Aspects of Negotiations with the USSR," in *Psychological Aspects of United States Strategy*, November 1955, Psychological Aspects of United States Strategy: Source Book of Individual Papers, Box 10, White House office, Office of the Special Assistant for National Security Affairs, NSC Series, Eisenhower Presidential Library, Abilene, KS (hereafter DEPL). For a discussion and contextualization of the document, see Kenneth Osgood, *Total Cold War* (Lawrence: University Press of Kansas, 2006), 181–183.

15. Kissinger, quoted in Niall Ferguson, *Kissinger: 1923–1968: The Idealist* (New York: Penguin, 2015), 356.

16. Historians such as Walter McDougall, Stephen Ambrose, and Robert Divine have made this claim. For a historiographical analysis, see Kenneth Osgood, "Before Sputnik: National Security and the Formation of U.S. Outer Space Policy," in *Spaceflight and the Myth of Presidential Leadership*, ed. Roger Launius and Howard McCurdy (Urbana: University of Illinois Press, 1997), 197–199.

17. Jean Edward Smith, *Eisenhower: In War and Peace* (New York: Random House, 2012), 640.

18. Smith, *Eisenhower*, 641.

19. Smith, *Eisenhower*, 643.

20. Osgood, *Total Cold War*, 71.

21. Osgood, *Total Cold War*, 71–75; David Callahan and Fred I. Greenstein, "Eisenhower and U.S. Space Policy," in *Spaceflight and the Myth of Presidential Leadership*, ed. Roger Launius and Howard McCurdy (Urbana: University of Illinois Press, 1997), 19–20.

22. Before 1953, a host of government agencies and private institutions took part in propaganda and psychological warfare studies and operations planning. The Central Intelligence Agency, the Department of Defense,

the State Department, and the White House, along with private institutions and universities, took part in immediate postwar propaganda and psychological warfare practices. The State Department handled the majority of overt and public propaganda, and the CIA organized covert propaganda programming.

23. Although the Jackson Committee recommended consolidating information activities under State Department jurisdiction, Eisenhower created an independent agency. The USIA officially came into being on August 1, 1953. Osgood, *Total Cold War*, 54–55, 88–89; Nicholas Cull, *The Cold War and the United States Information Agency: American Propaganda and Public Diplomacy, 1945–1989* (Cambridge: Cambridge University Press, 2010), 94–96; Callahan and Greenstein, "Eisenhower and U.S. Space Policy," 20.

24. A public affairs officer (PAO), working under the direction of the US ambassador, ran each post. Information officers, who primarily dealt with the local press and radio, and cultural affairs officers, who were in charge of working with local educational and cultural leaders, assisted these PAOs. Thomas Sorensen, remarks before the NASA Office of Public Information Staff Conference, June 27, 1961, Box 5, Entry P 243RG 306, NARA.

25. Sorensen remarks.

26. Osgood, *Total Cold War*, 98.

27. Cull, *The Cold War*, 101.

28. Osgood, *Total Cold War*, 74–75; Callahan and Greenstein, "Eisenhower and U.S. Space Policy," 19–20.

29. Francis H. Clauser et al., "Preliminary Design of an Experimental World-Circling Spaceship," Report No. SM-11827 (Santa Monica, CA: Douglas Aircraft Company, Santa Monica Plant Engineering Division, May 2, 1946).

30. McDougall, *The Heavens and the Earth*, 108.

31. Paul Kecskemeti, "The Satellite Rocket Vehicle: Political and Psychological Problems," RAND RM-567, October 4, 1950. See also McDougall, *The Heavens and the Earth*, 108.

32. Walter Sullivan, "James A. Van Allen, Discoverer of Earth-Circling Radiation Belts, Is Dead at 91," *New York Times*, August 10, 2006, C14.

33. "L. V. Berkner Dies," *New York Times*, June 5, 1967, 43.

34. *Teasel Muir-Harmony, "Tracking Diplomacy: The International Geophysical Year and American Scientific and Technical Exchange with East Asia, 1955–1973," in Globalizing Polar Science: Reconsidering the International Polar and Geophysical Year, ed. Roger Launius, James Fleming, and David DeVorkin (New York: Palgrave, 2010).*

35. Lloyd V. Berkner, Douglas Merritt Whitaker, and the National Research Council, *Science and Foreign Relations: International Flow of*

Scientific and Technological Information, Federal Foreign Policy Series 30 (Washington, DC: US Department of State, 1950); Allan A. Needell, *Science, Cold War, and the American State: Lloyd V. Berkner and the Balance of Professional Ideals* (Amsterdam: Harwood Academic, 2000), 299; Audra J. Wolfe, *Freedom's Laboratory: The Cold War Struggle for the Soul of Science* (Baltimore: Johns Hopkins University Press, 2018), 44–45.

36. In the end, sixty-seven nations employed some four thousand research stations around the world during the IGY. The United States National Committee for the IGY fell under the auspices of the National Academy of Sciences (NAS), an honorific society of American scientists. The United States funded roughly one-fourth of the $2 billion cost of the entire program, with some of this financial support coming from the CIA. And Berkner and Chapman, who both attended Van Allen's dinner party, served as president and vice president of the US National Committee, respectively. At first, the IGY organizing committee targeted nine areas of scientific research: meteorology, latitude and longitude determinations, geomagnetism, the ionosphere, aurora and airglow, solar activity, cosmic ray, glaciology, and oceanography. The committee added the launching of artificial satellites to the program for the support of geodetic and atmospheric studies of the Earth. Allan Needell, "Lloyd Berkner and the International Geophysical Year Proposal in Context," in *Globalizing Polar Science: Reconsidering the International Polar and Geophysical Years,* ed. Roger Launius, James Fleming, and David DeVorkin (New York: Palgrave, 2010), 217; Needell, *Science,* 299. Roger Launius, "Toward the Poles," in *Globalizing Polar Science: Reconsidering the International Polar and Geophysical Years,* ed. Roger Launius, James Fleming, and David DeVorkin (New York: Palgrave, 2010), 67; W. Patrick McCray, *Keep Watching the Skies! The Story of Operation Moonwatch & the Dawn of the Space Age* (Princeton, NJ: Princeton University Press, 2008), 58-63; McDougall, *The Heavens and the Earth,* 118-120; Osgood, "Before Sputnik," 205.

37. Grosse to Quarles, August 25, 1953, in *Exploring the Unknown: Selected Documents in the History of the U.S. Civil Space Program,* ed. John M. Logsdon et al. (Washington, DC: Government Printing Office, 1995), 1:236–244, 1:267–269; Osgood, "Before Sputnik," 202–203.

38. Frederick I. Ordway II and Mitchell R. Sharpe, *The Rocket Team* (New York: Crowell, 1979), 376.

39. McDougall, *The Heavens and the Earth,* 119.

40. Osgood, "Before Sputnik," 203; Dwayne A. Day, "Cover Stories and Hidden Agendas: Early American Space and National Security Policy," in *Reconsidering Sputnik: Forty Years Since the Soviet Satellite,* ed. Roger Launius, John Logsdon, and Robert W. Smith (Amsterdam: Harwood Academic, 2000), 163-165.

41. MacDougall, *The Heavens and the Earth*, 108-111.

42. Historian Kenneth Osgood makes this point in Osgood, "Before Sputnik," 205. See also Zuoyue Wang, *In Sputnik's Shadow: The President's Science Advisory Committee and Cold War America* (New Brunswick, NJ: Rutgers University Press, 2009), 49-54.

43. "Donald A. Quarles," *Washington Post*, May 9, 1959, A8.

44. Jack Raymond, "Quarles Dies in Sleep at 64; McElroy May Now Stay On," *New York Times*, May 9, 1969, 1.

45. "Mild Man Up in Arms: Donald Aubrey Quarles," *New York Times*, March 27, 1957, 20.

46. "Donald A. Quarles," A8.

47. Day, "Cover Stories," 166.

48. Siddiqi, *Red Rockets' Glare*, 320-322.

49. Henry Kissinger, *White House Years* (Boston: Little, Brown, 1979), 4.

50. Richard Norton Smith, *On His Own Terms: A Life of Nelson Rockefeller* (New York: Random House, 2014), 29.

51. "Memorandum of Discussion at the 250th Meeting of the National Security Council, Washington, May 26, 1955," *FRUS*, 1955-1957, vol. XI, United Nations and General International Matters, Document 341.

52. Annex B, Memorandum from the President's Special Assistant (Rockefeller) to the Executive Secretary of the National Security Council (Lay), May 17, 1955, draft in the White House; Memorandum of Discussion. See also Osgood, "Before Sputnik," 208.

53. McDougall, *The Heavens and the Earth*, 120-121.

54. The report's guidelines for a US scientific satellite program were in line with many of Rockefeller's recommendations. In addition to not interfering with military ballistic missile and reconnaissance satellite development, the scientific satellite program should demonstrate the United States' peaceful intentions in space, according to the council. To accomplish this, the satellites should be launched during the IGY and collect valuable scientific data. National Security Council Report 5520, May 20, 1955, *FRUS*, 1955-1957, vol. XI, United Nations and General International Matters, Document 340. See also Osgood, *Total Cold War*, 328-329.

55. "Memorandum of Discussion."

56. William I. Hitchcock, *The Age of Eisenhower: America and the World in the 1950s* (New York: Simon & Schuster, 2018), 268-272.

57. Smith, *On His Own Terms*, 239-242.

58. Quoted in Osgood, *Total Cold War*, 192-198.

59. Siddiqi, *Red Rockets' Glare*, 322-323.

60. R. Cargill Hall, "The Origins of U.S. Space Policy: Eisenhower, Open Skies, and Freedom of Space," *Colloquy* 14 (December 1993): 17. Press

release, statement by White House Press Secretary James C. Hagerty on earth-circling satellites as part of IGY program, July 29, 1955, Records of the President, Official Files, Box 624, OF 146-E International Geophysical Year (1), NAID #16646172.

61. Research and Reference Service West European Public Opinion Barometer Study, September 29, 1955, Box 4, Folder "WE-17," RG 306, Entry 1010, NARA.

62. John Hillaby, "Soviet Planning Early Satellite," *New York Times*, August 3, 1955, 8; "Russians Claim They'll Launch First Satellite," *Los Angeles Times*, August 3, 1955, 11; Siddiqi, *Red Rockets' Glare*, 323.

63. Siddiqi, *Red Rockets' Glare*, 324.

64. Asif Siddiqi, "Korolev, Sputnik and the IGY," in *Reconsidering Sputnik: Forty Years Since the Soviet Satellite*, ed. Roger Launius, John Logsdon, and Robert W. Smith (Amsterdam: Harwood Academic, 2000), 49–50.

65. Siddiqi, *Red Rockets' Glare*, 324–331.

66. The Stewart Committee's decision is treated in depth in Michael J. Neufeld, "Orbiter, Overflight, and the First Satellite: New Light on the Vanguard Decision," in *Reconsidering Sputnik: Forty Years Since the Soviet Satellite*, ed. Roger D. Launius, John M. Logsdon, and Robert W. Smith (Amsterdam: Harwood Academic, 2000), 231–257. On the Eisenhower administration's broader approach to early space policy, see McDougall, *The Heavens and the Earth*, 122.

67. "Memorandum of Discussion at the 322nd Meeting of the National Security Council, Washington, DC, May 10, 1957," *FRUS*, 1955–1957, United Nations and General International Matters, vol. XI.

68. Osgood, "Before Sputnik," 206, 219.

69. Korolev, quoted in Siddiqi, "Korolev, Sputnik and the IGY," 56.

70. Siddiqi, "Korolev, Sputnik and the IGY," 56–57.

71. The quality of Sputnik's surface was important for reflecting heat and ground tracking, but polishing the mock-up's surface was an aesthetic priority, not an engineering one. James J. Harford, "Korolev's Triple Play: Sputniks 1, 2, and 3," in *Reconsidering Sputnik: Forty Years Since the Soviet Satellite*, ed. Roger Launius, John M. Logsdon, and Robert W. Smith (Amsterdam: Harwood Academic, 2000), 81–82.

CHAPTER 2: SPUTNIK AND THE POLITICS OF SPACEFLIGHT, 1957

1. Benjamin Fine, "Arkansas Troops Bar Negro Pupils," *New York Times*, September 5, 1957, 1; Dudziak, *Cold War Civil Rights*, 119; National Park Service, "Little Rock Central High School: National Historic

Site," www.nps.gov/articles/little-rock-central-high-school-501428.htm #4/35.46/-98.57.

2. Odd Arne Westad, *The Cold War: A World History* (New York: Basic Books, 2017), 277; Dudziak, *Cold War Civil Rights*, 118–125; Cull, *The Cold War*, 147.

3. John Foster Dulles, quoted in Dudziak, *Cold War Civil Rights*, 118.

4. Robert Divine, *The Sputnik Challenge: Eisenhower's Response to the Soviet Satellite* (New York: Oxford University Press, 1993).

5. Paul Dickson, *Sputnik: The Shock of the Century* (New York: Walker, 2007), 12.

6. Anton L. Hales, "Lloyd Viel Berkner," in *Biographical Memoirs* (Washington, DC: National Academy Press, 1992), 61:3.

7. Walter Sullivan, "Soviet Embassy Guests Hear of Satellite from an American as Russians Beam," *New York Times*, October 5, 1957, 3; Dickson, *Sputnik*, 12–13.

8. Lyndon B. Johnson, *Vantage Point: Perspectives of the Presidency, 1963–1969* (New York: Random House, 1971), 272; Robert Caro, *Master of the Senate: The Years of Lyndon Johnson* (New York: Knopf, 2002), 1021; Dickson, *Sputnik*, 18.

9. Caro, *Master of the Senate*, 1022.

10. Divine, *The Sputnik Challenge*, xiv.

11. Dickson, *Sputnik*, 9.

12. "The Impact of 'Sputnik' upon the Press of Western Europe," October 18, 1957, Box 9, Entry P 243, RG 306, NARA.

13. McDougall, *The Heavens and the Earth*, 143.

14. Quoted in Cull, *The Cold War*, 135.

15. Vance C. Pace oral history interview by David Reuther, May 15, 2015, the Association of Diplomatic Studies and Training Foreign Affairs Oral History Project, 13.

16. Cathleen Lewis, "The Red Stuff: A History of the Public and Material Culture of Early Human Spaceflight in the U.S.S.R." (PhD diss., George Washington University, 2008), 45.

17. Khrushchev, quoted in Taubman, *Khrushchev*, 378.

18. Khrushchev, quoted in Taubman, *Khrushchev*, 378.

19. Oliver M. Gale, "Post-Sputnik Washington from an Inside Office," *Cincinnati Historical Society Bulletin* 31, no. 4 (1973): 226.

20. Both quotations are from Robert Dallek, *Lone Star Rising: Lyndon Johnson and His Times* (New York: Oxford University Press, 1991), 529–534.

21. Robert Divine, *The Johnson Years* (Lawrence: University of Kansas Press, 1987), 2:223.

22. Caro, *Master of the Senate*, 1024.

23. Most historical scholarship that recounts the early space race centers on questions of whether or not Sputnik surprised and shook the world. Did Sputnik mark or usher in a change? Was Sputnik the shock of the century? Was it a "Pearl Harbor"? These are questions that drive many accounts. But the answers are not essential to this book or, as I will argue, twentieth-century history. The critical element of this story for how and why spaceflight assumed national priority in the late 1950s and 1960s is revealed in how politicians within the United States, Soviet Union, and around the world used Sputnik for political ends. Robert Divine examines the response of Eisenhower to Sputnik and suggests that he failed to meet this test of his presidential leadership. Matthew Bille reviews US efforts to launch a satellite and explains that little was known of the Soviet missile capability before Sputnik was launched. Paul Dickson argues that "just when Americans were feeling self-confident and optimistic about the future, along came the crude, kerosene-powered Sputnik launch." Matthew Brzezinski's *Red Moon Rising* looks at both US and Soviet perspectives, but the book is centered on the question of whether or not the "shock" was justified. Matthew Bille, Erika Lishock, and James Van Allen, *The First Space Race: Launching the World's First Satellites* (College Station: Texas A&M Press, 2004); Matthew Brzezinski, *Red Moon Rising: Sputnik and the Hidden Rivalries That Ignited the Space Age* (New York: Times Books, 2007); Michael D'Antonio, *A Ball, a Dog, and a Monkey: 1957—The Space Race Begins* (New York: Simon & Schuster, 2007); Dickson, *Sputnik*; Divine, *The Sputnik Challenge*; Roger Launius, John Logsdon, and Robert W. Smith, eds., *Reconsidering Sputnik: Forty Years Since the Soviet Satellite* (Amsterdam: Harwood Academic, 2000). Asif Siddiqi's account of Sputnik reveals that it was not technologically determined, nor was it a "shock" to American scientists and the media. Scientists, engineers, government officials, and the public were at least minimally informed about the space programs in each country. Siddiqi, *Red Rockets' Glare*.

24. Charles J. V. Murphy, "The White House Since Sputnik," *Fortune*, January 1958, 100.

25. The NSF briefed Eisenhower first thing on Monday, October 7; on Tuesday he met with key members of his staff about Vanguard; he held a press conference on Wednesday; Thursday it was the NSC meeting; and then on Friday his Cabinet discussed the federal budget in the new post-Sputnik climate. The following week, in a meeting with the Science Advisory Committee he discussed funding scientific research, military preparedness, the status of science education, science collaboration with allies, and how to create a more fertile environment for the development of US science. Even though Sputnik dominated Eisenhower's agenda in the weeks following the launch, to many—especially within the opposing

political party—the president was not doing enough. "Memorandum of Conference with the President on American science education and Sputnik, October 15, 1957" (dated October 16), DDE's Papers as President, DDE Diary Series, Box 27, October '57 Staff Notes (2); NAID #12043792, DEPL.

26. "Memo of Conference with the President on October 8, 1957," 8:30 a.m. (dated October 9), DDE's Papers as President, DDE Diary Series, Box 27, October '57 Staff Notes (2); NAID #12043774, DEPL; "The Weather," *Washington Post*, October 8, 1957, A1.

27. "Memo of Conference with the President on October 8, 1957," 5:00 p.m. (dated October 9), DDE's Papers as President, DDE Diary Series, Box 27, October '57 Staff Notes (2); NAID #12043783, DEPL.

28. Official White House transcript of President Eisenhower's Press and Radio Conference #123 concerning the development by the U.S. of an earth satellite, October 9, 1957, DDE's Papers as President, Press Conference Series, Box 6, Press Conference Oct. 9, 1957; NAID #12086488, DEPL.

29. W. H. Lawrence, "President Voices Concern on U.S. Missiles Program, but Not on the Satellite," *New York Times*, October 10, 1957, 1; Edward T. Folliard, "U.S. Sets Tests in December and March: Sputnik Fails to Alarm Ike," *Washington Post*, October 10, 1957, A1.

30. During the Eisenhower administration the NSC was made up of the secretary of state, the secretary of defense, the chairman of the joint chiefs, the director of central intelligence, the director of defense mobilization, the vice president, the director of the Bureau of the Budget, the secretary of the Treasury, and the director of foreign aid. Smith, *Eisenhower*, 568.

31. Hitchcock, *The Age of Eisenhower*, 148–153.

32. "All his propaganda guns" is a reference to the combination of Sputnik and the announcement of an ICBM test as well as a large-scale hydrogen bomb test in Novaya Zemla. National Security Council, "Discussion at the 339th Meeting of the National Security Council, Thursday, October 10, 1957," October 11, 1957, NSC Series, Box 9, Eisenhower Papers, 1953–1961 (Ann Whitman File), DEPL. Allen Dulles, quoted in James Schwoch, *Global TV: New Media and the Cold War, 1946–69* (Urbana: University of Illinois Press, 2009), 51.

33. National Security Council, "Discussion at the 339th Meeting."

34. Bruce Lambert, "L. Arthur Larson Is Dead at 82; Top Eisenhower Aide and Writer," *New York Times*, April 1, 1993, 24; George Dixon, "Washington Scene . . . : 'Be-Beastly-to-USIA Week,'" *Washington Post*, June 10, 1957, A13.

35. National Security Council, "Discussion at the 339th Meeting."

36. Divine, *The Sputnik Challenge*, 21.

37. Lewis, "The Red Stuff," 83.

38. Siddiqi, *Red Rockets' Glare*, 360; Lewis, "The Red Stuff," 84.

39. Georgy Grechko, interview with James J. Harford in Harford, "Korolev's Triple Play," 86.

40. "World Opinion and the Soviet Satellite," October 17, 1957, Box 9, Folder "Satellites—Sputnik," Entry P 243, RG 306, NARA. See also Cull, *The Cold War*, 135.

41. NSC 5501, titled "Basic National Security Policy," January 6, 1955. Not printed. See *FRUS*, 1955–1957, vol. IX, Foreign Economic Policy; Foreign Information Program, Document 190; James Schwoch, "The Cold War, the Space Race, and the Globalization of Public Opinion Polling" (paper presented at the International Studies Association Conference, New Orleans, Louisiana, March 24–28, 2002); "Memorandum of Discussion at the 235th Meeting of the National Security Council, Thursday, February 3, 1955," Foreign Relations of the United States, 1955–1957, Korea, volume XXIII, part 2.

42. Mark Haefele, "John F. Kennedy, USIA, and World Public Opinion," *Diplomatic History* 25, no. 1 (2001): 67.

43. "World Opinion and the Soviet Satellite," October 17, 1957. See also Cull, *The Cold War*, 135.

44. "World Opinion and the Soviet Satellite."

45. By April, the postmaster general included a seat on the stamp advisory committee for a representative from the USIA.

46. Cull, *The Cold War*, 144–145.

47. Statement by George V. Allen before the House Science and Astronautics Committee, January 22, 1960, Box 12, Folder "USIA: Background Material from Media for Dir. Allen Appearance Before House Science and Astronautics Committee, 1960," Entry P 243, RG 306, NARA.

48. The OCB, created by Eisenhower in 1953, reported directly to the National Security Council and was staffed by the under secretary of state, deputy director of defense, director of the CIA, and the president's special assistant for psychological warfare, among other high-level officials. Eisenhower had tasked the OCB with implementing national security policies while also initiating proposals to the NSC. William Traum, letter to Abbott Washburn, October 21, 1957, Box 9, P 243, RG 306, NARA.

49. Dwight D. Eisenhower, "Executive Order 10483—Establishing the Operations Coordinating Board," American Presidency Project, www.presidency.ucsb.edu/node/234438.

50. Operations Coordinating Board, "Guide Lines for Public Information on the U.S. and Soviet Scientific Earth Satellite Programs," November 13, 1957, Box 7, Folder "OCB—Course of Action Relating to the

Public Information Aspects of Soviet Man-in-Space Program," RG 306, Entry P 243, NARA.

51. "Atomium," the major centerpiece of the fair, symbolized this enthusiasm and optimism for the role of science in modern life in the late 1950s. Standing tall above the exposition grounds, the metallic atomic-inspired structure housed a restaurant and panoramic tower.

52. Robert C. Hickok, "Comments on Brussels World's Fair," November 8, 1957, Box 1, Program and Media Studies, RG 306, NARA; Susan Reid, "The Soviet Pavilion at Brussels '58: Convergence, Conversion, Critical Assimilation, or Transculturation?," in *The Cold War International History Project*, ed. Christian F. Ostermann (Washington, DC: Woodrow Wilson International Center for Scholars, 2010).

53. Robert H. Haddow, *Pavilions of Plenty: Exhibiting American Culture Abroad in the 1950s* (Washington, DC: Smithsonian Institution Press, 1997), 85–88.

54. Osgood, *Total Cold War*, 343–344.

55. Osgood, *Total Cold War*, 344.

56. Gregory Kulacki and Jeffrey G. Lewis, "A Place for One's Mat: China's Space Program, 1956–2003," research paper, American Academy of Arts & Sciences, 2009, 13.

57. Amy Nelson, "Cold War Celebrity and the Courageous Canine Scout: The Life and Times of Soviet Space Dogs," in *Into the Cosmos: Space Exploration and Soviet Culture*, ed. James T. Andrews and Asif Siddiqi (Pittsburgh: University of Pittsburgh Press, 2011), 139–141; Colin Burgess and Chris Dubbs, *Animals in Space: From Research Rockets to the Space Shuttle* (New York: Springer, 2007), 156–161.

58. Walter Briggs, "U.S. Prestige in Asia Continues to Decline," *Washington Post*, November 6, 1957, A3.

59. "Barbarism with Sputniks," *New York Times*, November 4, 1957, 28.

60. "A Birthday Flexing of Red Biceps," *Life*, November 18, 1957, 35.

61. Affiliates of International Research Associates of New York conducted the survey. Research and Reference Service, "Recent Trends in Latin American Opinion Toward the United States and the Soviet Union, Report #16, October, 1958," RG 306, Box 2, Entry 1010, NARA; Office of Research and Intelligence, "Trends in Japanese Attitudes Following the Launching of Sputniks I and II and Explorer I," Report #16, July, 1968, Box 1, Entry 1010, RG 306, NARA; Research and Reference Service, "Post-Sputnik Attitudes Toward NATO and Western Defense, February 1958," RG 306, Box 5, Entry 1010, NARA; *USIA 9th Review of Operations, July 1–December 31, 1957*, 2; Cull, *The Cold War*, 150.

62. William J. Jorden, "Khrushchev Asks East-West Talks to End 'Cold War,'" *New York Times*, November 7, 1957, 1.

63. Khrushchev, *Nikita Khrushchev*, 261.

64. Taubman, *Khrushchev*, 300.

65. "A Proposal for a 'Giant Step,'" *Life*, November 18, 1957, 53.

66. George V. Allen stepped in to direct the USIA in Larson's place. Cull, *The Cold War*, 148-149.

67. David L. Stebenne, *Modern Republican: Arthur Larson and the Eisenhower Years* (Bloomington; Indiana University Press, 2006), 202-203.

68. Divine, *The Sputnik Challenge*, 45.

69. "Moment in History: President Speaks as the Citizens Listen," *Life*, November 18, 1957, 40.

70. Wang, *In Sputnik's Shadow*, 13.

71. Stebenne, *Modern Republican*, 202-203.

72. Caro, *Master of the Senate*, 1027-1028.

73. Day, "Cover Stories," 180; McDougall, *The Heavens and the Earth*, 154.

74. Quoted in Schwoch, *Global TV*, 52.

75. Cull, *The Cold War*, 150.

76. Harford, "Korolev's Triple Play," 87.

77. "Capitol Dismayed at Test Failure," *New York Times*, December 7, 1957.

78. Constance McLaughlin Green and Milton Lomask, *Vanguard: A History, NASA SP-4202* (Washington, DC: Smithsonian Institution Press, 1971), chapter 11.

CHAPTER 3: A SPACE PROGRAM FOR ALL HUMANKIND, 1958–1960

1. Oliver M. Gale was the special assistant to the secretary of defense and was responsible for coordinating interviews of Pentagon officials for Johnson's hearing. Gale, "Post-Sputnik Washington," 227.

2. Oliver M. Gale Papers, 1957-1960, 1971, 1974, DEPL.

3. Caro, *Master of the Senate*, 842.

4. Note, Reedy for LBJ, October 17, 1957, cited in McDougall, *Heavens and the Earth*, 148-149.

5. Note, Reedy for LBJ.

6. Johnson and Eisenhower, quoted in Caro, *Master of the Senate*, 1022.

7. Gerald Siegel oral history interview by John Logsdon, "Proceedings of an Oral History Workshop Conducted April 3, 1992 on the Legislative Origins of the National Aeronautics and Space Act of 1958," *Monographs in Aerospace History* 8 (Washington, DC: NASA History Office, 1998); Caro, *Master of the Senate*, 304-350.

8. Caro, *Master of the Senate*, 1022–1023.

9. Galloway first took a position on the Hill in 1941, beginning in the Legislative Reference Service and then moving to the Congressional Research Service. Interview of Eileen Galloway by Rebecca Wright, Johnson Space Center Oral History Project, August 7, 2000, https://history collection.jsc.nasa.gov/JSCHistoryPortal/history/oral_histories/NASA _HQ/Herstory/GallowayE/EG_8-7-00.pdf.

10. John W. Finney, "President and Aides Study Reports on Soviet's Feat," *New York Times*, November 5, 1957, 1.

11. William M. Blair, "Senators to Open Wide Inquiry on Missile and Satellite Program," *New York Times*, November 6, 1957, 12.

12. Divine, *The Johnson Years*, 2:223; "Excerpts from the Comments of Senator Johnson, Dr. Teller and Dr. Bush," *New York Times*, November 26, 1957, 20; Michael J. Neufeld, *Von Braun: Dreamer of Space, Engineer of War* (New York: Vintage, 2007), 317.

13. Barry M. Horstman, "Neil McElroy; He Served at P&G, Pentagon," *Cincinnati Post*, July 7, 1999, 1B; Gale, "Post-Sputnik Washington," 232.

14. Gale, "Post-Sputnik Washington," 228–232.

15. Caro, *Master of the Senate*, 1024–1025.

16. Galloway interview.

17. Dallek, *Lone Star Rising*, 532.

18. Hitchcock, *The Age of Eisenhower*, 376; "Man of the Year," *Time*, January 6, 1958.

19. Divine, *The Johnson Years*, 2:224.

20. Caro, *Master of the Senate*, 1020.

21. Johnson speech to Democratic Caucus, January 7, 1958, "Statement of Democratic Leader Lyndon B. Johnson to the Meeting of the Democratic Conference," Statements of LBJ, Box 23, Lyndon Baines Johnson Library, Austin, TX (hereafter LBJL).

22. George Reedy, quoted in Caro, *Master of the Senate*, 1028.

23. Galloway interview.

24. Johnson, *Vantage Point*, 274.

25. Quoted in Roger Launius, *NASA: The History of the US Civil Space Program* (Malabar, FL: Krieger, 2001), 29. See also "Inquiry into Satellite and Missile Programs: Hearings Before the Preparedness Investigating Subcommittee of the Committee on Armed Services, United States Senate, Eighty-Fifth Congress, First and Second Sessions" (Washington, DC: United States Government Printing Office, 1958); Thomas M. Gaskin, "Senator Lyndon B. Johnson, the Eisenhower Administration and U.S. Foreign Policy, 1957–60," *Presidential Studies Quarterly* 24, no. 2 (1994): 348.

26. Most coverage around the world highlighted German rocketry contributions to the first US satellite while downplaying von Braun's Nazi past. "Daily Report: Foreign Radio Broadcasts Supplement: World Reaction Series, No. 3—1958: World Radio and Press Reaction to Launching of American Earth Satellite," Box 9, Folder "Satellites—Explorer," RG 306, Entry P 243, NARA; United States Aeronautics and Space Activities, "1st Annual Report to Congress," House Document Number 71, 86th Congress, 1st Session, February 2, 1959.

27. "A Report on United States Foreign Operations," Box 9, Entry P 243, RG 306, NARA.

28. Nedville E. Nordness, letter to USIA Washington, February 14, 1958, Box 9, Entry P 243, RG 306, NARA.

29. Newspapers that drew on USIS material included *Illustrierte Berliner Zeitschrift, Telegraf-Wochenspiegel, Berliner Morgenpost, Koelnische Rundschau, Sueddeutsche Zeitung, Muenchner Merkur, Passaur Neue Presse,* and *Revue.* Nordness letter.

30. Nordness letter.

31. USIS Tehran to USIA Washington, June 11, 1958, Box 9, Entry P 243, RG 306, NARA.

32. Office of Research and Intelligence, "Trends in Japanese Attitudes"; Research and Reference Service, "Recent Trends in Latin American Opinion."

33. Gaskin, "Senator Lyndon B. Johnson."

34. Nixon is paraphrased in L. A. Minnich Jr.'s notes, "Legislative Leadership Meeting, Supplementary Notes," February 4, 1958, in *Exploring the Unknown: Selected Documents in the History of the U.S. Civil Space Program,* ed. John M. Logsdon et al. (Washington, DC: Government Printing Office, 1995), 1:631.

35. The American Rocket Society and the Rocket and Satellite Research Panel also pushed for a new separate civilian space agency. See Cull, *The Cold War,* 152; President's Advisory Committee on Government Organization, Executive Office of the President, Memorandum for the President, "Organization for Civil Space Programs," March 5, 1958, 3. Eisenhower expressed concern that locating US space activities within the military might increase competition among the branches for funding. Glen P. Wilson, "Lyndon Johnson and the Legislative Origins of NASA," *Prologue: Quarterly of the National Archives* 25 (Winter 1993): 362–372; Hitchcock, *The Age of Eisenhower,* 393.

36. Joseph M. Goldsen and Leon Lipson, "Some Political Implications of the Space Age," RAND P-1435, February 24, 1958.

37. Vanguard 1 would help prove that the Earth was slightly pear-shaped, not a perfect sphere. USIA Press and Publications Service Monthly

Report, March 1958, Box 4, Entry P 243, RG 306, NARA; Loyd S. Swenson Jr., James M. Grimwood, and Charles C. Alexander, *This New Ocean: A History of Project Mercury* (Washington, DC: NASA Special Publication, 1989), 30.

38. USIS Santiago to USIA Washington, March 31, 1958, Box 9, Entry P 243, RG 306, NARA.

39. "Four Objects Reported in Sputnik Orbit," *Aviation Week*, May 26, 1958, 28–29.

40. McDougall, *The Heavens and the Earth*, 250–251.

41. Taubman, *Khrushchev*, 378–380.

42. Divine, *The Sputnik Challenge*, 191.

43. In the current version of the act the word "mankind" has been replaced with the more inclusive "humankind." National Aeronautics and Space Act, Public Law No. 111-314, 124 Stat. 3328 (December 18, 2010); original version Sec. 102 (a), "The National Aeronautics and Space Act of 1958," Public Law No. 85-568.

44. Franklin Roosevelt's attempt to internationalize the New Deal through the Atlantic Charter reflected the beginning stages of the new position in international affairs, one in which the United States would establish New Deal–style policies and multilateral institutions outside its national borders. Although the New Deal laid the groundwork for America's relationship to the world after World War II, it was the experience of the war, and the way the war ended in particular, that shaped postwar international relations. After World War II, as the United States assumed a position of world economic, cultural, and military power, and after having fought the "good war," many Americans became convinced of their nation's essential goodness, benevolence, and duty in world affairs. This outlook infused US foreign policy in this period and can be observed in lines such as the Space Act's "for all mankind." Elizabeth Borgwardt, *A New Deal for the World: America's Vision for Human Rights* (Cambridge, MA: Harvard University Press, 2005), 6; Odd Arne Westad, *The Global Cold War* (New York: University of Cambridge Press, 2006), 20; Gary Gerstle, *American Crucible: Race and Nation in the Twentieth Century* (Princeton, NJ: Princeton University Press, 2001), 187.

45. John Krige, "NASA as an Instrument of U.S. Foreign Policy," in *Societal Impact of Spaceflight*, ed. Steven J. Dick and Roger D. Launius (Washington, DC: NASA, 2007), 207–218.

46. These employees came from the following NACA laboratories: Langley Aeronautical Laboratory at Hampton, Virginia; Ames Aeronautical Laboratory at Moffett Field, California; the Flight Research Center near Muroc Dry Lake, California; and the Lewis Flight Propulsion Laboratory in Cleveland, Ohio. See Sylvia K. Kraemer, "Organizing for Exploration,"

in *Exploring the Unknown: Selected Documents in the History of the U.S. Civil Space Program*, ed. John M. Logsdon et al. (Washington, DC: Government Printing Office, 1995), 1:661.

47. *Report to the Congress from the President of the United States*, U.S. Aeronautics and Space Activities, January 1 to December 31, 1959 (Washington, DC: U.S. Government Printing Office, 1960).

48. Office of Program Planning and Evaluation, "The Long-Range Plan of the National Aeronautics and Space Administration," December 16, 1969, NASA Historical Reference Collection, NASA History Office, NASA Headquarters, Washington, DC.

49. The director of the USIA was a member of the Operations Coordinating Board of the Outer Space Working Group as well as the National Security Council and participated in matters concerning space policy in these forums. "USIA Policy on Space," undated, Box 12, Folder "USIA: 1962 Proposals & Suggestions from all elements on Science Prog.," Entry P 243, RG 306, NARA; "USIA Participation in Space Policy Activities," December 21, 1960, Folder "US-USSR Outer Space Race," Box 13, Entry P 243, RG 306, NARA.

50. Goodwin had also discussed this approach with NASA Administrator T. Keith Glennan and his special assistant Frank Phillips the week before he met with Bonney. Harold Goodwin to Walter Bonney, November 26, 1958, Box 6, Entry P 243, RG 306, NARA.

51. Harold L. Goodwin paper on the USIA science program, June 22, 1958, Box 4, Entry P 243, RG 306, NARA.

52. Cull, *The Cold War*, 152.

53. Goodwin paper on the USIA science program.

54. "U.S. Information Agency Basic Guidance and Planning, Paper No. 4," November 18, 1958, Box 12, Folder "USIA: 1962 Proposals & Suggestions from All Elements on Science Prog.," Entry P 243, RG 306, NARA.

55. "U.S. Information Agency Basic Guidance and Planning, Paper No. 4."

56. Goodwin paper on the USIA science program.

57. "U.S. Information Agency Basic Guidance and Planning, Paper No. 4."

58. Hitchcock, *The Age of Eisenhower*, 406.

59. Dallek, *Lone Star Rising*, 531.

60. In 1959 the USIA distributed more than a hundred news releases on space activities in addition to more than a million pamphlets and posters. USIS posts received negatives, prints, lithographs, and plastic printing plates. These materials were complemented by daily news roundups, a series of television documentaries, and newsreel clips. The USIA filled its satellite libraries with books and pamphlets on space

exploration. Statement by George V. Allen before the House Science and Astronautics Committee, 22 January 1960, Box 12, Folder "USIA: Background Material from Media for Dir. Allen Appearance before House Science and Astronautics Committee, 1960," Entry P 243, RG 306, NARA. *Report to the Congress from the President of the United States.*

61. Robert Sivard to Harold Goodwin, April 7, 1960, Box 11, Folder "Space Exploration," Entry P 243, RG 306, NARA.

62. Allen statement.

63. The USIA produced nineteen copies of this exhibit in 1958 and continued to tour them around the world through 1960. In 1959 "Space Unlimited" was shown in Italy, Chile, Brazil, Uruguay, Bolivia, Spain, Pakistan, Afghanistan, Israel, Greece, Morocco, Mexico, Guatemala, Cuba, Finland, Denmark, The Hague, Laos, Malaya, Burma, the Philippines, Thailand, Sweden, Norway, and Japan; Allen statement.

64. *Report to the Congress from the President of the United States*; Sivard to Goodwin, April 7, 1960.

65. Dulles, quoted in Jason C. Parker, *Hearts, Minds, Voices: US Cold War Public Diplomacy and the Formation of the Third World* (New York: Oxford University Press, 2016), 107.

66. USIA to USIS Khartoum, August 15, 1960, Box 5, Entry P 243, RG 306, NARA.

67. Mr. Duggan to Secretary of State, February 27, 1961, Box 5, Entry P 243, RG 306, NARA.

68. "Project Mercury Installations in Africa: Bloc, Cairo and African Reactions," July 15, 1960, Box 5, Entry P 243, RG 306, NARA.

69. Department of State to the US Embassy Lagos, January 11, 1961, Box 5, Entry P 243, RG 306, NARA.

70. Mr. Palmer joint Embassy-USIS message to the Secretary of State, January 5, 1961, Box 5, Entry P 243, RG 306, NARA.

71. Sunny Tsiao, *"Read You Loud and Clear!" The Story of NASA's Spaceflight Tracking and Data Network* (Washington, DC: NASA History Office, 2008), 87.

72. Tsiao, *"Read You Loud and Clear!,"* 87.

73. The committee included USIA Director George Allen as well as State Department and Defense Department staff, an international officer from the Ford Foundation, and Eisenhower's political advisor C. D. Jackson. *Conclusions and Recommendations of the President's Committee on Information Activities Abroad, December 1960*, Box 4, Entry P 243, RG 306, NARA. See also Cull, *The Cold War*, 180-184; James Schwoch, *Global TV*, 53-56.

74. Henry Kissinger, "Psychological and Pressure Aspects"; Osgood, *Total Cold War*, 181-183.

75. *Conclusions and Recommendations of the President's Committee on Information Activities Abroad, December 1960*, 1.

76. Hugh Odishaw, "The Meaning of the International Geophysical Year," December 4, 1959, Box 6, A83-10, U.S. President's Committee on Information Activities Abroad (Sprague Committee) Records, 1959-1961, DEPL.

77. Hannah Arendt, *The Human Condition* (Chicago: University of Chicago Press, 1958), 1.

78. Goodwin left the USIA shortly after he prepared this report to become the director of NASA's Office of Program Development. See C. P. Snow, *The Two Cultures and the Scientific Revolution* (New York: Cambridge University Press, 1959); Harold L. Goodwin, "USIA and the Scientific Revolution," 1961, Box 5, Entry P 243, RG 306, NARA.

79. USIA officials saw science as a useful public diplomacy vehicle because, according to one USIA official, it "inherently rouses little or no suspicion and hostility, [so] it can be an effective means for communicating other aspects of the total American image." James Halsema to Mr. Wilson, undated (1961), Box 1, Entry P 243, RG 306, NARA.

CHAPTER 4: IF WE ARE TO WIN THE BATTLE, 1960–1961

1. Alan Brinkley, *Liberalism and Its Discontents* (Cambridge, MA: Harvard University Press, 1998), 216.

2. Quoted in Robert Dallek, *An Unfinished Life: John F. Kennedy 1917-1963* (New York: Little, Brown, 2003), 225.

3. Brinkley, *Liberalism*, 216.

4. Dallek, *An Unfinished Life*, 285.

5. John A. Farrell, *Richard Nixon: The Life* (New York: Doubleday, 2017), 284.

6. Theodore C. Sorensen, *Kennedy* (New York: Harper & Row, 1965), 196; Farrell, *Richard Nixon*, 270.

7. Dallek, *An Unfinished Life*, 286.

8. Sorensen, *Kennedy*, 198.

9. Dallek, *An Unfinished Life*, 285.

10. Sorensen, *Kennedy*, 197.

11. Donald M. Wilson, recorded interview by James Greenfield, September 2, 1964, John F. Kennedy Oral History Program, JFKL.

12. Hugh Sidey, *John F. Kennedy, President* (New York: Atheneum, 1964), 98.

13. Logsdon, *John F. Kennedy*, 10-12.

14. Remarks of Senator John F. Kennedy at Mormon Tabernacle, Salt Lake City, Utah, September 23, 1960, Papers of John F. Kennedy,

Pre-presidential Papers, Senate Files, Series 12, Speeches and the Press, Box 911, JFKL.

15. Stephen G. Rabe, "John F. Kennedy and the World," in *Debating the Kennedy Presidency*, ed. James N. Giglio and Stephen G. Rabe (Lanham, MD: Rowman & Littlefield, 2003), 9–10.

16. Papers of John F. Kennedy, Pre-presidential papers, Senate Files, Speeches and the Press, Speech Files, 1953–1960, "U.S. Military Power," Senate Floor, August 14 1958, JFK SEN-0901-022, JFKL; Douglas Brinkley, *American Moonshot: John F. Kennedy and the Great Space Race* (New York: Harper, 2019), 166–167.

17. Sorensen, *Kennedy*, 177–178.

18. Papers of John F. Kennedy, Pre-presidential papers, Senate Files, Speeches and the Press, Speech Files, 1953–1960, "Municipal Auditorium, Canton, Ohio," September 27, 1960, JFK SEN-0912-006, JFKL.

19. Sidey, *John F. Kennedy*, 95–96.

20. John F. Kennedy, "Address of Senator John F. Kennedy Accepting the Democratic Party Nomination for the Presidency of the United States—Memorial Coliseum, Los Angeles," July 15, 1960, American Presidency Project, www.presidency.ucsb.edu/ws/?pid=25966.

21. Robert A. Caro, *The Passage of Power: The Years of Lyndon Johnson* (New York: Alfred A. Knopf, 2012), 109–110.

22. Sorensen, *Kennedy*, 176.

23. Logsdon, *John F. Kennedy*, 12–13.

24. Nixon, quoted in Brinkley, *American Moonshot*, 201.

25. Nixon, quoted in Haefele, "John F. Kennedy," 69.

26. Sorensen, *Kennedy*, 203; Cull, *The Cold War*, 182.

27. "U.S. Survey Finds Others Consider Soviets Mightiest," *New York Times*, October 25, 1960, 1.

28. Nixon, quoted in William Jorden, "Campaign Issues—V: Debate on Status of Prestige Rouses Sharp Conflicts at Home and Abroad," *New York Times*, October 31, 1960.

29. Logsdon, *John F. Kennedy*, 12.

30. Haefele, "John F. Kennedy," 70; Logsdon, *John F. Kennedy*, 13.

31. "Donald Hornig, Last to See First A-Bomb, Dies at 92," *New York Times*, January 27, 2013, A20.

32. President's Science Advisory Committee, "Report of the Ad Hoc Panel on Man-in-Space," December 16, 1960, NASA Historical Reference Collection, History Office, NASA Headquarters, Washington, DC.

33. T. Keith Glennan, *The Birth of NASA: The Diary of T. Keith Glennan* (Washington, DC: NASA History Office, 1993), 292.

34. It is worth noting that there is some ambiguity about the decision process that went into giving Johnson the chairmanship of the Space

Council. John Logsdon details the discussions and the players involved in *John F. Kennedy*, 29–30. See also Caro, *The Passage of Power*, 172; Logsdon, *John F. Kennedy*, 22–23; Robert Dallek, *Flawed Giant: Lyndon Johnson and His Times, 1961–1973* (New York: Oxford University Press, 1998), 8–9.

35. President's Science Advisory Committee, "Report of the Ad Hoc Panel on Man-in-Space."

36. Eisenhower, quoted by Associate Administrator Robert Seamans in Logsdon, *John F. Kennedy*, 28.

37. Glennan, *The Birth of NASA*, 292.

38. Eisenhower, paraphrased by Glennan, *The Birth of NASA*, 292.

39. Logsdon, *John F. Kennedy*, 28.

40. Hubert Humphrey, paraphrased by Tam Dalyell, "Obituary: Jerome Wiesner," *Independent*, October 27, 1994.

41. "Report to the President-Elect of the Ad Hoc Committee on Space," January 10, 1961, NASA Historical Reference Collection.

42. "Report to the President-Elect."

43. Sorensen, *Kennedy*, 240–248.

44. Cull, *The Cold War*, 189–191.

45. President Kennedy's memorandum to the director of the USIA explained that "the mission of the United States Information Agency is to help achieve United States foreign policy objectives by (a) influencing public attitudes in other nations, and (b) advising the President, his representatives abroad, and the various departments and agencies on the implications of foreign opinion for present and contemplated United States policies, programs and official statements." See "Memorandum from President Kennedy to the Director of the U.S. Information Agency (Murrow)," January 25, 1963, Foreign Relations of the United States, 1961–1963, vol. XXV, Organization of Foreign Policy; Information Policy, United Nations; Scientific Matters. See also Cull, *The Cold War*, 192–193.

46. Sorensen, *Kennedy*, 240–248.

47. Sorensen, *Kennedy*, 292–293.

48. Edward Murrow to McGeorge Bundy, April 3, 1961, Papers of President Kennedy, National Security Files, Subjects, Box 307, "Space Activities, General, 4/61–6/61," JFKL.

49. Hugh Sidey, "How the News Hits Washington—with Some Reactions Overseas," *Life*, April 21, 1961, 26; Sidey, *John F. Kennedy*, 92–93.

50. Slava Gerovitch, "The Human Inside a Propaganda Machine: The Public Image and Professional Identity of Soviet Cosmonauts," in *Into the Cosmos: Space Exploration and Soviet Culture*, ed. James T. Andrews and Asif A. Siddiqi (Pittsburgh: University of Pittsburgh Press, 2011), 77–78.

51. Sidey, *John F. Kennedy*, 93.

52. In many ways, Gagarin was the ideal choice for the first Soviet human spaceflight mission. He came from a peasant family and had lived under Nazi occupation during World War II. Unlike his American counterparts, Gagarin had very little flying experience: just 230 hours. Gerovitch, "The Human Inside a Propaganda Machine," 81; Asif Siddiqi, *Sputnik and the Soviet Space Challenge* (Gainesville: University of Florida Press, 2003), 273–274.

53. Gerovitch, "The Human Inside a Propaganda Machine," 83.

54. Sidey, *John F. Kennedy*, 94.

55. Theodore C. Sorensen, oral history interview, March 25, 1964, National Archives and Records Administration, Office of Presidential Libraries, John F. Kennedy Library, JFKOH-TCS-01.

56. John F. Kennedy, "The President's News Conference," American Presidency Project, www.presidency.ucsb.edu/node/234594.

57. All quotations are from John M. Logsdon, *The Decision to Go to the Moon: Project Apollo and the National Interest* (Cambridge, MA: MIT Press, 1970), 103–104. See also U.S. Congress, House, Committee on Science and Astronautics, Discussion of Soviet Man-in-Space Shot, 87th Cong. 1st sess., 1961, 7–13, 375–378.

58. Robert Kennedy, quoted in Tim Weiner, "Theodore C. Sorensen, 1928–2010; Kennedy Wordsmith and More," *New York Times*, November 1, 2010, A1.

59. Sorensen, *Counselor: A Life at the Edge of History* (New York: Harper, 2008), 335.

60. Cabinet Room, April 21, 1961, Robert Knudsen, White House Photographs, JFKL.

61. Sidey, *John F. Kennedy*, 101.

62. Sidey, *John F. Kennedy*, 100.

63. Sidey, "How the News Hits Washington," 27.

64. Sorensen, quoted in Logsdon, *The Decision to Go to the Moon*, 107.

65. Rabe, "John F. Kennedy and the World," 13–14.

66. McGeorge Bundy, recorded interview by Richard Neustadt, March 1964, JFKL Oral History Program.

67. Sorensen, *Kennedy*, 310, 327.

68. Bundy interview.

69. Donald Wilson to McGeorge Bundy on the initial world reaction to Gagarin's flight, April 21, 1961, Papers of President Kennedy, National Security Files, Subjects, Box 307, JFKL.

70. Wilson to Bundy on the initial world reaction to Gagarin's flight.

71. Sidey, "How the News Hits Washington," 26.

72. "A Man in Space," *Washington Post*, April 13, 1961, A18; Harry Schwartz, "Moscow: Flight Is Taken as Another Sign That Communism

Is the Conquering Wave," *New York Times*, April 16, 1961, E3; Hanson Baldwin, "Flaw in Space Policy: U.S. Is Said to Lack Sense of Urgency in Drive for New Scientific Conquests," *New York Times*, April 17, 1961, 5.

73. Sorensen, *Kennedy*, 294.

74. John F. Kennedy, "Memorandum for Vice President Lyndon B. Johnson," April 20, 1961, Presidential Files, JFKL.

75. Logsdon, *John F. Kennedy*, 80-84.

76. Logsdon, *John F. Kennedy*, 84-85.

77. Robert S. McNamara, Secretary of Defense, Memorandum for the Vice President, "Brief Analysis of Department of Defense Space Program Efforts," April 21, 1961, LBJL.

78. Von Braun stressed that he was sharing his recommendations as a private citizen, not as the director of the Marshall Space Flight Center. Wernher von Braun to the Vice President of the United States, April 29, 1961, NASA Historical Reference Collection, History Office, NASA Headquarters, Washington, DC.

79. Walter Rostow memo to President Kennedy, "The Problems We Face," April 24, 1961, Papers of John F. Kennedy, Presidential Papers, National Security Files, Meetings and Memoranda, Staff memoranda: Rostow, Walt W., December 1960–June 1961, JFKNSF-323-006, JFKL.

80. Sorensen interview, 16.

81. Sorensen, quoted in John Logsdon, *John F. Kennedy*, 79.

82. With input from the secretary and deputy secretary of defense; General Schriever (Air Force); Admiral Hayward (Navy); Wernher von Braun (NASA); the administrator, deputy administrator, and other top officials at NASA; the special assistant to the president on science and technology; and representatives from the Bureau of the Budget. Johnson had also sought advice from George Brown (Brown & Root), Donald Cook (American Electric Power Service), and Frank Stanton (Columbia Broadcasting System), three businessmen and old friends who had helped his political career in one way or another. Lyndon B. Johnson, Vice President, Memorandum for the President, "Evaluation of Space Program," April 28, 1961, 2, Presidential Papers, JFKL.

83. Logsdon, *John F. Kennedy*, 92–94.

84. Swenson, Grimwood, and Alexander, *This New Ocean*, 350; Logsdon, *John F. Kennedy*, 96.

85. Sorensen, *Counselor*, 338; Logsdon, *John F. Kennedy*, 97.

86. Thomas C. Sorensen, *The Word War: The Story of American Propaganda* (New York: Harper & Row, 1968), 180-181.

87. Merrill Mueller to Lawrence O'Brien, May 10, 1961, Box 655, John F. Kennedy Presidential Papers, White House Central Subject Files, JFKL.

88. Sorensen, *The Word War*, 181; Cull, *The Cold War*, 198.

89. Press release, "Remarks of the President to Astronaut Comdr. Alan B. Shepard in the Rose Garden and the Latter's Reply," May 8, 1961, Box 654, John F. Kennedy Presidential Papers, White House Central Subject Files, JFKL.

90. The report also answers the question of sending robotic probes versus sending people: because "it is man, not merely machines, in space that captures the imagination of the world." James E. Webb, NASA administrator, and Robert S. McNamara, secretary of defense, to the vice president, May 8, 1961, with attached: "Recommendations for Our National Space Program: Changes, Policies, Goals," NASA Historical Reference Collection, History Office, NASA Headquarters, Washington, DC.

91. John F. Kennedy, "Special Message to the Congress on Urgent National Needs," May 25, 1961, American Presidency Project, www.presidency.ucsb.edu/ws/?pid=8151.

92. President John F. Kennedy, excerpts from "Urgent National Needs," speech to a Joint Session of Congress, May 25, 1961, Presidential Files, JFKL.

93. Sorensen, *Counselor*, 337.

94. Sorensen, *Kennedy*, 526.

95. McDougall, *The Heavens and the Earth*, 305.

96. John M. Logsdon et al., eds., *Exploring the Unknown: Selected Documents in the History of the U.S. Civil Space Program* (Washington, DC: Government Printing Office, 1995), 1:381; Logsdon, *John F. Kennedy*, 116.

97. Sorensen, *Counselor*, 337.

CHAPTER 5: JOHN GLENN AND *FRIENDSHIP 7*'S "FOURTH ORBIT," 1961–1963

1. Edward R. Murrow, quoted in "Shepard's Capsule Flown to Paris Show," *Baltimore Sun*, May 28, 1961, 11.

2. USIA, *16th Report to Congress, January 1–June 30, 1961* (Washington, DC, 1961), 7.

3. Dallek, *An Unfinished Life*, 396.

4. Memorandum of Conversation, June 3, 1961, Vienna, drafted by Akalovsky and approved by the White House on June 23. The meeting was held at the American ambassador's residence. Kennedy Library, President's Office Files, USSR, Secret; Eyes Only, JFKL.

5. Sorensen, *Kennedy*, 543–544.

6. *Report to the Congress from the President of the United States, US Aeronautics and Space Activities for 1961*.

7. Teasel Muir-Harmony, *Apollo to the Moon: A History in 50 Objects* (Washington, DC: National Geographic, 2018), 46–49.

8. Frederic O. Bundy and Harry Kendall to Thomas Sorensen, June 15, 1961, Box 5, Entry P 243, RG 306, NARA.

9. Cull, *The Cold War*, 205.

10. John Glenn and Nick Taylor, *John Glenn: A Memoir* (New York: Bantam, 1999), 250–275.

11. "The Nation," *Time*, March 2, 1962.

12. "Glenn Puts U.S. Back into the Space Race," *Los Angeles Times*, February 2, 1962, F4. In Japan alone, more than two hundred radio and TV stations carried coverage of the flight. *Report to the Congress from the President of the United States on United States Aeronautics and Space Activities, 1962*, 103–104.

13. Sorensen, *Kennedy*, 528.

14. *Report to the Congress from the President*, 103–104.

15. "Glenn Puts U.S. Back," F4.

16. Edward Murrow to John F. Kennedy, February 27, 1962, Box WH-23, "United States Information Agency 8/15/61–1/25/63," Papers of Arthur M. Schlesinger Jr., White House Files, JFKL.

17. Edward Murrow to John F. Kennedy, February 27, 1962, Box 91, Folder "USIA, 1/62–6/62," Papers of President Kennedy, National Security Files, Meetings and Memoranda, JFKL.

18. For a discussion of the significance of the *Friendship 7* mission in relation to the early space program, the extensive resources and labor that went into the flight, the scientific and engineering information gained from the flight, and the flight's political implications in the context of the cold war and the space race, see Glenn and Taylor, *John Glenn: A Memoir*; John Catchpole, *Project Mercury: NASA's First Manned Space Programme* (Chichester, UK: Springer-Praxis, 2001); Philip Scranton, "Behind the Icon: NASA's Mercury Capsules as Artefact, Process and Practice," in *Showcasing Space*, ed. Martin Collins (London: Science Museum, 2005); McDougall, *The Heavens and the Earth*.

19. Slava Gerovitch, *Soviet Space Mythologies: Public Images, Private Memories, and the Making of Cultural Identity* (Pittsburgh: University of Pittsburgh Press, 2015), 141–143; Andrew L. Jenks, *The Cosmonaut Who Couldn't Stop Smiling: The Life and Legend of Yuri Gagarin* (De Kalb, IL: Northern Illinois University Press, 2012), 151.

20. Murrow to Kennedy, February 27, 1962, Box 91, Folder "USIA, 1/62–6/62."

21. "Free World Media Treatment of First U.S. Orbital Flight," March 5, 1962, Box 7, Folder Office of Research, "R" Reports, 1960–32, "R-20-62,"

Research Notes (hereafter Entry 1029) RG 306, NARA; Osgood, *Total Cold War*, 176–180.

22. Ronald E. Doel and Zuoyue Wang, "Science and Technology in American Foreign Policy," in *Encyclopedia of American Foreign Policy*, 2nd ed., ed. Alexander DeConde, Richard Dean Burns, and Fredrik Logevall (New York: Scribner, 2001), 253.

23. For discussions on modernization theory and cold war foreign policy, see Gabrielle Hecht and Paul Edwards, "The Technopolitics of Cold War: Toward a Transregional Perspective," in *Essays on Twentieth-Century History*, ed. Michael Adas (Philadelphia: Temple University Press, 2010); David Engerman, Nils Gilman, Mark Haefele, and Michael Latham, eds., *Staging Growth* (Amherst: University of Massachusetts Press, 2003); Michael E. Latham, *Modernization as Ideology: American Social Science and "Nation Building" in the Kennedy Era* (Chapel Hill: University of North Carolina Press, 2000); Nils Gilman, *Mandarins of the Future: Modernization Theory in Cold War America* (Baltimore: Johns Hopkins University Press, 2004). For a thorough analysis of the history of science diplomacy in the cold war, see Wolfe, *Freedom's Laboratory*.

24. USIA Microfilm Microcopy No. NK-10A, Roll No. 14, USIA Press Releases: Africa, JFKL; USIA Microfilm Microcopy No. NK-10A, Roll No. 1, USIA Press Releases, Far East, JFKL.

25. This chapter follows the voyage of the capsule to exhibit sites and examines the precedents that this major tour set for future approaches to space diplomacy throughout the 1960s. As historian Stuart McCook has proposed, "Following something, as it moves around the world, in and out of particular places, and doing a contextually rich analysis of what happens as it moves" is a methodologically rich tool for doing global histories of science. Following the "fourth orbit" tour provides the opportunity to integrate the history of official planning and decision making with the experiences of those at whom public diplomacy was aimed. McCook, "Introduction: Global Currents in the National Histories of Science: The 'Global Turn' and the History of Science in Latin America," *Isis* 104, no. 4 (2013): 776; USIA Acting Director Donald Wilson to NASA Administrator James Webb, October 9, 1962, Box 37, Folder "Records Concerning Exhibits in Foreign Countries, 1955–67," Records Concerning Exhibits in Foreign Countries, 1955–1967, RG 306, NARA.

26. Handwritten internal correspondence, April 24, 1962, Nominal File 655, "Special Exhibition: Col. John Glenn's Capsule," Science Museum London, London, United Kingdom (hereafter SML).

27. Plans for the arrangement of the *Friendship 7* exhibit, Nominal File 655, "Special Exhibition: Col. John Glenn's Capsule," SML; USIA Press Release, May 9, 1962, Nominal File 655, "Special Exhibition: Col. John

Glenn's Capsule," SML. See correspondence in Nominal File 655, "Special Exhibition: Col. John Glenn's Capsule," SML.

28. Scholars of the history of public diplomacy often focus on US government elites and decision makers while minimizing the role of foreign participants in these programs. See Laura Belmonte, *Selling the American Way: U.S. Propaganda and the Cold War* (Philadelphia: University of Pennsylvania Press, 2008); Cull, *The Cold War*; Wilson Dizard Jr., *Inventing Public Diplomacy* (Boulder, CO: Lynne Rienner, 2004); Osgood, *Total Cold War*.

29. USIS London to USIA, May 1, 1962, Box 257, Folder "Outer Space, 14.B.5, Outer Space Exhibits, Jan-May, 1962, Part 2 of 2," Office of the Special Assistant to the Secretary of State for Atomic Energy and Outer Space, 1960-62, RG 59, NARA; USIS London to USIA, May 15, 1962, Box 257, Folder "Outer Space, 14.B.5, Outer Space Exhibits, Jan-May, 1962, Part 2 of 2," Entry A1 3008-A, RG 59, NARA.

30. USIS Paris to USIA, May 21, 1962, Box 257, Folder "Outer Space, 14.B.5, Outer Space Exhibits, Jan-May, 1962, Part 2 of 2," Entry A1 3008-A, RG 59, NARA; USIS Paris to USIA, May 20, 1962, Box 257, Folder "Outer Space, 14.B.5, Outer Space Exhibits, Jan-May, 1962, Part 2 of 2," Entry A1 3008-A, RG 59, NARA.

31. Edward Murrow to Madrid USITO, May 17, 1962, Box 257, Folder "Outer Space, 14.B.5, Outer Space Exhibits, Jan-May, 1962, Part 2 of 2," Entry A1 3008-A, RG 59, NARA.

32. Cathleen Lewis, "The Birth of the Soviet Space Museums: Creating the Earthbound Experience of Space Flight During the Golden Years of the Soviet Space Programme, 1957-68," in *Showcasing Space*, ed. Martin Collins (London: Science Museum, 2005), 142-158.

33. See Swenson, Grimwood, and Alexander, *This New Ocean*, 436; Wilson to Webb, October 9, 1962, Box 37, Entry A1 1039, RG 306, NARA.

34. Memorandum from the Apollo 11 Operations Office to Mr. Loomis, August 6, 1969, Box 15, Entry P 243, RG 306, NARA.

35. USIS Paris to USIA, May 20, 1962; Philip Hopkins, Director, Smithsonian National Air Museum, to Shelby Thompson, Director, Office of Technical Information and Education at NASA, April 10, 1962, Box 38, Folder "General Correspondence Jan.-June, 1962," Entry A1 1039, RG 306, NARA; Wilson to Webb, October 9, 1962.

36. The USIA had been sending jazz musicians around the world to mollify negative impressions of American civil rights progress while US political elites pushed for domestic civil rights reforms in part to win over the hearts and minds of the international public. See Dudziak, *Cold War Civil Rights*; Penny M. Von Eschen, *Satchmo Blows Up the World: Jazz Ambassadors Play the Cold War* (Cambridge, MA: Harvard University

Press, 2004); Thomas Borstelmann, *The Cold War and the Color Line: American Race Relations in the Global Arena* (Cambridge, MA: Harvard University Press, 2001); Brenda Gayle Plummer, ed., *Window on Freedom: Race, Civil Rights, and Foreign Affairs, 1945–1988* (Chapel Hill: University of North Carolina Press, 2003); Westad, *The Global Cold War*. On USIA space-related programming and race, see Cull, *The Cold War*, 212.

37. Dudziak, *Cold War Civil Rights*, 159.

38. Cull, *The Cold War*, 212.

39. USIA Microfilm Microcopy No. NK-10A, Roll No. 14.

40. Edward R. Murrow, Director of the USIA, to USITO Lagos, May 25, 1962, Box 257, Folder "Outer Space, 14.B.5, Outer Space Exhibits, Jan–May, 1962, Part 2 of 2," Entry A1 3008-A, RG 59, NARA.

41. USIA Microfilm Microcopy No. NK-10A, Roll No. 14.

42. "Glenn's Capsule Convinces Cairo," *Washington Post*, June 11, 1962, A24.

43. Egypt had been a battleground of Soviet and American aid since the mid-1950s. Because it was a neutral country, Egyptian leaders had played the United States against the Soviet Union for years in order to reap the greatest amount of development support without committing to political alliance. But the United States' support of Israel was widely unpopular in Egypt, which complicated the subject of American foreign aid for many people in the country. Quotation from *Al-Akhbar*. Edward R. Murrow, Director of the USIA, to President John F. Kennedy, February 27, 1962, Box 91, Folder "USIA, 1/62–6/62," Papers of President Kennedy, National Security Files, Departments and Agencies, JFKL; Robert J. McMahon, "The Illusion of Vulnerability: American Reassessments of the Soviet Threat, 1955–1956," *International History Review* 18, no. 3 (1996): 591–619.

44. USIS Tokyo to USIA Washington, June 18, 1962, Box 257, Folder "Outer Space, 14.B.5, Outer Space Exhibits, June–December, 1962, Part 1 of 2," Entry A1 3008-A, RG 59, NARA.

45. "'Friendship 7' Capsule," *Times of India*, June 27, 1962, 9.

46. USIS Bombay to USIA Washington, July 12, 1962, Box 257, Folder "Outer Space, 14.B.5, Outer Space Exhibits, June–December, 1962, Part 1 of 2," Entry A1 3008-A, RG 59, NARA.

47. Microcopy No. NK-10A, Roll No. 1.

48. Microcopy No. NK-10A, Roll No. 1; Paul Dench and Alison Gregg, *Carnarvon and Apollo: One Giant Leap for a Small Australian Town* (Kenthurst: Rosenberg, 2010), 18–24; Hamish Lindsay, *Tracking Apollo to the Moon* (Singapore: Springer-Verlag, 2001), 72–75.

49. USIS Tokyo to USIA Washington, June 4, 1962, Box 18, Entry A1 1039, RG 309, NARA.

50. Microcopy No. NK-10A, Roll No. 1; USIS Tokyo to USIA Washington, June 18, 1962; USIS Tokyo to USIA Washington, September 4, 1962, Box 258, Folder "Outer Space, 14.B.5, Outer Space Exhibits, June–December, 1962, Part 2 of 2," Entry A1 3008-A, RG 59, NARA; USIS Tokyo to USIA Washington, June 4, 1962.

51. Microcopy No. NK-10A, Roll No. 1.

52. "30,000 Koreans View 'Friendship 7' Exhibit," July 7, 1962, Microcopy No. NK-10A, Roll No 1.

53. "Seoul Makes Plans for 'Friendship 7' Exhibit," July 9, 1962, Microcopy No. NK-10A, Roll No. 1.

54. Wilson to Webb, October 9, 1962.

55. Dizard, *Inventing Public Diplomacy*, 5.

56. Although the agency was criticized for not devising a satisfactory way to measure its influence, no sufficient measures had been taken within the first twenty years of its operation to change this fact. See Report to the Congress, "Telling America's Story to the World—Problems and Issues," United States Information Agency by the comptroller general for the United States, March 25, 1974.

57. *Report to the Congress from the President of the United States, United States Aeronautics and Space Activities, 1962.*

58. Quoted in Osgood, *Total Cold War*, 176.

59. Based on the extraordinary turnout at all the exhibit sites, the extensive coverage in the press, and the enthusiastic comments that appeared in the USIA reports and in local press, it is clear that the display of *Friendship 7* was a significant event. Although the USIA reports are undeniably positive, they should be read with a critical eye. Even though these reports do not discuss any opposition to the exhibition of *Friendship 7*, it does not mean that none existed. Although these observations may not provide us with an unbiased picture of the reception of these exhibits, it is clear that they became popular events where people came together and were exposed to American scientific and technological information. See Osgood, *Total Cold War*, 170-180.

60. John Glenn Jr. to McGeorge Bundy, November 4, 1963, Box 308, Folder "Space Activities, General 10/63–11/63," Papers of President Kennedy, National Security Files, Subject Files, JFKL.

61. Harold McConeghey to Mr. Battey, October 17, 1962, Box 12, Folder "USIA: 1962—Proposals & Suggestions from all elements on Science prog," Entry P 243, RG 306, NARA.

62. The *Friendship 7* spacecraft was not a traditional instrument of power: its power lay in a particular interplay between US government officials and world publics, between producers and consumers of American

public diplomacy. Gabrielle Hecht and Michael Thad Allen, eds., *Technologies of Power: Essays in Honor of Thomas Parke Hughes and Agatha Chipley Hughes* (Cambridge, MA: MIT Press, 2001), 10-11.

63. Department of State memo, August 17, 1965, Box 28, Folder "SP 10 Astronaut Travel," Entry P 243, RG 306, NARA.

64. Simon Bourgin report on the astronaut overseas tour, September 1, 1965, Box 28, Folder "SP 10 Astronaut Travel," Entry P 243, RG 306, NARA.

65. Department of State memo, August 17, 1965; Bourgin report on the astronaut overseas tour; Burnett Anderson to Donald Wilson, December 4, 1963, Folder "SP 10 Astronaut Travel," Entry P 243, RG 306, NARA.

CHAPTER 6: "THE NEW EXPLORERS," 1963–1967

1. "Elton Stepherson, Jr.," *Washington Post*, December 5, 1987, B7.

2. For a history of the State Department's jazz tour program, see Von Eschen, *Satchmo Blows Up the World*.

3. Arthur Bardos to USIA Washington, July 7, 1964, Box 29, Entry P 243, RG 306, NARA.

4. Art Simmons, "Paris Scratchpad," *Jet*, September 10, 1970, 33; Jacob Gillespie, oral history interview by Charles Stuart Kennedy on February 4, 2010, Association for Diplomatic Studies and Training Foreign Affairs Oral History Project.

5. Michael Krenn, *Black Diplomacy: African Americans and the State Department, 1945–69* (Abingdon, UK: Routledge, 2015), 140.

6. Investigating space exploration and foreign relations activities in Africa, Asia, and South America offers a lens into some of the ramifications of the reorientation of the cold war front and a corrective of claims that science played a minimal role in the Johnson administration's foreign policy. See Ronald E. Doel and Kristine C. Harper, "Prometheus Unleashed: Science as a Diplomatic Weapon in the Lyndon B. Johnson Administration," *Osiris* 21, no. 1 (2006): 66. Studies that treat the political history of Project Apollo and situate it within the context of other 1960s government programs established to achieve "good ends" focus primarily on the domestic implications of the broadening of US government power. Walter McDougall's history of space exploration is one notable example. See McDougall, *The Heavens and the Earth*. See also Roger D. Launius, "Interpreting the Moon Landings: Project Apollo and the Historians," *History and Technology* 22, no. 3 (2006): 225-255.

7. *Report to the Congress from the President of the United States, United States and Aeronautics Space Activities, 1964*, 72.

8. USIS Tananarive Field Message no. 18, undated, Box 29, Entry P 243, RG 306, NARA.

9. USIS Tananarive Field Message no. 18.

10. USIS Tananarive Field Message no. 18.

11. USIS Tananarive Field Message no. 18.

12. USIS Tananarive Field Message no. 18.

13. USIS Tananarive to USIA Washington, June 29, 1964, Box 29, Entry P 243, RG 306, NARA.

14. USIS Tananarive Field Message no. 18.

15. USIS Tananarive Field Message no. 18.

16. "A Guide to the United States' History of Recognition, Diplomatic, and Consular Relations: Burundi," State Department, Office of the Historian, https://history.state.gov/countries/burundi.

17. Jacob Gillespie, interviewed by Charles Stuart Kennedy, February 4, 2010, Association for Diplomatic Studies and Training, Foreign Affairs Oral History Project, https://adst.org/wp-content/uploads/2013/12/Gillespie-Jacob-1.pdf.

18. Gillespie interview.

19. Robert Dallek, "Johnson, Project Apollo, and the Politics of Space Program Planning," in *Spaceflight and the Myth of Presidential Leadership*, ed. Roger D. Launius and Howard E. McCurdy (Urbana: University of Illinois Press, 1997), 71-72.

20. Johnson, *Vantage Point*, 285-286; Dallek, "Johnson, Project Apollo, and the Politics," 68-91.

21. Doel and Harper, "Prometheus Unleashed," 66-85.

22. Borstelmann, *The Cold War and the Color Line*, 174; David Ekbladh, *The Great American Mission: Modernization and the Construction of an American World Order* (Princeton, NJ: Princeton University Press, 2010).

23. Hubert Humphrey speech on space policy, March 19, 1965, *FRUS*, 1964-1968, vol. XXXIV, Energy Diplomacy and Global Issues, Document 31.

24. Dudziak, *Cold War Civil Rights*, 206–211.

25. USIS Tananarive to USIA Washington, June 29, 1964, Box 29, Entry P 243, RG 306, NARA.

26. NASA established Project Ranger in December 1959. The program was plagued by a series of failures, but on July 31, 1964, the Ranger 7 spacecraft successfully photographed the moon before a planned crash into the lunar surface. These images became important to scientists planning the Apollo missions and in geological studies of the moon. Don E. Wilhelms, *To a Rocky Moon: A Geologist's History of Lunar Exploration* (Tucson: University of Arizona Press, 1993), 94-101.

27. The USIA also produced a film on Ranger 7 that was shown with great results to university students in Pakistan. The American Embassy in Karachi observed that "it is believed that films such as this appeal to all groups, particularly the young, and make a lasting impression of U.S. leadership in the field of space research." US Embassy Karachi to State Department, November 13, 1964, Box 3147, Entry 1613, RG 59, NARA; *Report to the Congress from the President of the United States, United States Aeronautics and Space Activities, 1964.*

28. Lyndon Baines Johnson to Rear Admiral Ramon Castro Jijon, August 6, 1964, Box 3147, Entry 1613, RG 59, NARA. For the most part, respondents in nineteen countries and major cities believed that Soviet Union space activities, nuclear strength, and scientific development outpaced those of the United States. The one exception was in Turkey, where US superiority held. Dean Rusk to Hubert Humphrey, April 29, 1965, Box 9, Entry 3008D, RG 59, NARA.

29. Asif Siddiqi, *Challenge to Apollo: The Soviet Union and the Space Race, 1945-1974* (Military Bookshop, 2011), 407–408.

30. Taubman, *Khrushchev*, 617.

31. Henry Tanner, "Khrushchev Ouster: Reaction in Moscow," *New York Times*, October 25, 1964, E4.

32. Siddiqi, *Sputnik*, 421–460.

33. Deputy Vasily Mishin, quoted in Siddiqi, *Sputnik*, 447.

34. Voskhod 1, launched on October 12, 1964, carried three crew members, which made it the first multipassenger human spaceflight mission. *Report to the Congress from the President of the United States, United States Aeronautics and Space Activities, 1964.*

35. John Finney, "US Now Leads in Two Major Aspects of Space Race: Rendezvous and Endurance: But It Still Lags in Rocket Power," *New York Times*, December 16, 1965, 29.

36. For a comprehensive history of Project Gemini, see Barton C. Hacker and James M. Grimwood, *On the Shoulders of Titans: A History of Project Gemini* (Washington, DC: NASA, 1977).

37. "Gemini 4," *Times of India*, June 5, 1965, 6.

38. *Report to the Congress from the President of the United States, United States Aeronautics and Space Activities, 1965.*

39. Research and Reference Service, "World Press Reaction to Gemini IV Space Flight," R-74-65, June 11, 1965, Box 25, Entry P 142, RG 306, NARA.

40. United States Information Agency, "World Press Reactions to Gemini IV Space Flight," June 11, 1965, Box 25, Office of Research: Research Reports, 1960-1999 (hereafter Entry P 142), RG 306, NARA.

41. Rusk to Humphrey, April 29, 1965.

42. "Spaceman Jr. Romp, Snooze in White House," *Chicago Tribune*, June 19, 1965, 2; "A Conversation About the U.S. Space Program with Former President Lyndon B. Johnson as Broadcast During the CBS News Coverage of *Man on the Moon: The Epic Journey of Apollo*, CBS Television Network, July 21, 1969, Box 4, Entry A1 42, RG 306, NARA.

43. Art Buchwald, "Capitol Punishment . . . To Paris and Back," *Washington Post*, June 22, 1965, A17.

44. "Vostok Revealed," *Spaceflight*, July 1965, 161, quoted in Lewis, "The Red Stuff," 194.

45. US Embassy Paris to Secretary of State, June 19, 1965, Box 3153, Entry 1613, RG 59, NARA; *Report to the Congress from the President of the United States on United States Aeronautics and Space Activities, 1965*.

46. *Report to the Congress*.

47. US Embassy Lagos to Secretary of State, August 27, 1965, Box 3153, Entry 1613, RG 59, NARA. See also US Embassy Rio de Janeiro to Secretary of State, August 31, 1965, Box 3153, Entry 1613, RG 59, NARA.

48. US Embassy Nairobi to Secretary of State, September 11, 1965, Box 3153, Entry 1613, RG 59, NARA.

49. W. E. Weld Jr., Deputy Assistant Director (Africa) to the USIA Director, May 26, 1965, Box 50, Entry A1 1016, RG 306, NARA.

50. US Embassy Paris to Secretary of State, June 19, 1965, Box 3153, Entry 1613, RG 59, NARA; *Report to the Congress from the President of the United States on United States Aeronautics and Space Activities, 1965*.

51. *Report to the Congress*.

52. US Embassy Addis Ababa to Secretary of State, September 23, 1965, Box 3153, Entry 1613, RG 59, NARA.

53. Officials from the State Department, the USIA, and NASA had to work together to plan and execute these tours. At times, tensions ran high between agency representatives. The issue of which agency should pay for the tours, or portions of the tours, proved contentious as well. NASA Administrator James Webb argued that his agency did not have the budget for such tours. The State Department could cover many of the costs, but transportation often proved too expensive. In a number of instances, the Department of Defense or the White House stepped in and financed transportation. See Hugh Robinson to President Lyndon B. Johnson, October 12, 1965, Folder "Astronauts—Contracts-Life, etc.," RN 12917, Robert Sherrod Apollo Collection, National Aeronautics and Space Administration Headquarters, History Office, Washington, DC (hereafter NASA); Joe Califano, memorandum for the record, January 22, 1966, RN 12917, Robert Sherrod Apollo Collection, NASA.

54. American Ambassador C. Vaughan Ferguson, Jr., to the Secretary of State, September 23, 1965, Box 3153, Entry 1613, RG 59, NARA.

55. US Embassy Lagos to Secretary of State, October 9, 1965, Box 3153, Entry 1613, RG 59, NARA.

56. US Embassy Lagos to Secretary of State, September 28, 1965, Box 3153, Entry 1613, RG 59, NARA.

57. US Embassy Lagos to Secretary of State, October 9, 1965.

58. Muir-Harmony, "Tracking Diplomacy."

59. The USIA cosponsored the exhibit with the Argentine Air Force. Argentina covered a large portion of the expense, the exhibit was located at the Air Force base, and the Air Force provided twenty-four-hour security as well as personnel to manage crowds. John P. McKnight to USIA Washington, March 29, 1966, Box 2, Entry A1 1039, RG 306, NARA.

60. Harry Kendall, *A Farm Boy in the Foreign Service: Telling America's Story to the World* (Bloomington, IN: AuthorHouse, 2003).

61. Spacemobiles were large vehicles outfitted with displays and educational tools. Kendall, *A Farm Boy*, 110.

62. In October 1965 Glenn visited Frankfurt, Munich, Bonn, Bremen, Berlin, Hamburg, London, Amsterdam, The Hague, Rotterdam, Genoa, Rome, Naples, Madrid, and Lisbon. *Report to the Congress from the President of the United States on United States Aeronautics and Space Activities, 1965.*

63. G. A. Ewing to the Department of State, November 11, 1965, Box 3153, Entry 1613, RG 59, NARA.

64. Leonard H. Marks letter to President Lyndon B. Johnson, May 26, 1966, Box 75, National Security File: Agency File, LBJL.

65. Garland C. Routt to USIA Washington, March 4, 1966, Box 25, Entry P 243, RG 306, NARA.

66. USIS Rangoon to USIA Washington, March 4, 1966, Box 25, Entry 243, RG 306, NARA.

67. U Win Tin, editor of the *Mirror*, in USIS report by Peter Boog, Associated Press Bureau Chief Burma, Box 5, Entry A1 1039, RG 306, NARA.

68. US Embassy Rangoon to the Secretary of State, February 25, 1966, Box 3153, RG 59, NARA.

69. Frank Borman to President Lyndon B. Johnson, March 31, 1966, RN 12917, Robert Sherrod Apollo Collection, NASA.

70. Leonard Marks to Lyndon Johnson, March 1, 1966, Box 135, Papers of Lyndon Baines Johnson, President, 1963-1969, Confidential File; Agency Reports: US Information Agency, LBJL.

71. T. Nesbitt to Herman Pollack, April 14, 1966, Box 23, Entry 3008D, RG 59, NARA.

72. George Low, "Latin American Tour with Astronauts Armstrong and Gordon, October 7-31, 1966," RN 859, "Gordon, Richard F., Jr.," NASA.

73. James R. Hansen, *First Man: The Life of Neil A. Armstrong* (New York: Simon & Schuster, 2005), 300.

74. Low, "Latin American Tour."

75. Walter Rostow sent the president a draft of the letter. Each letter was slightly different, given the level of cooperation with the country, but each included sentiments about sharing information with the world scientific community. Walter Rostow to Lyndon Johnson, October 4, 1966, Box 1, National Security File: Special Head of State Correspondence File, LBJL.

76. Low, "Latin American Tour."

77. Kendall, *A Farm Boy*, 116.

78. Simon Bourgin to William Green, July 12, 1967, Box 26, Entry P 243, RG 306, NARA.

79. Jenifer Van Vleck, *Empire of the Air* (Cambridge: Harvard University Press, 2013), 11.

80. Dean Rusk to Hubert Humphrey, draft, March 18, 1965, Box 9, Entry 3008D, RG 59, NARA.

81. "Dean Rusk Dies; Vietnam War–Era Secretary of State," *Los Angeles Times*, December 22, 1994; Eric Pace, "Dean Rusk, Secretary of State in Vietnam War, Is Dead at 85," *New York Times*, December 22, 1994, 1.

82. Rusk noted that staff within the State Department viewed advanced space technology—space vehicle propulsion, the exploration of nearby planets, and the extension of space applications—as key to serving long-term foreign relations objectives. Rusk to Humphrey, draft, March 18, 1965.

83. In 1965 the Space Council consisted of Vice President Hubert Humphrey as chairman, Secretary of State Dean Rusk, Secretary of Defense Robert McNamara, NASA Administrator James Webb, and Atomic Energy Commission Director Glenn Seaborg. Herman Pollack to Dean Rusk, April 9, 1965, Box 9, Entry 3008D, RG 59, NARA.

84. Rusk articulated an elemental feature of what distinguished the process of globalization in this period: the consciousness of the unity of humankind. Akira Iriye, "Introduction," in *Global Interdependence: The World After 1945*, ed. Akira Iriye (Cambridge, MA: Harvard University Press, 2014), 5.

85. E. C. Welsh to Dean Rusk, Robert McNamara, James Webb, and Glenn Seaborg, April 14, 1965, Box 9, Entry 3008D, RG 59, NARA.

86. Rusk to Humphrey, April 29, 1965.

87. Dallek, "Johnson, Project Apollo, and the Politics," 81.

88. President Lyndon Baines Johnson, "Remarks at the Signing of the Treaty on Outer Space, January 27, 1967," *Public Papers of the Presidents of the United States: Lyndon B. Johnson, 1967.* vol. I, entry 18, 91–92

(Washington, DC: Government Printing Office, 1968); "Treaty on the Principles Governing the Activities of States in the Exploration and Use of Outer Space, Including the Moon and Other Celestial Bodies," Box 2, Executive Office Files on Outer Space, LBJL.

89. Brooks, Grimwood, and Swenson, *Chariots for Apollo*, chapter 9.

90. Trevor Rockwell, "Space Propaganda 'For All Mankind': Soviet and American Responses to the Cold War, 1957-1977" (PhD diss., University of Alberta, 2012), 153.

91. Simon Bourgin to Julian Scheer, March 3, 1967, Box 26, Entry P 243, RG 306, NARA; Kristen Amanda Starr, "NASA's Hidden Power: NACA/ NASA Public Relations and the Cold War, 1945-1967" (PhD diss., Auburn University, 2008), 297.

92. "Worldwide Treatment of Current Issues," January 30, 1967, Box 21, Entry P 243, RG 306, NARA.

93. Johnson, *Vantage Point*, 270-271.

94. Draft of presidential reply to messages of condolence for deaths of astronauts Grissom, White, and Chaffee, January 30, 1967, Box 18, National Security File: Files of Charles E. Johnson, LBJL.

95. Johnson, quoted in Dallek, "Johnson, Project Apollo, and the Politics," 84.

96. Brooks, Grimwood, and Swenson, *Chariots for Apollo*, 227.

CHAPTER 7: "RIDERS ON THE EARTH TOGETHER," 1968–1969

1. Don Oberdorfer, *Tet! The Turning Point in the Vietnam War* (Baltimore: Johns Hopkins University Press, 1971), 159.

2. Walter Cronkite, *CBS News Special Report*, originally aired on February 27, 1968, transcript, "Final Words: Cronkite's Vietnam Commentary," NPR, July 18, 2009, www.npr.org/templates/story/story.php?storyId=106775685.

3. William Bundy, *A Tangled Web: The Making of Foreign Policy in the Nixon Presidency* (New York: Hill and Wang, 1998), 20; Sönke Kunkel, *Empire of Pictures: Global Media and the 1960s Remaking of American Foreign Policy* (New York: Berghahn, 2016), 163.

4. Leonard Marks, quoted in Cull, *The Cold War*, 289.

5. Douglas Brinkley, *Cronkite* (New York: Harper Perennial, 2012), 400.

6. "USIA 31st Report to Congress," 7-12, 1968, Box 1, RG 306, Entry P 180, NARA.

7. "USIA 31st Report to Congress"; Kunkel, *Empire of Pictures*, 163.

8. Frank Borman, oral history interview with author, December 10, 2012.

9. Quoted in Robert Poole, *Earthrise: How Man First Saw the Earth* (New Haven, CT: Yale University Press, 2008), 17.

10. Lindsay, *Tracking Apollo*, 110-112.

11. Collins, *Carrying the Fire*, 144.

12. At the time of the fire, the command module was named AS-204. Later it became known as Apollo 1. Frank Borman with Robert J. Serling, *Countdown: An Autobiography* (New York: Silver Arrow, 1988), 171-189.

13. Richard Jurek, *The Ultimate Engineer: The Remarkable Life of NASA's Visionary Leader George M. Low* (Lincoln: University of Nebraska Press, 2019), 125-133.

14. Dwayne A. Day and Asif Siddiqi, "The Moon in the Crosshairs: CIA Intelligence on the Soviet Manned Lunar Progamme, Part 1—Launch Complex J," *Spaceflight* 45 (November 2003): 468.

15. Director of Central Intelligence, "The Soviet Space Program," *Central Intelligence Agency Historical Review Program*, April 4, 1968, 1.

16. Roger Launius, "NASA Looks to the East: American Intelligence Estimates of Soviet Capabilities and Project Apollo," *Air Power History* 48, no. 3 (2001): 10.

17. Borman, *Countdown*, 194.

18. Bourgin, quoted in Billy Watkins, *Apollo Moon Missions: The Unsung Heroes* (Lincoln: University of Nebraska Press, 2007), 70.

19. Simon Bourgin to Frank Borman, December 13, 1968, Folder: "Bourgin, Simon CB-497500-01," Smithsonian National Air and Space Museum Archives, Washington, DC.

20. Simon Bourgin to Frank Borman, December 15, 1968, Box 7, Simon Bourgin Collection, Boston University Archives, Boston, MA (hereafter BUA).

21. Bourgin to Borman, December 13, 1968.

22. Bourgin to Borman, December 15, 1968; Bourgin to Borman, December 13, 1968.

23. Bourgin to Borman, December 13, 1968.

24. Bourgin to Borman, December 15, 1968.

25. Lyndon B. Johnson, Daily Diary, December 21, 1968, LBJL.

26. Susan Borman, quoted in Poole, *Earthrise*, 19.

27. *Astronautics and Aeronautics, 1968: Chronology on Science, Technology, and Policy* (Washington, DC: NASA, 1969), 319.

28. "USIA 31st Report to Congress."

29. US Embassy Sofia to State Department, January 8, 1969, Box 3012, RG 59, NARA.

30. Consulate Martinique to Department of State, January 7, 1969, Box 3012, RG 59, Central Foreign Policy Files, 1967, 1969, NARA.

31. Jennifer Levasseur, *Through Astronaut Eyes: Photographing Early Human Spaceflight* (West Lafayette, IN: Purdue University Press, 2020), 56.

32. Bill Anders, *Apollo 8 Flight Journal*, mission elapsed time 075:47:30, corrected transcript and commentary, ed. W. David Woods and Frank O'Brien.

33. Borman, quoted in Poole, *Earthrise*, 2.

34. Levasseur, *Through Astronaut Eyes*, 62–63; Anders, *Apollo 8 Flight Journal*.

35. Poole, *Earthrise*, 24–25.

36. Borman, *Countdown*, 199–216.

37. Anders, *Apollo 8 Flight Journal*, Apollo 8 Transcript, Day 4: Lunar Orbit 9.

38. *Astronautics and Aeronautics, 1968*, 319.

39. "Pope and Apollo 8 on Satellite Debut," *New York Times*, December 25, 1968, 38.

40. US Embassy Lisbon to Department of State, January 3, 1969, Box 3012, RG 59, Central Foreign Policy Files, 1967, 1969, NARA.

41. "USIA 31st Report to Congress," 7–12, 1968, Box 1, RG 306, Entry P 180, NARA.

42. Simon Bourgin notes on the Apollo 8 tour, Box 4, Simon Bourgin Collection, BUA.

43. Borman interview; Bourgin to Borman, December 15, 1968.

44. Quoted in Poole, *Earthrise*, 135.

45. "USIA 31st Report to Congress"; *Astronautics and Aeronautics, 1968*, 327.

46. The most vocal critique of the reading came from Madalyn Murray O'Hair, who sued NASA. However, the court ruled that "the first amendment does not require the state to be hostile to religion, but only neutral." See Poole, *Earthrise*, 136–137.

47. Archibald MacLeish, "A Reflection: Riders on Earth Together, Brothers in Eternal Cold," *New York Times*, December 25, 1968, 1.

48. Arthur C. Clarke and Isidor Rabi, quoted in *Astronautics and Aeronautics, 1969: Chronology on Science, Technology, and Policy* (Washington, DC: NASA, 1970).

49. *Astronautics and Aeronautics, 1968*, 326.

50. Lyndon B. Johnson, Daily Diary, December 27, 1968, LBJL; telephone conversation #13825, sound recording, LBJ and SUSAN BORMAN, 12/27/1968, 11:15AM, Recordings and Transcripts of Telephone Conversations and Meetings, LBJL.

51. *Astronautics and Aeronautics, 1968*, 327.

52. "USIA 31st Report to Congress."

53. US Embassy Sofia to State Department, January 8, 1969.

54. US Embassy Kampala to Department of State, January 4, 1969, Box 3012, RG 59, Central Foreign Policy Files, 1967, 1969, NARA.

55. US Embassy Warsaw to Department of State, January 30, 1969, Box 3012, RG 59, Central Foreign Policy Files, 1967, 1969, NARA; *Astronautics and Aeronautics, 1968*, 330.

56. US Embassy Madrid to Department of State, January 11, 1969, Box 3012, RG 59, Central Foreign Policy Files, 1967, 1969, NARA; "USIA 31st Report to Congress."

57. *Times of India*, December 26, 1968, 8.

58. American Ambassador Walter Stoessel to Department of State, April 3, 1969, RG 59, Central Foreign Policy Files, 1967, 1969, NARA.

59. US Embassy Lisbon to Department of State, January 3, 1969, Box 3012, RG 59, Central Foreign Policy Files, 1967, 1969, NARA; US Embassy Madrid to Department of State, January 11, 1969; "USIA 31st Report to Congress."

60. US Embassy Fort Lamy to State Department, April 15, 1969, Box 3012, RG 59, Central Foreign Policy Files, 1967, 1969, NARA; author's stamp collection.

61. Lloyd Garrison, "De Gaulle Calls Borman 'a Very Nice Young Man,'" *New York Times*, February 7, 1969, 3; US Embassy Tehran to State Department, February 27, 1969, RG 59, Entry 1613, Box 3012, NARA.

62. "USIA 31st Report to Congress."

63. Walter Rostow to Lyndon Johnson, January 6, 1969, Box 1, National Security File: Special Head of State Correspondence File, LBJL.

64. Poole, *Earthrise*, 28.

65. Memorandum from Benjamin H. Read, Executive Secretary of State, to Mr. Walt W. Rostow, January 14, 1969, Box 1, National Security File: Special Head of State Correspondence File, LBJL.

66. Josip Broz Tito to Lyndon Johnson, February 19, 1969, Box 3012, Entry 1613, RG 59, NARA.

67. Lyndon Johnson, quoted in *Man on the Moon: The Epic Journey of Apollo 11*, as broadcast over the CBS Television Network on July 21, 1969.

68. Robert C. Maynard, "Gold Medals Presented by Johnson: Thousands Welcome Apollo Heroes," *Washington Post*, January 10, 1969, A1; John Noble Wilford, "Crew of Apollo 8 Is Saluted by President and Congress," *New York Times*, January 10, 1969, 1; *Astronautics and Aeronautics, 1969*, 7.

69. Congressional Record, House of Representatives, January 9, 1969, 367–368; *Astronautics and Aeronautics, 1969*, 7; Wilford, "Crew of Apollo," 1.

70. Poole, *Earthrise*, 33–34.

71. US Embassy Port au Prince to Secretary of State, March 3, 1969, Box 3012, RG 59, Central Foreign Policy Files, 1967, 1969, NARA; US Consulate Nassau to State Department, March 6, 1969, Box 3012, RG 59, Central Foreign Policy Files, 1967, 1969, NARA.

72. US Embassy Yaoundé to Secretary of State, January 24, 1969, Box 3012, RG 59, Central Foreign Policy Files, 1967, 1969, NARA.

73. US Embassy Warsaw to Department of State, January 30, 1969; US Embassy Warsaw to Secretary of State, February, 13, 1969, Box 3012, RG 59, Central Foreign Policy Files, 1967, 1969, NARA; US Embassy Warsaw to State Department, March 13, 1969, Box 3012, RG 59, Central Foreign Policy Files, 1967, 1969, NARA; US Embassy Warsaw to Department of State, May 1, 1969, Box 3012, RG 59, Central Foreign Policy Files, 1967, 1969, NARA.

74. US Embassy Prague to Secretary of State, February 19, 1969, Box 3012, RG 59, Central Foreign Policy Files, 1967, 1969, NARA; US Embassy Prague to State Department, March 17, 1969, Box 3012, RG 59, Central Foreign Policy Files, 1967, 1969, NARA.

75. Even the Siberian branch of the Soviet Academy of Sciences screened USIA films *Apollo 8* and *Project Apollo* at its May 1969 meeting, with the academy's vice president commenting on how Apollo 8 was an achievement of all mankind. US Embassy Moscow to Department of State, May 19, 1969, Box 3012, RG 59, Central Foreign Policy Files, 1967, 1969, NARA. See also Philip Arnold to Mr. Bardos, April 11, 1969, Box 17, Entry P 243, RG 306, NARA.

76. US Embassy Tananarive to State Department, March 15, 1969, Box 3012, RG 59, Central Foreign Policy Files, 1967, 1969, NARA.

77. US Consulate Poznan to Department of State, May 6, 1969, Box 3012, RG 59, Central Foreign Policy Files, 1967, 1969, NARA.

78. Borman, *Countdown*, 226; Nixon, Inaugural Address, American Presidency Project, www.presidency.ucsb.edu/node/239549; John Logsdon, *After Apollo? Richard Nixon and the American Space Program* (New York: Palgrave Macmillan, 2015), 9.

79. Nixon, Inaugural Address.

80. Robert Dallek, *Nixon and Kissinger: Partners in Power* (New York: HarperCollins, 2007), 95; Richard Nixon, *The Memoirs of Richard Nixon* (New York: Grosset & Dunlap, 1978), 366.

81. Dallek, *Nixon and Kissinger*, 99.

82. Dallek, *Nixon and Kissinger*, 99.

83. Borman, *Countdown*, 260.

84. Borman, *Countdown*, 226.

85. William Safire, *Before the Fall: An Inside View of the Pre-Watergate White House* (Garden City, NY: Doubleday, 1975), 147.

86. William Rogers to Frank Borman, March 27, 1969, RG 59, Central Foreign Policy Files, 1967–1969, Science, Box 3017, NARA. Borman's itinerary: London, Feb. 2–5; Paris, Feb. 5–7; Brussels, Feb. 7–10; The Hague, Feb. 10–11; Bonn, Feb. 11–12; West Berlin, Feb. 12–13; Rome, Feb. 13–17; Madrid, Feb. 17–19; and Lisbon, Feb. 19–21. *Astronautics and Aeronautics, 1969*, 37.

87. Richard Moose to Herb Klein, January 27, 1969, National Security Council Files: Names Files, Box 808, Folder "Borman, Frank," RNPL.

88. W. E. Weld Jr. (IAE) to Robert W. Akers, December 19, 1968, RG 306, Entry A1 42, Box 4, NARA; Thomas Paine, Acting Administrator NASA, to Charles E. Bohlen, Deputy Under Secretary for Political Affairs at the Department of State, January 3, 1969, RG 306, Entry A1 42, Box 4, NARA.

89. Borman interview; Borman, *Countdown*, 227.

90. John M. Lee, "Borman Blasts Off Smoothly as Envoy to Britain," *New York Times*, February 4, 1969, 4.

91. "Queen Welcomes Colonel Borman: Astronaut Who Saw Moon Sees Buckingham Palace," *Sun*, February 5, 1969, 2; Borman, *Countdown*, 228.

92. Bourgin notes on the Apollo 8 tour, Box 4; *Astronautics and Aeronautics, 1969*, 38.

93. Borman, *Countdown*, 229.

94. Simon Bourgin to Mariada Bourgin, February 5, 1969, Box 4, Simon Bourgin Collection, BUA.

95. Simon Bourgin to Mariada Bourgin, February 7, 1969, Box 4, Simon Bourgin Collection, BUA; Garrison, "De Gaulle Calls Borman," 3; Patricia Pullan, "Parisians Agog over Astronaut: Crowds Queue to Get Look at Borman, Apollo 8 Hero," *Sun*, February 7, 1969, A1; "Borman Sees Grandson of Jules Verne," *Los Angeles Times*, February 6, 1969, A12; Poole, *Earthrise*, 33.

96. Nixon, *Memoirs*, 370.

97. US Ambassador Sargent Shriver to State Department, February 9, 1969, RG 59, Entry 1613, Box 3017, NARA.

98. Bourgin notes on the Apollo 8 tour, Box 4; *Astronautics and Aeronautics, 1969*, 47.

99. US Embassy Brussels to State Department, February 21, 1969, RG 59, Entry 1613, Box 3017, NARA.

100. US Embassy Brussels to State Department, February 21, 1969; Bourgin notes on the Apollo 8 tour, Box 4; US Embassy Brussels to State Department, February 21, 1969, RG 59, Entry 1613, Box 3017, NARA.

101. US Embassy Brussels to State Department, February 21, 1969; *Astronautics and Aeronautics, 1969*, 46; Bourgin notes on the Apollo 8 tour, Box 4.

102. Simon Bourgin to Mariada Bourgin, undated, Box 4, Simon Bourgin Collection, BUA; US Embassy Brussels to State Department, February 21, 1969.

103. Th. H. Joekes, "Hero Anno 1969," *Nieuwe Rotterdamsche Courant*, February 11, 1969, translated by the US Embassy Brussels, US Embassy Brussels to State Department, February 21, 1969, RG 59, Entry 1613, Box 3017, NARA.

104. Simon Bourgin to Mariada Bourgin, Thursday night ???, February 1969, Box 4, Simon Bourgin Collection, BUA.

105. Poole, *Earthrise*, 132.

106. US Embassy Rome to State Department, February 7, 1969, RG 59, Entry 1613, Box 3017, NARA; "Pope, Praising Bravery, Greets Borman and Family," *New York Times*, February 16, 1969, 20.

107. Bourgin notes on the Apollo 8 tour, Box 4.

108. USIS Madrid to USIA and NASA, February 17, 1969, Box 4, Simon Bourgin Collection, BUA.

109. *Astronautics and Aeronautics, 1969*, 52.

110. Bourgin notes on the Apollo 8 tour, Box 4; *Astronautics and Aeronautics, 1969*, 54.

111. Bourgin notes on the Apollo 8 tour, Box 4.

112. Nixon, *Memoirs*, 370; H. R. Haldeman, *The Haldeman Diaries: Inside the Nixon White House* (New York: G.P. Putnam's Sons, 1994), 31, 34.

113. Haldeman, *The Haldeman Diaries*, 31–34; Richard Reeves, *President Nixon: Alone in the White House* (New York: Simon and Schuster, 2001), 48; Robert J. McMahon, *The Limits of Empire: The United States and Southeast Asia Since World War II* (New York: Columbia University Press, 1999), 161.

114. *Atlantic Consultation: President Nixon in Europe, February 23–March 2, 1969* (Washington, DC: Department of State, Office of Media Services, Bureau of Public Affairs, 1969), 3, 5.

115. *Atlantic Consultation*, 5; Safire, *Before the Fall*, 125.

116. Haldeman, *Haldeman Diaries*, 33–34; McMahon, *The Limits of Empire*, 161; Lien-Hang T. Nguyen, "Waging War on All Fronts: Nixon, Kissinger, and the Vietnam War, 1969-1972," in *Nixon in the World: American Foreign Relations, 1969-1977*, ed. Fredrik Logevall and Andrew Preston (New York: Oxford University Press, 2008), 188–189.

117. Nixon, *Memoirs*, 374.

CHAPTER 8: MAKING APOLLO 11 FOR ALL HUMANKIND, 1969

1. Reinhardt began his education in Tennessee's segregated school system, graduating from Knoxville College in 1939. After serving in WWII, he obtained a PhD in American Literature from the University of Wisconsin on the GI Bill. Martin Weil, "John E. Reinhardt, First Career Diplomat to Lead USIA, Dies at 95," *Washington Post*, February 24, 2016; Richard T. Arndt, *The First Resort of Kings: American Cultural Diplomacy in the Twentieth Century* (Dulles, VA: Potomac Books, 2006), 500–501.

2. John E. Reinhardt to USIA Public Affairs Officers, June 10, 1969, Box 18, Entry P 243, RG 306, NARA.

3. Stanley Moss to Simon Bourgin, March 7, 1969, Entry 243, Box 28, RG 306, NARA.

4. William Kling, "Moon Landing Trio Named," *Chicago Tribune*, January 10, 1969, 1.

5. Frank Borman to Richard Nixon, July 14, 1969, "EX OS 3-1 Astronauts Begin 7/31/69," Box 11, Subject Files: Outer Space, White House Central Files, RNPL.

6. *Apollo 9 Mission Report*, MSC-PA-R-69-2, National Aeronautics and Space Administration (Houston, TX: Manned Spacecraft Center, May 1969).

7. Collins, *Carrying the Fire*, 321.

8. Albert E. Hemsing to Mr. Ryan, February 27, 1969, Box 18, Entry P 243, RG 306, NARA.

9. "USIA 32nd Report to Congress 1–6/1969," Box 1, Reports to Congress (hereafter Entry P 180), RG 306, NARA.

10. In his essay on Project Apollo, Michael L. Smith argues that the "media coverage of Apollo *was* the event . . . never before had so ambitious an undertaking depended so thoroughly on its public presentation for significance." Building on Smith's observation about Project Apollo media coverage within the United States, this chapter sets out to examine public presentation of the flight within a global context. See Michael L. Smith, "Selling the Moon: The U.S. Manned Space Program and the Triumph of Commodity Scientism," in *The Culture of Consumption*, ed. Richard Wightman Fox and Jackson Lears (New York: Pantheon, 1983), 177.

11. USIS Iran to USIA Washington, June 3, 1969, Box 17, Entry P 243, RG 306, NARA.

12. American Consulate, Istanbul to the Secretary of State, June 23, 1969, Box 17, Entry P 243, RG 306, NARA.

13. USIA to All USIS Posts, May 13, 1969, Box 18, Entry P 243, RG 306, NARA.

14. Willis Shapley to George Mueller, April 19, 1969, Document II-70 in *Exploring the Unknown: Selected Documents in the History of the U.S. Civil Space Program*, ed. John Logsdon et al. (Washington, DC: National Aeronautics and Space Administration, 2008), 730.

15. Charles Spencer (IOP/RA) to Mr. Ryan, May 19, 1969, Box 18, Entry P 243, RG 306, NARA.

16. Mr. Bardos to Mr. Hedges, April 9, 1969, Box 18, Entry P 243, RG 306, NARA; Anne M. Platoff, "Where No Flag Has Gone Before: Political and Technical Aspects of Placing a Flag on the Moon," NASA Contractor Report 188251, 1993.

17. John P. Walsh, Acting Executive Secretary Department of State, memorandum for Henry Kissinger, the White House, June 8, 1969, Box 3013, Entry 1613, RG 59, NARA. For a discussion of the history and significance of planting a US flag on the moon, see Anne M. Platoff, "Where No Flag Has Gone Before: Political and Technical Aspects of Placing a Flag on the Moon," NASA Contractor Report 188251, www.jsc.nasa.gov /history/flag/flag.htm; Daniel Immerwahr, "Twilight of Empire," *Modern American History* 1, no. 1 (2018): 129–133.

18. Charles McC. Mathias Jr. to William P. Rogers, June 19, 1969, Box 3013, Entry 1613, RG 59, NARA.

19. Simon Bourgin to Mr. Ryan, April 14, 1969, Box 18, Entry P 243, RG 306, NARA.

20. Muir-Harmony, *Apollo to the Moon*, 182–185.

21. Collins, *Carrying the Fire*, 336.

22. All of the messages on the disk are reproduced in Tahir Rahman, *We Came in Peace for All Mankind: The Untold Story of the Apollo 11 Silicon Disc* (Overland Park, KS: Leathers, 2008).

23. NASA etched these messages alongside statements made by Eisenhower, Kennedy, Johnson, and Nixon; the names of leaders in Congress and a list of members of the four committees of the House and Senate that participated in the NASA legislation; and the names of NASA past and present administrators, deputy administrators, and other high-level managers. The words "From Planet Earth—July 1969" banded the top of the disk while the inscription "Goodwill messages from around the world brought to the Moon by the astronauts of Apollo 11" reemphasized the international character of the mission. The messages were photographed and reduced 200 times. Silicon was chosen because of its stability and the ability to withstand the extreme temperature range on the moon. NASA Press Release 69-83F, "Apollo 11 Goodwill Messages," July 13, 1969, NASA Headquarters History Office, Washington, DC;

Thomas Paine to U. Alexis Johnson, July 11, 1969, Box 3013, Entry 1613, RG 59, NARA.

24. Willis H. Shapley to Dr. Mueller, "Symbolic Activities for Apollo 11," July 2, 1969, Document II-71 in *Exploring the Unknown: Selected Documents in the History of the U.S. Civil Space Program*, ed. John Logsdon et al. (Washington, DC: National Aeronautics and Space Administration, 2008), 733.

25. John Reinhardt to Hewson Ryan, July 25, 1969, Box 20, Entry P 243, RG 306, NARA.

26. CBS Television Network, *10:56:20 PM EDT 7/20/1969* (New York: Columbia Broadcasting System, 1970), 93.

27. Muir-Harmony, *Apollo to the Moon*, 108–111; Collins, *Carrying the Fire*, 333–334.

28. *Apollo 10 Mission Report*, MSC-00126, National Aeronautics and Space Administration (Houston, TX: Manned Spacecraft Center, August 1969), https://history.nasa.gov/afj/ap10fj/pdf/a10-mission-report.pdf.

29. *Sunday Telegraph*, quoted in Apollo 11 Operations Center to Frank Shakespeare, June 6, 1969, Box 3, Entry A1 42, RG 306, NARA.

30. M. V. Kamath, "First Colour Telecast of the Moon," *Times of India*, May 23, 1969, 11.

31. USIA memo for all USIS posts; Apollo 11 Operations Center to Frank Shakespeare, June 6, 1969.

32. "VOA Coverage on the Flight of Apollo XI, 1969," September 11, 1969, Box 18, Entry P 243, RG 306, NARA; Apollo 11 Operations Office to Frank Shakespeare, June 27, 1969, Box 3, Entry A1 42, RG 306, NARA.

33. Collins, *Carrying the Fire*, 344.

34. Collins, *Carrying the Fire*, 344–347.

35. Frank Shakespeare to John L. McClellan, August 28, 1969, Box 4, Entry A1 42, RG 306, NARA; Apollo 11 Operations Center to Frank Shakespeare, June 6, 1969.

36. Apollo Operation Center to Henry Loomis, August 6, 1969, Box 15, Entry P 243, RG 306, NARA; Henry Dunlap to Harry Loomis, November 18, 1968, Box 4, Entry A1 42, RG 306, NARA; "USIA 32nd Report to Congress."

37. Apollo 11 Operations Office to Mr. Shakespeare, June 27, 1969, Box 3, Entry A1 42, RG 306, NARA.

38. The kiosk exhibits included a six-foot-tall, three-sided structure outfitted with blinking lights, music, photogelatin transparencies, and posters. Henry Loomis to William Rogers, June 30, 1969, Box 1, Office of Policy and Plans: Subject Files, 1966–1971 (hereafter Entry P 12), RG 306, NARA; *Report to the Congress from the President of the United States, US Aeronautics and Space Activities for 1969*.

39. "USIA 32nd Report to Congress"; Apollo 11 Operations Office to Frank Shakespeare, July 23, 1969, Box 3, Entry A1 42, RG 306, NARA; John Reinhardt to Hewson Ryan, July 25, 1969, Box 20, Entry P 243, RG 306, NARA.

40. USIA Circular, May 19, 1969, Box 17, Entry P 243, RG 306, NARA; Apollo Operation Center to Loomis, August 6, 1969; Karl E. Meyer, "Poland Parades Guns on 25th Birthday," *Washington Post*, July 23, 1969, A12.

41. Apollo Operation Center to Loomis, August 6, 1969.

42. "Special International Exhibitions 7th Annual Report FY 1969," Box 3, Entry P 173, RG 306, NARA.

43. "Special International Exhibitions 8th Annual Report FY 1970," Box 3, Entry P 173, RG 306, NARA.

44. "Special International Exhibitions 7th Annual Report FY 1969"; "Special International Exhibitions 8th Annual Report FY 1970."

45. "USIA 32nd Report to Congress."

46. Apollo Operation Center to Loomis, August 6, 1969.

47. "USIA 32nd Report to Congress."

48. David J. Whalen, "For All Mankind: Societal Impact of Application Satellites," in *Societal Impact of Spaceflight*, ed. Steven J. Dick and Roger D. Launius (Washington, DC: National Aeronautics and Space Administration, 2007), 289–312.

49. Schwoch, *Global TV*, 147.

50. An estimated forty million American homes tuned in to coverage of John Glenn's flight. Schwoch, *Global TV*, 127.

51. Schwoch, *Global TV*, 145.

52. Val Adams, "Pastore Predicts TV Links for All," *New York Times*, April 22, 1965, 67.

53. Schwoch, *Global TV*, 149.

54. Tsiao, *"Read You Loud and Clear!,"* chapter 5.

55. "Worldwide Treatment of Current Issues," July 25, 1969, Box 21, Folder "INF 7-6 Apollo 11 Worldwide Treatment of Current Issues," Entry P 243, RG 306, NARA.

56. Apollo 11 Operations Office to Frank Shakespeare, June 13, 1969, Box 3, Entry A1 42, RG 306, NARA.

57. Erik Tandberg, interview with author, May 29, 2017, Oslo, Norway.

58. Apollo Operation Center to Loomis, August 6, 1969.

59. "USIA 32nd Report to Congress."

60. Apollo 11 Operations Office to Shakespeare, June 27, 1969.

61. Apollo 11 Operations Center to Loomis, August 6, 1969.

CHAPTER 9: ONE GIANT LEAP, JULY 16–JULY 24, 1969

1. "Million Visitors to Witness Apollo 11 Take-Off," *Times of India*, July 15, 1969, 12; Norman Ferguson, *Project Apollo: The Moon Odyssey Explained* (Stroud: History Press, 2019), 109; Bernard Weinraub, "Tourists Crowd Cocoa Beach as Apollo Countdown Begins," *New York Times*, July 11, 1969, 1.

2. Jerry E. Bishop, "Localities Near Site of Moon Launch Gird for Onlooker Influx," *Wall Street Journal*, July 14, 1969, 11; William Greider, "Protestors, VIPs Flood Cape Area," *Washington Post*, July 16, 1969, A1.

3. Thomas Paine memorandum for record, July 17, 1969, NASA Historical Reference Collection. They sang "We Shall Overcome" while walking beside wagons drawn by mules they had named Jim Eastland and George Wallace after pro-segregationist southern politicians. The protest sought to draw attention to poverty in America, specifically that one-fifth of Americans lacked adequate shelter, food, clothing, and medical care. NASA Administrator Paine met them and responded with the message that Lyndon Johnson emphasized during his presidency: the space program could be a launchpad to a greater society. "Man on the Moon: Communist Reactions to the Voyage of Apollo 11," July 25, 1969, Box 20, Entry P 243, RG 306, NARA; Muir-Harmony, *Apollo to the Moon*, 129–131.

4. Thomas O'Toole, "Astronauts Poised for Historical Lunar Voyage," *Washington Post*, July 16, 1969, A1; Collins, *Carrying the Fire*, 356; Ferguson, *Project Apollo*, 38–44.

5. "Inside TV," *Los Angeles Times*, September 9, 1969, E19; US Embassy London to USIA Washington, June 30, 1969, Box 21, Entry P 243, RG 306, NARA; US Embassy Panama to USIA Washington, July 15, 1969, Box 21, Entry P 243, RG 306, NARA; US Embassy Saigon to USIA Washington, July 10, 1969, Box 21, Entry P 243, RG 306, NARA; David Meerman Scott and Richard Jurek, *Marketing the Moon: The Selling of the Apollo Lunar Program* (Cambridge, MA: MIT Press, 2014), 78–90.

6. *Report to the Congress from the President of the United States, US Aeronautics and Space Activities for 1969*.

7. Leonard Miall, "Obituary: Ruggero Orlando," *Independent*, May 6, 1994; Jean R. Hailey, "Louis Deroche, French Newsman, Dies," *Washington Post*, July 24, 1975, B10.

8. CBS, *Man on the Moon*, July 16, 1969.

9. Mailer, *Of a Fire on the Moon*, 98; Ferguson, *Project Apollo*, 37.

10. Apollo 11 Lunar Landing Mission Press Kit, Release No. 69-83K, July 6, 1969, NASA.

11. Lyndon Johnson, quoted in CBS, *Man on the Moon*, July 16, 1969.

12. "VOA Coverage on the Flight of Apollo XI, 1969," September 11, 1969.

13. Frank Shakespeare to Representative Edward J. Derwinski, August 4, 1969, Box 8, Entry A1 42, RG 306, NARA.

14. Apollo Operation Center to Loomis, August 6, 1969; Apollo 11 Operations Office to Shakespeare, July 3, 1969, Box 3, Entry A1 42, RG 306, NARA; Apollo 11 Operations Office to Shakespeare, July 23, 1969.

15. Apollo 11 Operations Office to Shakespeare, July 3, 1969; CBS Television Network, *10:56:20 PM EDT 7/20/1969*; Meerman Scott and Jurek, *Marketing the Moon*, 80.

16. CBS Television Network, *10:56:20 PM EDT 7/20/1969*, 61–62.

17. CBS Television Network, 63.

18. Collins, *Carrying the Fire*, 393; Armstrong, quoted in "Apollo 11, Day 4, part 1: Approaching the Moon," *Apollo Flight Journal*, corrected transcript and commentary by W. David Woods, Kenneth D. MacTaggart, and Frank O'Brien, 2019, https://history.nasa.gov/afj/ap11fj/12day4-loi1 .html.

19. "Apollo 11, Day 4, part 1: Approaching the Moon."

20. "The First Lunar Landing," *Lunar Surface Journal*, corrected transcript and commentary by Eric M. Jones, Apollo 11, 2018, https:// history.nasa.gov/alsj/a11/a11.landing.html; Chaikin, *A Man on the Moon*, 189–200.

21. "Brazil Cheers Lunar Landing," *Sun*, July 21, 1969, A4; "Japanese Apollo 11 Ovation Drowns Out Friction Issues," *Christian Science Monitor*, July 22, 1969, 4. The dance step created for the event resembled the "Lindy Hop," which had become popular after Charles Lindbergh's 1927 transatlantic flight. American Consulate Nice to the State Department, July 31, 1969, Box 3014, Entry 1613, RG 59, NARA.

22. Apollo Operation Center to Loomis, August 6, 1969; Apollo 11 Operations Office to Shakespeare, July 23, 1969; CBS Television Network, *10:56:20 PM EDT 7/20/1969*, 94; "USIA 32nd Report to Congress."

23. CBS Television Network, *10:56:20 PM EDT 7/20/1969*, 63, 86–89.

24. Memo for Frank Shakespeare, July 24, 1969, Box 4, Entry A1 42, RG 306, NARA.

25. USIA estimated that 650 million people watched the live television broadcast. "General Wrap-Up from USIA Apollo 11 Center," July 22, 1969, Box 4, Entry A1 42, RG 306, NARA; Muir-Harmony, *Apollo to the Moon*, 137; "The First Lunar Landing," https://history.nasa.gov/alsj/a11 /a11.landing.html; Chaikin, *A Man on the Moon*, 207–209.

26. It is possible that in the excitement and gravity of the moment, Armstrong did not fully enunciate the "a" in his statement. Some later

acoustic analysis of the recording suggests that he did not drop the article, but the issue has caused controversy over the years. Either way, the meaning Armstrong intended is clear. For a discussion of the line, see Hansen, *First Man*, 494–496.

27. Neil A. Armstrong, interviewed by Stephen Ambrose and Douglas Brinkley, Houston, TX, September 19, 2001, NASA Johnson Space Center Oral History Project.

28. Meerman Scott and Jurek, *Marketing the Moon*, 69; Hansen, *First Man*, 519–520.

29. In response to the 1967 Six-Day War, Egypt was attempting to force Israel out of the Sinai. On the night of the moon landing, however, the Israeli Air Force severely damaged the Egyptian forces. David A. Korn, *Stalemate: The War of Attrition and Great Power Diplomacy in the Middle East, 1967–1970* (New York: Routledge, 1992), 165–188; William Borders, "Even in Hostile Nations, the Feat Inspires Awe: All the World's in . . . ," *New York Times*, July 22, 1969, 1; "Chileans Dance, Soviets Scream 'Hooray' on Word of Landing," *Hartford Courant*, July 21, 1969, 9A; "General Wrap-Up from USIA Apollo 11 Center"; Robert Donovan, "Moon Shot Helps U.S. Image on Nixon's Tour," *Los Angeles Times*, July 29, 1969, 17.

30. Lella Scalia, "July 20th, 1969: Remembering When the Man Walked Where Nobody Had Gone Before," *Vogue Italia*, July 20, 2010.

31. Scalia, "July 20th, 1969."

32. The USIA reported that "school children in Bavaria and students in Mexico were excused from classes [the next] day . . . church bells rang out to announce the moon landing in various Latin American cities. . . . Laplanders followed the flight on their transistor radios while pasturing their reindeer." "General Wrap-Up from USIA Apollo 11 Center"; US Embassy Mogadiscio to State Department, July 24, 1969, Box 3013, Entry 1613, RG 59, NARA; "What's in a Name? Plenty if It's Neil," *Hartford Courant*, July 22, 1969, 5.

33. VOA broadcast live coverage of the flight in English, Portuguese, Spanish, Russian, Arabic, French, Greek, Japanese, Turkish, Chinese, and additional languages. The VOA coverage of John Glenn's flight in 1962 broke records with an estimated world audience of 300 million people; in comparison, the Apollo 11 audience was more than twice that size. *Report to the Congress from the President of the United States, US Aeronautics and Space Activities for 1969*; Apollo Operation Center to Loomis, August 6, 1969; "VOA Coverage on the Flight of Apollo XI, 1969," September 11, 1969; Patrick Buchanan to the White House, July 23, 1969, Box 3, Entry A1 42, RG 306, NARA.

34. This was an important indicator to the USIA of the significance and impact of the agency's Apollo 11 coverage and framing of the mission.

Arthur Bardos to Henry Loomis, July 24, 1969, Box 4, Director's Subject Files, 1968–1972, RG 306, NARA.

35. Bardos to Loomis, July 24, 1969.

36. Apollo 11 Operations Office to Shakespeare, July 23, 1969.

37. *Report to the Congress from the President of the United States, US Aeronautics and Space Activities for 1969.*

38. "Foreign Media Reaction: Apollo 11," July 21, 1969, Box 21, Entry P 243, RG 306, NARA; "USIA 32nd Report to Congress"; Bardos to Loomis, July 24, 1969.

39. "USIA 32nd Report to Congress."

40. Apollo Operation Center to Loomis, August 6, 1969.

41. US Embassy Caracas to State Department, July 21, 1969, Box 3013, Entry 1613, RG 59, NARA; US Embassy Santiago to State Department, July 18, 1969, Box 3013, Entry 1613, RG 59, NARA.

42. "Foreign Media Reaction: Apollo 11," July 21, 1969.

43. John Reinhardt to Hewson Ryan, July 25, 1969, Box 20, Entry P 243, RG 306, NARA.

44. "Man on the Moon: Communist Reactions to the Voyage of Apollo 11," July 25, 1969, Box 20, Entry P 243, RG 306, NARA.

45. Reinhardt to Ryan, July 25, 1969.

46. Apollo 11 Operations Office to Shakespeare, July 23, 1969; Embassy Tokyo to Secretary of State, September 22, 1969, Box 3015, Folder "SP 10 US 9/1/69," Entry 1613, RG 59, NARA.

47. The USIS post in Tokyo reported that "one somewhat misguided youth thought his message would carry more meaning if written in blood.—it was!—it did!" Apollo 11 Operations Office to Frank Shakespeare, October 10, 1969, Box 3, Entry A1 42, RG 306, NARA; "Worldwide Treatment of Current Issues," July 25, 1969.

48. US Embassy Tokyo to Secretary of State, September 22, 1969.

49. "Foreign Media Reaction: Apollo 11," July 21, 1969; CBS Television Network, *10:56:20 PM EDT 7/20/1969*, 54, 91.

50. Reinhardt to Ryan, July 25, 1969; American Embassy, Benghazi, to Department of State, August 1, 1969, Box 2014, Entry 1613, RG 59, NARA.

51. Asif A. Siddiqi, *The Soviet Space Race with Apollo* (Gainesville: University Press of Florida, 2000), 693–697.

52. "Man on the Moon: Communist Reactions," July 25, 1969.

53. Apollo Operations Center to Shakespeare, July 23, 1969.

54. "Man on the Moon: Communist Reactions," July 25, 1969; "USIA 32nd Report to Congress."

55. Henry Tasca to the Secretary of State, July 26, 1969, Box 3014, Entry 1613, RG 59, NARA.

56. Apollo Operation Center to Loomis, August 6, 1969.

57. Apollo Operation Center to Loomis, August 6, 1969.

58. USIS Rio de Janeiro to USIA Washington, August 6, 1969, Box 52, Entry A1 1016, RG 306, NARA.

59. "Relationships Among Opinions About the United States and U.S. Foreign Policy," April 1, 1973, Box 53, Entry A1 1016, RG 306, NARA.

60. Apollo 11 Operations Office to Frank Shakespeare, August 29, 1969, Box 3, Entry A1 42, RG 306, NARA; "Moon Probe via TV Costly to S. Africans," *Chicago Tribune*, July 14, 1969, 5.

61. "Japanese Apollo 11 Ovation," 4.

62. Nan-Shih Ho, chairman of the United Poet Association of China, to Richard Nixon, July 27, 1969, Box 3015, Entry 1613, RG 59, NARA.

63. USIS Tel Aviv to USIA Washington, July 24, 1969, Box 4, Entry A1 42, RG 306, NARA.

64. "Foreign Media Reaction: Apollo 11 Sunday Report," July 20, 1969, Box 4, Entry A1 42, RG 306, NARA.

65. Leonard A. Cheever, "The Spacecraft of Pablo Neruda and W. H. Auden," in *Flashes of the Fantastic: Selected Essays from the War of the Worlds Centennial, Nineteenth International Conference on the Fantastic in the Arts*, ed. David Ketterer (Westport, CT: Praeger, 2004), 239–246.

66. Francois Duvalier to Richard Nixon, July 26, 1969, Box 5, White House Central Files: Subject Files: Outer Space, RNPL; Jomo Kenyatta to Richard Nixon, July 22, 1969, Box 5, White House Central Files: Subject Files: Outer Space, RNPL; Josip Tito to Richard Nixon, July 22, 1969, Box 5, White House Central Files: Subject Files: Outer Space, RNPL; Hassan II to Richard Nixon, July 22, 1969, Box 5, White House Central Files: Subject Files: Outer Space, RNPL.

67. The use of "for all mankind" reflects scholarship on contact language in trading zones. Understandings of what "for all mankind" means likely differed, but the phrase provided a means to coordinate and connect. See Peter Galison, *Image and Logic: A Material Culture of Microphysics* (Chicago: University of Chicago Press, 1997); Lisa Messeri, "The Problem with Pluto: Conflicting Cosmologies and the Classification of Planets," *Social Studies of Science* 40, no. 2 (2010): 189–190.

68. "Nixoning the Moon," *New York Times*, July 19, 1969, 24.

69. Kathryn Cramer Brownell, "Nixoning the Moon," *Modern American History* 1, no. 1 (2018): 139.

70. Haldeman, *The Haldeman Diaries*, 73–74.

CHAPTER 10: OPERATION MOONGLOW, AUGUST 1969

1. "Apollo Operations Handbook Block II Spacecraft," Washington, DC, NASA, 1969; Collins, *Carrying the Fire*, 441.

2. Bob Fish, *Hornet Plus Three: The Story of the Apollo 11 Recovery* (Reno: Creative Minds, 2009), 113-119.

3. Fish, *Hornet Plus Three*, 113-119.

4. Nixon, *Memoirs*, 429; Haldeman, *The Haldeman Diaries*, 75.

5. Fish, *Hornet Plus Three*, 121.

6. Nixon, *Memoirs*, 429.

7. Richard Nixon, "Remarks to Apollo 11 Astronauts Aboard the U.S.S. Hornet Following Completion of Their Lunar Mission," American Presidency Project, www.presidency.ucsb.edu/node/239653; Nixon, *Memoirs*, 429.

8. "Nixon Cites Peace Aim in Manila," *Washington Post*, July 26, 1969; Robert Donovan, "Moon Shot Helps U.S. Image on Nixon's Tour," *Los Angeles Times*, July 29, 1969, 17.

9. Nixon, *Memoirs*, 394.

10. "Worldwide Treatment of Current Issues," July 25, 1969.

11. Robert Chalmers, "Apollo Gives U.S. a Big Boost in International Relations," *Washington Post*, July 23, 1969, A23.

12. Nixon, *Memoirs*, 393-394.

13. Dallek, *Nixon and Kissinger*, 146-147; Jussi M. Hanhimaki, "An Elusive Grand Design," in *Nixon in the World: American Foreign Relations, 1969-1977*, ed. Fredrik Logevall and Andrew Preston (New York: Oxford University Press, 2008), 35-37.

14. Memorandum from the President's Deputy Assistant (Butterfield) to the President's Assistant for National Security Affairs (Kissinger), June 2, 1969, National Archives, Nixon Presidential Materials, NSC Files, Box 672, Country Files, Europe, Czechoslovakia, vol. I, no classification marking, NARA.

15. Kissinger, *White House Years*, 156.

16. Kissinger, *White House Years*, 156.

17. Kissinger, *White House Years*, 156.

18. Helmut Sonnenfeldt memo for the record, June 23, 1969, "HAK Conversation with Romanian Ambassador Bogdan," Digital National Security Archive (DNSA) collection: Kissinger Conversations: Supplement I, 1969-1977.

19. Kissinger, *White House Years*, 156.

20. Memorandum of telephone conversation, July 2, 1969, between Henry Kissinger and John Ehrlichman, DNSA collection: Kissinger Telephone Conversations, 1969-1977.

21. Richard Nixon, "Information Remarks in Guam with Newsmen," July 25, 1969, American Presidency Project, www.presidency.ucsb.edu/ws/?pid=2140; Nixon, *Memoirs*, 394-395.

22. Kissinger, *White House Years*, 223–224.

23. "Worldwide Treatment of Current Issues," July 27, 1969, Box 21, Folder "INF 7-6 Apollo 11 Worldwide Treatment of Current Issues," Entry P 243, RG 306, NARA; "Worldwide Treatment of Current Issues," July 25, 1969.

24. Al Haig to Henry Kissinger, July 17, 1969, Box 464, Folder "East Asian Trip 1969 Part 1," National Security Council Files: Subject Files, RNPL.

25. "Worldwide Treatment of Current Issues," July 25, 1969.

26. Donovan, "Moon Shot Helps U.S. Image," 17.

27. "Worldwide Treatment of Current Issues," July 25, 1969; "Worldwide Treatment of Current Issues," July 27, 1969.

28. Memorandum for the Director of Current Intelligence, "Nerve Gas Incident on Okinawa," July 18, 1969, General CIA Records, CIA-RDP80B01439R000500090021-7, Central Intelligence Agency Digital Library.

29. Jon Mitchell, "Red Hat's Lethal Okinawa Smokescreen," *Japan Times*, July 27, 2013; Robert Kently, "Nerve Gas Accident: Okinawa Mishap Bares Overseas Deployment of Chemical Weapons," *Wall Street Journal*, July 18, 1969, 1; Jonathan B. Tucker, *War of Nerves: Chemical Warfare from World War I to Al-Qaeda* (New York: Pantheon, 2006), 215; Charles Mohr, "Few Issues Expected to Arise in Nixon's Brief Talks in Manila," *New York Times*, July 26, 1969, 7.

30. President's Arrival Remarks for Manila, July 21, 1969, Box 464, Folder "Manila, Djakarta, Bangkok, India, Pakistan, and Romania—Departure notes, toasts, etc. 18 July 1969," National Security Council Files: Subject Files, RNPL.

31. President's Suggested Toast for Manila, July 18, 1969, Box 464, Folder "Manila, Djakarta, Bangkok, India, Pakistan, and Romania—Departure notes, toasts, etc. 18 July 1969," National Security Council Files: Subject Files, RNPL.

32. President's Departure Remarks for Manila, July 18, 1969, Box 464, Folder "Manila, Djakarta, Bangkok, India, Pakistan, and Romania—Departure notes, toasts, etc. 18 July 1969," National Security Council Files: Subject Files, RNPL.

33. Arrival Remarks for Djakarta, undated, Box 50, President's Personal Files: President's Speech File, RNPL; "Nixon Offers Help to Asians," *Los Angeles Times*, July 28, 1969, 1; "Nixon Offers to Send Gift to Suharto: Piece of Moon," *Sun*, July 28, 1969, A1.

34. Donovan, "Moon Shot Helps U.S. Image," 17.

35. "Nixon Offers Help to Asians," 1.

36. Donovan, "Moon Shot Helps U.S. Image," 17.

37. "Worldwide Treatment of Current Issues," July 28, 1969, Box 21, Folder "INF 7-6 Apollo 11 Worldwide Treatment of Current Issues," Entry P 243, RG 306, NARA.

38. Kissinger, *White House Years*, 276-277; Haldeman, *The Haldeman Diaries*, 76.

39. "The Texts of the Statements by President Nixon and President Thieu," *New York Times*, July 31, 1969, 16.

40. Robert B. Semple Jr., "Nixon Sees Thieu, Talks to Troops in Vietnam Visit," *New York Times*, July 31, 1969, 1.

41. Special Memorandum, Foreign Radio and Press Reaction to President Nixon's Trip to Asia and Romania, July 23-August 3, 1969, Box 464, Folder "Manila, Djakarta, Bangkok, India, Pakistan, and Romania—Departure notes, toasts, etc. 18 July 1969," National Security Council Files: Subject Files, RNPL.

42. "Worldwide Treatment of Current Issues," July 24, 1969, Box 21, Folder "INF 7-6 Apollo 11 Worldwide Treatment of Current Issues," Entry P 243, RG 306, NARA.

43. Toast at Pakistan dinner, undated, Box 50, President's Personal Files: President's Speech File, RNPL.

44. Kissinger, *White House Years*, 180-181.

45. M. V. Kamath, "Nixon Home After 'Quest for Peace,'" *Times of India*, August 5, 1969, 11; Nixon, *Memoirs*, 395; Karl E. Meyer, "Moon Shot Warmed Up Reception: Apollo 11 Sparked Rumanian Spirit," *Washington Post*, August 3, 1969, 1.

46. Meyer, "Moon Shot Warmed Up Reception," 1; "Nixon's Trip to Rumania Held Significant," *Times of India*, July 27, 1969, 11.

47. Meyer, "Moon Shot Warmed Up Reception," 1; Robert B. Semple, "Nixon, in Rumania, Stresses Desire for World Peace," *New York Times*, August 3, 1969, 1.

48. Kissinger, *White House Years*, 157.

49. President's Toast for Bucharest, July 21, 1969, Box 464, Folder "Manila, Djakarta, Bangkok, India, Pakistan, and Romania—Departure notes, toasts, etc. 18 July 1969," National Security Council Files: Subject Files, RNPL.

50. Haldeman, *The Haldeman Diaries*, 77-78.

51. Nixon, *Memoirs*, 395.

52. Memorandum of conversation, private meeting between President Nixon and Ceaușescu, August 3, 1969, DNSA collection: Kissinger Transcripts, 1968-1977.

53. Nguyen, "Waging War on All Fronts," 189.

54. Memorandum of conversation, private meeting between President Nixon and Ceaușescu, August 3, 1969.

55. "Nixon's Trip to Rumania," 11.

56. Special Memorandum, Foreign Radio and Press Reaction to President Nixon's Trip to Asia and Romania, July 23–August 3, 1969.

57. Henry Kissinger to Richard Nixon, undated, Box 464, Folder "East Asian Trip 1969, Part 1," National Security Council Files: Subject Files, RNPL.

58. Nixon, *Memoirs*, 396.

59. Kissinger, *White House Years*, 278.

60. "Talks in Apartment," *New York Times*, October 12, 1972, 15; Kissinger, *White House Years*, 278.

61. "Xuan Thuy, Hanoi Envoy at Paris Talks, Dies," *New York Times*, June 20, 1985, 16.

62. Memorandum from the President's Assistant for National Security Affairs (Kissinger) to President Nixon, August 6, 1969, Nixon Presidential Materials, NSC Files, Box 863, For the President's File, Vietnam Negotiations, Camp David Memcons, 1969–1970, NARA.

63. Kissinger, *White House Years*, 281–282.

64. "Nixon, Returning, Hails Friendship He Found on Trip," *New York Times*, August 4, 1969, 1.

65. A few days later, Kissinger presented Nixon with options for how the administration could handle Vietnam. Haldeman noted that the president recognized he had to be "prepared for the heat" and potential blowback to his decision. "This is at least part of the reason for the efforts to build strong nationalism with space thing." Haldeman, *The Haldeman Diaries*, 81. See also Steven V. Robert, "Nixon Is Host in Los Angeles at a State Dinner for 3 Men," *New York Times*, August 14, 1969, 1; Don Oberdorfer, "Apollo Astronauts Hailed by Millions," *Washington Post*, August 14, 1969, A1; Logsdon, *After Apollo?*, 26.

CHAPTER 11: GIANTSTEP: THE APOLLO 11 DIPLOMATIC TOUR, 1969

1. Peter Flanigan to Thomas Paine, August 15, 1969, Box 12, Folder "[EX] OS 3-1 8/1/69-9/30/69," White House Central Files: Subject Files: Outer Space, RNPL.

2. Douglas Martin, "Peter M. Flanigan, Banker and Nixon Aide, Dies at 90," *New York Times*, August 1, 2013, A22.

3. See correspondence in Box 12, Folder "[EX] OS 3-1 8/1/69-9/30/69."

4. Oral history interview with Paul Findley conducted by Mark DePue, February 8, 2013, Jacksonville, IL, Interview #IS-A-L-2013-002, Abraham Lincoln Presidential Library, Springfield, IL.

5. Paul Findley to William Rogers, July 31, 1969, Box 17, Folder "Astronaut's Tour 69," Entry 243, RG 306, NARA.

6. John E. Reinhardt to Mr. Ryan, July 29, 1969, Box 17, Folder "Astronaut's Tour 69," Entry 243, RG 306, NARA; Weil, "John E. Reinhardt."

7. Richard Philips to U. Alexis Johnson, August 11, 1969, Box 17, Entry P 243, RG 306, NARA.

8. Logsdon, *After Apollo?*, 27–28.

9. US Embassy Rawalpindi to State Department, undated, Box 307, National Security Council Files: Subject Files, RNPL; US Embassy Tehran to Secretary of State, September 3, 1969, Box 307, National Security Council Files: Subject Files, RNPL; US Embassy Mexico City to Secretary of State, September 6, 1969, Box 307, National Security Council Files: Subject Files, RNPL.

10. William Thompson to Arthur Bardos, August 1, 1969, Box 17, Folder "Astronaut's Tour 69," Entry 243, RG 306, NARA.

11. Richard Nixon to John Glenn, July 14, 1969, Box 11, Folder "[EX] OS 3-1 Astronauts Begin 7/31/69," White House Central Files, Subject Files, Outer Space, RNPL.

12. Dan Oleksiw to Frank Shakespeare, June 1969, Box 4, Entry A1 42, RG 306, NARA.

13. Peter Flanigan to Julian Scheer, August 23, 1969, Box 11, Folder "[EX] OS 3-1 8/1/69-9/30/69," White House Central Files, Subject Files, Outer Space, RNPL; Nixon, quoted in Steve Wolfe, "Moonglow: Space Diplomacy in the Nixon Administration," *Quest* 18, no. 2 (2011): 43.

14. Al Haig to Henry Kissinger, September 10, 1969, Box 307, National Security Files, Subject Files, RNPL.

15. Hewson Ryan to Frank Shakespeare, September 4, 1969, Box 4, RG 306, Entry A1 42, NARA; Secretary of State to USINT Cairo, September 1969, Box 307, National Security Council Files: Subject Files, RNPL.

16. US Embassy The Hague to Department of State, October 22, 1969, Box 22, Entry P 243, RG 306, NARA.

17. Don Lesh to Henry Kissinger, September 3, 1969, Box 307, National Security Council Files: Subject Files, RNPL; Henry Kissinger to Richard Nixon, undated (early September), Box 307, National Security Council Files: Subject Files, RNPL.

18. US Embassy Budapest to Secretary of State, September 13, 1969, Box 307, National Security Council Files: Subject Files, RNPL.

19. Henry A. Kissinger to William Rogers, September 19, 1969, Box 307, National Security Council Files: Subject Files, RNPL.

20. Richardson gave examples of "cooling relations": "Secretary, for example, will not receive Under Secretary Puja in New York. So far as further bilateral talks, visits, and exchanges are concerned, you should await

Hungarian initiative and seek specific instructions on whether and how to proceed. We intend to limit bilateral talks to issues involving clear-cut, demonstrable and concert advantages to the interests of the US. We have also considered canceling visit of AEC Chairman Seaborg to Budapest." Elliot Richardson to Embassy in Hungary, September 20, 1969, NSC Files, Box 693, Country Files—Europe, Hungary, vol. I, Secret; RNPL.

21. After Tito split with Soviet leader Joseph Stalin in 1948, Yugoslavia pursued a policy of neutrality and by 1969 had a long history of diplomatic relations with the United States. Henry Kissinger to Richard Nixon, undated (September 1969), Box 307, National Security Council Files: Subject Files, RNPL; US Embassy Budapest to Secretary of State, September 13, 1969, Box 307, National Security Council Files: Subject Files, RNPL; Wolfe, "Moonglow," 43.

22. The itinerary released on September 29 listed the following stops: Mexico City, Mexico (Sept. 29-30); Bogotá, Colombia (Sept. 30–Oct. 1); Brasilia, Brazil (Oct. 1); Buenos Aires, Argentina (Oct. 1-2); Rio de Janeiro, Brazil (Oct. 2-4); Las Palmas, Canary Islands (Oct. 4-6); Madrid, Spain (Oct. 6–8); Paris, France (Oct. 8–9); Amsterdam, Holland (Oct. 9); Brussels, Belgium (Oct. 9–10); Oslo, Norway (Oct. 10–12); Cologne/Bonn and Berlin, Germany (Oct. 12–14); London, England (Oct. 14–15); Rome, Italy (Oct. 15–18); Belgrade, Yugoslavia (Oct. 18–20); Ankara, Turkey (Oct. 20–22); Kinshasa, Zaire (Oct. 22–24); Tehran, Iran (Oct. 24–26); Bombay, India (Oct. 26–27); Dacca, East Pakistan (Oct. 27–28); Bangkok, Thailand (Oct. 28–31); Perth, Australia (Oct. 31); Sydney, Australia (Oct. 31–Nov. 2); Agana, Guam (Nov. 2–3); Seoul, South Korea (Nov. 3–4); Tokyo, Japan (Nov. 4–5); Elmendorf, Alaska (Nov. 5); and Ottawa and Montreal, Canada (Dec. 2-3). Apollo 11 Operations Office memo to Frank Shakespeare, "Astronauts' World Tour," September 18, 1969, Box 17, Entry P 243, RG 306, NARA.

23. Logsdon, *After Apollo?*, 28–29.

24. "Japan Greets Astronauts as World Tour Nears End," *New York Times*, November 5, 1969, 93.

25. Simon Bourgin to Mariada Bourgin, end of September 1969, Box 4, #1633, Folder "Astronaut Tours/Letters to Ely Echo," Simon Bourgin Collection, BUA.

26. Simon Bourgin, *Simon Bourgin: An Odyssey That Began in Ely* (Ely, MN: Ely-Winton Historical Society, 2010), 100-101; Simon Bourgin to Neil Armstrong, February 26, 2000, Box 7, #1633, Folder "General Correspondence 1953-2001," Simon Bourgin Collection, BUA.

27. "'Conquistadores' of Moon Hailed," *Sun*, September 30, 1969, A1.

28. Michael Collins, interview with the author, July 18, 2019, Washington, DC.

29. "Astronauts Get Keys to Mexico City," *News* (Mexico City), September 30, 1969.

30. According to one report, the president chided the crew "on what he described as the injustice of the Nixon administration's intensive hunt for dope smugglers at Mexican border crossings" and "voiced his nation's resentment of 'operation intercept,'" the crackdown on marijuana that had bottlenecked crossing points and caused severe economic damage to Mexican border towns. "3 Astronauts Leave Mexico for Colombia," *Chicago Tribune*, October 1, 1969.

31. This designation for the Apollo 11 crew could have been taken as either a criticism of the United States as an imperial power or as local association of explorers; it is unclear from the context. The crew was often compared to European colonialist explorers, both by US government officials as well as by the media in many foreign countries. Examples in Portugal suggest a positive correlation, as one would expect. *Sun*, "Apollo 11 Crew Rests in Norway," October 12, 1969, 4; Simon Bourgin to Mariada Bourgin, end of September 1969; US Embassy Mexico to USIA, October 1969, Box 22, Entry P 243, RG 306, NARA.

32. Robert H. McBride, Ambassador to Mexico, to Frank Shakespeare, October 1, 1969, Box 4, "SP—Space and Astronautics," Entry A1 42, RG 306, NARA.

33. USIS Bogotá to USIA Washington, "Astronaut Tour Stops," October 22, 1969, RG 306, Entry P 243, Box 21, NARA.

34. Robert Amerson to Apollo Task Force, September 19, 1969, RG 306, Entry P 243, Box 17, NARA.

35. Apollo 11 Operations Office to Shakespeare, October 10, 1969, RG 306, Entry P 243, Box 17, Folder "Astronaut's Tour 69," NARA; Walter Bastian Jr. to USIA Washington, October 9, 1969, Box 22, Entry P 243, RG 306, NARA. To keep "the warmth engendered by the visit alive," the USIS hosted a series of events, distribution of pamphlets, film screenings, and radio broadcasts. USIS Bogotá to USIA Washington, "Astronaut Tour Stops."

36. USIS Buenos Aires to USIA Washington, October 22, 1969, Box 15, Entry P 243, RG 306, NARA; "Marriott Plaza Hotel Buenos Aires Celebrates 100 Years of Sophistication and Service," PR Newswire, July 21, 2009.

37. Simon Bourgin to Mariada Bourgin, October 2, 1969, Box 4, Simon Bourgin Collection, BUA.

38. Simon Bourgin to Mariada Bourgin, October 1969, Box 4, #1633, Folder "Astronaut Tours/Letters to Ely Echo," Simon Bourgin Collection, BUA.

39. Simon Bourgin to Mariada Bourgin, October 1969.

40. Albert Hemsing to Mr. Bardos and Mr. Bourgin, July 28, 1969, Box 17, Folder "Astronaut's Tour 69," Entry 243, RG 306, NARA.

41. "Apollo Moon Men Cheered by Spaniards," *Chicago Tribune*, October 8, 1969; Simon Bourgin to Mariada Bourgin, October 1969.

42. Hemsing to Bardos and Bourgin, July 28, 1969; "Apollo Moon Men Cheered by Spaniards," *Chicago Tribune*, October 8, 1969; Albert Harkness Jr., Counselor for Public Affairs US Embassy in Madrid, to Frank Shakespeare, October 17, 1969, Box 4, Folder "SP—Space and Astronautics," Entry A1 42, RG 306, NARA.

43. Simon Bourgin to Mariada Bourgin, October 1969.

44. Edward Rohrbach, "Paris Throng Hails Apollo 11 Astronauts," *Chicago Tribune*, October 9, 1969, S13; USIS Paris to USIA Washington, November 12, 1969, Box 21, Folder "Giant Steps," Entry 243, RG 306 NARA; Apollo 11 Operations Office to Frank Shakespeare, October 17, 1969, Box 3, Folder "INF 2-3 Weekly Reports to Director," Entry A1 42, RG 306, NARA.

45. US Embassy The Hague to Department of State, October 22, 1969; "Astronauts Wow Brussels," *Washington Daily News*, October 10, 1969, 2; Apollo 11 Operations Office to Shakespeare, October 17, 1969.

46. Simon Bourgin to Mariada Bourgin, October 1969; USIS Oslo to USIA Washington, October 31, 1969, Box 21, Folder "Giant Steps," Entry 243, RG 306, NARA.

47. Tandberg interview; Radio broadcast, "Apollo XI astronauts in Oslo," Program Leader Berit Griebenow, October 12, 1969, NRK (Norwegian Broadcasting Corporation) (2008/1837.P), Norwegian National Library Research Archives, Oslo, Norway.

48. Gwen Morgan, "Astronauts Hailed During Visit to Queen," *Chicago Tribune*, October 15, 1969.

49. In Rome, according to the *New York Times*, "At the city hall, the arrival of the group coincided with a demonstration by a group of irate mothers demanding more schools. But, for a moment, the local problem was set aside as the mothers joined in the applause." "Rome Welcomes Apollo 11 Crew," *New York Times*, October 16, 1969, 14.

50. US Embassy Belgrade to US Embassy Rome, October 1969, Box 23, Entry P 243, RG 306, NARA; Apollo 11 Operations Office to Frank Shakespeare, October 24, 1969, Box 1, Folder "INF 2-3 Weekly Reports—IOR Director to Agency Director 1969," Entry A1 42, RG 306, NARA; "400,000 Greet Apollo 11 Crew," *Chicago Tribune*, October 19, 1969, 3.

51. Collins interview.

52. US Embassy Belgrade to US Embassy Ankara, October 1969, Box 23, Entry P 243, RG 306, NARA; Collins interview.

53. Simon Bourgin to Mariada Bourgin, October 23, 1969; USIS Belgrade to USIA Washington, November 20, 1969, Box 21, Folder "Giant Steps," Entry 243, RG 306, NARA. The US Embassy in Belgrade described Tito as an avid outdoorsman, so it is not surprising that the astronauts went hunting with the Yugoslav leader to encourage friendship between their two countries, even though Armstrong may not have been an avid hunter himself. US Embassy in Belgrade to USIA Washington, October 18, 1969, Box 21, Folder "Giant Steps," Entry 243, RG 306, NARA; US Embassy Belgrade to US Embassy Rome, October 1969.

54. "Turkey's First and Only English Daily," *Daily News* (Turkey), undated; Jemima Kallas, "American Astronauts Arrive in Ankara," Box 21, Folder "Giant Steps," Entry P 243, RG 306, NARA.

55. Kallas, "American Astronauts Arrive in Ankara."

56. Henry L. Davis to Simon Bourgin, September 24, 1969, Box 22, Entry P 243, RG 306, NARA.

57. US Embassy Congo to the White House, October 25, 1969, Box 21, Folder "Giant Steps," Entry 243 RG 306, NARA; Apollo Operations Office to Frank Shakespeare, October 24, 1969, Box 4, Folder "SP—Space and Astronautics," Entry A1 42, RG 306, NARA.

58. "Briefing Paper for Astronauts: Iran," undated but likely September 1969, Box 17, RG 306, Entry P 243, NARA; US Embassy Tehran to USIA, October 26, 1969, Box 22, RG 306, Entry P 243, NARA.

59. Geneva B. Barnes, interviewed by Glenn Swanson, March 26, 1999, Washington, DC, NASA Johnson Space Center Oral History Project.

60. Asif Siddiqi, "Making Space for the Nation: Satellite Television, Indian Scientific Elites, and the Cold War," *Comparative Studies of South Asia, Africa and the Middle East* 35, no. 1 (2015): 35–49.

61. William F. Thompson to Mr. Bardos, August 1, 1969, Box 17, Entry P 243, RG 306, U.S. Information Agency Files, NARA.

62. Siddiqi, "Making Space for the Nation," 41.

63. "Briefing Paper for Astronauts," October 1969, Box 17, Entry P 243, RG 306, U.S. Information Agency Files, NARA.

64. Kenneth B. Keating, quoted in "Moon Men Here Oct. 26," *American Reporter—Bombay*, October 25, 1969, 1; USIS Bombay to USIA Washington, November 3, 1969, Box 22, Entry P 243, RG 306, U.S. Information Agency Files, NARA; Simon Bourgin notes on Apollo 11 tour, undated, Record Number 7093, Series Biographies—Government Officials, Folder: Bourgin, Simon, NASA Headquarters, Washington, DC; *Indian Express*, quoted in USIS Bombay to USIA Washington, November 3, 1969, Box 22, Entry P 243, RG 306, U.S. Information Agency Files, NARA.

65. Simon Bourgin to Mariada Bourgin, October 25, 1969, Box 4, Simon Bourgin Collection, BUA.

66. "Moon Men Touch Down to Biggest-Ever Welcome in City," *Times of India*, October 27, 1969, 1.

67. Simon Bourgin notes on Apollo 11 tour; *Free Press Journal*, quoted in USIS Bombay to USIA Washington, November 3, 1969, Box 22, Entry P 243, RG 306, U.S. Information Agency Files, NARA.

68. Keating, quoted in "Moon Men Touch Down," 1.

69. US Embassy New Delhi to US Embassy Belgrade, October 1969, Box 22, Entry P 243, RG 306, U.S. Information Agency Files, NARA.

70. Aldrin, quoted in "Moon Men Touch Down," 1.

71. US Embassy New Delhi to US Embassy Belgrade, October 1969; Bourgin notes on Apollo 11 tour, undated.

72. US Embassy New Delhi to US Embassy Belgrade, October 1969; Bourgin notes on Apollo 11 tour, undated.

73. Simon Bourgin to Mariada Bourgin, October 25, 1969.

74. USIS Bombay to USIA Washington, November 3, 1969, Box 22, Entry P 243, RG 306, U.S. Information Agency Files, NARA; Simon Bourgin notes on Apollo 11 tour, undated.

75. Buzz Aldrin with Wayne Warga, *Return to Earth* (New York: Random House, 1973), 79; Simon Bourgin to Mariada Bourgin, "Wednesday or maybe Thursday, end of October," 1969, Box 4, Simon Bourgin Collection, BUA.

76. Apollo 11 Operations Office to Shakespeare, October 24, 1969; USIA Washington to USIS Dacca, October 24, 1969, Box 15, Folder "SP 10 Apollo 11 [Folder 2/2]," Entry P 243, RG 306, NARA; USIS Rawalpindi to USIA Washington, October 22, 1969, Box 21, Folder "Giant Steps," Entry P 243, RG 306, NARA.

77. USIS Bangkok to USIA Washington, November 10, 1969, Box 21, Folder "Giant Steps," Entry P 243, RG 306, NARA.

78. US Embassy Canberra to USIA Washington, November 1969, Box 21, Folder "Giant Steps," Entry P 243, RG 306, NARA; US Embassy Seoul to USIA Washington, October 22, 1969, Box 21, Folder "Giant Steps," Entry P 243, RG 306, NARA.

79. "Japan Greets Astronauts as World Tour Nears End," *New York Times*, November 5, 1969, 93; USIS Tokyo to USIA Washington, November 1969, Box 21, Folder "Giant Steps," Entry P 243, RG 306, NARA; William Weathersby to Henry Loomis, December 15, 1969, Box 4, Folder "SP—Space and Astronautics," Entry A1 42, RG 306, NARA.

80. Hansen, *First Man*, 579.

81. Stephen E. Ambrose, *Nixon: The Triumph of a Politician, 1962–1972* (New York: Simon and Schuster, 1989), 303–304, 310.

82. R. Young, "Nixon Greets Moon Trio on End of Tour," *Chicago Tribune*, November 6, 1969.

83. Nan Robertson, "Apollo 11 Crew Feted by Nixons on Returning from World Tour," *New York Times*, November 6, 1969, 42.

84. Richard Nixon: "Remarks Welcoming the Apollo 11 Astronauts Following Their Goodwill Tour," November 5, 1969, American Presidency Project, www.presidency.ucsb.edu/ws/?pid=2309.

85. "NASA Authorization for Fiscal Year 1971, 'Hearings Before the Committee on Aeronautical and Space Sciences, United States Senate,'" Ninety-First Congress, second session on S. 3374, March 11, 1970.

86. Walter Bastian Jr. to USIA Washington, October 9, 1969, Box 22, Entry P 243, RG 306, NARA.

87. Aldrin, *Return to Earth*, 84.

88. Collins, *Carrying the Fire*, 465.

89. Hansen, *First Man*, 579.

90. Wolfe, "Moonglow," 43; Henry Kissinger to Mr. Ruwe, November 11, 1969.

91. Dwight L. Chapin to Lucy Winchester, September 29, 1969, Box 14, White House Central Files, Staff Member Office Files, Sanford Fox, RNPL; M. Smith, "'Ambassadors for Peace' Honored," *Washington Post*, November 6, 1969.

92. Aldrin, *Return to Earth*, 85.

AFTERGLOW

1. Eugene Cernan, quoted in the *Apollo Flight Journal*, Apollo 17 Day 1: "A regular human weather satellite," Corrected Transcript and Commentary, edited by W. David Woods and Ben Feist, 2018.

2. Many historians and scholars have examined the cultural, environmental, and historical significance of the *Blue Marble* image but have not traced its connections to 1960s American public diplomacy. Denis Cosgrove examines the long history of imagining the whole earth in *Apollo's Eye: A Cartographic Genealogy of the Earth in the Western Imagination* (Baltimore: Johns Hopkins University Press, 2001), and "Contested Global Visions: One-World, Whole-Earth, and the Apollo Space Photographs," *Annals of the Association of American Geographers* 84, no. 2 (1994): 270–294. The environmental context of the photographs is discussed in Neil Maher, "Shooting the Moon," *Environmental History* 9, no. 3 (2004): 526–531, and Sheila Jasanoff, "Image and Imagination: The Formation of Global and Environmental Consciousness," in *Changing the Atmosphere*, ed. Paul Edwards and Clark Miller (Cambridge, MA: MIT Press, 2001). Robert Poole considers the cultural significance of *Blue Marble* in *Earthrise*.

3. Eugene Cernan oral history interview by Rebecca Wright, December 11, 2007, Houston, Texas, NASA Johnson Space Center Oral History Project.

4. Brand would later use the *Blue Marble* image on the cover of his *Whole Earth Catalog*, a counterculture how-to manual. Stewart Brand, "Why Haven't We Seen the Whole Earth Yet?," in *The Sixties: The Decade Remembered Now, by the People Who Lived It Then*, ed. Lynda Obst (New York: Random House, 1977), 168. See also Maher, "Shooting the Moon," 526–531.

5. Chaikin, *A Man on the Moon*, 119.

6. Sheila Jasanoff, "Heaven and Earth," in *Earthly Politics: Local and Global in Environmental Governance*, ed. Sheila Jasanoff and Marybeth Long Martello (Cambridge, MA: MIT Press, 2004), 210. As historian Robert Poole argued, "All this amounted to a paradigm shift, along the lines of Thomas Kuhn's model of scientific revolutions." See Poole, *Earthrise*, 198. For further discussions of the idea of the revolution of human self-perception brought about by Project Apollo, see Sheila Jasanoff, "Image and Imagination."

7. Carl Sagan, *Pale Blue Dot: A Vision of the Human Future in Space* (New York: Random House, 1994), 215.

8. "Impact of U.S. Space Program on Domestic and Foreign Opinion," August 20, 1969, Box 4, Entry A1 42, RG 306, NARA.

9. "Impact of U.S. Space Program on Domestic and Foreign Opinion."

10. US Embassy Fort Lamy to State Department.

11. "Impact of U.S. Space Program on Domestic and Foreign Opinion," August 20, 1969, Box 4, Entry A1 42, RG 306, NARA.

12. "Impact of U.S. Space Program on Domestic and Foreign Opinion."

13. This type of shared experience, historians have argued, can contribute to social cohesion and transnational consciousness. See David Nye, *American Technological Sublime* (Cambridge, MA: MIT Press, 1994); Akira Iriye, "Making of a Transnational World," in *Global Interdependence: The World After 1945*, ed. Akira Iriye (Cambridge, MA: Harvard University Press, 2014); "Impact of U.S. Space Program on Domestic and Foreign Opinion."

14. Eighteenth-century print capitalism, according to Anderson, provided a common language, experience, and imagined association between people who might never meet. In the second edition of his book, Anderson analyzed additional institutions that began to flourish in the age of mechanical reproduction: museums, maps, and the census. Modern states employed these institutions to influence the content and shape of "imagined communities," Anderson explained. See Benedict Anderson,

Imagined Communities: Reflections on the Origin and Spread of Nationalism, rev. ed. (New York: Verso, 2006), 6.

15. In some ways, Anderson's description of the formation of nationalism resembles David Nye's and Michael Smith's discussions of technological display and national image making. In Nye's study, Americans' experience with awe-inspiring technologies has led to social cohesion. Communal events, according to Nye, contribute to American national identity. Smith focuses on the promotion, presentation, and iconography of Apollo, or what Anderson might call the "logoization" of spaceflight. Project Apollo, according to Smith, was "an agent of national self-definition." Both Nye and Smith are concerned with American experience and national identity making. But Project Apollo was not a merely domestic story. The image making and social cohesion that each author discusses apply to the role of Project Apollo in America's relationship to the world as well. Nye, *American Technological Sublime*; Smith, "Selling the Moon," 180.

16. "Impact of U.S. Space Program on Domestic and Foreign Opinion."

17. The decades following Apollo have been called the Age of Fracture because of the disintegration of shared values and collective purpose. For an analysis of the relationship between the US space program and the social and political movements of the 1960s, and the longer consequences, see Neil Maher, *Apollo in the Age of Aquarius* (Cambridge, MA: Harvard University Press, 2017). On the Age of Fracture, see Daniel T. Rodgers, *Age of Fracture* (Cambridge, MA: Belknap Press of Harvard University Press, 2012). For an estimate of the cost of US involvement in the Vietnam War, see "US Spent $141 Billion in Vietnam in 14 Years," *New York Times*, May 1, 1975, 20. For an overview of US foreign relations in the cold war, see Walter L. Hixson, *American Foreign Relations: A New Diplomatic History* (New York: Routledge, 2015).

18. Collins interview with author, July 18, 2019.

INDEX